ENCYCLOPEDIA OF MATHEMATICS AND ITS APPLICATIONS

Volume 107
Multiple Scattering

The interaction of waves with obstacles is an everyday phenomenon in science and engineering, cropping up for example in acoustics, electromagnetism, seismology and hydrodynamics. The mathematical theory and technology needed to understand the phenomenon is known as multiple scattering, and this book is the first devoted to the subject. The author covers a variety of techniques, for example separation of variables, T-matrix and integral equation methods, describing first the single-obstacle method and then extending it to the multiple-obstacle case. A key ingredient in many of these extensions is an appropriate addition theorem: a coherent, thorough exposition of these theorems is given, and computational and numerical issues around them are explored.

The application of these methods to different types of problems is also explained. In particular, sound waves, electromagnetic radiation, waves in solids and water waves. A comprehensive reference list of some 1400 items rounds off the book, which will be an essential reference on the topic for applied mathematicians, physicists and engineers.

ENCYCLOPEDIA OF MATHEMATICS AND ITS APPLICATIONS

All the titles listed below can be obtained from good booksellers or from Cambridge University Press. For a complete series listing visit
http://publishing.cambridge.org/stm/mathematics/eom/

88. Teo Mora *Solving Polynomial Equation Systems I*
89. Klaus Bichteler *Stochastic Integration with Jumps*
90. M. Lothaire *Algebraic Combinatorics on Words*
91. A.A. Ivanov & S.V. Shpectorov *Geometry of Sporadic Groups, 2*
92. Peter McMullen & Egon Schulte *Abstract Regular Polytopes*
93. G. Gierz et al. *Continuous Lattices and Domains*
94. Steven R. Finch *Mathematical Constants*
95. Youssef Jabri *The Mountain Pass Theorem*
96. George Gasper & Mizan Rahman *Basic Hypergeometric Series 2nd ed.*
97. Maria Cristina Pedicchio & Walter Tholen *Categorical Foundations*
98. Mourad Ismail *Classical and Quantum Orthogonal Polynomials in One Variable*
99. Teo Mora *Solving Polynomial Equation Systems II*
100. Enzo Olivieri & Maria Eulalia Vares *Large Deviations and Metastability*
101. A. Kushner, V. Lychagin & V. Roubtsov *Contact Geometry and Nonlinear Differential Equations*
102. R.J. Wilson & L. Beineke *Topics in Algebraic Graph Theory*
103. Olof Johan Staffans *Well-Posed Linear Systems*
104. John Lewis, S. Lakshmivaraham, Sudarshan Dhall *Dynamic Data Assimilation*
105. M. Lothaire *Applied Combinatorics on Words*

Multiple Scattering

Interaction of Time-Harmonic Waves with *N* Obstacles

P.A. MARTIN
Colorado School of Mines

CAMBRIDGE
UNIVERSITY PRESS

CAMBRIDGE UNIVERSITY PRESS
Cambridge, New York, Melbourne, Madrid, Cape Town, Singapore, São Paulo

Cambridge University Press
The Edinburgh Building, Cambridge CB2 2RU, UK

Published in the United States of America by Cambridge University Press, New York

www.cambridge.org
Information on this title: www.cambridge.org/9780521865548

First published 2006

Printed in the United Kingdom at the University Press, Cambridge

A catalogue record for this publication is available from the British Library

ISBN-13 978-0-521-86554-8 hardback
ISBN-10 0-521-86554-9 hardback

To Ruth, Richard, Frances and, last but not least, Ann

Contents

Preface

It has been said that 'a true scholar publishes only one book in his lifetime and that posthumously' [105, p. 435]. In fact, this book has not taken a lifetime to complete, although it has had a very long gestation period: there have been many distractions, most of which have not been unpleasant.

The book is concerned with the scattering of time-harmonic waves by obstacles; the words 'multiple scattering' in the title signify that there are at least two obstacles. The problems considered come from acoustics (sound waves, Helmholtz equation), electromagnetics (Maxwell's equations), elastodynamics (waves in solids) and hydrodynamics (surface water waves). The book describes a variety of mathematical techniques for solving such problems: the main techniques involve separation of variables, integral equations and T-matrices. Most of the book is concerned with exact methods, although the last chapter discusses several effective approximate methods. There are also two chapters on addition theorems; these are useful in other contexts as well as for multiple scattering. For detailed information on the topics covered, see the Table of Contents and Section 1.7; for a list of topics *not* covered, see Section 1.2.

The mathematics used is classical: separation of variables, special functions, Green's functions, Fourier methods, asymptotics. The reader should also have some familiarity with simple uses of boundary integral equations. Operator notation is used when convenient.

As far as I know, there is no other book that treats all four of the main physical domains: acoustics, electromagnetics, elastodynamics and hydrodynamics. The reader will see many connections between these domains. As far as I know, there is no other book that focuses on multiple scattering. (Of course, it is inevitable that we begin by considering scattering by one obstacle.) In addition, the book has an extensive reference list. This should be useful to future workers, and may help reduce duplication of effort.

The book contains a sprinkling of quotations from the literature. These are intended to be illuminating, amusing or both. Some may help the reader to not lose heart when the calculations become complicated; some may hint that great men are human; and some may simply reflect the author's English sense of humour!

I cannot give thanks individually to everybody who has helped me over the years, perhaps by sending me reprints or answering my questions. However, I must thank four people for shaping my outlook and taste in applied mathematics: Ralph Kleinman, Frank Rizzo, Fritz Ursell and Gerry Wickham. I also thank Chris Linton for his detailed comments on an early draft. No doubt some errors remain, for which I must take sole responsibilty. Please let me know if you find errors. In particular, if you think that I should have cited one of your papers, send me a copy!

1
Introduction

Privately, [Rayleigh] *often quoted with relish a saying attributed to Dalton when in the chair at a scientific meeting: "Well, this is a very interesting paper for those that take any interest in it".*

(Strutt [1157, p. 320])

1.1 What is 'multiple scattering'?

The mathematics of the full treatment may be altogether beyond human power in a reasonable time; nevertheless…

(Heaviside [489, p. 324])

'Multiple scattering' means different things to different scientists, but a general definition might be 'the interaction of fields with two or more obstacles'. For example, a typical multiple-scattering problem in classical physics is the scattering of sound waves by two rigid spheres. Further examples, such as the scattering of spherical electron waves by a cluster of atoms, can be found in condensed-matter physics [1379, 168, 422, 424, 423]. Many other examples will be discussed in this book.

The waves scattered by a single obstacle can be calculated in various well-known ways, such as by the method of separation of variables, T-matrix methods or integral-equation methods. All of these methods will be discussed in detail later.

If there are several obstacles, the field scattered from one obstacle will induce further scattered fields from all the other obstacles, which will induce further scattered fields from all the other obstacles, and so on. This recursive way of thinking about how to calculate the total field leads to another notion of multiple scattering; it can be used to actually compute the total scattered field – each step is called an *order of scattering*. In 1893, Heaviside [489, p. 323] gave a clear qualitative description of this 'orders-of-scattering' process.

1.1.1 Single scattering and independent scattering

In his well-known book on electromagnetic scattering, van de Hulst [1233, §1.2] considers two classifications, namely *single scattering* and *independent scattering*. Let us review his definitions of these ideas.

1.1.1.1 Single scattering

This is the simplest approximation, in which the effects of multiple scattering are ignored completely: 'the total scattered field is just the sum of the fields scattered by the individual [obstacles], each of which is acted on by the [incident] field in isolation from the other [obstacles]' [111, p. 9]. This approximation is used widely; it is only expected to be valid when the spacing is large compared with both the size of the obstacles and the length of the incident waves. Indeed, with these assumptions, higher-order approximations can be derived [1382, 1383, 1381] and these can be effective [511]. However, there are many instances where multiple scattering is important; for some natural examples, see Bohren's fascinating book [109] and his related paper [110]. Thus, in atmospheric physics, the single-scattering approximation is not justified, 'for example, by clouds, where *multiple scattering* can be appreciable' [111, p. 9].

1.1.1.2 Independent scattering

When waves interact with several obstacles, a 'cooperative effect' may occur. This could be constructive interference, leading to unexpectedly large fields, such as can happen with a periodic arrangement of identical scatterers as in a diffraction grating or a crystal lattice. Alternatively, there could be destructive interference, leading to unexpectedly small fields, such as can happen with a random arrangement of scatterers. These are examples of *dependent scattering*: in theory, one 'has to investigate in detail the phase relations between the waves scattered by neighboring [scatterers]' [1233, p. 4]. Thus, the 'assumption of independent scattering implies that there is no systematic relation between these phases' [1233, p. 5].

The notions of single scattering and independent scattering need not be separated. For example, the authors of [866] consider

> *only independent scattering, randomly positioned particles. This means that particles are separated widely enough, so that each particle scatters light in exactly the same way as if all other particles did not exist. Furthermore, there are no systematic phase relations between partial electromagnetic waves scattered by different particles, so that the intensities . . . of the partial waves can be added without regard to phase. In other words, we will assume that each particle is in the far-field zone of all other particles, and that scattering by different particles is incoherent.*
>
> (Mishchenko et al. [866, p. 4])

The authors go on to quantify what 'separated widely enough' means: 'Exact scattering calculations for randomly oriented two-sphere clusters composed of identical wavelength-sized spheres suggest that particles can scatter independently when the distance between their centers is as small as four times their radius [868]' [866, p. 5]. This is consistent with van de Hulst [1233, p. 5]: 'Early estimates have shown that a mutual distance of 3 times the radius is a sufficient condition for independence'.

1.1.2 Scattering by N obstacles

Suppose that we have N disjoint obstacles, B_i, $i = 1, 2, \ldots, N$. The boundary of B_i is S_i. A given wave is incident upon the N obstacles, and the problem is to calculate the scattered waves.

We assume that we know everything about every obstacle: location, shape, orientation and boundary condition; if the obstacles are penetrable, so that waves can travel through them, we assume that we know the internal composition. There are many situations where all of this information is not available; for example, the obstacles might be located randomly.

Mathematically, the exact (deterministic) multiple-scattering problem is easily formulated: it is an exterior boundary-value problem (with a radiation condition at infinity) where the boundary is not simply-connected. However, the problem is not easy to solve, due mainly to the complicated geometry: hence Heaviside's pessimistic comment. Another comment, in a similar vein, was made by van de Hulst [1233]:

> *Multiple scattering does not involve new physical problems, ... Yet the problem of finding the intensities inside and outside the cloud* [of N scatterers] *is an extremely difficult mathematical problem.*
>
> (van de Hulst [1233, p. 6])

This attitude led naturally to single-scattering approximations, as mentioned above. One scatters the incident wave from the ith obstacle (ignoring the presence of the other obstacles), and then sums over i. Indeed, van de Hulst's book and [276] are devoted entirely to single scattering.

At the other extreme, one may attempt to solve the N-body scattering problem directly, perhaps by setting up a boundary integral equation over

$$S = \bigcup_{j=1}^{N} S_j; \tag{1.1}$$

see Chapter 5. Analytically, although 'it would be esthetically preferable to treat the [N-body] configuration as a unit, this approach seems limited to certain special problems' [1196, p. 42]. Computationally, this direct approach can be expensive, especially for problems involving many three-dimensional obstacles.

In the first comprehensive review of the literature on multiple scattering, Twersky opined that

> *it is convenient in considering multiple scattering, to assume that solutions for the component scatterers when isolated are known, and that they may be regarded as "parameters" in the more general problem.*
>
> *Thus, one seeks representations for scattering by many objects in which the effects of the component scatterers are "separated" from the effects of the particular configuration (or statistical distribution of configurations) in the sense that the forms of the results are to hold independently of the type of scatterers involved.*
>
> (Twersky [1198, p. 715])

Similarly, in the context of hydrodynamics (where water waves interact with immersed structures, such as neighbouring ships, wave-power devices or elements of a single larger structure), Ohkusu wrote:

> *For the purpose of calculating hydrodynamic forces . . . , it is essential that only the hydrodynamic properties of each element be given. A method having such a merit will facilitate the calculation for a body having many elements and may be applied to the design arrangement of the elements.*
>
> (Ohkusu [927, p. 107])

In other words, assuming that we know everything about scattering by each obstacle in isolation, how can we use this knowledge to solve the multi-obstacle problem? The best way is to use a 'self-consistent' method. In the next section, we describe such a method in general terms.

1.1.3 Self-consistent methods

A self-consistent method

> *assumes that a wave is emitted by each scatterer of an amount and directionality determined by the radiation incident on that scatterer (the effective field). The latter is to be determined by adding to the incident beam the waves emitted by all other scatterers, and the waves emitted by those scatterers are in turn influenced by the radiation emitted by the scatterer in question. . . . The self-consistent procedure is not an expansion in primary, secondary, tertiary waves, etc. The field acting on a given scatterer, or emitted by it includes the effects of all orders of scattering.*
>
> (Lax [687, pp. 297–298])

Specifically, write the total field as

$$u = u_{\text{inc}} + \sum_{j=1}^{N} u_{\text{sc}}^{j}, \tag{1.2}$$

where u_{inc} is the given incident field and u_{sc}^{j} is the field scattered ('emitted') by the jth scatterer. Define the 'effective' or 'external' or 'exciting field' by

$$u_n \equiv u - u_{\text{sc}}^{n} = u_{\text{inc}} + \sum_{\substack{j=1 \\ j\neq n}}^{N} u_{\text{sc}}^{j}; \tag{1.3}$$

it is the 'radiation incident on [the nth] scatterer' in the presence of all the other scatterers.

Now, as the problem is linear, it must be possible to write

$$u_{\text{sc}}^{j} = \mathcal{T}_j u_j, \tag{1.4}$$

where $\mathcal{T}_j$ is an operator relating the field incident on the jth scatterer, u_j, to the field scattered by the jth scatterer, u^j_{sc}. Hence, (1.3) gives

$$u_n = u_{\text{inc}} + \sum_{\substack{j=1 \\ j\neq n}}^{N} \mathcal{T}_j u_j, \qquad n = 1, 2, \ldots, N, \tag{1.5}$$

or, equivalently,

$$u^n_{\text{sc}} = \mathcal{T}_n \left\{ u_{\text{inc}} + \sum_{\substack{j=1 \\ j\neq n}}^{N} u^j_{\text{sc}} \right\}, \qquad n = 1, 2, \ldots, N. \tag{1.6}$$

If one could solve (1.5) for u_n or (1.6) for u^n_{sc}, $n = 1, 2, \ldots, N$, the total field would then be given by

$$u = u_{\text{inc}} + \sum_{j=1}^{N} \mathcal{T}_j u_j \tag{1.7}$$

or (1.2), respectively.

The derivation of (1.5) given here follows [1193, Chapter 6, §3]; see also [1191, Chapter 7, §2]. Its simplicity is somewhat illusory, because we have not clearly defined the operator $\mathcal{T}_j$; also, we have not indicated *where* (1.5) or (1.6) is required to hold in space. Nevertheless, we have given an abstract framework within which a variety of concrete methods can be developed.

The general scheme leading to (1.5) and (1.7) is often called the *Foldy–Lax self-consistent method*. Foldy [354] used a special case of the method for 'isotropic point scatterers'; see Section 8.3 for a detailed description. Lax [687] used the general scheme, with a certain prescription for $\mathcal{T}_j$; see [687, §III]. We will see several specific realisations later, including the T-matrix methods developed in Chapter 7.

For simple geometries, such as circular cylinders or spheres, a self-consistent method is easily realised. One combines separated solutions of the Helmholtz equation (multipoles); a necessary ingredient is an addition theorem for expanding multipoles centred at one origin in terms of similar multipoles centred on a different origin. This old but useful method will be developed in detail in Chapter 4. The method itself goes back to a paper of Lord Rayleigh, published in 1892; we discuss this next.

1.1.4 Rayleigh's paper of 1892

In his paper 'On the influence of obstacles arranged in rectangular order upon the properties of a medium' [1009], Rayleigh considered potential flow through a periodic rectangular array of identical circular cylinders. As a special case of his analysis, let us consider an infinite square array of rigid cylinders of radius a with centres at $(x, y) = (mb, nb)$, where m and n are integers and $b > 2a$; see Fig. 1.1.

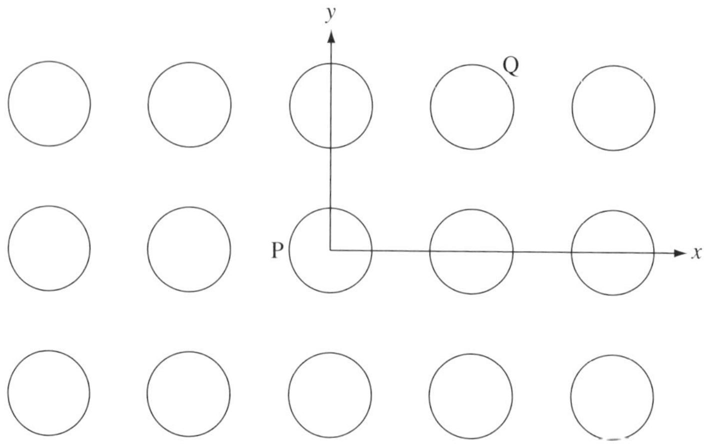

Fig. 1.1. Rayleigh's problem: an infinite square array of circles.

The ambient flow has potential $V = Hx$. In Rayleigh's words [1009, p. 482]: 'If we take the centre of one of the cylinders P as origin of polar coordinates, the potential external to the cylinder may be expanded in the series

$$V = A_0 + (A_1 r + B_1 r^{-1}) \cos\theta + (A_3 r^3 + B_3 r^{-3}) \cos 3\theta + \cdots, \text{'} \qquad (1.8)$$

where θ is measured from the x-axis. Symmetry implies that $V - A_0$ must be an odd function of x and an even function of y; these conditions lead to the form of the expansion (1.8). Imposing the boundary condition $\partial V/\partial r = 0$ on $r = a$ gives

$$B_n = a^{2n} A_n, \quad n = 1, 3, 5, \ldots. \qquad (1.9)$$

Next [1009, p. 483]: 'The values of the coefficients A_1, B_1, A_3, $B_3 \ldots$ are necessarily the same for all the cylinders, and each may be regarded as a similar multiple source of potential. The first term A_0, however, varies from cylinder to cylinder, as we pass up or down the stream'.

At this stage, we have obtained one condition relating A_n and B_n, namely (1.9), but we need another. To find it, Rayleigh begins as follows [1009, p. 483]: 'The potential V at any point may be regarded as due to external sources at infinity (by which the flow is caused) and to multiple sources situated on the axes of the cylinders. The first part may be denoted by Hx'.

Then, Rayleigh proceeds [1009, p. 484] 'by equating two forms of the expression for the potential at a point x, y near P. The part of the potential due to Hx and to the multiple sources Q (P not included) is

$$A_0 + A_1 r \cos\theta + A_3 r^3 \cos 3\theta + \ldots;$$

or, if we subtract Hx, we may say that the potential at x, y due to multiple sources at Q is the real part of'

$$A_0 + (A_1 - H)z + A_3 z^3 + A_5 z^5 + \cdots, \quad \text{with } z = x + \mathrm{i}y. \tag{1.10}$$

Continuing: 'But if x', y' are the coordinates of the same point when referred to the centre of one of the Q's, the same potential may be expressed by'

$$\Sigma\{B_1 z'^{-1} + B_3 z'^{-3} + \cdots\} \quad \text{with } z' = x' + \mathrm{i}y', \tag{1.11}$$

'the summation being extended over all the Q's. If ξ, η be the coordinates of a Q referred to P, $x' = x - \xi$, $y' = y - \eta$; so that' $B_n z'^{-n} = B_n(z - z_0)^{-n}$ with $z_0 = \xi + \mathrm{i}\eta$. Then, the binomial theorem gives

$$z'^{-n} = (-z_0)^{-n}\left\{1 + n(z/z_0) + \frac{1}{2}n(n+1)(z/z_0)^2 + \cdots\right\}. \tag{1.12}$$

Hence [1009, p. 484]: 'Since (1.10) is the expansion of (1.11) in rising powers of $x + \mathrm{i}y$ $[= z]$, we obtain, equating term to term,'

$$\left.\begin{aligned} H - A_1 &= B_1\Sigma_2 + 3B_3\Sigma_4 + 5B_5\Sigma_6 + \cdots \\ -3!\,A_3 &= 3!\,B_1\Sigma_4 + \frac{1}{2}5!\,B_3\Sigma_6 + \cdots \\ -5!\,A_5 &= 5!\,B_1\Sigma_6 + \frac{1}{2}7!\,B_3\Sigma_8 + \cdots \end{aligned}\right\} \tag{1.13}$$

'and so on, where

$$\Sigma_{2n} = \Sigma(\xi + \mathrm{i}\eta)^{-2n}, \text{'} \tag{1.14}$$

'the summation extending over all the Q's.' (As $\Sigma_{2s+1} = 0$, we also obtain $A_0 = 0$.)

Thus, the system comprising (1.9) and (1.13) can now be solved, in principle, for A_n and B_n. Note that, for two-dimensional potential flow (Laplace's equation), the addition theorem amounts to an application of the binomial theorem, (1.12). Note also that the situation becomes more complicated when the periodicity is destroyed, because then the coefficents A_n and B_n will vary from cylinder to cylinder.

Rayleigh [1009] also considered flow past a rectangular three-dimensional array of identical spheres, and (briefly) low-frequency sound waves through rectangular arrays of rigid cylinders or spheres. For further comments, see [1205, §(6)], [648, §2] and [888, §3.1].

1.1.5 Kasterin, KKR and the electronic structure of solids

Shortly after Rayleigh's paper [1009] *was published, a graduate student at Moscow University, N.P. Kasterin, set out to apply* [Rayleigh's] *ideas to a genuine scattering problem. He chose the relatively simple phenomenon of reflection and refraction of low-frequency sound by an orthorhombic grid of hard*

> *spheres. ... Kasterin's results were published in his 1903 Moscow thesis. A preliminary report ... came out in 1898.*
>
> (Korringa [648, p. 346])

As a special case of Kasterin's analysis, consider an infinite planar square array of spheres of radius a with centres at $(x, y, z) = (mb, nb, 0)$, where m and n are integers and $b > 2a$. We take the incident field as a plane wave at normal incidence to the array, $u_{\text{inc}} = \mathrm{e}^{\mathrm{i}kz}$.

Generalising (1.8), the total field at $\boldsymbol{r}$ near the jth sphere can be expanded as

$$u(\boldsymbol{r}) = \sum_{n,m} \left\{ d_{nj}^m \hat{\psi}_n^m(\boldsymbol{r}_j) + c_{nj}^m \psi_n^m(\boldsymbol{r}_j) \right\}, \tag{1.15}$$

where $\boldsymbol{r} = \boldsymbol{r}_j + \boldsymbol{b}_j$ and $\boldsymbol{r} = \boldsymbol{b}_j$ is the sphere's centre. Here, $\psi_n^m(\boldsymbol{r}_j)$ are outgoing multipoles (separated solutions of the Helmholtz equation in spherical polar coordinates), singular at $\boldsymbol{r}_j = \boldsymbol{0}$ $(\boldsymbol{r} = \boldsymbol{b}_j)$ and $\hat{\psi}_n^m(\boldsymbol{r}_j)$ are regular spherical solutions. (Precise definitions will be given later.) The coefficients d_{nj}^m and c_{nj}^m correspond to Rayleigh's A_n and B_n, respectively. Evidently, the periodic geometry and the simple incident field imply that

$$d_{nj}^m \equiv d_n^m \quad \text{and} \quad c_{nj}^m \equiv c_n^m :$$

the coefficients are the same for every sphere.

Applying the boundary condition $\partial u/\partial r = 0$ on $r = a$ yields one relation between d_n^m and c_n^m, namely

$$d_n^m = \gamma_n c_n^m, \tag{1.16}$$

where γ_n is a known constant (see Section 4.6).

The effective field incident on the jth sphere is

$$\sum_{n,m} d_n^m \hat{\psi}_n^m(\boldsymbol{r}_j). \tag{1.17}$$

This must be the same as the sum of the actual incident field and the scattered fields emitted by all the other spheres, namely

$$\mathrm{e}^{\mathrm{i}kz} + \sum_{\substack{l \\ l \neq j}} \sum_{n,m} c_n^m \psi_n^m(\boldsymbol{r}_l). \tag{1.18}$$

Equating (1.17) and (1.18) in a neighbourhood of the jth sphere gives a second relation between d_n^m and c_n^m. This solves the problem, in principle.

To proceed further, suppose that we have the expansions

$$\mathrm{e}^{\mathrm{i}kz} = \sum_{n,m} e_n^m \hat{\psi}_n^m(\boldsymbol{r}_j) \tag{1.19}$$

and

$$\psi_n^m(\boldsymbol{r}_l) = \sum_{\nu,\mu} S_{n\nu}^{m\mu}(\boldsymbol{b}_j - \boldsymbol{b}_l)\, \hat{\psi}_\nu^\mu(\boldsymbol{r}_j); \tag{1.20}$$

we will discuss e_n^m and $S_{n\nu}^{m\mu}(\boldsymbol{b})$ shortly. Then, equating (1.17) and (1.18), using (1.16), (1.19) and (1.20), we obtain

$$\gamma_n c_n^m - \sum_{\nu,\mu} c_\nu^\mu \sum_{\substack{l \\ l\neq j}} S_{\nu n}^{\mu m}(\boldsymbol{b}_j - \boldsymbol{b}_l) = e_n^m.$$

In this linear system of algebraic equations, we can take $j = 0$ without loss of generality.

The Rayleigh–Kasterin method, described above, is rigorous, and it can be generalised in various ways. It has been used to obtain numerical solutions for many related problems with (infinite) periodic geometries. For example, see [727] for acoustic scattering by a single periodic row of circles, and see [984] for two-dimensional elastic waves around a square array of circular cavities; see also [888, Chapter 3].

The Rayleigh–Kasterin method was also adapted to problems in solid-state physics. In that context, it is known as the *KKR* (Korringa–Kohn–Rostoker) *method*; see, for example, [1379, §10.3], [775, 424] or [423, §6.8]. For a clear presentation of the two-dimensional KKR method (for sound waves around an infinite square array of soft circles), see [90].

For historical background, including a detailed description of Kasterin's work, see [648].

To generalise Rayleigh's method to a *non-periodic* configuration, consider the problem of acoustic scattering by two spheres (see Section 1.3 for background information). Suppose that the spheres are centred at O_1 and O_2. Write the scattered field u_{sc} as a superposition of outgoing multipoles ψ_n^m, one set singular at O_1 and the other set singular at O_2:

$$u_{sc} = \sum_{n,m} \{a_n^m\, \psi_n^m(\boldsymbol{r}_1) + b_n^m\, \psi_n^m(\boldsymbol{r}_2)\}.$$

Then, determine the coefficients a_n^m and b_n^m by applying the boundary condition on each sphere in turn: this requires the expansion of $\psi_n^m(\boldsymbol{r}_2)$ in terms of regular spherical solutions centred on O_1, $\hat{\psi}_n^m(\boldsymbol{r}_1)$. Thus, we need the *addition theorem*

$$\psi_n^m(\boldsymbol{r}_2) = \sum_{\nu,\mu} S_{n\nu}^{m\mu}(\boldsymbol{b})\, \hat{\psi}_\nu^\mu(\boldsymbol{r}_1),$$

which is valid for $r_1 < b$, where $r_1 = |\boldsymbol{r}_1|$, $\boldsymbol{r}_2 = \boldsymbol{r}_1 + \boldsymbol{b}$ and $b = |\boldsymbol{b}|$ is the distance between O_1 and O_2. The matrix $\mathbf{S} = (S_{n\nu}^{m\mu})$ is called the *separation matrix* or the *translation matrix* or the *propagator matrix*. It is an important ingredient in several exact theories of multiple scattering. We will give much attention to various methods for calculating $\mathbf{S}$, with emphasis on acoustic problems (Helmholtz equation) in two (Chapter 2) and three (Chapter 3) dimensions.

We also need expansions of the incident field, similar to (1.19); these will be derived too.

Kasterin did not have explicit expressions for the matrix $\mathbf{S}$: we can see that the expansions (1.19) and (1.20) are analogous to Taylor expansions about $\boldsymbol{r}_j = \mathbf{0}$, and so the coefficients could be obtained by applying appropriate differential operators. This is one of several methods for constructing **S** that we shall develop later.

1.2 Narrowing the scope: previous reviews and omissions

Multiple scattering is a huge subject with a huge literature. For an extensive review up to 1964, see [1198] and the supplement [163]. For a collection of articles surveying many aspects of scattering (including theory, computation and application), see the 957-page volume edited by Pike & Sabatier [977].

There is a 1981 survey by Oguchi on 'multiple scattering of microwaves or millimeter waves by an assembly of hydrometeors' [924, p. 719]. Two approaches are reviewed. One is the *Foldy–Lax–Twersky integral equation method*, introduced by Foldy in 1945 [354] and generalised by Lax [687, 688] and Twersky [1200, 1201]. The second approach is based on the *radiative transfer equation*; this may be regarded as the final stage in a larger calculation:

> *the treatment of light scattering by a cloud of randomly positioned, widely separated particles can be partitioned into three steps:* (i) *computation of the far-field scattering and absorption properties of an individual particle ...* (ii) *computation of the scattering and absorption properties of a small volume element containing a tenuous particle collection by using the single-scattering approximation; and* (iii) *computation of multiple scattering by the entire cloud by solving the radiative transfer equation supplemented by appropriate boundary conditions.*
>
> (Mishchenko *et al.* [870, p. 7])

We do not consider radiative transfer further, but see [1190] for more information.

In 2000, Tourin *et al.* [1182] reviewed a variety of applications, including theory and experiment: in one example of note, sound waves in water are scattered by a random collection of 1000 identical parallel steel rods.

Major areas *not* covered in this book include the following.

(i) Scattering by an infinite number of identical obstacles arranged in some periodic manner, such as in a row or in a regular lattice. For plane-wave scattering, problems of this type can be reduced to a problem in a single 'unit cell' (for lattices) or to waveguide problems (for a row of equally-spaced obstacles). The prototype for this reduction, of course, is Rayleigh's paper [1009], discussed in Section 1.1.4.

Larsen [684] gave an early review of scattering by periodic rows of identical cylinders. For scattering by a semi-infinite periodic row of cylinders, see [859, 502, 501, 861, 729].

(ii) Scattering by an infinite rough surface. For such problems, one has to solve a governing partial differential equation (such as the Helmholtz equation) in the region $y > f(x)$, where $y = f(x)$, $-\infty < x < \infty$, is the rough surface with

$0 \le f(x) \le h$, say, and f and h are given. There is an extensive literature on this topic; see, for example, the reviews [280, 1286, 1049, 279] or the books by Ogilvy [923] and Voronovich [1268].

(iii) Scattering by unbounded obstacles. We consider bounded scatterers in three dimensions (such as spheres) or cylindrical scatterers with bounded cross-sections in two dimensions. Thus, we do not consider problems such as diffraction by parallel semi-infinite rigid planes or by wedges and cones. For early work on such problems with one scatterer, see [135].

(iv) Scattering by random arrangements of scatterers. Problems of this type will be discussed in Section 8.6, although we do not develop probabilistic techniques in detail. In particular, we do not consider the *localisation* of waves by random arrays of identical scatterers; see, for example, [508, 615, 228, 1132, 1100].

(v) Transient problems. We exclusively consider time-harmonic problems. Transient problems may be reduced to time-harmonic problems using Fourier transforms. Laplace transforms may be used for initial-value problems. There is also an extensive (and classical) literature on the use of retarded potentials, leading to integral equations of Volterra type. For more information, see the books by Friedlander [365] and by Baker & Copson [53]. Numerical aspects are discussed in [60, §10.5], [116, Chapter 7], and [367, 773, 1040].

(vi) Non-linear problems. In practice, non-linear effects are usually negligible or they are ignored.

(vii) Variational (weak) formulations. We do not use such formulations, so that, in particular, we do not discuss finite-element methods (except that certain hybrid methods are mentioned) or the use of Sobolev spaces. For some information in this direction, see [526, 836, 903, 879].

1.3 Acoustic scattering by N obstacles

May not Music be described as the Mathematic of sense, Mathematic as Music of the reason? the soul of each the same! Thus, the musician feels Mathematic, the mathematician thinks Music, – Music the dream, Mathematic the working life – each to receive its consummation from the other when the human intelligence, elevated to its perfect type, shall shine forth glorified in some future Mozart–Dirichlet or Beethoven–Gauss – a union already not indistinctly foreshadowed in the genius and labours of Helmholtz!

(J.J. Sylvester, quoted in [331, p. 133])

Sound waves in a homogeneous, compressible fluid are governed by the wave equation,

$$\nabla^2 U = \frac{1}{c^2}\frac{\partial^2 U}{\partial t^2},$$

where ∇^2 is the Laplacian, c is the constant speed of sound, t is time, and U is a physical quantity such as the (excess) pressure or the velocity potential; see, for example, [885, Chapter 6], [714, Chapter 1] or [974, Chapter 1]. We consider time-harmonic motions exclusively, so that

$$U = \mathrm{Re}\left\{u\,\mathrm{e}^{\ \mathrm{i}\omega t}\right\}, \tag{1.21}$$

where ω is the radian frequency. Hence, the complex-valued function u satisfies the Helmholtz equation,

$$(\nabla^2 + k^2)u = 0 \quad \text{in } B_\mathrm{e}, \tag{1.22}$$

where B_e is the unbounded exterior region occupied by the fluid and $k = \omega/c$. We assume that the wavenumber k is real and positive. Then, the problem is to solve (1.22), subject to a boundary condition on S and a radiation condition at infinity.

Nowadays, the use of complex-valued functions u, as in (1.21), is so familiar that it usually goes unremarked. However, the following quotation suggests that this was not always the case.

> *There was Lamb, who had only recently gone to Adelaide* [in 1875], *and whose book on hydrodynamics* [682, 514] *(then a slight volume, being an exposition of lectures he had given at Trinity) was the first English book that revealed a use of the complex variable in mathematical physics: let me add that it was an age when the use of* $\sqrt{-1}$ *was suspect at Cambridge even in trigonometrical formulae.*
>
> (Forsyth [357, p. 164])

1.3.1 The Sommerfeld radiation condition

The radiation condition serves two purposes. First, it ensures that the scattered waves are not incoming at infinity. Thus, imagine enclosing the N obstacles by a large sphere, and then sending an incident wave into this sphere; then, we suppose that scattered waves can leave the large sphere but not enter. This is physically reasonable.

Second, the radiation condition (usually) serves a mathematical purpose. Thus, for reasonable boundary conditions, the radiation condition ensures *uniqueness*: in the absence of an incident wave, the only solution of the boundary-value problem is $u \equiv 0$ in B_e.

For example, in three dimensions, the scattered field u_sc is required to satisfy

$$r\left(\frac{\partial u_\mathrm{sc}}{\partial r} - \mathrm{i}k u_\mathrm{sc}\right) \to 0 \quad \text{as } r \to \infty, \tag{1.23}$$

uniformly in θ and ϕ, where r, θ and ϕ are spherical polar coordinates with respect to a chosen origin in the vicinity of the scatterers. Equation (1.23) is the *Sommerfeld radiation condition*; it implies that ru_sc is $O(1)$ as $r \to \infty$ [223, Remark 3.4].

A consequence of (1.23) is that

$$u_\mathrm{sc}(r, \theta, \phi) \sim \frac{\mathrm{e}^{\mathrm{i}kr}}{\mathrm{i}kr} f(\theta, \phi) \quad \text{as } r \to \infty, \tag{1.24}$$

where $f(\theta, \phi)$ is known as the *far-field pattern* [223, Corollary 3.7]. The function f is not known but it may be calculated by solving the boundary-value problem for u_{sc}. Note that the factor of $(ik)^{-1}$ on the right-hand side of (1.24) is sometimes absorbed into f; the definition (1.24) is convenient because, if u_{sc} is dimensionless, then so is f.

In two dimensions, the Sommerfeld radiation condition is

$$r^{1/2}\left(\frac{\partial u_{sc}}{\partial r} - iku_{sc}\right) \to 0 \quad \text{as } r \to \infty, \tag{1.25}$$

uniformly in θ, where r and θ are plane polar coordinates.

For further discussion on the Sommerfeld radiation condition, see, for example, [1149, §9.1], [1144], [564, §1.27], [53, Chapter I, §4.2] and [974, §4-5].

1.3.2 Boundary conditions

For a typical scattering problem, we have a given incident field u_{inc}. We then write the total field as

$$u = u_{inc} + u_{sc} \quad \text{in } B_e,$$

where u_{sc} is the (unknown) scattered field. The boundary condition is usually specified in terms of u, but it can then be rewritten in terms of u_{sc}, if convenient. There are three standard boundary conditions:

Dirichlet condition ('sound-soft' or 'pressure-release')

$$u = 0 \quad \text{or} \quad u_{sc} = -u_{inc} \quad \text{on } S.$$

Neumann condition ('sound-hard' or 'rigid')

$$\frac{\partial u}{\partial n} = 0 \quad \text{or} \quad \frac{\partial u_{sc}}{\partial n} = -\frac{\partial u_{inc}}{\partial n} \quad \text{on } S.$$

Here, $\partial/\partial n$ denotes normal differentiation on S. The (unknown) boundary values of u can be used to calculate the far-field pattern directly; see (5.81).

Robin condition [457] (impedance condition)

$$\frac{\partial u}{\partial n} + \lambda u = 0 \quad \text{on } S,$$

where λ is given; λ may be real or complex, it may be a constant, or it may vary with position on S.

Uniqueness theorems for these three problems are proved by Colton & Kress [223, Theorems 3.13 and 3.37]. Their proofs are for one obstacle, but the number of obstacles is irrelevant to the argument.

For multiple-scattering problems, there is the possibility that we have a different boundary condition on each scatterer. For example, we could have N_1 sound-hard obstacles and N_2 sound-soft obstacles, with $N_1 + N_2 = N$. We shall refer to these as *mixture problems*. For more information and a uniqueness theorem, see [657].

1.3.3 Transmission problems

The three standard boundary conditions given above all suppose that the obstacles B_i are impenetrable: there is no transmission through the interfaces S_i. There are situations where this is unrealistic. For example, consider the scattering of a wave in a compressible fluid by blobs of another fluid; the wave will be partially reflected and partially transmitted through S into the blobs. Then, mathematically, we have to solve the following *transmission problem*: given u_{inc}, find u and u_0, where u satisfies (1.22), u_0 satisfies

$$(\nabla^2 + k_0^2)u_0 = 0 \quad \text{in} \quad B = \bigcup_{j=1}^{N} B_j,$$

$u - u_{\text{inc}}$ satisfies the radiation condition (1.23), and u and u_0 satisfy a pair of *transmission conditions* on the interfaces,

$$u = u_0 \quad \text{and} \quad \frac{\partial u}{\partial n} = \gamma \frac{\partial u_0}{\partial n} \quad \text{on } S. \tag{1.26}$$

Here, $k_0 = \omega/c_0$ is the wavenumber in B, c_0 is the speed of sound in B and γ may be interpreted as a density ratio (when u is regarded as the excess pressure); see (1.31) and [262, §1.A.3].

This is the simplest transmission problem, where k, k_0 and γ are all constants. There are situations where k_0 and γ are different constants for each obstacle. There are also important situations where k_0 varies with position, in which case the obstacles are *inhomogeneous*; see Section 1.3.4.

If k, k_0 and γ are all constants, we have uniqueness provided that their values are restricted suitably [627]; in particular, we have uniqueness if k, k_0 and γ are all real and positive [223, Theorem 3.40].

1.3.4 Inhomogeneous obstacles

Time-harmonic waves in a compressible, inhomogeneous fluid are governed by [885, p. 408]

$$\rho \operatorname{div} \left(\rho^{-1} \operatorname{grad} p\right) + \kappa^2 p = 0, \tag{1.27}$$

where ρ is the density, p is the acoustic pressure, $\kappa = \omega/C$ and C is the speed of sound; in general, ρ and C are functions of position. According to Pierce [975], (1.27) was first given by Bergmann in 1946 [87]; a derivation is given in [793].

If the density is constant, (1.27) reduces to

$$\nabla^2 p + k^2 n(\boldsymbol{r})\, p = 0, \tag{1.28}$$

where $n(\boldsymbol{r}) = [c/C(\boldsymbol{r})]^2$ is the (square of the) *refractive index* at position $\boldsymbol{r}$, $k = \omega/c$ and c is a constant sound speed. For our discussion here, we assume that $n(\boldsymbol{r}) \to 1$ as $r \to \infty$ in all directions; this excludes media that are layered at infinity, for example.

There is a considerable literature on (1.28), especially in the context of quantum mechanics; some of this will be mentioned below. In addition, several point-source solutions (Green's functions) are known for various functional forms of $n(\boldsymbol{r})$; see [711] for a review.

For scattering problems in acoustics, there are essentially three cases, depending on properties of n. First, suppose that $[1 - n(\boldsymbol{r})]$ has compact support, so that $n(\boldsymbol{r}) \equiv 1$ for $r = |\boldsymbol{r}| > a$, say. Suppose further that $n(\boldsymbol{r})$ is smooth for all $\boldsymbol{r}$ in three-dimensional space. Then, one can reduce the scattering problem to an integral equation over B, the finite region in which $n(\boldsymbol{r}) \not\equiv 1$. One such is the Lippmann–Schwinger equation; see Section 6.3.3 and, for example, [14], [225, §8.2] and [911, §10.3]. Asymptotic approximations are available for $ka \ll 1$ [14, 660]; the paper [660] also discusses (1.27) briefly.

Second, we could have situations in which $[1 - n(\boldsymbol{r})]$ does *not* have compact support, but is such that $n(\boldsymbol{r}) \to 1$ as $r \to \infty$. The corresponding scattering problems are uncommon in acoustics.

Third, we could have $n(\boldsymbol{r}) \equiv 1$ outside B with n discontinuous across S, the boundary of B. The corresponding scattering problem will require transmission conditions across the interface S. (If the material in B is actually homogeneous, so that $n(\boldsymbol{r}) = n_0$, a constant, in B, we recover the standard transmission problem discussed in Section 1.3.3.) Here, we are mainly concerned with this third class of problem: acoustic scattering by N bounded inhomogeneous obstacles B_j, $j = 1, 2, \dots, N$, surrounded by an unbounded homogeneous fluid occupying B_e. In B_e, we denote the density, speed of sound and pressure by ρ_e, c and p_e, respectively, where ρ_e and c are constants. For scattering problems, we write

$$p_e = p_{inc} + p_{sc} \quad \text{in } B_e,$$

where p_{inc} is the given incident field and p_{sc} is the corresponding scattered field. From (1.27), the governing equation for p_{sc} is

$$(\nabla^2 + k^2) p_{sc} = 0 \quad \text{in } B_e, \tag{1.29}$$

where $k = \omega/c$ is a positive constant. We assume that p_{inc} satisfies (1.29) everywhere, except possibly at some places in B_e. We require that p_{sc} satisfies the Sommerfeld radiation condition (1.23).

Inside the inhomogeneous obstacles, we denote the density, speed of sound and pressure by ρ_0, c_0 and p_0, respectively, so that (1.27) becomes

$$\rho_0 \,\text{div}\left(\rho_0^{-1} \,\text{grad}\, p_0\right) + k_0^2 p_0 = 0 \quad \text{in } B, \tag{1.30}$$

where $k_0 = \omega/c_0$; in general, ρ_0 and k_0 vary within B. Across the interface S, we require continuity of pressure and normal velocity. These transmission conditions reduce to

$$p_e = p_0 \quad \text{and} \quad \frac{1}{\rho_e}\frac{\partial p_e}{\partial n} = \frac{1}{\rho_0}\frac{\partial p_0}{\partial n} \quad \text{on } S. \tag{1.31}$$

This defines the scattering problem for inhomogeneous obstacles.

1.3.4.1 Reduced equations

We can reduce Bergmann's equation to an equation without first derivatives by introducing a new dependent variable [87]; thus, define

$$p_0 = \rho_0^{1/2} v_0 \tag{1.32}$$

whence v_0 is found to satisfy

$$\nabla^2 v_0 + (k_0^2 + K)v_0 = 0, \tag{1.33}$$

where

$$K = \frac{1}{2}\rho_0^{-1}\nabla^2\rho_0 - \tfrac{3}{4}\rho_0^{-2}\,|\text{grad}\,\rho_0|^2 \tag{1.34}$$

$$= -\rho_0^{1/2}\nabla^2\left(\rho_0^{-1/2}\right). \tag{1.35}$$

Equations (1.32)–(1.34) (but not (1.35)), can be found in [138, p. 171].

In a homogeneous region, ρ_0 and c_0 are constants. Then, $K \equiv 0$, k_0 is a constant and (1.33) reduces to the standard Helmholtz equation. If ρ_0 is constant but c_0 is not, we still have $K \equiv 0$ and then (1.33) is usually written as (1.28); see, for example, [225, Chapter 8].

Finally, we can write (1.33) as

$$\nabla^2 v_0 + (k^2 - V)v_0 = 0, \tag{1.36}$$

where k^2 is a constant and $V = k^2 - k_0^2 - K$. In this form, (1.36) is recognised as *Schrödinger's equation* with potential V [911, eqn (10.59)].

For scattering problems, it is natural to mimic (1.32) in B_e, defining v_e by $p_e = \rho_e^{1/2} v_e$. Then, the transmission conditions (1.31) become

$$\gamma^{1/2} v_e = v_0 \quad \text{and} \quad \gamma^{-1/2}\frac{\partial v_e}{\partial n} = \frac{\partial v_0}{\partial n} + \frac{v_0}{2\rho_0}\frac{\partial \rho_0}{\partial n} \quad \text{on } S,$$

where $\gamma = \rho_e/\rho_0$ evaluated on S.

1.4 Multiple scattering of electromagnetic waves

In the past 25 years, numerical techniques for scattering and absorption by variously-shaped objects have proliferated like weeds.

(Wiscombe & Mugnai [1330, p. 119])

The scattering of electromagnetic waves is governed by Maxwell's equations. Thus, in the exterior domain B_e, we have

$$\operatorname{curl} \boldsymbol{E} - \mathrm{i}\mu\omega \boldsymbol{H} = \mathbf{0} \quad \text{and} \quad \operatorname{curl} \boldsymbol{H} + \mathrm{i}\varepsilon\omega \boldsymbol{E} = \mathbf{0}, \tag{1.37}$$

where $\boldsymbol{E}$ is the electric field, $\boldsymbol{H}$ is the magnetic field, ε is the electric permittivity, μ is the magnetic permeability and the time-dependence of $\mathrm{e}^{-\mathrm{i}\omega t}$ is suppressed. We always assume that ε and μ are constants. Eliminating $\boldsymbol{H}$, say, from (1.37) gives

$$\operatorname{curl}\operatorname{curl} \boldsymbol{E} - k^2 \boldsymbol{E} = \mathbf{0},$$

where $k = \omega\sqrt{\varepsilon\mu}$; $\boldsymbol{H}$ satisfies the same equation. Also, (1.37) implies that $\operatorname{div} \boldsymbol{E} = 0$ and $\operatorname{div} \boldsymbol{H} = 0$. It follows that

$$(\nabla^2 + k^2)\boldsymbol{E} = \mathbf{0} \quad \text{and} \quad (\nabla^2 + k^2)\boldsymbol{H} = \mathbf{0},$$

so that all the Cartesian components of $\boldsymbol{E}$ and $\boldsymbol{H}$ satisfy the same Helmholtz equation.

A given electromagnetic wave, $\{\boldsymbol{E}_{\text{inc}}, \boldsymbol{H}_{\text{inc}}\}$, is incident upon the obstacles B_i, $i = 1, 2, \ldots, N$. Write

$$\boldsymbol{E} = \boldsymbol{E}_{\text{inc}} + \boldsymbol{E}_{\text{sc}} \quad \text{and} \quad \boldsymbol{H} = \boldsymbol{H}_{\text{inc}} + \boldsymbol{H}_{\text{sc}} \quad \text{in } B_e, \tag{1.38}$$

where $\{\boldsymbol{E}_{\text{sc}}, \boldsymbol{H}_{\text{sc}}\}$ is the scattered field. Then, the problem is to solve Maxwell's equations (1.37) in B_e subject to appropriate boundary conditions on S and a radiation condition at infinity.

1.4.1 The Silver–Müller radiation conditions

The generalisation of the Sommerfeld radiation condition to electromagnetics was given by Silver and Müller in the late 1940s. There are two conditions, namely

$$\sqrt{\mu}\,\hat{\boldsymbol{r}} \times \boldsymbol{H}_{\text{sc}} + \sqrt{\varepsilon}\,\boldsymbol{E}_{\text{sc}} \to \mathbf{0} \tag{1.39}$$

and

$$\sqrt{\varepsilon}\,\hat{\boldsymbol{r}} \times \boldsymbol{E}_{\text{sc}} - \sqrt{\mu}\,\boldsymbol{H}_{\text{sc}} \to \mathbf{0} \tag{1.40}$$

as $r = |\boldsymbol{r}| \to \infty$, uniformly in $\hat{\boldsymbol{r}}$, where $\boldsymbol{r} = r\hat{\boldsymbol{r}}$. In fact, it is sufficient to impose just one of (1.39) and (1.40), as either one implies the other [223, Corollary 4.6]. Also, Maxwell's equations can be used to deduce radiation conditions for $\boldsymbol{E}_{\text{sc}}$ and $\boldsymbol{H}_{\text{sc}}$ separately; for example, (1.39) and (1.37) give

$$\hat{\boldsymbol{r}} \times \operatorname{curl} \boldsymbol{E}_{\text{sc}} + \mathrm{i}k\boldsymbol{E}_{\text{sc}} \to \mathbf{0} \quad \text{as } r \to \infty.$$

In particular, we can deduce that $\hat{\boldsymbol{r}}\cdot\boldsymbol{E}_{\text{sc}}$ and $\hat{\boldsymbol{r}}\cdot\boldsymbol{H}_{\text{sc}}$ vanish as $r\to\infty$. For more information, see [223, §4.2] or [262, §1.EM.4].

1.4.2 Boundary conditions

In electromagnetic problems, the most important boundary condition is that corresponding to a perfectly-conducting surface. This condition is

$$\boldsymbol{n}\times\boldsymbol{E}=\boldsymbol{0} \quad \text{or} \quad \boldsymbol{n}\times\boldsymbol{E}_{\text{sc}}=-\boldsymbol{n}\times\boldsymbol{E}_{\text{inc}} \quad \text{on } S, \tag{1.41}$$

where $\boldsymbol{n}$ is the unit normal pointing from S into B_{e}; see [223, §4.1] or [225, §6.4].

1.4.3 Transmission problems

If the regions B_i, $i=1,2,\ldots,N$, are filled with a different material, we are led to an electromagnetic transmission problem: given $\{\boldsymbol{E}_{\text{inc}},\boldsymbol{H}_{\text{inc}}\}$, find $\{\boldsymbol{E},\boldsymbol{H}\}$ and $\{\boldsymbol{E}_0,\boldsymbol{H}_0\}$, where $\{\boldsymbol{E},\boldsymbol{H}\}$ satisfies (1.37), $\{\boldsymbol{E}_0,\boldsymbol{H}_0\}$ satisfies

$$\operatorname{curl}\boldsymbol{E}_0-\mathrm{i}\mu_0\omega\boldsymbol{H}_0=\boldsymbol{0} \quad \text{and} \quad \operatorname{curl}\boldsymbol{H}_0+\mathrm{i}\varepsilon_0\omega\boldsymbol{E}_0=\boldsymbol{0} \quad \text{in } B, \tag{1.42}$$

$\{\boldsymbol{E}-\boldsymbol{E}_{\text{inc}},\boldsymbol{H}-\boldsymbol{H}_{\text{inc}}\}$ satisfies the Silver–Müller radiation conditions, and $\{\boldsymbol{E},\boldsymbol{H}\}$ and $\{\boldsymbol{E}_0,\boldsymbol{H}_0\}$ satisfy

$$\boldsymbol{n}\times\boldsymbol{E}=\boldsymbol{n}\times\boldsymbol{E}_0 \quad \text{and} \quad \boldsymbol{n}\times\boldsymbol{H}=\boldsymbol{n}\times\boldsymbol{H}_0 \quad \text{on } S. \tag{1.43}$$

Here, ε_0 and μ_0 are the electric permittivity and the magnetic permeability, respectively, of the material in B. For more information on this problem, see [482, 799]. For the transmission problem when the material in B is *chiral*, see [47, 48].

1.4.4 Inhomogeneous obstacles

Suppose that the obstacles B are inhomogeneous, with electric permittivity $\varepsilon_0(\boldsymbol{r})$ and magnetic permeability $\mu_0(\boldsymbol{r})$ at position $\boldsymbol{r}$. Maxwell's equations, (1.42), continue to hold. Eliminating $\boldsymbol{H}_0$ gives

$$\mu_0\operatorname{curl}(\mu_0^{-1}\operatorname{curl}\boldsymbol{E}_0)-k_0^2\boldsymbol{E}_0=\boldsymbol{0}, \tag{1.44}$$

where $k_0^2(\boldsymbol{r})=\omega^2\mu_0\varepsilon_0$. Similarly, eliminating $\boldsymbol{E}_0$ gives

$$\varepsilon_0\operatorname{curl}(\varepsilon_0^{-1}\operatorname{curl}\boldsymbol{H}_0)-k_0^2\boldsymbol{H}_0=\boldsymbol{0}.$$

These equations are discussed in [196, §1.3], together with their generalisation to anisotropic media for which $\varepsilon_0(\boldsymbol{r})$ and $\mu_0(\boldsymbol{r})$ are replaced by (invertible) matrices. For problems where μ_0 is constant, see [225, Chapter 9].

1.4.5 Two-dimensional problems

For electromagnetic scattering by perfectly-conducting cylinders, where the fields, boundary conditions and geometry do not vary with z, say, we can reduce the problem to two uncoupled problems for the scalar Helmholtz equation [564, §1.34]. They are

$$(\nabla^2 + k^2)E_z = 0 \quad \text{with } E_z = 0 \text{ on } S \tag{1.45}$$

and

$$(\nabla^2 + k^2)H_z = 0 \quad \text{with } \partial H_z/\partial n = 0 \text{ on } S. \tag{1.46}$$

Solutions of (1.45) and (1.46) are known as *TM waves* and *TE waves*, respectively [549, §8.2].

1.5 Multiple scattering of elastic waves

In this section, we formulate some problems involving the scattering of elastic waves by obstacles. Thus, we begin by supposing that the exterior domain B_e is filled with homogeneous elastic material, with elastic stiffnesses c_{ijkl} and mass density ρ. We assume, as usual, that

$$c_{ijkl} = c_{jikl} = c_{klij}. \tag{1.47}$$

For an isotropic solid,

$$c_{ijkl} = \lambda\delta_{ij}\delta_{kl} + \mu\left(\delta_{ik}\delta_{jl} + \delta_{il}\delta_{jk}\right), \tag{1.48}$$

where λ and μ are the Lamé moduli. Hooke's law gives the stresses as

$$\tau_{ij} = c_{ijkl}(\partial/\partial x_k)u_l,$$

where u_i are the components of the total displacement $\boldsymbol{u}$ and we have used the standard Einstein summation convention. The equations of motion for $\boldsymbol{u}$ are

$$\frac{\partial}{\partial x_j}\tau_{ij} + \rho\omega^2 u_i = 0, \quad i = 1, 2, 3, \quad \text{in } B_e. \tag{1.49}$$

For an isotropic material, these reduce to

$$k^{-2}\text{grad div}\,\boldsymbol{u} - K^{-2}\text{curl curl}\,\boldsymbol{u} + \boldsymbol{u} = \boldsymbol{0} \quad \text{in } B_e, \tag{1.50}$$

where the two wavenumbers k and K are defined by

$$\rho\omega^2 = (\lambda + 2\mu)k^2 = \mu K^2. \tag{1.51}$$

We assume that k and K are real and positive, and, as usual, the time dependence $e^{-i\omega t}$ is suppressed.

A given stress wave, $\boldsymbol{u}_{\text{inc}}$, is incident upon the obstacles B_i, $i = 1, 2, \ldots, N$. Write

$$\boldsymbol{u} = \boldsymbol{u}_{\text{inc}} + \boldsymbol{u}_{\text{sc}} \quad \text{in } B_{\text{e}},$$

where $\boldsymbol{u}_{\text{sc}}$ is the scattered field. Then, the problem is to solve (1.49) or (1.50) subject to a boundary condition on S and a radiation condition at infinity.

There are many good textbooks on elastodynamics. See, for example, [2, 429, 520, 83, 250].

1.5.1 The Kupradze radiation conditions

Kupradze generalised the Sommerfeld radiation to elastodynamics, assuming that the (homogeneous) material in B_{e} is isotropic; see [671, Chapter III, §2], [673, pp. 124–130] or [520, §6.5]. Thus, decompose the scattered field as

$$\boldsymbol{u}_{\text{sc}} = \boldsymbol{u}_{\text{sc}}^{P} + \boldsymbol{u}_{\text{sc}}^{S} \quad \text{in } B_{\text{e}},$$

where

$$\boldsymbol{u}_{\text{sc}}^{P} = -k^{-2} \operatorname{grad} \operatorname{div} \boldsymbol{u}_{\text{sc}} \quad \text{and} \quad \boldsymbol{u}_{\text{sc}}^{S} = K^{-2} \operatorname{curl} \operatorname{curl} \boldsymbol{u}_{\text{sc}}.$$

Then, in three dimensions, the elastodynamic radiation conditions are

$$r\left(\frac{\partial \boldsymbol{u}_{\text{sc}}^{P}}{\partial r} - \mathrm{i}k\boldsymbol{u}_{\text{sc}}^{P}\right) \to \boldsymbol{0} \quad \text{and} \quad r\left(\frac{\partial \boldsymbol{u}_{\text{sc}}^{S}}{\partial r} - \mathrm{i}K\boldsymbol{u}_{\text{sc}}^{S}\right) \to \boldsymbol{0}$$

as $r \to \infty$, uniformly in $\hat{\boldsymbol{r}}$. Note that the fields $\boldsymbol{u}_{\text{sc}}^{P}$ and $\boldsymbol{u}_{\text{sc}}^{S}$ are the longitudinal and transverse parts, respectively, of the scattered field; they satisfy

$$(\nabla^2 + k^2)\boldsymbol{u}_{\text{sc}}^{P} = \boldsymbol{0} \quad \text{and} \quad (\nabla^2 + K^2)\boldsymbol{u}_{\text{sc}}^{S} = \boldsymbol{0}$$

and correspond to radiated P-waves and S-waves, respectively. Note also that $\operatorname{curl} \boldsymbol{u}_{\text{sc}}^{P} = \boldsymbol{0}$ and $\operatorname{div} \boldsymbol{u}_{\text{sc}}^{S} = 0$.

1.5.2 Boundary conditions

In elastodynamics, the most important boundary condition is that corresponding to a cavity or hole. This condition is

$$T\boldsymbol{u} = \boldsymbol{0} \quad \text{or} \quad T\boldsymbol{u}_{\text{sc}} = -T\boldsymbol{u}_{\text{inc}} \quad \text{on } S, \tag{1.52}$$

where the traction operator is defined by

$$(T\boldsymbol{u})_m = n_i c_{imkl} \frac{\partial u_k}{\partial x_l} = \lambda n_m \frac{\partial u_j}{\partial x_j} + \mu n_j \left(\frac{\partial u_m}{\partial x_j} + \frac{\partial u_j}{\partial x_m}\right) \tag{1.53}$$

and $\boldsymbol{n}$ is the unit normal vector pointing from S into B_{e}.

Equation (1.52) may be regarded as the elastodynamic analogue of the Neumann condition. The corresponding Dirichlet condition is

$$\boldsymbol{u} = \boldsymbol{0} \quad \text{or} \quad \boldsymbol{u}_{\text{sc}} = -\boldsymbol{u}_{\text{inc}} \quad \text{on } S.$$

This condition is usually said to characterise a rigid surface S. In fact, if B contains a rigid material (which cannot deform, by definition), it will move, and this should be taken into account [936].

1.5.3 Transmission problems

If the regions B_i, $i = 1, 2, \ldots, N$, are filled with a different homogeneous isotropic elastic material, we are led to an elastodynamic transmission problem (or *inclusion problem*): given $\boldsymbol{u}_{\text{inc}}$, find $\boldsymbol{u}$ and $\boldsymbol{u}_0$, where $\boldsymbol{u}$ satisfies (1.50), $\boldsymbol{u}_0$ satisfies

$$k_0^{-2}\text{grad div}\,\boldsymbol{u}_0 - K_0^{-2}\text{curl curl}\,\boldsymbol{u}_0 + \boldsymbol{u}_0 = \boldsymbol{0} \quad \text{in } B,$$

$\boldsymbol{u} - \boldsymbol{u}_{\text{inc}}$ satisfies the Kupradze radiation conditions, and $\boldsymbol{u}$ and $\boldsymbol{u}_0$ satisfy

$$\boldsymbol{u} = \boldsymbol{u}_0 \quad \text{and} \quad T\boldsymbol{u} = T_0\boldsymbol{u}_0 \quad \text{on } S. \tag{1.54}$$

Here, k_0, K_0 and T_0 are defined by (1.51) and (1.53) with ρ, λ and μ replaced by ρ_0, λ_0 and μ_0, respectively, which are the material constants for the elastic solid occupying B. More generally, each inclusion could be made of a different anisotropic elastic material.

The transmission conditions (1.54) correspond to a 'welded' or 'perfect' interface. They can be modified to account for various imperfect interfaces [790].

The acoustic transmission problem (Section 1.3.3) can be modified so as to represent the problem of *fluid–solid interactions*: given the acoustic pressure field u_{inc}, find the acoustic scattered field u_{sc} in B_{e} and an elastic displacement $\boldsymbol{u}_0$ in B, subject to

$$\frac{\partial u}{\partial n} = \rho_{\text{f}}\omega^2 \boldsymbol{n} \cdot \boldsymbol{u}_0 \quad \text{and} \quad -u\boldsymbol{n} = T_0\boldsymbol{u}_0 \quad \text{on } S,$$

where $u = u_{\text{inc}} + u_{\text{sc}}$ and ρ_{f} is the density of the compressible inviscid fluid occupying B_{e}. For more information on this problem, see [750].

1.5.4 Inhomogeneous obstacles

Suppose that the obstacles are inhomogeneous and anisotropic, with elastic stiffnesses $c^0_{ijkl}(\boldsymbol{r})$ and density $\rho_0(\boldsymbol{r})$ at position $\boldsymbol{r}$. The equations of motion will now involve derivatives of the stiffnesses:

$$\frac{\partial}{\partial x_j}\tau^0_{ij} + \rho_0\omega^2 u^0_i = 0, \quad i = 1, 2, 3, \quad \text{in } B,$$

where Hooke's law gives the stresses as

$$\tau_{ij}^0 = c_{ijkl}^0(\partial/\partial x_k)u_l^0,$$

and u_i^0 are the components of $\boldsymbol{u}_0$, the displacement in B. The traction vector is $T_0\boldsymbol{u}_0$ with components $\tau_{ij}^0 n_j$.

If the stiffnesses are only piecewise-smooth functions of position within B, one may sometimes make progress by partitioning B into subregions, within each of which the stiffnesses are smooth. Then, one has to impose transmission conditions, similar to (1.54), across all the internal interfaces. This approach can be used, for example, if each scatterer is layered like an onion or the trunk of a tree.

1.5.5 Two-dimensional problems

Let (x, y, z) be Cartesian coordinates and suppose that each B_i is a cylinder with generators parallel to the z-axis. Then, if the incident field is independent of z, the three-dimensional scattering problems formulated above can be decomposed into two subproblems in a cross-sectional plane. Thus, let $\boldsymbol{u} = (u, v, w)$, where $u(x, y)$ and $v(x, y)$ are the in-plane components of $\boldsymbol{u}$ and $w(x, y)$ is the anti-plane component. The problem of finding w reduces to an acoustic problem, as $(\nabla^2 + K^2)w = 0$; solutions for w are usually referred to as *SH-waves*. Finding u and v remains as a genuinely coupled problem. For more information, see, for example, [2, §2.7] or [520, §1.8]. The two-dimensional Kupradze radiation conditions are discussed in [69, 1167] and [520, §6.9].

In-plane problems are often reduced to problems involving acoustic potentials. Thus, with $\boldsymbol{u}_2 \equiv (u, v)$ being the in-plane displacement, we can write

$$\boldsymbol{u}_2 = \operatorname{grad}\phi + \operatorname{curl}(\psi\hat{\mathbf{z}}), \tag{1.55}$$

where

$$(\nabla^2 + k^2)\phi = 0, \quad (\nabla^2 + K^2)\psi = 0$$

and $\hat{\mathbf{z}}$ is a unit vector in the z-direction [520, §2.7]. The representation (1.55) is convenient for constructing valid displacement fields. However, although ϕ and ψ satisfy separate Helmholtz equations, they are inevitably coupled through the boundary conditions on S, and this fact makes elastic problems more difficult to solve.

1.5.6 Half-space problems

There are several areas of application for which the most relevant problems involve obstacles in (or attached to) an elastic half-space, rather than an unbounded 'full-space'. Examples come from seismology [1123], soil-structure interaction [585, 269] and non-destructive evaluation.

Suppose that the half-space occupies the region $y > 0$. The boundary is usually taken to be free from tractions, so that $T\boldsymbol{u} = \boldsymbol{0}$ on (all or most of) $y = 0$. The obstacles could be grooves, cracks, canyons or other indentations in the free surface, they could be buried cavities, cracks or inclusions, or they could be elastic structures extending into $y < 0$. The flat free surface extends to infinity, and can support Rayleigh (surface) waves; see, for example, [1265], [2, §5.11] or [520, §3.4].

For problems of this kind, it is convenient to write $\boldsymbol{u} = \boldsymbol{u}_{\text{inc}} + \boldsymbol{u}_{\text{sc}}$ inside the half-space, where $\boldsymbol{u}_{\text{inc}}$ now consists of the total displacement *in the absence of the scattering obstacles*. If the incident wave is a plane P-wave, for example, then $\boldsymbol{u}_{\text{inc}}$ would consist of this wave together with the reflected P- and S-waves: in other words, $\boldsymbol{u}_{\text{inc}}$ is constructed so that $T\boldsymbol{u}_{\text{inc}} = \boldsymbol{0}$ on $y = 0$.

A uniqueness theorem for two-dimensional half-plane problems has been proved by Gregory [440].

1.6 Multiple scattering of water waves

Consider an infinite ocean of constant depth h. Specifically, let (x, y, z) be Cartesian coordinates, chosen so that $z = 0$ is the undisturbed free surface with z increasing with depth: the rigid bottom is at $z = h$. The water is assumed to be incompressible and inviscid. The motion is assumed to be irrotational, whence a velocity potential Φ exists. For time-harmonic motions, we have $\Phi = \text{Re}\{\phi\, \text{e}^{-\text{i}\omega t}\}$, where

$$\nabla^2 \phi = 0 \quad \text{in } B_{\text{e}}, \tag{1.56}$$

which is the region occupied by the water.

We suppose that there are N obstacles B_i $(i = 1, 2, \ldots, N)$ immersed in the water; these may be completely submerged or they may pierce the free surface. The wetted boundary of B_i is S_i.

A given wave with velocity potential ϕ_{inc} is incident upon the obstacles. Write

$$\phi = \phi_{\text{inc}} + \phi_{\text{sc}},$$

where ϕ_{sc} is the scattered potential. Then, the problem is to solve Laplace's equation subject to boundary conditions on S (defined by (1.1)), the bottom and the free surface, and a radiation condition.

The standard references on water waves are the books of Lamb [682] and Stoker [1145], and the encyclopaedic review by Wehausen & Laitone [1303]. More recent books include [102, 842, 731, 675].

For reviews on the interaction of water waves with multiple bodies, see [816, 909, 832].

1.6.1 Boundary conditions

In hydrodynamics, there is usually no flow through S or the bottom, whence

$$\frac{\partial \phi}{\partial n} = 0 \quad \text{or} \quad \frac{\partial \phi_{\text{sc}}}{\partial n} = -\frac{\partial \phi_{\text{inc}}}{\partial n} \quad \text{on } S, \tag{1.57}$$

and

$$\frac{\partial \phi}{\partial z} = 0 \quad \text{on the bottom, } z = h. \tag{1.58}$$

For ‘deep water’ ($h = \infty$), the bottom boundary condition (1.58) is replaced by

$$|\operatorname{grad} \phi| \to 0 \quad \text{as } z \to \infty.$$

On the free surface, we have a Robin condition,

$$K\phi + \frac{\partial \phi}{\partial z} = 0 \quad \text{on the free surface, } F,$$

where $K = \omega^2/g$ and g is the acceleration due to gravity. The boundary F is all of $z = 0$ if the obstacles are all submerged, otherwise it is the portion of $z = 0$ not intersected by the obstacles.

The boundary condition (1.57) is appropriate for scattering by *fixed* obstacles. More generally, we could consider floating, movable obstacles. The solution of such problems (with N three-dimensional obstacles) can be written as a linear combination of $6N$ (in general) radiation problems (with no incident wave) and the solution of the scattering problem (with all obstacles held fixed). The appropriate linear combination is determined using the equations of motion of the moving obstacles; see, for example, [753]. So, for one obstacle, one has to solve six radiation problems, in general; see, for example, [1303, §19β]. In this context, the scattering problem is sometimes known as the *diffraction problem* [1303, §18α], [906].

1.6.2 Radiation conditions

Water waves are surface waves with gravity as the restoring force. Typical velocity potentials for wave-like solutions are

$$e^{ikx} \cosh k(h-z) \quad \text{and} \quad J_0(kr) \cosh k(h-z), \tag{1.59}$$

where k is the unique positive real solution of

$$K = k \tanh kh, \tag{1.60}$$

J_0 is a Bessel function, and $r = (x^2 + y^2)^{1/2}$. For deep water, $k = K$.

The free-surface elevation is given by

$$-\frac{1}{g} \frac{\partial \Phi}{\partial t}\bigg|_{z=0} = \operatorname{Re}\left\{\eta(x,y)\, e^{-i\omega t}\right\},$$

so that $\eta(x,y) = i(\omega/g)\,\phi(x,y,0)$. Thus, for three-dimensional water waves, the waves propagate over a two-dimensional plane (the free surface, $z = 0$). Also, we see that the surface elevations corresponding to the wave-like solutions (1.59) satisfy two-dimensional Helmholtz equations so that it is natural to impose the two-dimensional Sommerfeld radiation condition, (1.25), on the scattered potential $\phi_{sc}(x,y,z)$; this condition is to hold uniformly in all horizontal directions and for all depths z. For more information on radiation conditions for water-wave problems, see [731, §1.3.1] or [675, p. 13].

1.6.3 Two-dimensional problems

For two-dimensional scattering problems, it is usual to express the far field (and hence, the radiation condition) in terms of reflection and transmission coefficients. Thus, we identify two problems, corresponding to waves incident from the right or the left. For deep water and a wave incident from $x = +\infty$, we have

$$\phi_{\text{inc}}(x, z) = \mathrm{e}^{-Kz-\mathrm{i}Kx} = \phi_{\text{inc}}^{-},$$

say, and then $\phi \equiv \phi_+$ satisfies

$$\phi_+(x, z) \sim \begin{cases} \mathrm{e}^{-Kz}(\mathrm{e}^{-\mathrm{i}Kx} + R_+\mathrm{e}^{\mathrm{i}Kx}) & \text{as } x \to \infty, \\ T_+\mathrm{e}^{-Kz-\mathrm{i}Kx} & \text{as } x \to -\infty. \end{cases}$$

Similarly, for deep water and a wave incident from $x = -\infty$, we have

$$\phi_{\text{inc}}(x, z) = \mathrm{e}^{-Kz+\mathrm{i}Kx} = \phi_{\text{inc}}^{+},$$

say, and then $\phi \equiv \phi_-$ satisfies

$$\phi_-(x, z) \sim \begin{cases} T_-\mathrm{e}^{-Kz+\mathrm{i}Kx} & \text{as } x \to \infty, \\ \mathrm{e}^{-Kz}(\mathrm{e}^{\mathrm{i}Kx} + R_-\mathrm{e}^{-\mathrm{i}Kx}) & \text{as } x \to -\infty. \end{cases}$$

The constants $R_\pm$ and $T_\pm$ are the (complex) reflection and transmission coefficients, respectively. They satisfy the following relations:

$$T_+ = T_- = T, \quad |R_+| = |R_-| = |R|,$$
$$|R|^2 + |T|^2 = 1 \quad \text{and} \quad T\overline{R_+} + \overline{T}R_- = 0,$$

where the first two equations define T and $|R|$, respectively, and the overbar denotes complex conjugation. Proofs of these relations and further references can be found in [787] and [842, §7.6.2]. Furthermore, if one applies Green's theorem to ϕ_+ and ϕ_{inc}^{+} in a region bounded by F, S and straight lines at $x = X$, $x = -X$ and $z = Z$ (where $X \to \infty$ and $Z \to \infty$), the result is a formula for T_+; repeating the calculation for the other three combinations of signs in $\phi_\pm$ and $\phi_{\text{inc}}^{\pm}$ gives formulae for the other three coefficients. Thus, we have

$$R_\pm = -\mathrm{i}\int_S \phi_\pm \frac{\partial}{\partial n}\phi_{\text{inc}}^{\mp}\,\mathrm{d}s \tag{1.61}$$

and

$$T_\pm = 1 - \mathrm{i}\int_S \phi_\pm \frac{\partial}{\partial n}\phi_{\text{inc}}^{\pm}\,\mathrm{d}s. \tag{1.62}$$

Such applications of Green's theorem (and other reciprocity relations) often yield useful formulae with minimal work; for a careful exposition, with many examples, see Achenbach's book [3].

The formulae (1.61) and (1.62) are valid for any (finite) number of scatterers. They are useful because they express the reflection and transmission coefficients in terms of the boundary values of the total potentials, $\phi_{\pm}$, and these may be computed by solving a boundary integral equation, for example.

1.6.4 Vertical cylinders

Suppose that every obstacle B_i is a vertical cylinder and that the water has constant finite depth, h. For the scattering problem, we start with an incident plane wave, given by $\phi_{\text{inc}}(x, y, z) = \mathrm{e}^{\mathrm{i}kx} \cosh k(h-z)$. Then, provided each cylinder extends from the bottom at $z = h$ through the free surface, we can separate out the dependence on z, writing $\phi_{\text{sc}}(x, y, z) = \varphi(x, y) \cosh k(h-z)$. As ϕ_{sc} satisfies Laplace's equation in three dimensions, we find that the reduced potential φ satisfies the two-dimensional Helmholtz equation,

$$(\nabla^2 + k^2)\varphi = 0.$$

Hence, the problem reduces to a two-dimensional acoustic scattering problem. For radiation problems (with prescribed normal velocity on S), we can still separate the variables, writing $\phi(x, y, z) = \varphi(x, y)\, Z(z)$, say, but now, in general, we obtain an infinite set of eigenfunctions for Z. Explicitly, we obtain

$$\phi(x, y, z) = \varphi(x, y)\, Z_0(z) + \sum_{n=1}^{\infty} \varphi_n(x, y)\, Z_n(z), \tag{1.63}$$

where

$$Z_0(z) = \cosh k(h-z), \quad Z_n(z) = \cos k_n(h-z) \quad \text{for } n = 1, 2, \ldots, \tag{1.64}$$

k_n $(n = 1, 2, \ldots)$ are positive real solutions of

$$K = -k_n \tan k_n h \tag{1.65}$$

and φ_n solves $(\nabla^2 - k_n^2)\varphi_n = 0$, the two-dimensional modified Helmholtz equation. Notice that the vertical eigenfunctions are orthogonal:

$$\int_0^h Z_n(z)\, Z_m(z)\, \mathrm{d}z = 0 \quad \text{if } n \neq m, \text{ with } n \geq 0 \text{ and } m \geq 0. \tag{1.66}$$

1.6.5 Uniqueness

Although the basic problems for the scattering and radiation of linear water waves were formulated long ago, the first general mathematical analysis was given by John in 1950 [559]. In particular, he studied the question of uniqueness. He proved that the boundary-value problems for bodies that are not completely submerged do have at most one solution, provided the bodies satisfy a certain geometrical condition. In the same year, Ursell [1225] proved uniqueness for a single horizontal circular cylinder, completely submerged beneath the free surface of deep water. Subsequently, there was much work in which the geometrical conditions were weakened: it was widely believed that the geometrical conditions were superfluous, and that a general uniqueness proof would be found one day. All this changed when a counter-example was found by McIver in 1996 [826]. She constructed a symmetric catamaran-type structure in two dimensions for which the corresponding homogeneous boundary-value problem has a non-trivial solution at one value of K. Further examples have been found since; for a good discussion, see [675].

1.7 Overview of the book

In this chapter, we have delineated the scope of the book. The basic problem to be studied is the scattering of time-harmonic waves by a finite number of bounded obstacles: let us call this the *basic multiple-scattering problem*. As a prototype, we consider acoustic scattering (governed by the Helmholtz equation) in detail. Results for two and three dimensions are given. Many extensions are also described, including those for elastic waves, electromagnetic waves and water waves.

The oldest methods for solving the basic multiple-scattering problem combine separated solutions (for example, in cylindrical or spherical polar coordinates) with appropriate addition theorems. Indeed, the general idea of combining multipole expansions with addition theorems has been used in many physical contexts, involving configurations of cylinders and spheres. Thus, we begin in Chapter 2 with two-dimensional separated solutions, including their associated addition theorems. The corresponding three-dimensional results are given in Chapter 3, where the emphasis is on the use of spherical polar coordinates. These two chapters are rather technical, but the material developed there is used extensively in the subsequent chapters.

Apart from cylindrical wavefunctions (involving Bessel and Hankel functions), Chapter 2 also contains new results for elliptical wavefunctions (Mathieu functions) and a discussion of multipole solutions for water-wave problems. Chapter 3 contains a wealth of formulae for spherical wavefunctions. The main justification for this development is that three-dimensional addition theorems are complicated, and new methods continue to be devised for the efficient computational use of these theorems. For example, we derive various formulae that have been used in 'fast multipole methods' (see Section 6.14). Chapter 3 also includes new addition theorems for water-wave multi-pole potentials.

Chapter 4 is dedicated to methods based on separation of variables. Results are given for circular cylindrical coordinates, elliptic cylindrical coordinates and spherical polar coordinates. These methods are exact, in principle, and they are fast in practice.

However, they are limited to scatterers with simple shapes. Consequently, more general methods are developed in the subsequent chapters.

Methods based on integral equations (mainly boundary integral equations) are developed in Chapters 5 and 6. The first of these covers the basic theory with simple acoustic applications. Chapter 6 considers many other problems and physical applications: penetrable and inhomogeneous scatterers; electromagnetic problems, including inhomogeneous obstacles and thin wires; elastodynamic problems, including inclusions and cracks; hydrodynamic problems; modified integral equations, designed to eliminate 'irregular frequencies'; and the use of exact Green's functions. We also discuss Twersky's method for the basic multiple-scattering problem. The chapter ends with a discussion of some analytical aspects of fast multipole methods.

T-matrix and related methods are developed in Chapter 7. These methods may be viewed as generalisations of the method of separation of variables, in which spherical multipoles are used for non-spherical scatterers, for example. They may also be viewed as a method for solving certain integral equations. Chapter 7 also considers several related topics: Kupradze's method and the 'method of fundamental solutions'; complete sets of functions, as used to represent functions defined on a curve or surface; and various properties of the T-matrix.

Chapter 8 is concerned with approximations. How do the exact methods in the rest of the book simplify if the scatterers are small or widely separated? Detailed results are given for acoustic scattering by one small obstacle in two and three dimensions; these lead to new low-frequency approximations for the T-matrix. An exact method due to Foldy is described: it assumes that each obstacle scatters isotropically, and so is appropriate for small soft obstacles, for example. We extend Foldy's method to non-isotropic scatterers by including acoustic dipoles of unknown strength and orientation. Various wide-spacing approximations are discussed in Section 8.5. Finally, we also give a brief discussion of problems in which the scatterers are located randomly. This is a large topic, with a large literature; we outline some of this work, with emphasis on multiple scattering.

The book ends with nine appendices and a list of references. The reference list is extensive, from Abramowitz & Stegun [1] to Zurk *et al.* [1386]. It is not intended to be exhaustive because, first, the subject is too old and too broad, and second, we only cite material that we have actually seen: according to one study [1103], 'only about 20% of citers read the original'!

2
Addition theorems in two dimensions

Bessel functions are not algebraic functions, ... they are not simply periodic functions, and ... they are not doubly periodic functions. Consequently, in accordance with a theorem of Weierstrass, it is not possible to express $J_\nu(Z+z)$ as an algebraic function of $J_\nu(Z)$ and $J_\nu(z)$. That is to say, that Bessel functions do not possess addition theorems in the strict sense of the term.

(Watson [1298, p. 358])

2.1 Introduction

Watson began his chapter on 'Addition Theorems' for Bessel functions with the remarks above. Nevertheless, he went on to give proofs of various formulae which are 'commonly described as addition theorems'.

An example of an addition theorem in the strict sense is

$$e^{i(x+y)} = e^{ix}\, e^{iy}.$$

In the wider sense, we have the formula

$$J_0(x+y) = \sum_{n=-\infty}^{\infty} (-1)^n J_n(x)\, J_n(y), \tag{2.1}$$

where J_n is a Bessel function. Further examples are given by Askey [46, Lecture 4].

Addition theorems are at the heart of most theories of multiple scattering. Roughly speaking, they are used to transform one expansion about some point in space into a similar expansion about a different point. We shall prove several theorems of this type, in both two and three dimensions. This chapter is concerned with two-dimensional addition theorems; these results are mainly classical. Analogous results in three dimensions are proved in Chapter 3. These are typically more complicated, and they are still the subject of current research.

2.2 Cartesian coordinates

Let $(x_1, x_2, \dots, x_q)$ be Cartesian coordinates. In practice, we are usually interested in $q=2$ or $q=3$; when convenient, we write x, y and z for x_1, x_2 and x_3, respectively.

Consider the general linear homogeneous partial differential equation

$$\sum_{i=1}^{q}\sum_{j=1}^{q} a_{ij}\frac{\partial^2 u}{\partial x_i \partial x_j} + \sum_{i=1}^{q} b_i \frac{\partial u}{\partial x_i} + cu = 0, \tag{2.2}$$

where a_{ij}, b_i and c are all constants. Typical examples are Laplace's equation,

$$\nabla_q^2 u = 0,$$

and the Helmholtz equation,

$$(\nabla_q^2 + k^2)u = 0,$$

where ∇_q^2 is the q-dimensional Laplacian,

$$\nabla_q^2 = \frac{\partial^2}{\partial x_1^2} + \frac{\partial^2}{\partial x_2^2} + \cdots + \frac{\partial^2}{\partial x_q^2}.$$

Solutions of Laplace's equation are called *harmonic functions*, whereas solutions of the Helmholtz equation are called *wavefunctions*.

It is obvious that if u solves (2.2), then so does any linear combination of derivatives of u with respect to any x_k. Thus, given one solution of (2.2), further solutions can be found by differentiation. The use of this idea for constructing harmonic functions in three dimensions (by repeated differentiation of $1/r$) is described at length by Thomson & Tait [1175, Appendix B].

In order to define multipole solutions, we introduce a radial coordinate,

$$r = \sqrt{\varphi_q}, \quad \text{where} \quad \varphi_q = x_1^2 + x_2^2 + \cdots + x_q^2.$$

Hobson's theorem, several variants of which are proved below, is convenient for applying differential operators in Cartesian coordinates to functions of φ_q. To be precise, let $f_n(x_1, x_2, \dots, x_q)$ be a homogeneous polynomial of degree n, where $n \geq 0$ is an integer. This means that

$$f_n(\lambda x_1, \lambda x_2, \dots, \lambda x_q) = \lambda^n f_n(x_1, x_2, \dots, x_q),$$

for any λ, and implies that

$$f_n(x_1, x_2, \dots, x_q) = \sum_{j=0}^{n} x_1^j g_{n-j}(x_2, x_3, \dots, x_q),$$

where g_m is a homogeneous polynomial of degree m in the $q-1$ variables $x_2, x_3, \dots, x_q$. Given f_n, we define the differential operator of order n, $f_n(\nabla_q)$ by

$$f_n(\nabla_q) = f_n\left(\frac{\partial}{\partial x_1}, \frac{\partial}{\partial x_2}, \dots, \frac{\partial}{\partial x_q}\right). \tag{2.3}$$

Hobson's theorem gives the result of applying $f_n(\nabla_q)$ to $F(\varphi_q)$, where F is any sufficiently smooth function of one variable. The result simplifies if f_n is a *harmonic* homogeneous polynomial, $\mathcal{H}_n$ say, so that

$$\nabla_q^2 \mathcal{H}_n(x_1, x_2, \ldots, x_q) = 0.$$

In this chapter, we take $q = 2$. In this case, there are exactly two linearly independent harmonic homogeneous polynomials of degree n for each positive n; they can be taken as

$$\mathcal{H}_n^{\pm}(x, y) = (x \pm \mathrm{i}y)^n = r^n \mathrm{e}^{\pm \mathrm{i}n\theta}, \tag{2.4}$$

where r and θ are plane polar coordinates ($x = r\cos\theta$ and $y = r\sin\theta$).

2.3 Hobson's theorem

We start with a simple one-dimensional version of Hobson's theorem; according to Hobson [507, p. 125], it is due to Schlömilch.

Theorem 2.1 *Let $\varphi_1 = x^2$. Let n be a positive integer. Then*

$$\frac{\mathrm{d}^n}{\mathrm{d}x^n} F(\varphi_1) = \sum_{l=0} \frac{2^{n-2l}}{l!} F^{(n-l)}(\varphi_1) \frac{\mathrm{d}^{2l}}{\mathrm{d}x^{2l}} x^n,$$

where $F^{(m)}$ is the mth derivative of F and the summation has a finite number of terms.

Proof We give an inductive proof. The case $n = 1$ is trivial. Assuming the result for n, we consider

$$\begin{aligned}
\frac{\mathrm{d}^{n+1}}{\mathrm{d}x^{n+1}} F(\varphi_1) &= \frac{\mathrm{d}}{\mathrm{d}x} \sum_{l=0} \frac{2^{n-2l}}{l!} F^{(n-l)}(\varphi_1) \frac{\mathrm{d}^{2l}}{\mathrm{d}x^{2l}} x^n, \\
&= \sum_{l=0} \frac{2^{n-2l}}{l!} \left\{ 2xF^{(n+1-l)} \frac{\mathrm{d}^{2l}}{\mathrm{d}x^{2l}} x^n + F^{(n-l)} \frac{\mathrm{d}^{2l+1}}{\mathrm{d}x^{2l+1}} x^n \right\} \\
&= 2^{n+1} x^{n+1} F^{(n+1)} + \sum_{l=1} \frac{2^{n+1-2l}}{l!} F^{(n+1-l)} \Lambda_n^l,
\end{aligned}$$

where

$$\Lambda_n^l = x \frac{\mathrm{d}^{2l}}{\mathrm{d}x^{2l}} x^n + 2l \frac{\mathrm{d}^{2l-1}}{\mathrm{d}x^{2l-1}} x^n = \frac{\mathrm{d}^{2l}}{\mathrm{d}x^{2l}} x^{n+1};$$

the last equality can be proved by direct calculation or by taking $P = x_1^n$ in Lemma 2.2. Thus, the result is true for $n + 1$, and so the theorem is proved. □

Lemma 2.2 *Let ∇_q^2 be the q-dimensional Laplacian, let $(x_1, x_2, \ldots, x_q)$ be Cartesian coordinates, and let l be a positive integer. Then*

$$\nabla_q^{2l}(x_1 P) - x_1 \nabla_q^{2l} P = 2l \nabla_q^{2l-2} \frac{\partial P}{\partial x_1}$$

for any sufficiently smooth function $P(x_1, x_2, \ldots, x_q)$, where, by definition, $\nabla_q^{2l} = (\nabla_q^2)^l$.

Proof The result is trivial for $l = 1$:

$$\nabla_q^2(x_1 P) - x_1 \nabla_q^2 P = 2\frac{\partial P}{\partial x_1}. \tag{2.5}$$

Assuming the result for l, we consider

$$\begin{aligned}\nabla_q^{2l+2}(x_1 P) - x_1 \nabla_q^{2l+2} P &= \nabla_q^{2l}\left\{\nabla_q^2(x_1 P)\right\} - x_1 \nabla_q^{2l}\nabla_q^2 P \\ &= \nabla_q^{2l}\left\{x_1 \nabla_q^2 P + 2\frac{\partial P}{\partial x_1}\right\} - x_1 \nabla_q^{2l}\nabla_q^2 P,\end{aligned}$$

using (2.5). Then, applying the assumed formula to $\nabla_q^2 P$, we see that the right-hand side becomes

$$2l\nabla_q^{2l-2}\frac{\partial}{\partial x_1}\nabla_q^2 P + 2\nabla_q^{2l}\frac{\partial P}{\partial x_1} = 2(l+1)\nabla_q^{2l}\frac{\partial P}{\partial x_1},$$

as required. □

We generalise Theorem 2.1 to two dimensions.

Theorem 2.3 (Hobson's theorem in two dimensions) *Let $\varphi_2 = x^2 + y^2$. Let $f_n(x, y)$ be a homogeneous polynomial of degree n, where n is a positive integer. Then, with $f_n(\nabla_2)$ defined by (2.3),*

$$f_n(\nabla_2)F(\varphi_2) = \sum_{l=0} \frac{2^{n-2l}}{l!} F^{(n-l)}(\varphi_2)\nabla_2^{2l} f_n(x, y),$$

where the summation has a finite number of terms.

Proof Again, we give an inductive proof. First of all, we note that the one-dimensional theorem, Theorem 2.1, is valid for functions of φ_2, so that we have

$$\frac{\partial^n}{\partial y^n} F(\varphi_2) = \sum_{l=0} \frac{2^{n-2l}}{l!} F^{(n-l)}(\varphi_2)\nabla_2^{2l} y^n, \tag{2.6}$$

for every positive n.

For the inductive argument, we start by noting that the result is true for $n = 1$. In this case, we have $f_1(x, y) = a_0 x + a_1 y$, and so

$$f_1(\nabla_2)F(\varphi_2) = 2F'(\varphi_2)f_1(x, y),$$

as required ($\nabla_2^{2l} f_1 = 0$ for $l \geq 1$).

Next, we assume that the result is true for n. We have

$$f_{n+1}(x, y) = x f_n(x, y) + a y^{n+1} \tag{2.7}$$

for some coefficient a. Thus,

$$\begin{aligned} f_{n+1}(\nabla_2) F(\varphi_2) &= \frac{\partial}{\partial x} f_n(\nabla_2) F(\varphi_2) + a \frac{\partial^{n+1}}{\partial y^{n+1}} F(\varphi_2) \\ &= \frac{\partial}{\partial x} \sum_{l=0} \frac{2^{n-2l}}{l!} F^{(n-l)}(\varphi_2) \nabla_2^{2l} f_n \\ &\quad + a \sum_{l=0} \frac{2^{n+1-2l}}{l!} F^{(n+1-l)}(\varphi_2) \nabla_2^{2l} y^{n+1}, \end{aligned}$$

using the assumed result and (2.6). The first term on the right-hand side is

$$\sum_{l=0} \frac{2^{n-2l}}{l!} \left\{ 2x F^{(n+1-l)} \nabla_2^{2l} f_n + F^{(n-l)} \frac{\partial}{\partial x} \nabla_2^{2l} f_n \right\},$$

whereas, by (2.7), the second term is

$$\sum_{l=0} \frac{2^{n+1-2l}}{l!} F^{(n+1-l)} \nabla_2^{2l} f_{n+1} - \sum_{l=0} \frac{2^{n+1-2l}}{l!} F^{(n+1-l)} \nabla_2^{2l} (x f_n).$$

Hence,

$$f_{n+1}(\nabla_2) F(\varphi_2) = \sum_{l=0} \frac{2^{n+1-2l}}{l!} F^{(n+1-l)}(\varphi_2) \nabla_2^{2l} f_{n+1}(x, y) + L_n,$$

where

$$L_n = \sum_{l=0} \frac{2^{n-2l}}{l!} \left(2F^{(n+1-l)} \left[x \nabla_2^{2l} f_n - \nabla_2^{2l} (x f_n) \right] + F^{(n-l)} \frac{\partial}{\partial x} \nabla_2^{2l} f_n \right).$$

We must show that $L_n \equiv 0$ to complete the proof. To do this, we note that the expression in square brackets vanishes when $l = 0$, whence

$$L_n = \sum_{l=1} \frac{2^{n+1-2l}}{l!} F^{(n+1-l)} \left\{ x \nabla_2^{2l} f_n - \nabla_2^{2l} (x f_n) + 2l \frac{\partial}{\partial x} \nabla_2^{2l-2} f_n \right\};$$

this vanishes by Lemma 2.2, as required. □

The theorems above are special cases of a general theorem proved by Hobson [507, §79]. Our proofs are less elegant but perhaps more transparent.

The result of Hobson's theorem simplifies if f_n is harmonic. Thus, we obtain the following corollary.

Corollary 2.4 *Let $r = \sqrt{x^2 + y^2}$. Let $\mathcal{H}_n(x, y)$ be a harmonic homogeneous polynomial of degree n, where n is a positive integer. Then*

$$\mathcal{H}_n(\nabla_2) F(r) = \mathcal{H}_n(x, y) \left(\frac{1}{r} \frac{\mathrm{d}}{\mathrm{d}r} \right)^n F(r).$$

Thus, if f_n is harmonic, the (finite) series in Theorem 2.3 reduces to just one term. A simple example follows; applications to wavefunctions will be given in the next section.

Example 2.5 Take $F(r) = \log r$ and $\mathcal{H}_n = \mathcal{H}_n^{\pm}$, defined by (2.4), giving

$$\left(\frac{\partial}{\partial x} \pm \mathrm{i}\frac{\partial}{\partial y}\right)^n \log r = 2^{n-1}(n-1)!\,\frac{\mathrm{e}^{\pm\mathrm{i}n\theta}}{r^n}.$$

The q-dimensional version of Corollary 2.4 can be found in [893, §12, Theorem 1].

2.4 Wavefunctions

Hobson applied his theorem to harmonic functions in three dimensions. We apply it here to wavefunctions (solutions of the Helmholtz equation), in two dimensions.

Let $Z_\nu(w)$ denote any cylinder function of order ν, that is

$$Z_\nu(w) = J_\nu(w),\ Y_\nu(w),\ H_\nu^{(1)}(w) \text{ or } H_\nu^{(2)}(w). \tag{2.8}$$

Then, we have ([1, eqn (9.1.30)], [1298, §3.2])

$$\left(\frac{1}{w}\frac{\mathrm{d}}{\mathrm{d}w}\right)^m \frac{Z_\nu(w)}{w^\nu} = (-1)^m \frac{Z_{\nu+m}(w)}{w^{\nu+m}} \tag{2.9}$$

for $m = 0, 1, 2, \ldots$ and any ν. In particular, setting $\nu = 0$ and $w = kr$ gives

$$\left(\frac{1}{kr}\frac{\mathrm{d}}{\mathrm{d}r}\right)^m Z_0(kr) = \frac{(-1)^m}{r^m} Z_m(kr). \tag{2.10}$$

Next, we show that wavefunctions can be generated by the application of certain differential operators.

Definition 2.6 *The differential operators $\mathcal{D}_m^{\pm}$ are defined by*

$$\mathcal{D}_m^{\pm} = \left[\frac{-1}{k}\left(\frac{\partial}{\partial x} \pm \mathrm{i}\frac{\partial}{\partial y}\right)\right]^m,$$

for $m = 0, 1, 2, \ldots$, where $\mathcal{D}_0^{\pm} = I$, the identity operator.

The action of these operators on cylindrical wavefunctions is described by the next two theorems.

Theorem 2.7

$$\mathcal{D}_m^{\pm}\{Z_0(kr)\} = Z_m(kr)\,\mathrm{e}^{\pm\mathrm{i}m\theta} \quad \textit{for } m = 0, 1, 2, \ldots.$$

First proof Apply Corollary 2.4 to $F(r) = Z_0(kr)$, using (2.10), with $\mathcal{H}_m = (-1/k)^m(x \pm \mathrm{i}y)^m$. □

We can also give a direct proof of Theorem 2.7 using an inductive argument.

Second proof The theorem is obviously true for $m=0$. For subsequent values, we use induction. We have, assuming the result is true for m,

$$\mathcal{D}_{m+1}^{\pm} Z_0 = \mathcal{D}_1^{\pm}\{\mathcal{D}_m^{\pm} Z_0\} = \mathcal{D}_1^{\pm}\{Z_m(kr)\,\mathrm{e}^{\pm\mathrm{i}m\theta}\}.$$

Now, in terms of plane polar coordinates, we have

$$\mathcal{D}_1^{\pm}(\boldsymbol{r}) = \frac{-1}{k}\,\mathrm{e}^{\pm\mathrm{i}\theta}\left(\frac{\partial}{\partial r} \pm \frac{\mathrm{i}}{r}\frac{\partial}{\partial\theta}\right).$$

Hence $\mathcal{D}_{m+1}^{\pm}\{Z_0(kr)\} = Z_{m+1}(kr)\,\mathrm{e}^{\pm\mathrm{i}(m+1)\theta}$, since [1, §9.1.27]

$$Z_\nu'(w) - (\nu/w)\,Z_\nu(w) = -Z_{\nu+1}(w),$$

and the result follows. □

Another interesting relation, similar to Theorem 2.7, is

$$\mathcal{T}_m\{Z_0(kr)\} = \mathrm{i}^m\,Z_m(kr)\cos m\theta, \tag{2.11}$$

where $\mathcal{T}_m = T_m((\mathrm{i}k)^{-1}\partial/\partial x)$ and $T_m(w)$ is a Chebyshev polynomial of the first kind [427, §8.94]. Formula (2.11), which is [83, eqn (2.6.14)], can be proved by induction, using the recurrence relation for T_n.

We can generalise Theorem 2.7, but first we state a trivial but useful property of $\mathcal{D}_m^{\pm}$.

Lemma 2.8 *If u is any solution of the two-dimensional Helmholtz equation,*

$$\mathcal{D}_m^{+}\mathcal{D}_m^{-}u = \mathcal{D}_m^{-}\mathcal{D}_m^{+}u = (-1)^m u \quad \textit{for } m = 0, 1, 2, \ldots.$$

We shall use this lemma in the proof of the next theorem.

Theorem 2.9

$$\mathcal{D}_m^{+}\{Z_n(kr)\,\mathrm{e}^{\mathrm{i}n\theta}\} = Z_{m+n}(kr)\,\mathrm{e}^{\mathrm{i}(m+n)\theta}, \tag{2.12}$$

$$\begin{aligned}\mathcal{D}_m^{-}\{Z_n(kr)\,\mathrm{e}^{\mathrm{i}n\theta}\} &= (-1)^n Z_{m-n}(kr)\,\mathrm{e}^{-\mathrm{i}(m-n)\theta},\\ &= (-1)^m Z_{n-m}(kr)\,\mathrm{e}^{\mathrm{i}(n-m)\theta},\end{aligned} \tag{2.13}$$

for $m = 0, 1, 2, \ldots$ and $n = 0, \pm 1, \pm 2, \ldots$.

Proof Let A denote the left-hand side of (2.12). If $n \ge 0$, we have

$$A = \mathcal{D}_m^{+}\mathcal{D}_n^{+}Z_0 = \mathcal{D}_{m+n}^{+}Z_0 = Z_{m+n}(kr)\,\mathrm{e}^{\mathrm{i}(m+n)\theta}.$$

If $n < 0$, set $n = -l$, whence

$$A = (-1)^l\mathcal{D}_m^{+}\mathcal{D}_l^{-}Z_0.$$

Now, there are two cases. First, if $m \ge l$, we use Lemma 2.8 to eliminate $\mathcal{D}_l^{-}$:

$$A = \mathcal{D}_{m-l}^{+}Z_0 = \mathcal{D}_{m+n}^{+}Z_0,$$

as before. Second, if $m < l$, we use Lemma 2.8 to eliminate $\mathcal{D}_m^+$:

$$A = (-1)^{l+m}\mathcal{D}_{l-m}^- Z_0 = (-1)^{l+m} Z_{l-m}(kr)\,\mathrm{e}^{-\mathrm{i}(l-m)\theta}$$

and again we obtain the desired result, after using $Z_{-n} = (-1)^n Z_n$. A similar argument succeeds for (2.13). □

We remark that similar results can be proved for modified Bessel functions $I_n(w)$ and $K_n(w)$.

2.5 Addition theorems

Consider two origins, O_1 and O_2. Let $\boldsymbol{r}_j$ be the position vector of a general point P with respect to O_j, for $j = 1, 2$. Let $\boldsymbol{b}$ be the position vector of O_1 with respect to O_2, so that $\boldsymbol{r}_2 = \boldsymbol{r}_1 + \boldsymbol{b}$. Let $\boldsymbol{r}_j = (r_j \cos\theta_j, r_j \sin\theta_j)$ and $\boldsymbol{b} = (b\cos\beta, b\sin\beta)$. See Fig. 2.1 for a sketch of the geometry. The simplest addition theorem was proved by C. Neumann in 1867; see [1298, §11.2], [884, p. 1371], [691, eqn (5.12.2)] or [31, eqn (4.10.2)].

Theorem 2.10 (Neumann's addition theorem)

$$J_0(kr_2) = \sum_{n=-\infty}^{\infty} (-1)^n J_n(kb)\, J_n(kr_1)\,\mathrm{e}^{\mathrm{i}n(\theta_1-\beta)} \tag{2.14}$$

$$= \sum_{n=0}^{\infty} \epsilon_n (-1)^n J_n(kb)\, J_n(kr_1)\, \cos n(\theta_1-\beta), \tag{2.15}$$

where $\epsilon_0 = 1$ *and* $\epsilon_n = 2$ *for* $n > 0$.

Note that (2.14) reduces to (2.1) when $\theta_1 = \beta$.

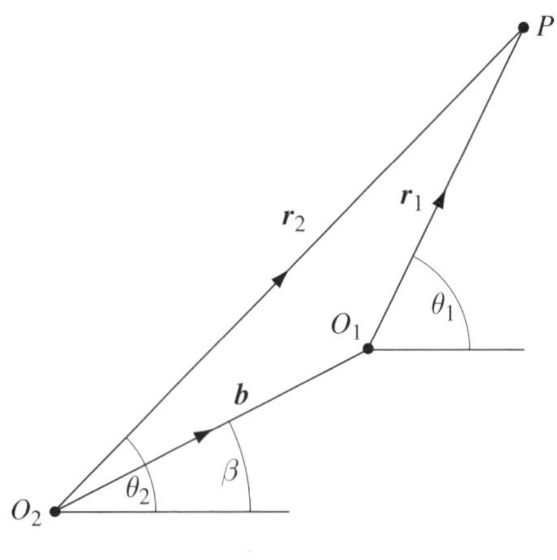

Fig. 2.1. Configuration for two-dimensional addition theorems.

First proof The Parseval integral for $J_0(kr_2)$ gives

$$J_0(kr_2) = \frac{1}{2\pi}\int_{-\pi}^{\pi} e^{ikr_2\cos\varphi}\,d\varphi = \frac{1}{2\pi}\int_{-\pi}^{\pi} e^{ikr_2\cos(\varphi-\theta_2)}\,d\varphi$$
$$= \frac{1}{2\pi}\int_{-\pi}^{\pi} e^{ikr_1\cos(\varphi-\theta_1)}\,e^{ikb\cos(\varphi-\beta)}\,d\varphi. \tag{2.16}$$

Now, from the generating function for Bessel functions [1298, p. 14],

$$\exp\left\{\frac{1}{2}w(t-t^{-1})\right\} = \sum_{n=-\infty}^{\infty} t^n J_n(w),$$

with the substitution $t = ie^{i\varphi}$, we obtain the Jacobi expansion [1298, p. 22]

$$e^{iw\cos\varphi} = \sum_{n=-\infty}^{\infty} i^n J_n(w)\,e^{in\varphi}. \tag{2.17}$$

Use this expansion twice in (2.16), once for each exponential, and then integrate over φ using

$$\int_{-\pi}^{\pi} e^{im\varphi}\,d\varphi = 2\pi\,\delta_{0m} \tag{2.18}$$

to complete the proof. □

Before generalising Neumann's theorem, we give an alternative proof. This is based on properties of the differential operators $\mathcal{D}_m^{\pm}$, defined by Definition 2.6.

Second proof of Neumann's addition theorem Since $J_0(kr_2)$ is a solution of the two-dimensional Helmholtz equation, it must have an expansion of the form

$$J_0(kr_2) = \sum_{n=-\infty}^{\infty} C_n J_n(kr_1)\,e^{in\theta_1}, \tag{2.19}$$

for some coefficients C_n. To find these coefficients, we apply the operator $\mathcal{D}_m^+(\boldsymbol{r}_2)$ to (2.19), with $m \geq 0$, keeping $\boldsymbol{b}$ fixed; this gives

$$J_m(kr_2)\,e^{im\theta_2} = \sum_{n=-\infty}^{\infty} C_n \mathcal{D}_m^+(\boldsymbol{r}_2)\left\{J_n(kr_1)\,e^{in\theta_1}\right\}.$$

Since the Cartesian axes are parallel, we have $\mathcal{D}_n^+(\boldsymbol{r}_2) = \mathcal{D}_n^+(\boldsymbol{r}_1)$. Hence, using Theorem 2.9, we obtain

$$J_m(kr_2)\,e^{im\theta_2} = \sum_{n=-\infty}^{\infty} C_n J_{m+n}(kr_1)\,e^{i(m+n)\theta_1}, \tag{2.20}$$

for $m = 0, 1, 2, \ldots$. Now, let $r_1 \to 0$, so that $r_2 \to b$ and $\theta_2 \to \beta$; since $J_m(0) = \delta_{0m}$, we obtain

$$J_m(kb)\,e^{im\beta} = C_{-m}, \tag{2.21}$$

for $m = 0, 1, 2, \ldots$. A similar argument, using $\mathcal{D}_m^-(\boldsymbol{r}_2)$, gives

$$J_m(kr_2)\,\mathrm{e}^{-\mathrm{i}m\theta_2} = \sum_{n=-\infty}^{\infty} C_n(-1)^n J_{m-n}(kr_1)\,\mathrm{e}^{-\mathrm{i}(m-n)\theta_1}, \tag{2.22}$$

for $m = 0, 1, 2, \ldots$. Letting $r_1 \to 0$ shows that (2.21) holds for $m < 0$ too. Then, substitution of (2.21) in (2.19) completes the proof. □

In the course of this proof, we have also proved a more general result. Thus, combining (2.20), (2.21) and (2.22), we obtain the following result.

Theorem 2.11 (Graf's addition theorem for $J_m(kr)\,\mathrm{e}^{\mathrm{i}m\theta}$) *For $m = 0, \pm 1, \pm 2, \ldots$, we have*

$$J_m(kr_2)\,\mathrm{e}^{\mathrm{i}m\theta_2} = \sum_{n=-\infty}^{\infty} J_n(kb)\,\mathrm{e}^{\mathrm{i}n\beta}\,J_{m-n}(kr_1)\,\mathrm{e}^{\mathrm{i}(m-n)\theta_1}$$
$$= \sum_{n=-\infty}^{\infty} J_{m-n}(kb)\,\mathrm{e}^{\mathrm{i}(m-n)\beta}\,J_n(kr_1)\,\mathrm{e}^{\mathrm{i}n\theta_1}.$$

This is a special case of an addition theorem published by Graf in 1893 [428]. Graf's theorem gives an expansion for $J_\nu(kr_2)\,\mathrm{e}^{\mathrm{i}\nu\theta_2}$ with arbitrary complex ν; see [1298, §11.3] or [1, eqn (9.1.79)]. We are mainly interested in integer values of ν. For this case, a simpler proof was given by Walker [1270]; we give it here, as it is a straightforward extension of the first proof of Neumann's addition theorem.

Second proof of Graf's addition theorem for $J_m(kr)\mathrm{e}^{\mathrm{i}m\theta}$ From (2.17), we have

$$\begin{aligned}
\mathrm{i}^m J_m(kr_2) &= \frac{1}{2\pi}\int_{-\pi}^{\pi} \mathrm{e}^{\mathrm{i}kr_2\cos\varphi}\,\mathrm{e}^{-\mathrm{i}m\varphi}\,\mathrm{d}\varphi \\
&= \frac{1}{2\pi}\int_{-\pi}^{\pi} \mathrm{e}^{\mathrm{i}kr_2\cos(\theta_2-\varphi)}\,\mathrm{e}^{-\mathrm{i}m(\theta_2-\varphi)}\,\mathrm{d}\varphi \\
&= \frac{1}{2\pi}\,\mathrm{e}^{-\mathrm{i}m\theta_2}\int_{-\pi}^{\pi} \mathrm{e}^{\mathrm{i}kr_1\cos(\theta_1-\varphi)}\,\mathrm{e}^{\mathrm{i}kb\cos(\beta-\varphi)}\,\mathrm{e}^{\mathrm{i}m\varphi}\,\mathrm{d}\varphi.
\end{aligned} \tag{2.23}$$

Now, use (2.17) twice and integrate over φ to give the result. □

Notice that we can rewrite (2.23) as

$$J_n(kr)\,\mathrm{e}^{\mathrm{i}n\theta} = \frac{(-\mathrm{i})^n}{2\pi}\int_{-\pi}^{\pi} \mathrm{e}^{\mathrm{i}k(x\cos\varphi + y\sin\varphi)}\mathrm{e}^{\mathrm{i}n\varphi}\,\mathrm{d}\varphi. \tag{2.24}$$

This formula expresses regular cylindrical wavefunctions as a superposition of plane waves. The three-dimensional analogue of (2.24) is known as the Funk–Hecke formula; see Theorem 3.12.

In our first proof of Graf's addition theorem for $J_m(kr)\,\mathrm{e}^{\mathrm{i}m\theta}$, we combined appropriate differential operators ($\mathcal{D}_m^+$ and $\mathcal{D}_m^-$) with a simpler addition theorem for $J_0(kr)$. This idea can be used for three-dimensional problems in spherical polar coordinates; see the second proof of Theorem 3.26. In fact, we used it to obtain the addition theorem for J_0 itself. One virtue of these proofs is that they extend easily to Y_m, $H_m^{(1)}$ and $H_m^{(2)}$. In particular, we have the following theorem.

Theorem 2.12 (Graf's addition theorem for $H_m^{(1)}(kr)\,\mathrm{e}^{\mathrm{i}m\theta}$) *For $m=0,\pm1,\pm2,\dots$, we have*

$$H_m^{(1)}(kr_2)\,\mathrm{e}^{\mathrm{i}m\theta_2} = \sum_{n=-\infty}^{\infty} H_{m-n}^{(1)}(kb)\,\mathrm{e}^{\mathrm{i}(m-n)\beta}\,J_n(kr_1)\,\mathrm{e}^{\mathrm{i}n\theta_1} \tag{2.25}$$

for $r_1 < b$, and

$$H_m^{(1)}(kr_2)\,\mathrm{e}^{\mathrm{i}m\theta_2} = \sum_{n=-\infty}^{\infty} J_{m-n}(kb)\,\mathrm{e}^{\mathrm{i}(m-n)\beta}\,H_n^{(1)}(kr_1)\,\mathrm{e}^{\mathrm{i}n\theta_1} \tag{2.26}$$

for $r_1 > b$.

First proof Assume that $r_1 < b$. As $H_0^{(1)}(kr_2)$ is a solution of the two-dimensional Helmholtz equation for $r_2 > 0$, it must have an expansion of the form

$$H_0^{(1)}(kr_2) = \sum_{n=-\infty}^{\infty} C_n J_n(kr_1)\,\mathrm{e}^{\mathrm{i}n\theta_1},$$

for some coefficients C_n, but only for $r_1 < b$ (due to the singularity on the left-hand side at $r_2 = 0$). Now, apply $\mathcal{D}_m^{\pm}(\boldsymbol{r}_2)$ and then let $r_1 \to 0$, exactly as in the second proof of Neumann's addition theorem. The result (2.25) follows. The result (2.26) follows by interchanging the roles of $\boldsymbol{r}_1$ and $\boldsymbol{b}$. □

Second proof Suppose that $r_1 > b$. Then, there must be an expansion of the form

$$H_m^{(1)}(kr_2)\,\mathrm{e}^{\mathrm{i}m\theta_2} = \sum_{n=-\infty}^{\infty} C_n H_n^{(1)}(kr_1)\,\mathrm{e}^{\mathrm{i}n\theta_1}. \tag{2.27}$$

Now, let $r_1 \to \infty$, with $\boldsymbol{b}$ fixed. We have $r_2^2 = r_1^2 + 2\boldsymbol{b}\cdot\boldsymbol{r}_1 + b^2$ and

$$\tan\theta_2 = \frac{b\sin\beta + r_1\sin\theta_1}{b\cos\beta + r_1\cos\theta_1}.$$

Hence, $r_2 \sim r_1 + b\cos(\theta_1 - \beta)$ and $\theta_2 \sim \theta_1$ as $r_1 \to \infty$. Also, it is known that (see, for example, [1298, p. 197] or [1, eqn (9.2.3)])

$$H_n^{(1)}(w) \sim \sqrt{2/(\pi w)}\,\exp\left\{\mathrm{i}\left(w - \frac{1}{2}n\pi - \frac{1}{4}\pi\right)\right\} \quad \text{as } w \to \infty. \tag{2.28}$$

So, letting $r_1 \to \infty$ in (2.27), we obtain

$$(-\mathrm{i})^m\,\mathrm{e}^{\mathrm{i}kb\cos(\theta_1-\beta)}\mathrm{e}^{\mathrm{i}m\theta_1} = \sum_{n=-\infty}^{\infty} C_n\,(-\mathrm{i})^n\,\mathrm{e}^{\mathrm{i}n\theta_1}.$$

The result (2.26) follows after using (2.17) and (2.18).

If $r_1 < b$, begin with the expansion

$$H_m^{(1)}(kr_2)\,\mathrm{e}^{\mathrm{i}m\theta_2} = \sum_{n=-\infty}^{\infty} C_n H_n^{(1)}(kb)\,\mathrm{e}^{\mathrm{i}n\beta}$$

and then let $b \to \infty$, with $\boldsymbol{r}_1$ fixed, making use of

$$r_2 \sim b + r_1 \cos(\theta_1 - \beta) \quad \text{and} \quad \theta_2 \sim \beta \quad \text{as } b \to \infty.$$

The result (2.25) follows readily. □

A special case of Graf's theorem, obtained by putting $m = 0$, is

$$H_0^{(1)}(kr_2) = \sum_{n=-\infty}^{\infty} (-1)^n H_n^{(1)}(kb)\, J_n(kr_1)\, e^{in(\theta_1 - \beta)} \tag{2.29}$$

for $r_1 < b$, and

$$H_0^{(1)}(kr_2) = \sum_{n=-\infty}^{\infty} (-1)^n J_n(kb)\, H_n^{(1)}(kr_1)\, e^{in(\theta_1 - \beta)} \tag{2.30}$$

for $r_1 > b$. These are useful as $H_0^{(1)}$ is a fundamental solution for the two-dimensional Helmholtz equation; see Chapter 5.

Another proof of Theorem 2.12 is given in Section 2.8.1. A fourth proof of (2.29) can be found in Section 6.13.2.

2.6 The separation matrices S and $\widehat{\mathrm{S}}$

Let us introduce a short-hand notation for the various wavefunctions.

Definition 2.13 *Let* $\mathbf{r} = (r\cos\theta, r\sin\theta)$. *The* outgoing cylindrical wavefunction ψ_n *is defined by*

$$\psi_n(\boldsymbol{r}) = H_n^{(1)}(kr)\, e^{in\theta}.$$

Similarly, the regular cylindrical wavefunction $\hat{\psi}_n$ *is defined by*

$$\hat{\psi}_n(\boldsymbol{r}) = J_n(kr)\, e^{in\theta}.$$

The wavefunction ψ_n is described as 'outgoing' because it corresponds to outgoing waves at infinity: ψ_n satisfies the Sommerfeld radiation condition in two dimensions, (1.25). To see this, use (2.28) to deduce that

$$\mathrm{Re}\left\{\psi_n(\boldsymbol{r})\, e^{-i\omega t}\right\} \sim \sqrt{\frac{2}{\pi k r}} \cos\left\{(kr - \omega t) + n\left(\theta - \frac{1}{2}\pi\right) - \frac{1}{4}\pi\right\}$$

as $r \to \infty$; the argument $(kr - \omega t)$ signifies a wave propagating outwards as t increases.

The wavefunctions $\hat{\psi}_n$ and ψ_n are easily seen to satisfy

$$\hat{\psi}_n(\boldsymbol{r}) = (-1)^n \hat{\psi}_n(-\boldsymbol{r}) = (-1)^n \overline{\hat{\psi}_{-n}(\boldsymbol{r})} \tag{2.31}$$

and

$$\psi_n(\boldsymbol{r}) = (-1)^n \, \psi_n(-\boldsymbol{r}), \tag{2.32}$$

respectively, where the overbar denotes complex conjugation.

Using Definition 2.13, we can rewrite Theorem 2.11 as

$$\hat{\psi}_m(\boldsymbol{r}_2) = \sum_{n=-\infty}^{\infty} \widehat{S}_{mn}(\boldsymbol{b}) \, \hat{\psi}_n(\boldsymbol{r}_1), \tag{2.33}$$

where $\boldsymbol{r}_2 = \boldsymbol{r}_1 + \boldsymbol{b}$ and

$$\widehat{S}_{mn}(\boldsymbol{b}) = \hat{\psi}_{m-n}(\boldsymbol{b}).$$

Similarly, we can rewrite Theorem 2.12 as

$$\psi_m(\boldsymbol{r}_2) = \sum_{n=-\infty}^{\infty} S_{mn}(\boldsymbol{b}) \, \hat{\psi}_n(\boldsymbol{r}_1) \tag{2.34}$$

for $r_1 < b$, and

$$\psi_m(\boldsymbol{r}_2) = \sum_{n=-\infty}^{\infty} \widehat{S}_{mn}(\boldsymbol{b}) \, \psi_n(\boldsymbol{r}_1)$$

for $r_1 > b$, where

$$S_{mn}(\boldsymbol{b}) = \psi_{m-n}(\boldsymbol{b}).$$

We call the matrices $\widehat{\mathbf{S}} = (\widehat{S}_{mn})$ and $\mathbf{S} = (S_{mn})$ *separation matrices* as they depend on $\boldsymbol{b}$, where b is the separation between the two origins, O_1 and O_2: $\boldsymbol{r}_2 = \boldsymbol{r}_1 + \boldsymbol{b}$. These matrices play a crucial role in multiple-scattering theories.

From (2.31), we obtain

$$\widehat{S}_{mn}(\boldsymbol{b}) = (-1)^{m+n} \, \widehat{S}_{mn}(-\boldsymbol{b}) = (-1)^{m+n} \, \overline{\widehat{S}_{nm}(\boldsymbol{b})} \, ,$$

whence

$$\widehat{\mathbf{S}}(-\boldsymbol{b}) = \overline{\widehat{\mathbf{S}}^{\mathrm{T}}(\boldsymbol{b})} \, , \tag{2.35}$$

where $\widehat{\mathbf{S}}^{\mathrm{T}}$ is the transpose of $\widehat{\mathbf{S}}$. For $\mathbf{S}$, we have merely

$$S_{mn}(\boldsymbol{b}) = (-1)^{m+n} \, S_{mn}(-\boldsymbol{b}).$$

Taking the shorthand notation further, define vectors $\boldsymbol{\psi} = (\psi_n)$ and $\hat{\boldsymbol{\psi}} = (\hat{\psi}_n)$. Then, Graf's addition theorem for $\hat{\psi}_m$, (2.33) (for all m), becomes

$$\hat{\boldsymbol{\psi}}(\boldsymbol{r}_2) = \widehat{\mathbf{S}}(\boldsymbol{b}) \, \hat{\boldsymbol{\psi}}(\boldsymbol{r}_1), \tag{2.36}$$

where $\boldsymbol{r}_2 = \boldsymbol{r}_1 + \boldsymbol{b}$. If we suppose further that $\boldsymbol{r}_1 = \boldsymbol{r}_3 + \boldsymbol{c}$, we obtain

$$\hat{\boldsymbol{\psi}}(\boldsymbol{r}_2) = \widehat{\mathbf{S}}(\boldsymbol{b})\,\widehat{\mathbf{S}}(\boldsymbol{c})\,\hat{\boldsymbol{\psi}}(\boldsymbol{r}_3) \tag{2.37}$$

after a second application of Graf's addition theorem for $\hat{\boldsymbol{\psi}}$. But we also have that $\boldsymbol{r}_2 = \boldsymbol{r}_3 + (\boldsymbol{b} + \boldsymbol{c})$, whence

$$\hat{\boldsymbol{\psi}}(\boldsymbol{r}_2) = \widehat{\mathbf{S}}(\boldsymbol{b} + \boldsymbol{c})\,\hat{\boldsymbol{\psi}}(\boldsymbol{r}_3). \tag{2.38}$$

If we compare (2.37) and (2.38), we see that the regular separation matrix satisfies

$$\widehat{\mathbf{S}}(\boldsymbol{b})\,\widehat{\mathbf{S}}(\boldsymbol{c}) = \widehat{\mathbf{S}}(\boldsymbol{b} + \boldsymbol{c}) = \widehat{\mathbf{S}}(\boldsymbol{c})\,\widehat{\mathbf{S}}(\boldsymbol{b}). \tag{2.39}$$

In particular, if we take $\boldsymbol{c} = -\boldsymbol{b}$, we obtain

$$\widehat{\mathbf{S}}(\boldsymbol{b})\,\widehat{\mathbf{S}}(-\boldsymbol{b}) = \widehat{\mathbf{S}}(-\boldsymbol{b})\,\widehat{\mathbf{S}}(\boldsymbol{b}) = \widehat{\mathbf{S}}(\mathbf{0}) = \mathbf{I},$$

the identity, using $J_n(0) = \delta_{n0}$. When this is combined with (2.35), we obtain

$$\widehat{\mathbf{S}}^{-1} = \overline{\widehat{\mathbf{S}}^{\mathrm{T}}};$$

this property means that $\widehat{\mathbf{S}}$ is *unitary*.

Next, consider Graf's addition theorem for ψ_m, (2.34), rewritten as

$$\boldsymbol{\psi}(\boldsymbol{r}_2) = \mathbf{S}(\boldsymbol{b})\,\hat{\boldsymbol{\psi}}(\boldsymbol{r}_1), \tag{2.40}$$

where $\boldsymbol{r}_2 = \boldsymbol{r}_1 + \boldsymbol{b}$ and $r_1 < b$. Recalling the notation at the beginning of Section 2.5, we see that (2.40) holds for any point P inside a circle C_1 with centre O_1 and radius b. Now, introduce a third origin O_3 inside this circle and set $\boldsymbol{r}_1 = \boldsymbol{r}_3 + \boldsymbol{c}$. Thus, P has position vector $\boldsymbol{r}_j$ with respect to O_j, for $j = 1, 2, 3$. Moreover, O_3 has position vectors $\boldsymbol{c}$ and $\boldsymbol{b} + \boldsymbol{c}$ with respect to O_1 and O_2, respectively. As O_3 is inside C_1, we have $b > c = |\boldsymbol{c}|$. See Fig. 2.2 for a sketch of the geometry.

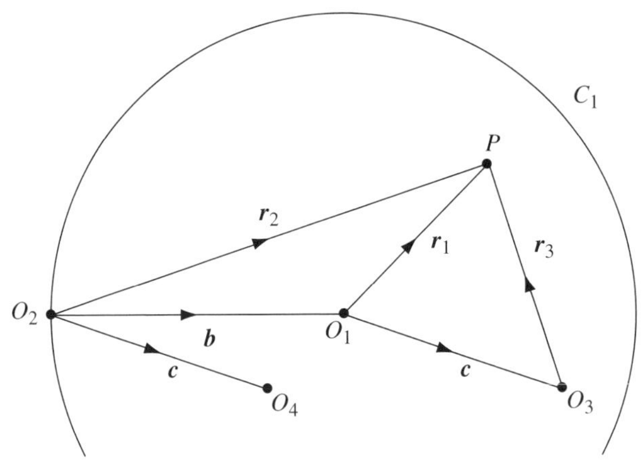

Fig. 2.2. Geometry for addition theorems.

Then, use of (2.36) gives $\hat{\boldsymbol{\psi}}(\boldsymbol{r}_1) = \widehat{\mathbf{S}}(\boldsymbol{c})\,\hat{\boldsymbol{\psi}}(\boldsymbol{r}_3)$, whence (2.40) gives

$$\boldsymbol{\psi}(\boldsymbol{r}_2) = \mathbf{S}(\boldsymbol{b})\,\widehat{\mathbf{S}}(\boldsymbol{c})\,\hat{\boldsymbol{\psi}}(\boldsymbol{r}_3). \tag{2.41}$$

But $\boldsymbol{r}_2 = \boldsymbol{r}_3 + (\boldsymbol{b}+\boldsymbol{c})$, whence (2.34) gives

$$\boldsymbol{\psi}(\boldsymbol{r}_2) = \mathbf{S}(\boldsymbol{b}+\boldsymbol{c})\,\hat{\boldsymbol{\psi}}(\boldsymbol{r}_3) \tag{2.42}$$

for $r_3 < |\boldsymbol{b}+\boldsymbol{c}|$. Comparing (2.41) and (2.42) in a neighbourhood of O_3, we obtain

$$\mathbf{S}(\boldsymbol{b})\,\widehat{\mathbf{S}}(\boldsymbol{c}) = \mathbf{S}(\boldsymbol{b}+\boldsymbol{c}) \quad \text{for } c < b. \tag{2.43}$$

In (2.41), we have used a regular expansion of $\boldsymbol{\psi}(\boldsymbol{r}_2)$ about O_3, obtained via a regular expansion about O_1. Instead, introduce a fourth origin O_4 with position vector $\boldsymbol{c}$ with respect to O_2; thus, $O_1O_2O_4O_3$ forms a parallelogram with sides of lengths b and c (see Fig. 2.2). If we expand $\boldsymbol{\psi}(\boldsymbol{r}_2)$ *outside* a circle C_4 with centre O_4 and radius c, we obtain

$$\boldsymbol{\psi}(\boldsymbol{r}_2) = \widehat{\mathbf{S}}(\boldsymbol{c})\,\boldsymbol{\psi}(\boldsymbol{r}_4)$$

for $r_4 > c$, where $\boldsymbol{r}_4$ is the position vector of P with respect to O_4. But $\boldsymbol{r}_4 = \boldsymbol{r}_3 + \boldsymbol{b}$ and, assuming that $b > c$, O_3 is outside C_4, and so we obtain

$$\boldsymbol{\psi}(\boldsymbol{r}_2) = \widehat{\mathbf{S}}(\boldsymbol{c})\,\mathbf{S}(\boldsymbol{b})\,\hat{\boldsymbol{\psi}}(\boldsymbol{r}_3).$$

Hence, comparing with (2.42), we see that

$$\widehat{\mathbf{S}}(\boldsymbol{c})\,\mathbf{S}(\boldsymbol{b}) = \mathbf{S}(\boldsymbol{b}+\boldsymbol{c}) \quad \text{for } c < b.$$

When this is combined with (2.43), we obtain

$$\mathbf{S}(\boldsymbol{b})\,\widehat{\mathbf{S}}(\boldsymbol{c}) = \mathbf{S}(\boldsymbol{b}+\boldsymbol{c}) = \widehat{\mathbf{S}}(\boldsymbol{c})\,\mathbf{S}(\boldsymbol{b}) \quad \text{for } c < b, \tag{2.44}$$

which should be compared with (2.39).

2.7 Use of rotation matrices

Let $Ox'y'$ be the Cartesian-coordinate axes obtained by rotating the axes Oxy by an angle α in the anti-clockwise direction. Thus

$$x' = x\cos\alpha + y\sin\alpha \quad \text{and} \quad y' = y\cos\alpha - x\sin\alpha.$$

In terms of the corresponding plane polar angles θ and θ', we have $\theta = \theta' + \alpha$.

Let $\hat{\mathbf{x}}$ be a unit vector in the x-direction. Then, we have

$$\widehat{S}_{mn}(b\hat{\mathbf{x}}) = J_{m-n}(kb) \quad \text{and} \quad S_{mn}(b\hat{\mathbf{x}}) = H^{(1)}_{m-n}(kb).$$

These separation matrices are used for translating along the x-axis, rather than in a general direction given by $\boldsymbol{b}$.

However, we can obtain the general case by combining two rotations with the special case. Thus, we make the following steps.

- Rotate the axes $O_2x_2y_2$ by β into $O_2x_2'y_2'$, so that O_1 is at $x_2' = b$, $y_2' = 0$, $\theta_2' = 0$.
- Translate from O_2 to O_1 using $\widehat{\mathbf{S}}(b\hat{\mathbf{x}})$ or $\mathbf{S}(b\hat{\mathbf{x}})$, as appropriate; at O_1, we have a set of parallel axes, $O_1x_1'y_1'$ (with $x_1' = x_2' - b$ and $y_1' = y_2'$).
- Rotate the axes $O_1x_1'y_1'$ by $-\beta$ into $O_1x_1y_1$.

Observe that rotating coordinates changes angles but not radial distances.

Explicitly, for example, we have

$$\begin{aligned} J_m(kr_2)\,\mathrm{e}^{\mathrm{i}m\theta_2} &= \mathrm{e}^{\mathrm{i}m\beta}\,J_m(kr_2)\,\mathrm{e}^{\mathrm{i}m\theta_2'} \\ &= \mathrm{e}^{\mathrm{i}m\beta}\sum_{n=-\infty}^{\infty}\widehat{S}_{mn}(b\hat{\mathbf{x}})\,J_n(kr_1)\,\mathrm{e}^{\mathrm{i}n\theta_1'} \\ &= \sum_{n=-\infty}^{\infty}\left\{\mathrm{e}^{\mathrm{i}m\beta}\,\widehat{S}_{mn}(b\hat{\mathbf{x}})\,\mathrm{e}^{-\mathrm{i}n\beta}\right\}J_n(kr_1)\,\mathrm{e}^{\mathrm{i}n\theta_1}. \end{aligned}$$

In matrix notation, we have

$$\widehat{\mathbf{S}}(\boldsymbol{b}) = \mathbf{D}(\beta)\,\widehat{\mathbf{S}}(b\hat{\mathbf{x}})\,\mathbf{D}(-\beta),$$

where

$$\mathbf{D}(\beta) = \mathrm{diag}\left\{\mathrm{e}^{\mathrm{i}m\beta}\right\}$$

is a diagonal matrix, the *rotation matrix*. Evidently,

$$\mathbf{D}(-\beta) = \overline{\mathbf{D}(\beta)} = \overline{[\mathbf{D}(\beta)]^T} = [\mathbf{D}(\beta)]^{-1}.$$

We also have

$$\mathbf{S}(\boldsymbol{b}) = \mathbf{D}(\beta)\,\mathbf{S}(b\hat{\mathbf{x}})\,\mathbf{D}(-\beta).$$

We have seen that we can decompose a general translation into a rotation, followed by a translation in a special direction, followed by a rotation back. This approach does not offer significant gains in two dimensions. However, it can be useful in three dimensions, where the separation matrices are much more complicated (but so are the rotation matrices); see Section 3.15.

2.8 Two-centre expansions

So far, we have considered situations in which a single point P is referred to two (or more) origins O_j. There are other situations in which *two* points, P and P', are of interest, and expansions are again sought via intermediate expansions around O_1 and O_2. For example, suppose that P is in the vicinity of O_1 and that P' is in the vicinity of O_2. Suppose further that there is a wave source at P and that one wants to expand this about P'. One way of doing this is to make a 'local expansion' about O_1, and then to re-expand about O_2. It turns out that this indirect approach is computationally efficient whenever one has sources at many points P_i and one wants to calculate their effects at many points P'_j: this is the basic idea behind so-called 'fast multipole methods' for solving the boundary integral equations of scattering theory.

We use the same notation as before, with $\boldsymbol{r}_j$ being the position vector of P with respect to O_j, $j = 1, 2$, and $\boldsymbol{b}$ being the position vector of O_1 with respect to O_2. Similarly, let $\boldsymbol{r}'_j$ be the position vector of P' with respect to O_j, $j = 1, 2$. Also, let $\boldsymbol{R}$ be the position vector of P with respect to P', so that

$$\boldsymbol{R} = \boldsymbol{r}_1 - \boldsymbol{r}'_1 = \boldsymbol{r}_2 - \boldsymbol{r}'_2 = \boldsymbol{r}_1 + \boldsymbol{b} - \boldsymbol{r}'_2. \tag{2.45}$$

The geometry is sketched in Fig. 2.3.

Theorem 2.12 gives

$$H_0^{(1)}(kR) = \sum_{m=-\infty}^{\infty} \hat{\psi}_{-m}(\boldsymbol{r}_1)\, \psi_m(-\boldsymbol{r}'_1) \quad \text{for } r_1 < r'_1 = |\boldsymbol{r}'_1|$$

and

$$\psi_m(-\boldsymbol{r}'_1) = \sum_{n=-\infty}^{\infty} S_{mn}(\boldsymbol{b})\, \hat{\psi}_n(-\boldsymbol{r}'_2) \quad \text{for } b > r'_2 = |\boldsymbol{r}'_2|,$$

where $R = |\boldsymbol{R}|$ and we have used $-\boldsymbol{r}'_1 = \boldsymbol{b} - \boldsymbol{r}'_2$. Hence, if both inequalities are satisfied, we obtain

$$H_0^{(1)}(kR) = \sum_{m=-\infty}^{\infty} \sum_{n=-\infty}^{\infty} (-1)^{m+n} S_{mn}(\boldsymbol{b})\, \overline{\hat{\psi}_m(\boldsymbol{r}_1)}\, \hat{\psi}_n(\boldsymbol{r}'_2),$$

where we have used (2.31). Thus, we have proved the following two-centre expansion.

Theorem 2.14 (Two-centre expansion of $H_0^{(1)}(kR)$) *Let $\boldsymbol{R} = \boldsymbol{a} + \boldsymbol{b} + \boldsymbol{c}$. Then*

$$H_0^{(1)}(kR) = \sum_{m=-\infty}^{\infty} \sum_{n=-\infty}^{\infty} (-1)^m S_{mn}(\boldsymbol{b})\, \overline{\hat{\psi}_m(\boldsymbol{a})}\, \hat{\psi}_n(\boldsymbol{c})$$

for $a = |\boldsymbol{a}| < |\boldsymbol{b} + \boldsymbol{c}|$ and $c = |\boldsymbol{c}| < b = |\boldsymbol{b}|$; both inequalities hold if $a + c < b$.

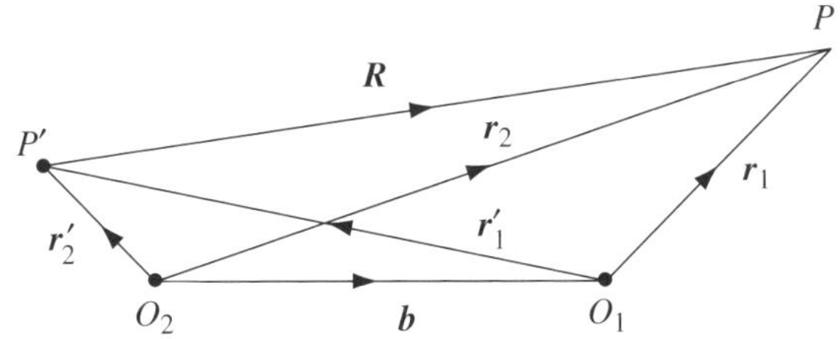

Fig. 2.3. Geometry for two-centre expansions.

2.8.1 Third proof of Graf's addition theorem for $H_m^{(1)}(kr)\,\mathrm{e}^{\mathrm{i}m\theta}$

We can use two-centre expansions to give another proof of Theorem 2.12 for any m, assuming that the result is known for $m = 0$; the proof also makes use of Graf's addition theorem for $\hat{\psi}_m$.

From $\boldsymbol{R} = \boldsymbol{r}_1 - \boldsymbol{r}_1'$, (2.29) gives

$$H_0^{(1)}(kR) = \sum_{l=-\infty}^{\infty} \hat{\psi}_l(\boldsymbol{r}_1')\,\psi_{-l}(-\boldsymbol{r}_1) \quad \text{for } r_1 > r_1'. \tag{2.46}$$

Then, as $\boldsymbol{r}_1' = \boldsymbol{r}_2' - \boldsymbol{b}$, Graf's addition theorem for $\hat{\psi}_l$, (2.33), gives

$$\hat{\psi}_l(\boldsymbol{r}_1') = \sum_{m=-\infty}^{\infty} \hat{\psi}_{l-m}(-\boldsymbol{b})\,\hat{\psi}_m(\boldsymbol{r}_2'),$$

whence (2.46) gives

$$H_0^{(1)}(kR) = \sum_{m=-\infty}^{\infty} (-1)^m \hat{\psi}_m(\boldsymbol{r}_2') \sum_{l=-\infty}^{\infty} \hat{\psi}_{l-m}(\boldsymbol{b})\,\psi_{-l}(\boldsymbol{r}_1), \tag{2.47}$$

where we have used (2.31) and (2.32).

Alternatively, assume that $r_2 > r_2'$ with $\boldsymbol{R} = \boldsymbol{r}_2 - \boldsymbol{r}_2'$, whence

$$H_0^{(1)}(kR) = \sum_{m=-\infty}^{\infty} \hat{\psi}_m(\boldsymbol{r}_2')\,\psi_{-m}(-\boldsymbol{r}_2). \tag{2.48}$$

Comparing (2.47) and (2.48), in a neighbourhood of O_2, we see that

$$\psi_{-m}(\boldsymbol{r}_2) = \sum_{l=-\infty}^{\infty} \hat{\psi}_{l-m}(\boldsymbol{b})\,\psi_{-l}(\boldsymbol{r}_1).$$

The result (2.26) (or (2.34)) follows after replacing m by $-m$ and l by $-n$.

2.9 Elliptical wavefunctions

We confine our formulas here to the minimum, though this minimum is not insignificant, owing to the unfortunate propensity of Mathieu function formulas for coming in fours . . .

(Morse & Feshbach [884, p. 1408])

Elliptic cylindrical coordinates, (ξ, η), are defined by

$$x = c\cosh\xi\,\cos\eta \quad \text{and} \quad y = c\sinh\xi\,\sin\eta, \tag{2.49}$$

where $0 \le \xi < \infty$, $-\pi \le \eta < \pi$ and c is a positive constant. The coordinate ξ is the radial-like variable and η is the angular variable. Also, ξ and η are related to plane polar coordinates, r and θ, by

$$r^2 = \frac{1}{2}c^2(\cosh 2\xi + \cos 2\eta) \quad \text{and} \quad \tan\theta = \tanh\xi\,\tan\eta,$$

so that

$$r \sim \frac{1}{2} c\, \mathrm{e}^{\xi} \quad \text{and} \quad \theta \sim \eta \quad \text{as } \xi \to \infty. \tag{2.50}$$

Choosing $\xi = \xi_0$, say, gives an ellipse

$$(x/A)^2 + (y/B)^2 = 1,$$

with $A = c \cosh \xi_0$ and $B = c \sinh \xi_0$. The foci are at $(x, y) = (\pm c, 0)$, and $\xi > \xi_0$ gives the exterior of the ellipse. In particular, the ellipse degenerates into a finite strip of width $2c$ when $\xi_0 \to 0$.

In terms of ξ and η, the two-dimensional Helmholtz equation becomes

$$\frac{\partial^2 u}{\partial \xi^2} + \frac{\partial^2 u}{\partial \eta^2} + (kc)^2 (\sinh^2 \xi + \sin^2 \eta) u = 0.$$

Separated solutions, $u(\xi, \eta) = R(\xi)\, \Theta(\eta)$, can be found, where

$$R'' - (a - 2q \cosh 2\xi) R = 0, \tag{2.51}$$

$$\Theta'' + (a - 2q \cos 2\eta) \Theta = 0 \tag{2.52}$$

and $q = (\frac{1}{2} kc)^2$; the parameter a is the separation constant. Equation (2.52) is *Mathieu's equation*, whereas (2.51) is the *modified Mathieu equation*.

Seeking 2π-periodic solutions of (2.52) leads to a countable set of values of a, denoted by $a_n(q)$; these eigenvalues are usually divided into those giving even functions of η (denoted by $\mathrm{ce}_n(\eta, q)$) and those giving odd functions of η (denoted by $\mathrm{se}_n(\eta, q)$).

For each $a_n(q)$, there are two independent solutions of (2.51); those associated with ce_n are $\mathrm{Mc}_n^{(j)}(\xi, q)$ with $j = 1, 2$, whereas those associated with se_n are $\mathrm{Ms}_n^{(j)}(\xi, q)$, $j = 1, 2$. The functions with $j = 1$ are regular for all ξ.

The notation for Mathieu functions is cumbersome and varied. We combine the definitions of [1, Chapter 20] with the shorthand ideas of [162], and define $e_{\sigma n}(\eta, q)$ ($\sigma = \mathrm{c}, \mathrm{s}$) by

$$e_{\mathrm{c}n}(\eta, q) = \mathrm{ce}_n(\eta, q) \quad \text{and} \quad e_{\mathrm{s}n}(\eta, q) = \mathrm{se}_n(\eta, q). \tag{2.53}$$

These functions satisfy $e_{\sigma n}(\eta + \pi, q) = (-1)^n e_{\sigma n}(\eta, q)$ and

$$\int_{-\pi}^{\pi} e_{\tau m}(\eta, q)\, e_{\sigma n}(\eta, q)\, \mathrm{d}\eta = \pi \delta_{mn} \delta_{\tau\sigma}. \tag{2.54}$$

Similarly, we define $M_{\sigma n}^{(j)}(\xi, q)$ by

$$M_{\mathrm{c}n}^{(j)}(\xi, q) = \mathrm{Mc}_n^{(j)}(\xi, q) \quad \text{and} \quad M_{\mathrm{s}n}^{(j)}(\xi, q) = \mathrm{Ms}_n^{(j)}(\xi, q).$$

The functions $e_{\sigma n}(\eta, q)$, $M_{\sigma n}^{(1)}(\xi, q)$ and $M_{\sigma n}^{(2)}(\xi, q)$, $\sigma = \mathrm{c}, \mathrm{s}$, are all real. The functions with $j = 3$, defined by

$$M_{\sigma n}^{(3)} = M_{\sigma n}^{(1)} + \mathrm{i} M_{\sigma n}^{(2)},$$

are analogous to Hankel functions of the first kind; their use ensures that the Sommerfeld radiation condition is satisfied: from [1, eqn (20.9.1)], we have

$$M^{(3)}_{\sigma n}(\xi, q) \sim (2/(\pi k r))^{1/2} \exp\left\{ \mathrm{i}\left(kr - \frac{1}{2} n\pi - \frac{1}{4}\pi \right) \right\} \quad \text{as } \xi \to \infty. \tag{2.55}$$

Let us now define elliptical wavefunctions.

Definition 2.15 *The* outgoing elliptical wavefunction $\psi_{\sigma n}$ *is defined by*

$$\psi_{\sigma n}(\xi, \eta, q) = M^{(3)}_{\sigma n}(\xi, q)\, e_{\sigma n}(\eta, q).$$

Both $\psi_{\mathrm{s}n}$ and $(\partial/\partial\xi)\psi_{\mathrm{c}n}$ are discontinuous across the interfocal line.

Similarly, the regular elliptical wavefunction $\hat{\psi}_{\sigma n}$ *is defined by*

$$\hat{\psi}_{\sigma n}(\xi, \eta, q) = M^{(1)}_{\sigma n}(\xi, q)\, e_{\sigma n}(\eta, q).$$

Notice that $\hat{\psi}_{\sigma n}$ is real.

From [1, eqns (20.7.34) & (20.7.35)], we have

$$\hat{\psi}_{\sigma n}(\xi, \eta, q) = \frac{(-\mathrm{i})^n}{2\pi} \int_{-\pi}^{\pi} \mathrm{e}^{\mathrm{i}k(x\cos\varphi + y\sin\varphi)}\, e_{\sigma n}(\varphi, q)\, \mathrm{d}\varphi, \tag{2.56}$$

which should be compared with (2.24). See also [835, §10.50] and [322, §16.8].

Formula (2.56) can be used to obtain an expansion of a plane wave:

$$\mathrm{e}^{\mathrm{i}k(x\cos\varphi + y\sin\varphi)} = 2\sum_{\sigma,n} \mathrm{i}^n\, \hat{\psi}_{\sigma n}(\xi, \eta, q)\, e_{\sigma n}(\varphi, q). \tag{2.57}$$

Here, we have used the shorthand notation

$$\sum_{\sigma,n} = \sum_{\sigma=\mathrm{c,s}} \sum_{n=0}^{\infty}. \tag{2.58}$$

To obtain (2.57), begin by noting that the left-hand side is a regular wavefunction, and so it must have an expansion of the form

$$\sum_{\sigma,n} C^{\sigma}_{n}\, \hat{\psi}_{\sigma n}(\xi, \eta, q).$$

The coefficients C^{σ}_{n} can be found by first observing that

$$x\cos\varphi + y\sin\varphi = \tilde{x}\cos\eta + \tilde{y}\sin\eta,$$

with $\tilde{x} = c\cosh\xi\,\cos\varphi$ and $\tilde{y} = c\sinh\xi\,\sin\varphi$, and then using (2.54) and (2.56).

The plane-wave expansion (2.57) is given in, for example, [886], [835, §10.53], [884, eqn (11.2.94)] and [845, §2.68].

We can give a bilinear expansion of $H^{(1)}_0(kr_2)$ using elliptic cylindrical coordinates, analogous to (2.29) and (2.30). Thus, let (ξ_1, η_1) and (ξ_b, η_b) be the coordinates of points with position vectors $\boldsymbol{r}_1$ and $\boldsymbol{b}$, respectively, and let $\boldsymbol{r}_2 = \boldsymbol{r}_1 + \boldsymbol{b}$. Then

$$H^{(1)}_0(kr_2) = 2\sum_{\sigma,n} (-1)^n\, \psi_{\sigma n}(\xi_1, \eta_1, q)\, \hat{\psi}_{\sigma n}(\xi_b, \eta_b, q) \quad \text{for } \xi_1 > \xi_b; \tag{2.59}$$

interchange ξ_1 and ξ_b if $\xi_1 < \xi_b$; see [884, eqn (11.2.93)]. To obtain (2.59), note first that the left-hand side is an outgoing wavefunction, and so it must have an expansion of the form

$$\sum_{\sigma,n} C_n^\sigma \psi_{\sigma n}(\xi_1, \eta_1, q) \quad \text{for } \xi_1 > \xi_b.$$

To find the coefficients C_n^σ, we let $\xi_1 \to \infty$. We have

$$r_2 = \{(\boldsymbol{r}_1 + \boldsymbol{b}) \cdot (\boldsymbol{r}_1 + \boldsymbol{b})\}^{1/2} \sim r_1 + \boldsymbol{b} \cdot \hat{\boldsymbol{r}}_1 \quad \text{as } r_1 \to \infty,$$

where $\hat{\boldsymbol{r}}_1 = \boldsymbol{r}_1 / r_1 \sim (\cos\eta_1, \sin\eta_1)$ as $\xi_1 \to \infty$, by (2.50). Thus

$$\begin{aligned} H_0^{(1)}(kr_2) &\sim (2/(\pi k r_1))^{1/2} \, \mathrm{e}^{\mathrm{i}(kr_1 - \pi/4)} \exp\{\mathrm{i}k(x_b \cos\eta_1 + y_b \sin\eta_1)\} \\ &= (2/(\pi k r_1))^{1/2} \, \mathrm{e}^{\mathrm{i}(kr_1 - \pi/4)} \, 2 \sum_{\sigma,n} \mathrm{i}^n \, \hat{\psi}_{\sigma n}(\xi_b, \eta_b, q) \, e_{\sigma n}(\eta_1, q) \end{aligned}$$

as $\xi_1 \to \infty$, where x_b and y_b are the Cartesian components of $\boldsymbol{b}$ and we have used (2.57). The result follows after using (2.55).

For more information on Mathieu functions, see [835], [1, Chapter 20], [884, pp. 562–568, pp. 633–644, pp. 1407–1432 & pp. 1568–1573], [845], [322, §§ 16.2–16.8], [427, §8.6] and [1140, 364].

2.9.1 Addition theorems

Consider two right-handed Cartesian coordinate systems, $O_1 x_1 y_1$ and $O_2 x_2 y_2$. O_1 is at $(x_2, y_2) = (x_b, y_b)$. The axis $O_1 x_1$ is rotated by an angle γ anticlockwise from $O_2 x_2$. Thus,

$$x_2 = x_b + x_1 \cos\gamma - y_1 \sin\gamma \quad \text{and} \quad y_2 = y_b + x_1 \sin\gamma + y_1 \cos\gamma; \tag{2.60}$$

see Fig. 2.4. Equivalently, we have

$$x_1 = (x_2 - x_b)\cos\gamma + (y_2 - y_b)\sin\gamma, \tag{2.61}$$

$$y_1 = (y_2 - y_b)\cos\gamma - (x_2 - x_b)\sin\gamma. \tag{2.62}$$

Now, introduce elliptic cylindrical coordinates (ξ_j, η_j) at O_j, with $2c_j$ as the interfocal distance and $q_j = (\frac{1}{2}kc_j)^2$, $j = 1, 2$. Then, we have the following addition theorem for regular elliptical wavefunctions.

Theorem 2.16

$$\hat{\psi}_{\sigma n}(\xi_2, \eta_2, q_2) = \sum_{\tau,m} \widehat{S}_{nm}^{\sigma\tau}(\xi_b, \eta_b, q_1, q_2, \gamma) \, \hat{\psi}_{\tau m}(\xi_1, \eta_1, q_1), \tag{2.63}$$

where $x_b = c_2 \cosh\xi_b \cos\eta_b$, $y_b = c_2 \sinh\xi_b \sin\eta_b$,

$$\widehat{S}_{nm}^{\sigma\tau} = \frac{2}{\pi} (-1)^n \sum_{\nu,l} (-1)^s \, I_{nml}^{\sigma\tau\nu} \, \hat{\psi}_{\nu l}(\xi_b, \eta_b, q_2), \tag{2.64}$$

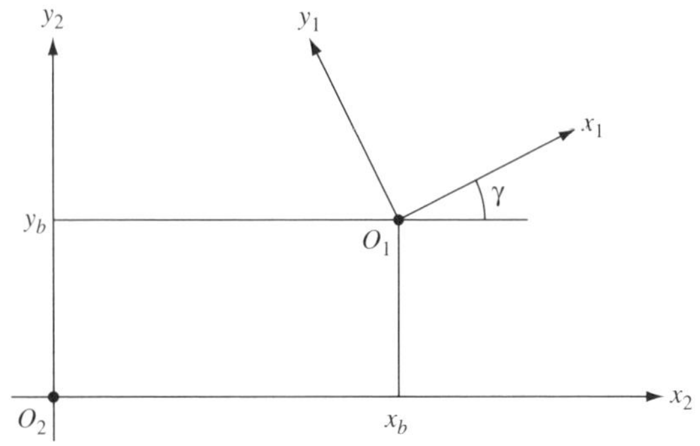

Fig. 2.4. Geometry for addition theorems with elliptical wavefunctions.

the sum over l is such that $n+m+l=2s$ is even, and

$$I_{nml}^{\sigma\tau\nu} = \int_{-\pi}^{\pi} e_{\sigma n}(\varphi, q_2)\, e_{\tau m}(\varphi-\gamma, q_1)\, e_{\nu l}(\varphi, q_2)\, \mathrm{d}\varphi = I_{lmn}^{\nu\tau\sigma}. \tag{2.65}$$

Similarly,

$$\hat{\psi}_{\sigma n}(\xi_1, \eta_1, q_1) = \sum_{\tau,m} \widehat{S}_{mn}^{\tau\sigma}(\xi_b, \eta_b, q_1, q_2, \gamma)\, \hat{\psi}_{\tau m}(\xi_2, \eta_2, q_2). \tag{2.66}$$

Proof For (2.63), use (2.60) in (2.56), and then use (2.57) twice. Similarly, for (2.66), use (2.61) and (2.62). Notice that (2.63) contains $\widehat{S}_{nm}^{\sigma\tau}$ whereas (2.66) contains the transpose, $\widehat{S}_{mn}^{\tau\sigma}$; the proof exploits two facts: $\widehat{S}_{nm}^{\sigma\tau}$ is real and $e_{\sigma n}(\varphi, q)$ is 2π-periodic in φ. □

The real integral (2.65) is elementary but tedious to evaluate: $e_{\sigma n}$ has a known Fourier series for $\sigma = \mathrm{c}, \mathrm{s}$ and $n = 0, 1, 2, \ldots$; there are, of course, four cases depending on the value of σ and on whether n is even or odd. Notice that $I_{nml}^{\sigma\tau\nu} = 0$ if $n+m+l$ is odd.

There are similar addition theorems for outgoing wavefunctions. We shall derive them in two ways. First, we begin with the method of Section 2.8.1. Thus, let $\boldsymbol{r}_j$ and $\boldsymbol{r}'_j$ be the position vectors of points P and P', respectively, with respect to O_j, $j = 1, 2$. Let $\boldsymbol{b}$ be the position vector of O_1 with respect to O_2. Let $\boldsymbol{R}$ be the position vector of P with respect to P', so that (2.45) holds. In particular, $\boldsymbol{R} = \boldsymbol{r}_1 - \boldsymbol{r}'_1$ and so (2.59) gives

$$\begin{aligned} H_0^{(1)}(kR) &= 2\sum_{\sigma,n} (-1)^n\, \psi_{\sigma n}(\xi_1, \eta_1, q_1)\, \hat{\psi}_{\sigma n}(\xi'_1, \eta'_1 + \pi, q_1) \\ &= 2\sum_{\sigma,n} \psi_{\sigma n}(\xi_1, \eta_1, q_1)\, \hat{\psi}_{\sigma n}(\xi'_1, \eta'_1, q_1), \quad \xi_1 > \xi'_1, \end{aligned} \tag{2.67}$$

where (ξ_j, η_j) and (ξ'_j, η'_j) are the elliptic cylindrical coordinates of P and P', respectively, with respect to O_j, $j = 1, 2$. But, from (2.66),

$$\hat{\psi}_{\sigma n}(\xi'_1, \eta'_1, q_1) = \sum_{\tau,m} \widehat{S}^{\tau\sigma}_{mn}(\xi_b, \eta_b, q_1, q_2, \gamma)\, \hat{\psi}_{\tau m}(\xi'_2, \eta'_2, q_2),$$

whence (2.67) gives

$$H_0^{(1)}(kR) = 2\sum_{\sigma,n} \hat{\psi}_{\sigma n}(\xi'_2, \eta'_2, q_2) \sum_{\tau,m} \widehat{S}^{\sigma\tau}_{nm}(\xi_b, \eta_b, q_1, q_2, \gamma)\, \psi_{\tau m}(\xi_1, \eta_1, q_1). \quad (2.68)$$

Alternatively, assume that $\xi_2 > \xi'_2$ with $\boldsymbol{R} = \boldsymbol{r}_2 - \boldsymbol{r}'_2$, whence (2.59) gives

$$H_0^{(1)}(kR) = 2\sum_{\sigma,n} \psi_{\sigma n}(\xi_2, \eta_2, q_2)\, \hat{\psi}_{\sigma n}(\xi'_2, \eta'_2, q_2).$$

Comparing this expansion with (2.68), in a neighbourhood of O_2, we see that

$$\psi_{\sigma n}(\xi_2, \eta_2, q_2) = \sum_{\tau,m} \widehat{S}^{\sigma\tau}_{nm}(\xi_b, \eta_b, q_1, q_2, \gamma)\, \psi_{\tau m}(\xi_1, \eta_1, q_1) \quad (2.69)$$

for $\xi_1 > \xi_1^+$, where ξ_1^+ will be defined shortly.

Similarly, if we reverse the roles of O_1 and O_2, beginning with $\boldsymbol{R} = \boldsymbol{r}_2 - \boldsymbol{r}'_2$ and ending with expansions in the neighbourhood of O_1, we obtain, for $\xi_2 > \xi_2^+$,

$$\psi_{\sigma n}(\xi_1, \eta_1, q_1) = \sum_{\tau,m} \widehat{S}^{\tau\sigma}_{mn}(\xi_b, \eta_b, q_1, q_2, \gamma)\, \psi_{\tau m}(\xi_2, \eta_2, q_2). \quad (2.70)$$

Notice that the real parts of (2.69) and (2.70) are precisely (2.63) and (2.66), respectively.

In order to specify ξ_1^+ and ξ_2^+, we first let $\mathcal{L}_j$ denote the interfocal line defined by $-c_j \le x_j \le c_j$, $y_j = 0$, $j = 1, 2$. Let $\xi_j = \xi_j^+$ and $\xi_j = \xi_j^-$ be two confocal ellipses, centred at O_j, with $\xi_j^+ > \xi_j^-$, $j = 1, 2$. Then, $\xi_1 = \xi_1^+$ is the smallest ellipse that contains $\mathcal{L}_2$ in its interior and $\xi_1 = \xi_1^-$ is the largest ellipse that does not contain any piece of $\mathcal{L}_2$ in its interior. The quantities ξ_2^+ and ξ_2^- are defined similarly, using $\mathcal{L}_1$.

Now, as the left-hand side of (2.69) is discontinuous across $\mathcal{L}_2$ (more precisely, see Definition 2.15) but the right-hand side is a wavefunction for $\xi_1 > 0$, we infer that the series (2.69) converges for $\xi_1 > \xi_1^+$. Similar reasoning shows that (2.70) holds for $\xi_2 > \xi_2^+$.

For our second proof, we adapt the argument used in the second proof of Theorem 2.12. Thus, in order to prove (2.69), we begin by asserting that there must be an expansion of the form

$$\psi_{\sigma n}(\xi_2, \eta_2, q_2) = \sum_{\tau,m} C^\tau_m\, \psi_{\tau m}(\xi_1, \eta_1, q_1) \quad \text{for } \xi_1 > \xi_1^+. \quad (2.71)$$

Now, let $\xi_1 \to \infty$ with $\boldsymbol{b}$ fixed. To do this, put $(x_b, y_b) = b(\cos\beta, \sin\beta)$ and $(x_j, y_j) = r_j(\cos\theta_j, \sin\theta_j)$, $j = 1, 2$. Then, (2.60) gives

$$r_2^2 = r_1^2 + 2r_1 b\cos(\theta_1 + \gamma - \beta) + b^2$$

and

$$\tan\theta_2 = \frac{b\sin\beta + r_1\sin(\theta_1+\gamma)}{b\cos\beta + r_1\cos(\theta_1+\gamma)}.$$

Thus, $r_2 \sim r_1 + b\cos(\theta_1+\gamma-\beta)$ and $\theta_2 \sim \theta_1+\gamma$ as $r_1 \to \infty$, which is equivalent to letting $\xi_1 \to \infty$. Also, $\theta_1 \sim \eta_1$, $\theta_2 \sim \eta_2$ and $\eta_2 \sim \eta_1+\gamma$ as $\xi_1 \to \infty$. So, letting $\xi_1 \to \infty$ in (2.71), using (2.55), we obtain

$$(-\mathrm{i})^n\, \mathrm{e}^{\mathrm{i}kb\cos(\eta_1+\gamma-\beta)}\, e_{\sigma n}(\eta_1+\gamma, q_2) = \sum_{\tau,m} C_m^\tau\, (-\mathrm{i})^m\, e_{\tau m}(\eta_1, q_1).$$

The result (2.69) follows after using (2.54) and (2.57).

Similarly, if $\xi_1 < \xi_1^-$, we begin with the expansion

$$\psi_{\sigma n}(\xi_2, \eta_2, q_2) = \sum_{\tau,m} C_m^\tau\, \psi_{\tau m}(\xi_b, \eta_b, q_2) \tag{2.72}$$

and then let $b \to \infty$ with ξ_1 and η_1 fixed, making use of

$$r_2 \sim b + r_1\cos(\theta_1+\gamma-\beta) \quad \text{and} \quad \theta_2 \sim \beta \quad \text{as } b \to \infty.$$

Also, $\eta_b \sim \eta_2$ and $\theta_2 \sim \eta_2$ as $b \to \infty$. So, letting $b \to \infty$ in (2.72), using (2.55), we obtain

$$(-\mathrm{i})^n\, \mathrm{e}^{\mathrm{i}kr_1\cos[\theta_1-(\eta_2-\gamma)]}\, e_{\sigma n}(\eta_2, q_2) = \sum_{\tau,m} C_m^\tau\, (-\mathrm{i})^m\, e_{\tau m}(\eta_2, q_2),$$

whence

$$C_l^\nu = \frac{2}{\pi}(-1)^n \sum_{\tau,m} (-1)^s\, I_{nml}^{\sigma\tau\nu}\, \hat{\psi}_{\tau m}(\xi_1, \eta_1, q_1).$$

Then, substitution in (2.72) gives

$$\psi_{\sigma n}(\xi_2, \eta_2, q_2) = \sum_{\tau,m} S_{nm}^{\sigma\tau}(\xi_b, \eta_b, q_1, q_2, \gamma)\, \hat{\psi}_{\tau m}(\xi_1, \eta_1, q_1) \quad \text{for } \xi_1 < \xi_1^-,$$

where

$$S_{nm}^{\sigma\tau} = \frac{2}{\pi}(-1)^n \sum_{\nu,l} (-1)^s\, I_{nml}^{\sigma\tau\nu}\, \psi_{\nu l}(\xi_b, \eta_b, q_2), \tag{2.73}$$

the sum over l is such that $n+m+l = 2s$ is even and $I_{nml}^{\sigma\tau\nu}$ is defined by (2.65). Notice that $\widehat{S}_{nm}^{\sigma\tau} = \mathrm{Re}\{S_{nm}^{\sigma\tau}\}$.

We summarise these results in the following theorem.

Theorem 2.17

$$\psi_{\sigma n}(\xi_2, \eta_2, q_2) = \sum_{\tau,m} \widehat{S}_{nm}^{\sigma\tau}\, \psi_{\tau m}(\xi_1, \eta_1, q_1) \quad \textit{for } \xi_1 > \xi_1^+, \tag{2.74}$$

$$\psi_{\sigma n}(\xi_2, \eta_2, q_2) = \sum_{\tau,m} S_{nm}^{\sigma\tau}\, \hat{\psi}_{\tau m}(\xi_1, \eta_1, q_1) \quad \textit{for } \xi_1 < \xi_1^-, \tag{2.75}$$

$$\psi_{\sigma n}(\xi_1, \eta_1, q_1) = \sum_{\tau,m} \widehat{S}_{mn}^{\tau\sigma} \psi_{\tau m}(\xi_2, \eta_2, q_2) \quad \textit{for } \xi_2 > \xi_2^+, \tag{2.76}$$

$$\psi_{\sigma n}(\xi_1, \eta_1, q_1) = \sum_{\tau,m} S_{mn}^{\tau\sigma} \hat{\psi}_{\tau m}(\xi_2, \eta_2, q_2) \quad \textit{for } \xi_2 < \xi_2^-. \tag{2.77}$$

Here, $\widehat{S}_{nm}^{\sigma\tau} = \widehat{S}_{nm}^{\sigma\tau}(\xi_b, \eta_b, q_1, q_2, \gamma)$ and $S_{nm}^{\sigma\tau} = S_{nm}^{\sigma\tau}(\xi_b, \eta_b, q_1, q_2, \gamma)$ are defined by (2.64) *and* (2.73)*, respectively.*

Notice that there are no expansions valid for $\xi_j^- < \xi < \xi_j^+$, $j = 1, 2$.

Addition theorems of the form (2.66) and (2.76) are given in [845, §2.54]. Subsequently, Særmark [1046] also obtained (2.77). His proof uses representations of $\hat{\psi}_{\sigma n}$ and $\psi_{\sigma n}$ in terms of cylinder functions (see (2.8)). In all cases, $\widehat{S}_{nm}^{\sigma\tau}$ and $S_{nm}^{\sigma\tau}$ are found as double infinite series of cylinder functions; our expressions are single infinite series of Mathieu functions.

2.10 Vector cylindrical wavefunctions

In Section 1.5.5, we saw that in-plane elastodynamic scattering problems may be solved using acoustic potentials; see (1.55). These representations can be used to construct vector wavefunctions. For example, we can define 'compressional multipoles' and 'shear multipoles' by

$$\boldsymbol{u}_n = \operatorname{grad}\{H_n(kr)\,\mathrm{e}^{\mathrm{i}n\theta}\} \quad \text{and} \quad \boldsymbol{v}_n = \operatorname{curl}\{\hat{\mathbf{z}}\,H_n(Kr)\,\mathrm{e}^{\mathrm{i}n\theta}\}, \tag{2.78}$$

respectively, where k and K are the compressional and shear wavenumbers, respectively. Both $\boldsymbol{u}_n$ and $\boldsymbol{v}_n$ are elastodynamic wavefunctions; they are singular at $r = 0$ and they satisfy the two-dimensional Kupradze radiation conditions. We can construct regular wavefunctions by substituting J_n for H_n.

For half-plane problems, one can construct compressional multipoles that are singular at a point $(x, y) = (0, h)$ within the half-plane. They have the form

$$\boldsymbol{u}_n = \operatorname{grad}\{H_n(kr)\,\mathrm{e}^{\mathrm{i}n\theta} + E_n\} + \operatorname{curl}\{\hat{\mathbf{z}}\,\chi_n\},$$

where $r = \{x^2 + (y - h)^2\}^{1/2}$ and the wavefunctions E_n and χ_n are to be chosen so that $T\boldsymbol{u}_n = \mathbf{0}$ on $y = 0$. This construction is given in detail in [438]. Shear multipoles can be constructed similarly.

One can also construct vector wavefunctions using scalar elliptic wavefunctions in place of $H_n\,\mathrm{e}^{\mathrm{i}n\theta}$ in (2.78), with appropriate choices for q. For example, we can use

$$\boldsymbol{u}_n = \operatorname{grad}\{M_{\sigma n}^{(3)}(\xi, q)\, e_{\sigma n}(\eta, q)\}$$

with $q = (\frac{1}{2}kc)^2$ for outgoing compressional waves, and

$$\boldsymbol{v}_n = \operatorname{curl}\{\hat{\mathbf{z}}\, M_{\sigma n}^{(3)}(\xi, Q)\, e_{\sigma n}(\eta, Q)\}$$

with $Q = (\frac{1}{2}Kc)^2$ for outgoing shear waves.

2.11 Multipoles for water waves

Consider two-dimensional water-wave problems in deep water. Introduce plane polar coordinates (r, θ), where

$$x = r\cos\theta, \quad z = r\sin\theta \quad \text{and} \quad 0 \le \theta \le \pi;$$

the free surface is part of $z = 0$ $(\theta = 0, \pi)$. See Fig. 2.5. We want to construct multipole solutions; these must satisfy Laplace's equation in the water $(z > 0)$ and the free-surface condition on $z = 0$. These are readily found: if $\chi(r, \theta)$ is harmonic and satisfies $\chi(r, 0) = \chi(r, \pi) = 0$, then

$$\left(K - \frac{\partial}{\partial z}\right)\chi = \left(K - \sin\theta\,\frac{\partial}{\partial r} - \frac{\cos\theta}{r}\frac{\partial}{\partial \theta}\right)\chi$$

is harmonic and satisfies the free-surface condition. For χ, we can choose any constant multiple of $r^\nu \sin\nu\theta$, where ν is any non-zero integer. Our specific choices are given next.

Definition 2.18 *Let* $\boldsymbol{r} = (r\cos\theta, r\sin\theta)$. *The cylindrical wavefree potential* Φ_n *is defined by*

$$\Phi_n(\boldsymbol{r}) = -\frac{(n-2)!}{K^n}\left(K - \frac{\partial}{\partial z}\right)\frac{\sin(n-1)\theta}{r^{n-1}}$$
$$= \frac{(n-1)!}{(Kr)^n}\cos n\theta - \frac{(n-2)!}{(Kr)^{n-1}}\sin(n-1)\theta, \quad n = 2, 3, 4, \dots.$$

The analogous regular potential $\widehat{\Phi}_n$ *is defined by*

$$\widehat{\Phi}_n(\boldsymbol{r}) = -\frac{K^n}{(n+1)!}\left(K - \frac{\partial}{\partial z}\right) r^{n+1}\sin(n+1)\theta$$
$$= \frac{(Kr)^n}{n!}\cos n\theta - \frac{(Kr)^{n+1}}{(n+1)!}\sin(n+1)\theta, \quad n = 0, 1, 2, \dots.$$

The functions Φ_n are called 'wavefree potentials' because they decay algebraically as $|x| \to \infty$. In order to treat scattering problems, we also need potentials that generate

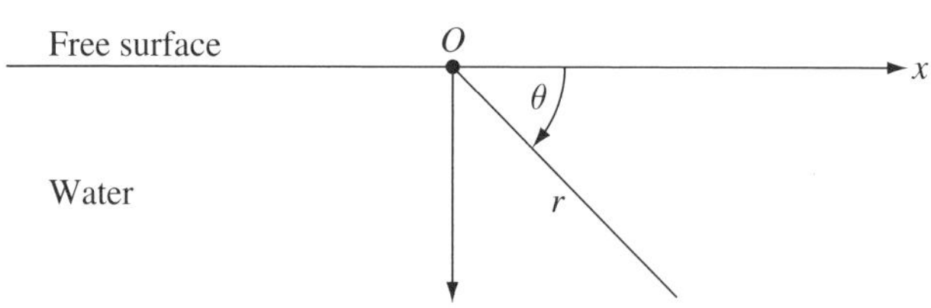

Fig. 2.5. Polar coordinates for water-wave problems.

waves; these are more complicated. They are defined by

$$\Phi_0(\boldsymbol{r}) = \int_0^\infty \mathrm{e}^{-kz} \cos kx \, \frac{\mathrm{d}k}{k-K} \quad \text{and} \quad \Phi_1(\boldsymbol{r}) = -\frac{1}{K}\frac{\partial}{\partial x}\Phi_0, \tag{2.79}$$

where the path of integration passes below the pole of the integrand at $k = K$. The function Φ_0 represents a wave source at the origin, O; it is an even function of x. The function Φ_1 represents a horizontal wave dipole at the origin; it is an odd function of x. For large $|x|$, we have

$$\Phi_0 \sim \pi \mathrm{i}\, \mathrm{e}^{-Kz} \mathrm{e}^{\pm \mathrm{i}Kx} \quad \text{and} \quad \Phi_1 \sim \pm\pi\, \mathrm{e}^{-Kz} \mathrm{e}^{\pm \mathrm{i}Kx} \quad \text{as } x \to \pm\infty, \tag{2.80}$$

so that the waves propagate away from the origin. We also have the following convergent expansion,

$$\begin{aligned} \Phi_0(\boldsymbol{r}) = &- \left\{ (\log Kr - \mathrm{i}\pi + \gamma) \cos Kx + \left(\frac{1}{2}\pi - \theta\right) \sin Kx \right\} \mathrm{e}^{-Kz} \\ &+ \sum_{m=1}^{\infty} \frac{(-Kr)^m}{m!} \left(\sum_{n=1}^{m} \frac{1}{n} \right) \cos \left\{ m \left(\frac{1}{2}\pi - \theta \right) \right\}, \end{aligned} \tag{2.81}$$

where $\gamma = 0.5772\ldots$ is Euler's constant. The expansion (2.81) is due to Yu & Ursell [1369]; it shows that $\Phi_0(\boldsymbol{r}) \sim -\log r$ as $r \to 0$.

The potentials $\{\Phi_n\} = \Phi_0, \Phi_1, \Phi_2, \ldots$ are known as *Ursell's multipole potentials* [1222, 1223, 1229]. They are harmonic, satisfy the free-surface and radiation conditions, and are singular at the origin. Multipole potentials can also be defined for water of constant finite depth, and with singularities at a submerged point, $(x, z) = (\xi, \zeta)$, instead of at $(x, z) = (0, 0)$. In particular, the wave source at (ξ, ζ) is recognised as the fundamental solution $G(x, z; \xi, \zeta)$; see Section 2.11.1. For more information on multipole potentials, see [1176] and [731, Appendix B]. For multipole potentials in elliptic cylindrical coordinates, see [1223].

An inductive argument shows that

$$\frac{\partial^m}{\partial x^m}\left(\frac{\sin\theta}{r}\right) = (-1)^m m! \, \frac{\sin(m+1)\theta}{r^{m+1}}, \qquad m = 0, 1, 2, \ldots,$$

whence

$$\Phi_{n+2}(\boldsymbol{r}) = \frac{(-1)^{n+1}}{K^{n+2}} \left(K - \frac{\partial}{\partial z} \right) \frac{\partial^n}{\partial x^n} \left(\frac{\sin\theta}{r} \right), \qquad n = 0, 1, 2, \ldots.$$

Thus, using

$$\frac{\sin\theta}{r} = \frac{z}{x^2+z^2} = \int_0^\infty \mathrm{e}^{-kz} \cos kx \, \mathrm{d}k, \qquad z > 0,$$

we obtain the following integral representations for the wavefree potentials, valid for $m = 0, 1, 2, \dots$:

$$\Phi_{2m+2}(\boldsymbol{r}) = \frac{(-1)^{m+1}}{K^{2m+2}} \int_0^\infty (k+K)\, k^{2m} \mathrm{e}^{-kz} \cos kx \, \mathrm{d}k, \tag{2.82}$$

$$\Phi_{2m+3}(\boldsymbol{r}) = \frac{(-1)^{m+1}}{K^{2m+3}} \int_0^\infty (k+K)\, k^{2m+1} \mathrm{e}^{-kz} \sin kx \, \mathrm{d}k. \tag{2.83}$$

We note the following elementary properties:

$$(\partial/\partial x)\Phi_1 = K(\Phi_0 - \Phi_2), \tag{2.84}$$

$$(\partial/\partial x)\Phi_n = -K\Phi_{n+1}, \quad n = 0 \quad \text{and} \quad n = 2, 3, \dots, \tag{2.85}$$

$$(\partial/\partial x)\widehat{\Phi}_n = K\widehat{\Phi}_{n-1}, \quad n = 1, 2, 3, \dots. \tag{2.86}$$

2.11.1 A fundamental solution for deep water

As we have already noted, the function Φ_0, defined by (2.79), represents a wave source at the origin. More generally, we can put the source at a point Q in the water, and then calculate the velocity potential at a different point, P. We denote this potential by $G(P, Q)$; it is a fundamental solution. In particular, for two-dimensional water waves on deep water, it is given by

$$G(P, Q) \equiv G(x, z; \xi, \zeta) = \frac{1}{\pi} \log \frac{R}{R'} - \frac{2}{\pi} \fint_0^\infty \mathrm{e}^{-k(z+\zeta)} \frac{\cos k(x-\xi)}{k-K} \, \mathrm{d}k, \tag{2.87}$$

where $R = |\boldsymbol{r}_P - \boldsymbol{r}_Q| = \{(x-\xi)^2 + (z-\zeta)^2\}^{1/2}$ and $R' = \{(x-\xi)^2 + (z+\zeta)^2\}^{1/2}$. In the far field, we have

$$G(x, z; \xi, \zeta) \sim -2\mathrm{i}\, \mathrm{e}^{-K(z+\zeta)} \mathrm{e}^{\mathrm{i}K|x-\xi|} \quad \text{as } |x-\xi| \to \infty. \tag{2.88}$$

Note that $G(P, O) = -(2/\pi)\Phi_0(\boldsymbol{r}_P)$.

The fundamental solution for water of constant finite depth is also available. For references and further details, see [559], [1303, pp. 479–483], [842, Appendix 7.A] and [675, §1.1].

The function G can be used to derive boundary integral equations, and it plays a role in null-field and T-matrix methods. There is also a bilinear expansion for G, which we give next. This expansion will also be used to prove addition theorems for Φ_n.

Theorem 2.19 **(Ursell** [1228]) *Suppose that* $R = |\boldsymbol{r}_P - \boldsymbol{r}_Q|$ *with* $|\boldsymbol{r}_P| < |\boldsymbol{r}_Q|$. *Then the two-dimensional deep-water fundamental solution* $G(P, Q)$, *defined by* (2.87), *has the expansion*

$$G(\boldsymbol{r}_P, \boldsymbol{r}_Q) = -\frac{2}{\pi} \sum_{n=0}^{\infty} \alpha_n(\boldsymbol{r}_P)\, \Phi_n(\boldsymbol{r}_Q), \tag{2.89}$$

where

$$\alpha_n(\boldsymbol{r}) = \sum_{s=0}^{\infty}(-1)^s \widehat{\Phi}_{n+2s}(\boldsymbol{r}), \quad n = 0, 1, 2, \ldots. \tag{2.90}$$

Proof Fix $P = (x, z)$. We choose $\alpha_0(x, z)$ and $\alpha_1(x, z)$ so that

$$G(x, z; \xi, \zeta) + (2/\pi)[\alpha_0(x, z)\,\Phi_0(\xi, \zeta) + \alpha_1(x, z)\,\Phi_1(\xi, \zeta)] \equiv H(x, z; \xi, \zeta),$$

say, is wavefree as $\xi \to \pm\infty$; using (2.80) and (2.88), we deduce that

$$\alpha_0(\boldsymbol{r}) = \mathrm{e}^{-Kz}\cos Kx \quad \text{and} \quad \alpha_1(\boldsymbol{r}) = \mathrm{e}^{-Kz}\sin Kx. \tag{2.91}$$

As H is wavefree, harmonic and satisfies the free-surface condition, it must have an expansion in terms of wavefree potentials. To realise this, we assume initially that $z < \zeta$. Making use of [427, 3.951(3)]

$$\log\frac{R}{R'} = \int_0^{\infty}\left(\mathrm{e}^{-k(\zeta+z)} - \mathrm{e}^{-k|\zeta-z|}\right)\cos k(x-\xi)\,\frac{\mathrm{d}k}{k},$$

we find that

$$H = \frac{2}{\pi}\int_0^{\infty}\mathrm{e}^{-k\zeta}\,F\,\mathrm{d}k, \tag{2.92}$$

where

$$\begin{aligned} F &= \frac{\mathrm{e}^{-Kz}}{k-K}\left\{\cos Kx\,\cos k\xi + \frac{k}{K}\sin Kx\,\sin k\xi\right\} \\ &\quad - \left\{\frac{\sinh kz}{k} + \frac{\mathrm{e}^{-kz}}{k-K}\right\}\cos k(x-\xi) \\ &= F_{\mathrm{c}}(x, z)\,\cos k\xi + F_{\mathrm{s}}(x, z)\,\sin k\xi, \end{aligned} \tag{2.93}$$

$$F_{\mathrm{c}}(x, z) = \frac{\mathrm{e}^{-Kz}\cos Kx - \mathrm{e}^{-kz}\cos kx}{k-K} - \frac{\sinh kz}{k}\cos kx,$$

$$F_{\mathrm{s}}(x, z) = \frac{\mathrm{e}^{-Kz}(k/K)\sin Kx - \mathrm{e}^{-kz}\sin kx}{k-K} - \frac{\sinh kz}{k}\sin kx$$

and we have used (2.87) and (2.79). To complete the proof, we expand F_{c} and F_{s} in powers of k, and then use the integral representations for the wavefree potentials, (2.82) and (2.83). We have

$$\begin{aligned} \mathrm{e}^{\pm Kz + \mathrm{i}Kx} &= \sum_{n=0}^{\infty}\frac{(\mathrm{i}Kr)^n}{n!}\,\mathrm{e}^{\mp\mathrm{i}n\theta} \\ &= \sum_{m=0}^{\infty}(-1)^m\frac{(Kr)^{2m}}{(2m)!}\left\{\mathrm{e}^{\mp 2m\mathrm{i}\theta} + \frac{\mathrm{i}Kr}{2m+1}\,\mathrm{e}^{\mp(2m+1)\mathrm{i}\theta}\right\}, \end{aligned}$$

whence

$$\mathrm{e}^{\pm Kz}\cos Kx = \sum_{m=0}^{\infty}(-1)^m\frac{(Kr)^{2m}}{(2m)!}\left\{\cos 2m\theta \pm \frac{Kr}{2m+1}\sin(2m+1)\theta\right\}$$

and

$$e^{\pm Kz}\sin Kx = \sum_{m=0}^{\infty}(-1)^m \frac{(Kr)^{2m+1}}{(2m+1)!}\left\{\cos(2m+1)\theta \mp \frac{Kr}{2m+2}\sin(2m+2)\theta\right\};$$

in particular, these show that (2.90) reduces to (2.91) when $n=0$ and $n=1$. Then, using the identities

$$k^n - K^n = (k-K)\sum_{q=1}^{n} K^{n-q}k^{q-1}, \quad n = 1, 2, \ldots, \tag{2.94}$$

$$\sum_{q=1}^{2m} K^{-q}k^{q-1} = (k+K)\sum_{q=0}^{m-1} K^{-2q-2}k^{2q}, \quad m = 1, 2, \ldots, \tag{2.95}$$

we obtain

$$F_c(\boldsymbol{r}) = \sum_{m=1}^{\infty}(-1)^{m+1}\widehat{\Phi}_{2m}(\boldsymbol{r})\,(k+K)\sum_{q=0}^{m-1} K^{-2q-2}k^{2q},$$

$$F_s(\boldsymbol{r}) = \sum_{m=1}^{\infty}(-1)^{m+1}\widehat{\Phi}_{2m+1}(\boldsymbol{r})\,(k+K)\sum_{q=0}^{m-1} K^{-2q-3}k^{2q+1}.$$

Substitution in (2.93) and (2.92), using (2.82) and (2.83), gives

$$H = \frac{2}{\pi}\sum_{m=1}^{\infty}\sum_{q=0}^{m-1}(-1)^{m+q}\left\{\widehat{\Phi}_{2m}(\boldsymbol{r}_P)\,\Phi_{2q+2}(\boldsymbol{r}_Q) + \widehat{\Phi}_{2m+1}\Phi_{2q+3}\right\}.$$

But, changing the order of summation, we have

$$\sum_{m=1}^{\infty}\sum_{q=0}^{m-1} F_{mq} = \sum_{q=0}^{\infty}\sum_{m=q+1}^{\infty} F_{mq} = \sum_{q=0}^{\infty}\sum_{s=0}^{\infty} F_{q+1+s,q},$$

for any F_{mq}. Hence

$$\begin{aligned} H &= \frac{2}{\pi}\sum_{q=0}^{\infty}\sum_{s=0}^{\infty}(-1)^{s+1}\Big\{\widehat{\Phi}_{2q+2+2s}(\boldsymbol{r}_P)\,\Phi_{2q+2}(\boldsymbol{r}_Q) \\ &\qquad + \widehat{\Phi}_{2q+3+2s}(\boldsymbol{r}_P)\,\Phi_{2q+3}(\boldsymbol{r}_Q)\Big\} \\ &= \frac{2}{\pi}\sum_{q=2}^{\infty}\sum_{s=0}^{\infty}(-1)^{s+1}\widehat{\Phi}_{q+2s}(\boldsymbol{r}_P)\,\Phi_q(\boldsymbol{r}_Q), \end{aligned}$$

which gives the desired result, assuming that $z < \zeta$. However, both sides of (2.89) are regular harmonic functions for $|\boldsymbol{r}_P| < |\boldsymbol{r}_Q|$, and so analytic continuation implies that the expansion (2.89) is valid for all points P and Q satisfying $|\boldsymbol{r}_P| < |\boldsymbol{r}_Q|$. □

As a special case of Theorem 2.19, put P at a point O_2 in the free surface; let $\boldsymbol{c}$ be the position vector of O_2 with respect to O. Then, as $G(O_2, Q) = -(2/\pi)\Phi_0(\boldsymbol{r}_Q - \boldsymbol{c})$, we obtain

$$\Phi_0(\boldsymbol{r}_Q - \boldsymbol{c}) = \sum_{n=0}^{\infty} \alpha_n(\boldsymbol{c})\, \Phi_n(\boldsymbol{r}_Q), \quad |\boldsymbol{c}| < |\boldsymbol{r}_Q|. \tag{2.96}$$

This formula will be used in our proofs of addition theorems for Ursell's multipole potentials; see Theorem 2.20.

2.11.2 Addition theorems

Let O_1 and O_2 be two origins in the free surface. Let $\boldsymbol{r}_j$ be the position vector of a general point P with respect to O_j, for $j = 1, 2$. Let $\boldsymbol{b}$ be the position vector of O_1 with respect to O_2, so that $\boldsymbol{r}_2 = \boldsymbol{r}_1 + \boldsymbol{b}$. See Fig. 2.6.

Our addition theorems for Φ_n involve the functions α_n, defined by (2.90); in particular, α_0 and α_1 are given by (2.91). The functions α_n are harmonic and regular, and they satisfy the free-surface boundary condition. We note that, as a consequence of (2.86) and (2.91),

$$(\partial/\partial x)\alpha_0 = -K\alpha_1 \quad \text{and} \quad (\partial/\partial x)\alpha_n = K\alpha_{n-1}, \tag{2.97}$$

for $n = 1, 2, 3, \ldots$. Also, the definition (2.90) yields

$$\alpha_n + \alpha_{n+2} = \widehat{\Phi}_n, \quad n = 0, 1, 2, \ldots. \tag{2.98}$$

The addition theorems also involve quantities such as $\alpha_n(-\boldsymbol{b})$. To define these explicitly, let (x_2, z) be Cartesian coordinates at O_2. Suppose that O_1 is at $(x_2, z) = (b\cos\beta, 0)$, where b is the distance between O_1 and O_2, and $\beta = 0$ or $\beta = \pi$. Then

$$\alpha_n(-\boldsymbol{b}) = (-\cos\beta)^n \sum_{s=0}^{\infty} (-1)^s \frac{(Kb)^{n+2s}}{(n+2s)!},$$

$\widehat{\Phi}_n(-\boldsymbol{b}) = (-\cos\beta)^n (Kb)^n/n!$ and $\Phi_n(-\boldsymbol{b}) = (-\cos\beta)^n (n-1)!/(Kb)^n$, the last formula holding for $n \geq 2$. For $\Phi_0(-\boldsymbol{b})$ and $\Phi_1(-\boldsymbol{b})$, we can use (2.81) and $(2.79)_2$.

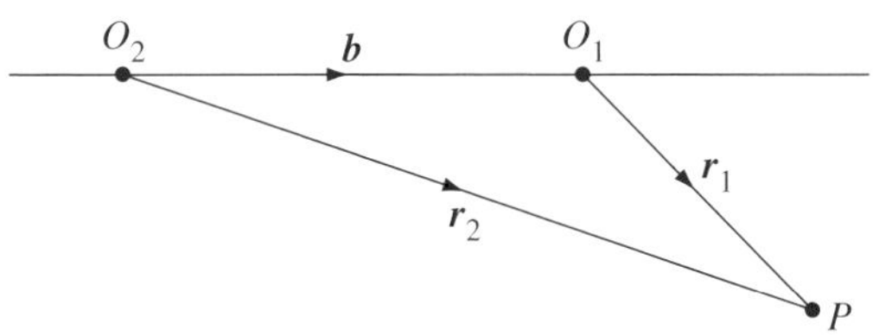

Fig. 2.6. Configuration for water-wave addition theorems.

Theorem 2.20 *Let $\boldsymbol{r}_2 = \boldsymbol{r}_1 + \boldsymbol{b}$. For $n = 0, 1, 2, \ldots$, we have*

$$\Phi_n(\boldsymbol{r}_2) = \sum_{m=0}^{\infty} \widehat{S}_{nm}(\boldsymbol{b})\,\Phi_m(\boldsymbol{r}_1) \quad \textit{for } r_1 > b, \tag{2.99}$$

$$\Phi_n(\boldsymbol{r}_2) = \sum_{m=0}^{\infty} S_{nm}(\boldsymbol{b})\,\alpha_m(\boldsymbol{r}_1) \quad \textit{for } r_1 < b. \tag{2.100}$$

The quantities $\widehat{S}_{nm}$ and S_{nm} are defined as follows:

$$\widehat{S}_{0m}(\boldsymbol{b}) = \alpha_m(-\boldsymbol{b}), \quad m \geq 0;$$

$$\widehat{S}_{10}(\boldsymbol{b}) = -\alpha_1(-\boldsymbol{b}), \quad \widehat{S}_{1m}(\boldsymbol{b}) = \alpha_{m-1}(-\boldsymbol{b}), \quad m \geq 1;$$

$$\widehat{S}_{nm}(\boldsymbol{b}) = \widehat{\Phi}_{m-n}(-\boldsymbol{b}), \quad m \geq n \geq 2;$$

$$\widehat{S}_{nm}(\boldsymbol{b}) = 0, \quad 0 \leq m < n, \quad n \geq 2;$$

$$S_{0m}(\boldsymbol{b}) = \Phi_m(-\boldsymbol{b}), \quad m \geq 0; \quad S_{10}(\boldsymbol{b}) = -\Phi_1(-\boldsymbol{b}),$$

$$S_{11}(\boldsymbol{b}) = \Phi_0(-\boldsymbol{b}) - \Phi_2(-\boldsymbol{b}), \quad S_{1m}(\boldsymbol{b}) = -\Phi_{m+1}(-\boldsymbol{b}), \quad m \geq 2;$$

$$S_{n0}(\boldsymbol{b}) = (-1)^n \Phi_n(-\boldsymbol{b}), \quad S_{n1}(\boldsymbol{b}) = (-1)^n \Phi_{n+1}(-\boldsymbol{b}), \quad n \geq 2;$$

$$S_{nm}(\boldsymbol{b}) = (-1)^n \{\Phi_{m+n-2}(-\boldsymbol{b}) + \Phi_{m+n}(-\boldsymbol{b})\}, \quad n \geq 2, \quad m \geq 2.$$

Proof When $n = 0$, (2.99) reduces to

$$\Phi_0(\boldsymbol{r}_2) = \sum_{m=0}^{\infty} \alpha_m(-\boldsymbol{b})\,\Phi_m(\boldsymbol{r}_1), \qquad r_1 > b. \tag{2.101}$$

This formula is (2.96); it is a special case of the bilinear expansion of G (Theorem 2.19). To prove (2.99) for $n > 0$, we merely differentiate (2.101) with respect to x_2, where $x_2 = x_1 + b\cos\beta$. As $\partial/\partial x_2 = \partial/\partial x_1$, we first obtain

$$(\partial/\partial x_2)\Phi_0(\boldsymbol{r}_2) = \sum_{m=0}^{\infty} \alpha_m(-\boldsymbol{b})\,(\partial/\partial x_1)\Phi_m(\boldsymbol{r}_1),$$

whence

$$\begin{aligned}\Phi_1(\boldsymbol{r}_2) &= \alpha_0\Phi_1 - \alpha_1(\Phi_0 - \Phi_2) + \sum_{m=2}^{\infty} \alpha_m \Phi_{m+1} \\ &= -\alpha_1\Phi_0 + \alpha_0\Phi_1 + \sum_{m=2}^{\infty} \alpha_{m-1}\Phi_m,\end{aligned}$$

which gives (2.99) for $n = 1$; here, we have used (2.84) and (2.85). Differentiating again gives

$$\Phi_2(\boldsymbol{r}_2) - \Phi_0(\boldsymbol{r}_2) = -\alpha_1\Phi_1 + \alpha_0(\Phi_2 - \Phi_0) + \sum_{m=2}^{\infty} \alpha_{m-1}\Phi_{m+1}.$$

Making use of (2.101), we obtain

$$\Phi_2(\boldsymbol{r}_2) = \sum_{m=0}^{\infty} \alpha_m \Phi_m - \alpha_0 \Phi_0 - \alpha_1 \Phi_1 + \sum_{m=2}^{\infty} \alpha_{m-2} \Phi_m$$
$$= \sum_{m=2}^{\infty} (\alpha_m + \alpha_{m-2}) \Phi_m(\boldsymbol{r}_1),$$

which gives the stated result for $n = 2$, once (2.98) has been used. For $n > 2$, differentiate repeatedly, making use of (2.85).

To prove (2.100), we start from

$$\Phi_0(\boldsymbol{r}_2) = \sum_{m=0}^{\infty} \Phi_m(-\boldsymbol{b})\, \alpha_m(\boldsymbol{r}_1), \qquad r_1 < b, \tag{2.102}$$

which also follows from Theorem 2.19. Equation (2.102) is (2.100) with $n = 0$. Differentiating once, using (2.97), gives

$$\Phi_1(\boldsymbol{r}_2) = \Phi_0 \alpha_1 - \sum_{m=1}^{\infty} \Phi_m \alpha_{m-1}$$
$$= -\Phi_1 \alpha_0 + (\Phi_0 - \Phi_2)\alpha_1 - \sum_{m=2}^{\infty} \Phi_{m+1} \alpha_m(\boldsymbol{r}_1),$$

which is (2.100) with $n = 1$. Differentiating again, using (2.84), (2.97) and (2.102), gives

$$\Phi_2(\boldsymbol{r}_2) = \Phi_2 \alpha_0 + \Phi_3 \alpha_1 + \sum_{m=2}^{\infty} (\Phi_m + \Phi_{m+2}) \alpha_m(\boldsymbol{r}_1), \tag{2.103}$$

which is (2.100) with $n = 2$. For $n > 2$, we use an inductive argument, based on

$$\Phi_n(\boldsymbol{r}_2) = (-1)^n (\Phi_n \alpha_0 + \Phi_{n+1} \alpha_1) + (-1)^n \sum_{m=2}^{\infty} (\Phi_{m+n-2} + \Phi_{m+n}) \alpha_m(\boldsymbol{r}_1). \tag{2.104}$$

Assume that this formula holds for some $n \geq 2$. Differentiate once with respect to x_2, using (2.85) and (2.97); elementary calculations then show that (2.104) holds with n replaced by $n+1$. As (2.104) is valid for $n = 2$ (when it reduces to (2.103)), it is valid for all $n \geq 2$. This completes the proof of (2.100). □

For $n \geq 2$, Φ_n is wavefree and satisfies $\Phi_n(\boldsymbol{r}) = O(r^{-n})$ as $r \to \infty$; this explains why $\widehat{S}_{nm} = 0$ for $0 \leq m < n$.

In [786], two other proofs of (2.100) are sketched. One makes use of the integral representations for Φ_n, whereas the other uses complex-variable techniques. The proof given here is simpler, and generalises to three dimensions.

3
Addition theorems in three dimensions

Part of the secret of success in studying and using special functions is to try to remember exactly what is necessary, and nothing more.

(Askey [46, p. 9])

3.1 Introduction

In this chapter, we shall derive several three-dimensional addition theorems. Before doing so, we have to discuss several special functions. We begin with spherical harmonics, Y_n^m, and then we exploit Hobson's theorem. We introduce spherical wavefunctions, defined using spherical Bessel functions and spherical harmonics. We define the Erdélyi operator $\mathcal{Y}_n^m$ in Section 3.6. This operator is used subsequently to derive general addition theorems from much simpler axisymmetric addition theorems; the latter are proved in Section 3.8. A composition formula for $\mathcal{Y}_n^m$ is given in Section 3.10. This formula is derived from the linearisation formula for spherical harmonics, expressing the product $Y_n^m Y_\nu^\mu$ as a linear combination of spherical harmonics with coefficients that are usually known as Gaunt coefficients. The next six sections are concerned with various forms of the addition theorems for spherical wavefunctions. Vector spherical wavefunctions are discussed briefly in Section 3.17. The last section contains results for water-wave multipole potentials.

3.2 Spherical harmonics

Some of the present generation are at a loss to explain why Thomson and Tait [1175] *perpetrated the ill-founded title "Spherical harmonics". Musicians and acousticians should be sole concessionaires of the word* harmonics!

(McLachlan [834, p. 260])

In terms of spherical polar coordinates (r, θ, ϕ), defined by

$$x = r \sin\theta \cos\phi, \quad y = r \sin\theta \sin\phi \quad \text{and} \quad z = r\cos\theta, \tag{3.1}$$

the three-dimensional Laplacian

$$\nabla_3^2 u = \frac{\partial^2 u}{\partial x^2} + \frac{\partial^2 u}{\partial y^2} + \frac{\partial^2 u}{\partial z^2} \tag{3.2}$$
$$= \frac{1}{r^2}\frac{\partial}{\partial r}\left(r^2 \frac{\partial u}{\partial r}\right) + \frac{1}{r^2 \sin\theta}\frac{\partial}{\partial \theta}\left(\sin\theta \frac{\partial u}{\partial \theta}\right) + \frac{1}{r^2 \sin^2\theta}\frac{\partial^2 u}{\partial \phi^2}.$$

Then, the method of separation of variables shows that Laplace's equation, $\nabla_3^2 u = 0$, has solutions

$$r^n P_n^m(\cos\theta)\,\mathrm{e}^{\pm im\phi} \quad \text{and} \quad r^{-n-1} P_n^m(\cos\theta)\,\mathrm{e}^{\pm im\phi} \tag{3.3}$$

for $n = 0, 1, 2, \ldots$ and $m = 0, 1, \ldots, n$, where P_n^m is an *associated Legendre function*. The definition of P_n^m varies from book to book. We use what is known as Ferrers' definition ([507, §57], [343, p. 75]):

Definition 3.1 *For* $m = 0, 1, \ldots, n$, $n = 0, 1, 2, \ldots$ *and* $-1 \le t \le 1$,

$$P_n^m(t) = \frac{(1-t^2)^{m/2}}{2^n n!}\frac{\mathrm{d}^{m+n}}{\mathrm{d}t^{m+n}}(t^2-1)^n. \tag{3.4}$$

This definition is used by Bateman [73, p. 39], Whittaker & Watson [1317, §15.5], Lamb [682, §86], Stratton [1149, §7.3], Morse & Feshbach [884, p. 1325], Messiah [846], Edmonds [305, p. 23], Zare [1371, §1.3], Colton & Kress [225, §2.3], Dahlen & Tromp [250, §B.4] and Nédélec [903, §2.4.3]. It differs by a factor of $(-1)^m$ from that used by Hobson [507, §55], Erdélyi *et al.* [321, §3.6.1], Jones [564, p. 79], Abramowitz & Stegun [1], Lebedev [691, §7.12], Jackson [549, p. 98] and Gradshteyn & Ryzhik [427]. Jeffreys & Jeffreys [556, p. 633] use a different notation and a different definition, having discussed various possibilities.

Definition 3.1 makes sense for negative values of m, with $|m| \le n$. It follows that

$$P_n^{-m}(t) = (-1)^m \frac{(n-m)!}{(n+m)!} P_n^m(t) \tag{3.5}$$

for $m = 0, 1, \ldots, n$ and $n \ge 0$.

Various properties of associated Legendre functions are summarised in Appendix A.

The restrictions of the harmonic functions (3.3) to a spherical surface $r = r_0$, say, are called *spherical harmonics* or *tesseral harmonics*. They are usually normalised in some way, but there is no agreement on the normalisation or the notation. We shall use both an unnormalised definition ($\tilde{Y}_n^m$) and a corresponding normalised definition (Y_n^m).

Definition 3.2 *For $m = 0, \pm 1, \pm 2, \ldots, \pm n$ and $n = 0, 1, 2, \ldots,$*

$$\tilde{Y}_n^m(\hat{\boldsymbol{r}}) = \tilde{Y}_n^m(\theta, \phi) = P_n^m(\cos\theta)\,\mathrm{e}^{\mathrm{i}m\phi},$$
$$Y_n^m(\hat{\boldsymbol{r}}) = Y_n^m(\theta, \phi) = (4\pi)^{-1/2} A_n^m\, \tilde{Y}_n^m(\theta, \phi),$$
$$A_n^m = (-1)^m \sqrt{2n+1}\sqrt{\frac{(n-m)!}{(n+m)!}}. \tag{3.6}$$

Here, $\hat{\boldsymbol{r}}$ is a unit vector with Cartesian components given by

$$\hat{\boldsymbol{r}} = (\sin\theta\cos\phi,\ \sin\theta\sin\phi,\ \cos\theta).$$

Note that, from (3.5), we have

$$Y_n^{-m}(\hat{\boldsymbol{r}}) = (-1)^m\, \overline{Y_n^m(\hat{\boldsymbol{r}})}, \tag{3.7}$$

where the overbar denotes complex conjugation. Also, replacing θ by $\pi - \theta$ and ϕ by $\phi \pm \pi$, and then using (A.6), we see that

$$Y_n^m(-\hat{\boldsymbol{r}}) = (-1)^n\, Y_n^m(\hat{\boldsymbol{r}}). \tag{3.8}$$

The unnormalised definition is most convenient for the statement of recurrence formulae. According to Edmonds [305, p. 21], the normalised definition above 'is the most commonly used convention'. It is convenient because it renders the spherical harmonics orthonormal. Specifically, let $\Omega = \{(\theta, \phi) : 0 \le \theta \le \pi,\ -\pi \le \phi < \pi\}$ be the surface of the unit ball. Then, it can be shown [225, Theorem 2.7] that the spherical harmonics form a complete orthonormal system (see Section 7.7) in $L^2(\Omega)$:

$$\int_\Omega Y_n^m \overline{Y_\nu^\mu}\,\mathrm{d}\Omega = \delta_{n\nu}\delta_{m\mu}; \tag{3.9}$$

this follows from two orthogonality results, namely (2.18) and

$$\int_0^\pi P_n^m(\cos\theta)\,P_\nu^m(\cos\theta)\,\sin\theta\,\mathrm{d}\theta = \frac{2}{2n+1}\,\frac{(n+m)!}{(n-m)!}\,\delta_{n\nu}. \tag{3.10}$$

We shall often expand functions defined on Ω using spherical harmonics. Thus, if $g(\hat{\boldsymbol{r}})$ is twice-continuously differentiable, we have (see, for example, [243, p. 513])

$$g(\hat{\boldsymbol{r}}) = \sum_{n=0}^\infty \sum_{m=-n}^n Y_n^m(\hat{\boldsymbol{r}}) \int_\Omega g(\hat{\boldsymbol{r}}_1)\,\overline{Y_n^m(\hat{\boldsymbol{r}}_1)}\,\mathrm{d}\Omega(\hat{\boldsymbol{r}}_1). \tag{3.11}$$

The normalised definition can be inconvenient, because of the cumbersome square-root factors. When we wish to compare contiguous spherical harmonics, the following simple relations satisfied by the normalisation factor A_n^m are useful:

$$A_{n-1}^m = \sqrt{\frac{(2n-1)(n+m)}{(2n+1)(n-m)}}\, A_n^m, \tag{3.12}$$

$$A_{n+1}^m = \sqrt{\frac{(2n+3)(n-m+1)}{(2n+1)(n+m+1)}}\, A_n^m, \tag{3.13}$$

$$A_{n-1}^{m+1} = -\sqrt{\frac{2n-1}{(2n+1)(n-m)(n-m-1)}}\, A_n^m, \tag{3.14}$$

$$A_{n+1}^{m+1} = -\sqrt{\frac{2n+3}{(2n+1)(n+m+2)(n+m+1)}}\, A_n^m, \tag{3.15}$$

$$A_{n-1}^{m-1} = -\sqrt{\frac{(2n-1)(n+m)(n+m-1)}{2n+1}}\, A_n^m, \tag{3.16}$$

$$A_{n+1}^{m-1} = -\sqrt{\frac{(2n+3)(n-m+2)(n-m+1)}{2n+1}}\, A_n^m. \tag{3.17}$$

Another simple relation is $A_n^m A_n^{-m} = 2n+1$.

As $r^n Y_n^m(\theta, \phi)$ solves $\nabla_3^2 u = 0$, (3.2) shows that Y_n^m satisfies

$$\frac{\partial}{\partial\theta}\left(\sin\theta\,\frac{\partial Y_n^m}{\partial\theta}\right) + \frac{1}{\sin\theta}\frac{\partial^2 Y_n^m}{\partial\phi^2} + n(n+1)\,Y_n^m\sin\theta = 0. \tag{3.18}$$

For the computation of Y_n^m using recurrence relations, see [810].

3.3 Legendre's addition theorem

Let $\hat{\boldsymbol{r}}_1$ and $\hat{\boldsymbol{r}}_2$ be two unit vectors with Cartesian components given by

$$\hat{\boldsymbol{r}}_1 = (\sin\theta_1\cos\phi_1,\ \sin\theta_1\sin\phi_1,\ \cos\theta_1),$$
$$\hat{\boldsymbol{r}}_2 = (\sin\theta_2\cos\phi_2,\ \sin\theta_2\sin\phi_2,\ \cos\theta_2).$$

The angle between $\hat{\boldsymbol{r}}_1$ and $\hat{\boldsymbol{r}}_2$ is γ, defined by

$$\hat{\boldsymbol{r}}_1\cdot\hat{\boldsymbol{r}}_2 = \cos\gamma = \sin\theta_1\sin\theta_2\cos(\phi_1-\phi_2) + \cos\theta_1\cos\theta_2.$$

Then, we have

$$P_n(\cos\gamma) = \sum_{m=0}^{n} \epsilon_m \frac{(n-m)!}{(n+m)!}\, P_n^m(\cos\theta_1)\, P_n^m(\cos\theta_2)\,\cos m(\phi_1-\phi_2), \tag{3.19}$$

where P_n is a Legendre polynomial, $\epsilon_0 = 1$ and $\epsilon_m = 2$ for $m > 0$. This is *Legendre's addition theorem*. For a proof, see [1317, p. 395], [507, p. 143], [1149, §7.5], [225, Theorem 2.8], [31, §9.6] or [903, Theorem 2.4.5].

We can give several variants of (3.19). Thus, writing the cosine as the sum of two exponentials, and using (3.5), we obtain

$$P_n(\cos\gamma) = \sum_{m=-n}^{n} \frac{(n-m)!}{(n+m)!} P_n^m(\cos\theta_1)\, P_n^m(\cos\theta_2)\, \mathrm{e}^{\mathrm{i}m(\phi_1-\phi_2)}$$

$$= \sum_{m=-n}^{n} (-1)^m P_n^m(\cos\theta_1)\, P_n^{-m}(\cos\theta_2)\, \mathrm{e}^{\mathrm{i}m(\phi_1-\phi_2)}.$$

In terms of spherical harmonics, we have

$$P_n(\hat{\boldsymbol{r}}_1 \cdot \hat{\boldsymbol{r}}_2) = \sum_{m=-n}^{n} (-1)^m \tilde{Y}_n^m(\hat{\boldsymbol{r}}_1)\, \tilde{Y}_n^{-m}(\hat{\boldsymbol{r}}_2)$$

$$= \frac{4\pi}{2n+1} \sum_{m=-n}^{n} (-1)^m Y_n^m(\hat{\boldsymbol{r}}_1)\, Y_n^{-m}(\hat{\boldsymbol{r}}_2)$$

$$= \frac{4\pi}{2n+1} \sum_{m=-n}^{n} Y_n^m(\hat{\boldsymbol{r}}_1)\, \overline{Y_n^m(\hat{\boldsymbol{r}}_2)}. \tag{3.20}$$

Using (3.20) twice, followed by (3.9) and a third use of (3.20) gives

$$\int_\Omega P_n(\hat{\boldsymbol{r}}_1 \cdot \hat{\boldsymbol{r}})\, P_\nu(\hat{\boldsymbol{r}}_2 \cdot \hat{\boldsymbol{r}})\, \mathrm{d}\Omega(\hat{\boldsymbol{r}}) = \frac{4\pi\,\delta_{n\nu}}{2n+1} P_n(\hat{\boldsymbol{r}}_1 \cdot \hat{\boldsymbol{r}}_2). \tag{3.21}$$

3.4 Cartesian coordinates

The advantages of the use of spherical harmonics expressed in Cartesian coordinates were recognised almost simultaneously by Thomson (Lord Kelvin) in England, and by Clebsch in Germany. The most complete development of the subject is contained in the well-known Appendix B of Thomson and Tait's Natural Philosophy [1175].

(Hobson [507, p. 119])

As in Chapter 2, we construct multipole solutions by differentiating one solution with respect to any Cartesian coordinate. The use of this idea for constructing harmonic functions in three dimensions (by repeated differentiation of $1/r$) is described at length by Thomson & Tait [1175, Appendix B].

In the next section, we give Hobson's theorem in three dimensions. This gives the result of applying certain differential operators in Cartesian coordinates to functions of φ_3, where $\varphi_3 = r^2 = x^2 + y^2 + z^2$. The operators are

$$f_n(\nabla_3) = f_n\left(\frac{\partial}{\partial x}, \frac{\partial}{\partial y}, \frac{\partial}{\partial z}\right), \tag{3.22}$$

where $f_n(x, y, z)$ is any homogeneous polynomial of degree $n \geq 0$.

Hobson's theorem simplifies if f_n is a harmonic homogeneous polynomial, $\mathcal{H}_n$ say, so that

$$\nabla_3^2 \mathcal{H}_n(x, y, z) = 0.$$

It is known that there are exactly $2n+1$ linearly independent harmonic homogeneous polynomials of degree n for each $n \geq 0$; they can be taken as

$$\mathcal{H}_n^m(x, y, z) = r^n Y_n^m(\theta, \phi) \quad \text{or} \quad \tilde{\mathcal{H}}_n^m(x, y, z) = r^n \tilde{Y}_n^m(\theta, \phi), \tag{3.23}$$

for $m = 0, \pm 1, \pm 2, \ldots, \pm n$, where r, θ and ϕ are spherical polar coordinates and the spherical harmonics, Y_n^m and $\tilde{Y}_n^m$, are defined by Definition 3.2. For example, we have

$$\mathcal{H}_1^0 = \sqrt{\frac{3}{4\pi}}\, z, \quad \mathcal{H}_1^1 = -\sqrt{\frac{3}{8\pi}}\,(x+\mathrm{i}y), \quad \mathcal{H}_1^{-1} = \sqrt{\frac{3}{8\pi}}\,(x-\mathrm{i}y). \tag{3.24}$$

Note that, from (3.7),

$$\mathcal{H}_n^{-m} = (-1)^m\, \overline{\mathcal{H}_n^m}. \tag{3.25}$$

Explicitly, using (A.2), we have

$$\tilde{\mathcal{H}}_n^m(x, y, z) = (x+\mathrm{i}y)^m \sum_{l=0}^{[(n-m)/2]} B_l^{n,m} z^{n-m-2l} r^{2l} \tag{3.26}$$

for $0 \leq m \leq n$, where $[n/2]$ is the integer part of $n/2$ (so that $[n/2] = N$ when $n = 2N$ or $2N+1$) and

$$B_l^{n,m} = \frac{(-1)^l (2n-2l)!}{2^n l!\,(n-l)!\,(n-2l-m)!}. \tag{3.27}$$

Also, using (3.5), we obtain

$$\tilde{\mathcal{H}}_n^{-m}(x, y, z) = (-1)^m \frac{(n-m)!}{(n+m)!} \overline{\tilde{\mathcal{H}}_n^m(x, y, z)} \quad \text{for } 0 \leq m \leq n. \tag{3.28}$$

Comparing (3.28) with (3.25) gives another reason for preferring normalised spherical harmonics.

3.5 Hobson's theorem

One thing Cambridge made almost inevitable for an analyst; intensive study of Legendre functions, and all that. Such "dictionary" subjects are utterly unsuitable for a good man. It was, however, the more inevitable in that a lecture was provided (by E. W. Hobson; he later wrote a standard text-book): I was the sole member of the class.

(Littlewood [734, p. 86])

We generalise the results of Section 2.3 to three dimensions.

Theorem 3.3 (Hobson's theorem in three dimensions) *Let $\varphi_3 = x^2 + y^2 + z^2$. Let $f_n(x, y, z)$ be a homogeneous polynomial of degree n, where n is a positive integer. Then, with $f_n(\nabla_3)$ defined by (3.22),*

$$f_n(\nabla_3) F(\varphi_3) = \sum_{l=0} \frac{2^{n-2l}}{l!} F^{(n-l)}(\varphi_3) \nabla_3^{2l} f_n(x, y, z),$$

where the summation has a finite number of terms.

Proof The proof is very similar to that for the two-dimensional case, Theorem 2.3. That theorem is valid for functions of φ_3, so that

$$f_n\left(\frac{\partial}{\partial y}, \frac{\partial}{\partial z}\right) F(\varphi_3) = \sum_{l=0} \frac{2^{n-2l}}{l!} F^{(n-l)}(\varphi_3) \nabla_3^{2l} f_n(y, z),$$

where, in this formula, $f_n(y, z)$ is a homogeneous polynomial in y and z. We then note that $f_{n+1}(x, y, z) = x f_n(x, y, z) + g_{n+1}(y, z)$, where g_{n+1} is a homogeneous polynomial of degree $n+1$ in two variables. The inductive argument then proceeds as before; we do not repeat it here. □

The special case of Theorem 3.3 with $F(\varphi_3) = \varphi_3^{-1/2}$ is given in [74, §6.22] and in [362, Theorem 2.3.1]. Specialising further gives the following result.

Example 3.4 (Hobson [507, p. 134]**)** Take $F(\varphi_3) = r^{-1}$ with $r = \sqrt{\varphi_3}$, and

$$f_n(x, y, z) = \frac{z^{n-m}(x \pm \mathrm{i}y)^m}{(n-m)!}, \qquad 0 \le m \le n.$$

Thus

$$2^l F^{(l)}(\varphi_3) = \left(\frac{1}{r}\frac{\mathrm{d}}{\mathrm{d}r}\right)^l \frac{1}{r} = \frac{(-1)^l (2l)!}{2^l\, l!} \frac{1}{r^{2l+1}}$$

and $\nabla_3^2 f_{l+2} = f_l$ for $l \ge m$, with $\nabla_3^2 f_m = \nabla_3^2 f_{m+1} = 0$. Then, Theorem 3.3 gives

$$f_n(\nabla_3)\frac{1}{r} = \sum_{l=0}^{[(n-m)/2]} (-1)^n B_l^{n,m} \frac{z^{n-2l-m}(x \pm \mathrm{i}y)^m}{r^{2n-2l+1}},$$

where $B_l^{n,m}$ is defined by (3.27). Introducing spherical polar coordinates (3.1) and comparing the sum with the definition of $P_n^m(\cos\theta)$, (A.2), we obtain the following result, for $0 \le m \le n$:

$$\frac{\partial^{n-m}}{\partial z^{n-m}} \left(\frac{\partial}{\partial x} \pm \mathrm{i}\frac{\partial}{\partial y}\right)^m \frac{1}{r} = \frac{(-1)^n (n-m)!}{r^{n+1}} P_n^m(\cos\theta)\, \mathrm{e}^{\pm \mathrm{i}m\phi}. \qquad (3.29)$$

Let us now assume that f_n is harmonic. Then, we obtain the following corollary to Hobson's theorem, Theorem 3.3.

Corollary 3.5 *Let $r = \sqrt{x^2 + y^2 + z^2}$. Let $\mathcal{H}_n(x, y, z)$ be a harmonic homogeneous polynomial of degree n, where n is a positive integer. Then*

$$\mathcal{H}_n(\nabla_3) F(r) = \mathcal{H}_n(x, y, z) \left(\frac{1}{r}\frac{\mathrm{d}}{\mathrm{d}r}\right)^n F(r).$$

Example 3.6 (Hobson [507, p. 127]**)** Take $F(r) = r^{-1}$ and $\mathcal{H}_n = \mathcal{H}_n^m$, defined by (3.23), giving

$$\mathcal{H}_n^m(-\nabla_3)\frac{1}{r} = \frac{(2n)!}{2^n n!} \frac{1}{r^{n+1}} Y_n^m(\hat{\boldsymbol{r}}).$$

This formula holds for $0 \le |m| \le n$, as a consequence of (3.25) and (3.7).

It is interesting to compare Example 3.4 and Example 3.6. Both give representations of the same spherical harmonic Y_n^m in terms of derivatives of r^{-1}, for $0 \le m \le n$. The two formulae agree for $n = 0$ and for $n = 1$, but they differ for $n \ge 2$. This lack of uniqueness is not surprising, as we know that $\nabla_3^{2l}(1/r) = 0$ for $l \ge 1$. Thus, if we can find one suitable homogeneous polynomial of even degree, $f_{2n}(x, y, z)$, we can always add any multiple of $(x^2 + y^2 + z^2)^n$; for odd degree, $2n + 1$, we can add any multiple of $z(x^2 + y^2 + z^2)^n$.

The following corollary to Theorem 3.3 is also useful.

Corollary 3.7 *Let $r = \sqrt{x^2 + y^2 + z^2}$ and let $x_i = x$, y or z. Let $\mathcal{H}_n(x, y, z)$ be a harmonic homogeneous polynomial of degree n, where n is a positive integer. Then*

$$\frac{\partial}{\partial x_i}\mathcal{H}_n(\nabla_3)F(r) = x_i\mathcal{H}_n(x, y, z)\left(\frac{1}{r}\frac{\mathrm{d}}{\mathrm{d}r}\right)^{n+1}F(r) + \frac{\partial}{\partial x_i}\mathcal{H}_n(x, y, z)\left(\frac{1}{r}\frac{\mathrm{d}}{\mathrm{d}r}\right)^{n}F(r).$$

Proof We apply Theorem 3.3 with $f_{n+1}(x, y, z) = x_i\mathcal{H}_n(x, y, z)$. The result follows by noting that if u and v are both harmonic, then

$$\nabla_3^2(uv) = 2\,\mathrm{grad}\,u \cdot \mathrm{grad}\,v, \tag{3.30}$$

whence $\nabla_3^2 f_{n+1} = 2(\partial/\partial x_i)\mathcal{H}_n$ and $\nabla_3^{2l} f_{n+1} = 0$ for $l \ge 2$. □

3.6 Wavefunctions and the operator $\mathcal{Y}_n^m$

[The] *common* [multipole] *potential ... is proportional to*

$$\frac{d^i}{dh_1\,dh_2 \ldots dh_i}\left(\frac{1}{r}\right),$$

where there are i differentiations of r^{-1} with respect to the axes h_1, h_2, &c., any number of which may in particular cases coincide. It might perhaps have been expected that a similar law would hold for the [acoustic] *velocity potential with the substitution of $r^{-1}e^{-ikr}$ for r^{-1}. This however is not the case; ... though of course the function $r^{-1}e^{-ikr}$ after any number of differentiations continues to satisfy the fundamental equation $(\nabla^2 + k^2)\,\psi = 0$.*

(Lord Rayleigh [1010, p. 249])

We have seen that spherical harmonics can be constructed by repeated differentiation of r^{-1} with respect to Cartesian coordinates. The corresponding construction for wavefunctions is not so straightforward, as the quotation above suggests.

Before giving this construction, let us define spherical wavefunctions. Thus, let $z_\nu(w)$ denote any spherical function of order ν, that is

$$z_\nu(w) = j_\nu(w),\ y_\nu(w),\ h_\nu^{(1)}(w) \text{ or } h_\nu^{(2)}(w); \tag{3.31}$$

by definition,

$$z_\nu(w) = \sqrt{\frac{1}{2}\pi/w}\, Z_{\nu+1/2}(w),$$

where Z_ν is the corresponding cylinder function, (2.8). In particular, $\mathrm{i}h_0^{(1)}(w) = \mathrm{e}^{\mathrm{i}w}/w$. From (2.9), we have

$$\left(\frac{1}{w}\frac{\mathrm{d}}{\mathrm{d}w}\right)^m \frac{z_\nu(w)}{w^\nu} = (-1)^m \frac{z_{\nu+m}(w)}{w^{\nu+m}} \tag{3.32}$$

for $m = 0, 1, 2, \dots$ and any ν, whence

$$\left(\frac{1}{kr}\frac{\mathrm{d}}{\mathrm{d}r}\right)^m z_0(kr) = \frac{(-1)^m}{r^m} z_m(kr). \tag{3.33}$$

Next, we define general spherical wavefunctions as follows.

Definition 3.8 *The* unnormalised spherical wavefunction *is defined by*

$$\tilde{\Omega}_n^m(x, y, z) = z_n(kr)\, \tilde{Y}_n^m(\theta, \phi) = z_n(kr)\, P_n^m(\cos\theta)\, \mathrm{e}^{\mathrm{i}m\phi}$$

for $0 \le |m| \le n$. *For* $|m| > n$, $\tilde{\Omega}_n^m \equiv 0$ (*since* $P_n^{|m|} \equiv 0$). *Similarly, the* normalised spherical wavefunction *is defined, for* $0 \le |m| \le n$, *by*

$$\Omega_n^m(x, y, z) = z_n(kr)\, Y_n^m(\theta, \phi);$$

in particular, the normalised regular spherical wavefunction *is defined by*

$$\widehat{\Omega}_n^m(x, y, z) = j_n(kr)\, Y_n^m(\theta, \phi).$$

Now, just as in the two-dimensional case, we distinguish certain useful differential operators.

Definition 3.9 *Define differential operators* $\mathcal{Y}_n^m$ *and* $\tilde{\mathcal{Y}}_n^m$ *by*

$$\mathcal{Y}_n^m = \mathcal{H}_n^m\left(\frac{-1}{k}\nabla_3\right) = \frac{(-1)^n}{k^n}\mathcal{H}_n^m\left(\frac{\partial}{\partial x}, \frac{\partial}{\partial y}, \frac{\partial}{\partial z}\right), \tag{3.34}$$

$$\tilde{\mathcal{Y}}_n^m = \tilde{\mathcal{H}}_n^m\left(\frac{-1}{k}\nabla_3\right) = \frac{(-1)^n}{k^n}\tilde{\mathcal{H}}_n^m\left(\frac{\partial}{\partial x}, \frac{\partial}{\partial y}, \frac{\partial}{\partial z}\right), \tag{3.35}$$

for $0 \le |m| \le n$, *where* $\mathcal{H}_n^m$ *and* $\tilde{\mathcal{H}}_n^m$ *are defined by* (3.23).

For example, if we use (3.24), we find that

$$\mathcal{Y}_1^0 = \sqrt{\frac{3}{4\pi}}\,\mathcal{D}_1^0, \quad \mathcal{Y}_1^1 = -\sqrt{\frac{3}{8\pi}}\,\mathcal{D}_1^+ \quad \text{and} \quad \mathcal{Y}_1^{-1} = \sqrt{\frac{3}{8\pi}}\,\mathcal{D}_1^-, \tag{3.36}$$

where the first-order differential operators $\mathcal{D}_1^\pm$ and $\mathcal{D}_1^0$ are defined by

$$\mathcal{D}_1^\pm = \frac{-1}{k}\left(\frac{\partial}{\partial x} \pm \mathrm{i}\frac{\partial}{\partial y}\right) \quad \text{and} \quad \mathcal{D}_1^0 = \frac{-1}{k}\frac{\partial}{\partial z}, \tag{3.37}$$

respectively; see Definition 2.6. Note that

$$\mathcal{Y}_n^{-m} = (-1)^m \overline{\mathcal{Y}_n^m}.$$

Explicitly, we have, for $0 \le m \le n$,

$$\tilde{\mathcal{Y}}_n^m = (\mathcal{D}_1^+)^m \sum_{l=0}^{[(n-m)/2]} B_l^{n,m} (\mathcal{D}_1^0)^{n-m-2l} (k^{-2}\nabla_3^2)^l, \tag{3.38}$$

where $B_l^{n,m}$ is given by (3.27). Hence, if u is a wavefunction, we have

$$\tilde{\mathcal{Y}}_n^m u = (\mathcal{D}_1^+)^m \sum_{l=0}^{[(n-m)/2]} (-1)^l B_l^{n,m} (\mathcal{D}_1^0)^{n-m-2l} u. \tag{3.39}$$

The main reason for introducing the operators $\mathcal{Y}_n^m$ and $\tilde{\mathcal{Y}}_n^m$ is contained in the next theorem.

Theorem 3.10 *For* $0 \le |m| \le n$,

$$\mathcal{Y}_n^m\{z_0(kr)\} = \Omega_n^m = z_n(kr)\, Y_n^m(\theta, \phi), \tag{3.40}$$

$$\tilde{\mathcal{Y}}_n^m\{z_0(kr)\} = \tilde{\Omega}_n^m = z_n(kr)\, P_n^m(\cos\theta)\, \mathrm{e}^{\mathrm{i}m\phi}. \tag{3.41}$$

Proof For $0 \le m \le n$, apply Corollary 3.5 to $F(r) = z_0(kr)$ with $\mathcal{H}_n = (-1/k)^n \mathcal{H}_n^m(x, y, z)$, using (3.33), giving (3.40). For negative m, we have

$$\begin{aligned} \mathcal{Y}_n^m z_0 &= (-1)^m \overline{\mathcal{Y}_n^{-m}} z_0 = (-1)^m (-1/k)^n \overline{\mathcal{H}_n^{-m}(\nabla_3)} z_0 \\ &= (-1)^m (-1/k)^n \overline{\mathcal{H}_n^{-m}(x, y, z)} (-k/r)^n z_n \\ &= z_n (-1)^m \overline{Y_n^{-m}} = z_n Y_n^m. \end{aligned}$$

Similarly, (3.41) follows by using $\mathcal{H}_n = (-1/k)^n \tilde{\mathcal{H}}_n^m(x, y, z)$. □

An operator similar to $\tilde{\mathcal{Y}}_n^0$ was first introduced by van der Pol in 1936 [1238]. The generalisation to non-axisymmetric wavefunctions was made by Erdélyi [319]. Thus, for $0 \le m \le n$, his operator is

$$\mathcal{P}_n^m = (\mathrm{i}\mathcal{D}_1^+)^m \sum_{l=0}^{[(n-m)/2]} B_l^{n,m} (\mathrm{i}\mathcal{D}_1^0)^{n-m-2l},$$

whereas $\tilde{\mathcal{Y}}_n^m$ is given by (3.38); note that if u is a wavefunction, then $\mathcal{P}_n^m u = \mathrm{i}^n \tilde{\mathcal{Y}}_n^m u$.

Erdélyi [319] gave a proof of (3.41). Other proofs have been given by Nozawa [919] and by Clercx & Schram [218]; see also [83, p. 81]. Similar results can be obtained for modified spherical Bessel functions.

Erdélyi [319] showed that $\mathcal{P}_n^m$ has a simple action on plane waves.

Theorem 3.11 *Let ξ be a constant. Then, for $0 \le |m| \le n$,*

$$\mathcal{Y}_n^m\{\exp(\mathrm{i}k\xi\boldsymbol{r}\cdot\hat{\boldsymbol{r}}_1)\} = (-\mathrm{i}\xi)^n Y_n^m(\hat{\boldsymbol{r}}_1)\exp(\mathrm{i}k\xi\boldsymbol{r}\cdot\hat{\boldsymbol{r}}_1).$$

Proof Let $\Phi = \exp(\mathrm{i}k\xi\boldsymbol{r}\cdot\hat{\boldsymbol{r}}_1)$. Then

$$\begin{aligned}\mathcal{Y}_n^m\Phi &= (-k)^{-n}\mathcal{H}_n^m(\nabla_3)\,\Phi = (-k)^{-n}\mathcal{H}_n^m(\mathrm{i}k\xi\hat{\boldsymbol{r}}_1)\,\Phi \\ &= (-\mathrm{i}\xi)^n\mathcal{H}_n^m(\hat{\boldsymbol{r}}_1)\,\Phi = (-\mathrm{i}\xi)^n Y_n^m(\hat{\boldsymbol{r}}_1)\,\Phi,\end{aligned}$$

as (3.23) gives $\mathcal{H}_n^m(\hat{\boldsymbol{r}}) = Y_n^m(\hat{\boldsymbol{r}})$. □

Combining this theorem with Theorem 3.10, we obtain the Funk–Hecke formula.

Theorem 3.12 (Funk–Hecke formula) *For $n = 0, 1, 2, \ldots$ and $0 \le |m| \le n$, we have*

$$j_n(kr)\,Y_n^m(\hat{\boldsymbol{r}}) = \frac{(-\mathrm{i})^n}{4\pi}\int_\Omega \exp(\mathrm{i}k\boldsymbol{r}\cdot\hat{\boldsymbol{r}}_1)\,Y_n^m(\hat{\boldsymbol{r}}_1)\,\mathrm{d}\Omega(\hat{\boldsymbol{r}}_1) \tag{3.42}$$

$$= \frac{\mathrm{i}^n}{4\pi}\int_\Omega \exp(-\mathrm{i}k\boldsymbol{r}\cdot\hat{\boldsymbol{r}}_1)\,Y_n^m(\hat{\boldsymbol{r}}_1)\,\mathrm{d}\Omega(\hat{\boldsymbol{r}}_1), \tag{3.43}$$

where Ω is the unit sphere.

Proof Direct calculation gives

$$\int_\Omega \exp(\mathrm{i}k\boldsymbol{r}\cdot\hat{\boldsymbol{r}}_1)\,\mathrm{d}\Omega(\hat{\boldsymbol{r}}_1) = 2\pi\int_0^\pi \mathrm{e}^{\mathrm{i}kr\cos\theta}\sin\theta\,\mathrm{d}\theta = 4\pi j_0(kr), \tag{3.44}$$

using $j_0(w) = w^{-1}\sin w$. Now apply the operator $\mathcal{Y}_n^m$ to this equation, using Theorem 3.10 on the right-hand side (with $z_0 = j_0$) and Theorem 3.11 on the left-hand side (with $\xi = 1$), giving (3.42). Equation (3.43) follows from (3.42), either by replacing k by $-k$, or by complex conjugation. □

This proof illustrates once more how the application of a Cartesian differential operator ($\mathcal{Y}_n^m$) to a spherically-symmetric formula yields a more general result. For the two-dimensional version of Theorem 3.12, see (2.24); for a q-dimensional version, see [31, Lemma 9.10.2].

Theorem 3.12 is useful as it gives a representation of a regular spherical wavefunction in terms of plane waves. In fact, it is a special case of a more general theorem ([892, Theorem 6, p. 20)], [362, §3.6]). Thus, if $f(\mu)$ is continuous for $-1 \le \mu \le 1$, then

$$\int_\Omega f(\hat{\boldsymbol{r}}\cdot\hat{\boldsymbol{r}}_1)\,Y_n^m(\hat{\boldsymbol{r}}_1)\,\mathrm{d}\Omega(\hat{\boldsymbol{r}}_1) = 2\pi\,Y_n^m(\hat{\boldsymbol{r}})\int_{-1}^1 f(\mu)\,P_n(\mu)\,\mathrm{d}\mu. \tag{3.45}$$

This can be proved by combining a Fourier–Legendre expansion of f with Legendre's addition theorem. For a q-dimensional version of (3.45), see [893, §4, Theorem 1] or [31, Theorem 9.7.1].

3.7 First derivatives of spherical wavefunctions

I often think about [how to explain an idea], *especially when I'm teaching some esoteric technique such as integrating Bessel functions. When I see equations, I see the letters in colors – I don't know why. As I'm talking, I see vague pictures of Bessel functions from Jahnke and Emde's book* [550], *with light-tan j's, slightly violet-bluish n's,* [called y_n here] *and dark brown x's flying around. And I wonder what the hell it must look like to the students.*

(Feynman [344, p. 223])

In this section, we prove three-dimensional analogues of Theorem 2.9. We start by calculating a z-derivative of spherical wavefunctions.

Theorem 3.13 *For* $0 \le |m| \le n$,

$$\mathcal{D}_1^0 \tilde{\Omega}_n^m = \frac{n-m+1}{2n+1}\,\tilde{\Omega}_{n+1}^m - \frac{n+m}{2n+1}\,\tilde{\Omega}_{n-1}^m, \tag{3.46}$$

$$\mathcal{D}_1^0 \Omega_n^m = \sqrt{\frac{(n+1)^2-m^2}{(2n+1)(2n+3)}}\,\Omega_{n+1}^m - \sqrt{\frac{n^2-m^2}{4n^2-1}}\,\Omega_{n-1}^m, \tag{3.47}$$

where $\mathcal{D}_1^0 = -k^{-1}(\partial/\partial z)$. *In particular, for* $n \ge 0$,

$$\mathcal{D}_1^0 \tilde{\Omega}_n^n = (2n+1)^{-1}\tilde{\Omega}_{n+1}^n, \quad \mathcal{D}_1^0 \tilde{\Omega}_n^{-n} = \tilde{\Omega}_{n+1}^{-n}, \quad \mathcal{D}_1^0 \Omega_n^{\pm n} = \frac{1}{\sqrt{2n+3}}\,\Omega_{n+1}^{\pm n}.$$

Proof By Theorem 3.10 and Corollary 3.7, we have

$$\begin{aligned}\mathcal{D}_1^0 \tilde{\Omega}_n^m &= \mathcal{D}_1^0 \tilde{\mathcal{Y}}_n^m\{z_0(kr)\}\\ &= \frac{z}{r^{n+1}}\,z_{n+1}(kr)\,\tilde{\mathcal{H}}_n^m(x,y,z) + \frac{1}{r^n}\,z_n(kr)\,\mathcal{D}_1^0\tilde{\mathcal{H}}_n^m(x,y,z).\end{aligned}$$

For $m = n = 0$, the second term vanishes, since $\tilde{\mathcal{H}}_0^0 = 1$; the result follows from $P_1^0(\mu) = \mu$. Otherwise, we expand the second term using

$$\frac{\partial}{\partial z} = \cos\theta\,\frac{\partial}{\partial r} - \frac{\sin\theta}{r}\,\frac{\partial}{\partial \theta},$$

whence

$$\frac{\partial}{\partial z}\tilde{\mathcal{H}}_n^m(x,y,z) = r^{n-1}\left[n\mu P_n^m(\mu) + (1-\mu^2)\frac{\mathrm{d}}{\mathrm{d}\mu}P_n^m(\mu)\right]\mathrm{e}^{\mathrm{i}m\phi},$$

where $\mu = \cos\theta$. Hence,

$$-\mathrm{e}^{-\mathrm{i}m\phi}\,\mathcal{D}_1^0 \tilde{\Omega}_n^m = \mu P_n^m(\mu)\left[\frac{n}{kr}\,z_n(kr) - z_{n+1}\right] + \frac{z_n(kr)}{kr}\,(1-\mu^2)\frac{\mathrm{d}P_n^m}{\mathrm{d}\mu}.$$

Finally, we obtain the desired result by using (A.13), (A.14) and

$$(2n+1)z_n(w) = w\{z_{n-1}(w) + z_{n+1}(w)\}. \tag{3.48}$$

The result (3.47) is obtained by multiplying (3.46) by the normalisation factor A_n^m (defined by (3.6)) and then using (3.12) and (3.13). □

Theorem 3.13 shows that differentiating with respect to z changes the degree (n) but does not affect the (azimuthal) order m; this is altered by differentiating with respect to x or y, as the next result shows.

Theorem 3.14 *For* $0 \le |m| \le n$,

$$\mathcal{D}_1^+ \tilde{\Omega}_0^0 = \tilde{\Omega}_1^1, \quad \mathcal{D}_1^- \tilde{\Omega}_0^0 = -2\tilde{\Omega}_1^{-1}, \tag{3.49}$$

$$\mathcal{D}_1^+ \tilde{\Omega}_n^m = \frac{1}{2n+1}\left(\tilde{\Omega}_{n-1}^{m+1} + \tilde{\Omega}_{n+1}^{m+1}\right), \tag{3.50}$$

$$\mathcal{D}_1^- \tilde{\Omega}_n^m = -\frac{(n+m)(n+m-1)}{2n+1}\tilde{\Omega}_{n-1}^{m-1} - \frac{(n-m+1)(n-m+2)}{2n+1}\tilde{\Omega}_{n+1}^{m-1}, \tag{3.51}$$

where the differential operators $\mathcal{D}_1^\pm$ *are defined by* (3.37). *For* Ω_n^m,

$$\mathcal{D}_1^+ \Omega_n^m = -\sqrt{\frac{(n+m+2)(n+m+1)}{(2n+1)(2n+3)}}\,\Omega_{n+1}^{m+1} - \sqrt{\frac{(n-m)(n-m-1)}{4n^2-1}}\,\Omega_{n-1}^{m+1}, \tag{3.52}$$

$$\mathcal{D}_1^- \Omega_n^m = \sqrt{\frac{(n-m+2)(n-m+1)}{(2n+1)(2n+3)}}\,\Omega_{n+1}^{m-1} + \sqrt{\frac{(n+m)(n+m-1)}{4n^2-1}}\,\Omega_{n-1}^{m-1}. \tag{3.53}$$

Proof By Theorem 3.10 and Corollary 3.7, we have

$$\frac{1}{k}\left(\frac{\partial}{\partial x} \pm \mathrm{i}\frac{\partial}{\partial y}\right)\tilde{\Omega}_n^m = -\frac{x \pm \mathrm{i}y}{r^{n+1}} z_{n+1}\tilde{\mathcal{H}}_n^m + \frac{z_n}{r^n}\frac{1}{k}\left(\frac{\partial}{\partial x} \pm \mathrm{i}\frac{\partial}{\partial y}\right)\tilde{\mathcal{H}}_n^m.$$

For $m = n = 0$, the second term vanishes, and the results (3.49) follow from $P_1^1(\cos\theta) = \sin\theta$ and $P_1^{-1}(\cos\theta) = -\frac{1}{2}\sin\theta$. Otherwise, we expand the second term using

$$\frac{\partial}{\partial x} \pm \mathrm{i}\frac{\partial}{\partial y} = \mathrm{e}^{\pm\mathrm{i}\phi}\left\{\sin\theta\,\frac{\partial}{\partial r} + \frac{\cos\theta}{r}\frac{\partial}{\partial\theta} \pm \frac{\mathrm{i}}{r\sin\theta}\frac{\partial}{\partial\phi}\right\},$$

whence

$$\left(\frac{\partial}{\partial x} \pm \mathrm{i}\frac{\partial}{\partial y}\right)\tilde{\mathcal{H}}_n^m(x, y, z) = r^{n-1}\left[n\sqrt{1-\mu^2}\,P_n^m(\mu) - \Lambda_0^\pm(\mu)\right]\mathrm{e}^{\mathrm{i}(m\pm1)\phi},$$

where

$$\Lambda_0^{\pm}(\mu) = \mu\sqrt{1-\mu^2}\frac{\mathrm{d}}{\mathrm{d}\mu}P_n^m(\mu) \pm \frac{m}{\sqrt{1-\mu^2}}P_n^m(\mu)$$

and $\mu = \cos\theta$. Hence, using (3.48),

$$\frac{1}{k}\left(\frac{\partial}{\partial x} \pm \mathrm{i}\frac{\partial}{\partial y}\right)\tilde{\Omega}_n^m = \left[\Lambda_1^{\pm} z_{n-1}(kr) + \Lambda_2^{\pm} z_{n+1}(kr)\right]\mathrm{e}^{\mathrm{i}(m\pm1)\phi},$$

where

$$(2n+1)\Lambda_1^{\pm} = n\sqrt{1-\mu^2}\,P_n^m(\mu) - \Lambda_0^{\pm}(\mu),$$
$$(2n+1)\Lambda_2^{\pm} = -(n+1)\sqrt{1-\mu^2}\,P_n^m(\mu) - \Lambda_0^{\pm}(\mu).$$

To simplify $\Lambda_0^{\pm}$, we use (A.17) and (A.18), whence

$$\Lambda_0^{+} = \mu P_n^{m+1}(\mu) + m\sqrt{1-\mu^2}\,P_n^m(\mu),$$
$$\Lambda_0^{-} = -(n-m+1)(n+m)\mu P_n^{m-1}(\mu) - m\sqrt{1-\mu^2}\,P_n^m(\mu).$$

Hence, using (A.13) and (A.20), we obtain

$$(2n+1)\Lambda_1^{+} = -P_{n-1}^{m+1}(\mu) \quad \text{and} \quad (2n+1)\Lambda_2^{+} = -P_{n+1}^{m+1}(\mu),$$

which proves (3.50). Similarly, using (A.13) and (A.21), we obtain

$$(2n+1)\Lambda_1^{-} = (n+m)(n+m-1)\,P_{n-1}^{m-1}(\mu),$$
$$(2n+1)\Lambda_2^{-} = (n-m+1)(n-m+2)\,P_{n+1}^{m-1}(\mu),$$

which proves (3.51). The results (3.52) and (3.53) follow by using (3.14)–(3.17). □

Another proof of Theorems 3.13 and 3.14 will be given later; see Theorem 3.25. Comments on the relevant literature will also be made then.

3.8 Axisymmetric addition theorems

After all our preparation, we are now ready to prove some addition theorems for spherical wavefunctions. Choose two points, O and O_2, distance b apart. Choose Cartesian coordinates (x, y, z) at O so that O_2 is at $(0, 0, b)$. Choose a similar system (x_2, y_2, z_2) at O_2, with axes parallel to the system at O. We also introduce two systems of spherical polar coordinates, (r, θ, ϕ) at O, and (r_2, θ_2, ϕ_2) at O_2: thus O_2 is at $r = b$, $\theta = 0$. See Fig. 3.1.

We are going to expand wavefunctions centred at O_2 about O. The simplest case is to expand $z_0(kr_2)$ as a function of r, θ and ϕ. Indeed, as $z_0(kr_2)$ is an axisymmetric

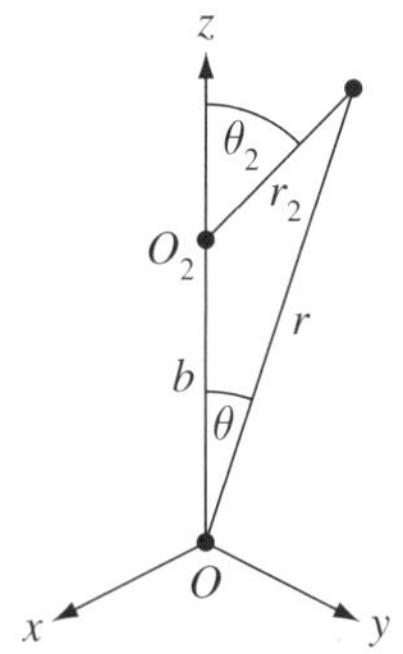

Fig. 3.1. Geometry for axisymmetric addition theorems.

wavefunction, it has an expansion of the form

$$z_0(kr_2) = \sum_{n=0}^{\infty} A_n\, j_n(kr)\, P_n(\cos\theta) \tag{3.54}$$

for $r < b$. It remains to determine the coefficients A_n; clearly, they depend on kb but not on θ. The result is as follows.

Theorem 3.15 *Let* $r_2 = \sqrt{r^2 + b^2 - 2rb\cos\theta}$. *Then*

$$z_0(kr_2) = \sum_{n=0}^{\infty} (2n+1)\, z_n(kb)\, j_n(kr)\, P_n(\cos\theta) \quad \textit{for } r < b.$$

We shall give two proofs of this theorem, but first we sketch a possible proof. Start by putting $\theta = 0$ and $w = kr$ in (3.54) giving

$$\sum_{n=0}^{\infty} A_n\, j_n(w) = f(w), \tag{3.55}$$

where $f(w) = z_0(kb - w)$. Then, putting $w = 0$ immediately gives

$$A_0 = f(0) = z_0(kb), \tag{3.56}$$

since $j_n(0) = \delta_{0n}$. Similarly, as $j_n'(0) = \frac{1}{3}\delta_{1n}$, differentiating with respect to w gives $\frac{1}{3}A_1 = f'(0)$, whence

$$A_1 = -3z_0'(kb) = 3z_1(kb). \tag{3.57}$$

Repeating this procedure (differentiating and setting $w = 0$) is equivalent to finding the Taylor series for f. Thus, assume that

$$f(w) = \sum_{m=0}^{\infty} f_m w^m,$$

where the coefficients f_m are known. Then, find a similar expansion for the left-hand side of (3.55); we have

$$j_n(w) = \sum_{l=0}^{\infty} c_l^{(n)} w^{n+2l} \quad \text{with} \quad c_l^{(n)} = \frac{(-1)^l \sqrt{\pi}}{2^{2l+1} l!\, \Gamma(n+l+3/2)}, \tag{3.58}$$

whence

$$\begin{aligned}\sum_{n=0}^{\infty} A_n j_n(w) &= \sum_{m=0}^{\infty} \left\{ A_{2m} \sum_{l=0}^{\infty} c_l^{(2m)} w^{2m+2l} \right. \\ &\qquad \left. + A_{2m+1} \sum_{l=0}^{\infty} c_l^{(2m+1)} w^{2m+2l+1} \right\} \\ &= \sum_{p=0}^{\infty} \left\{ w^{2p} \sum_{m=0}^{p} c_{p-m}^{(2m)} A_{2m} + w^{2p+1} \sum_{m=0}^{p} c_{p-m}^{(2m+1)} A_{2m+1} \right\}.\end{aligned}$$

Comparing powers of w then gives

$$f_{2n} = \sum_{m=0}^{n} c_{n-m}^{(2m)} A_{2m} \quad \text{and} \quad f_{2n+1} = \sum_{m=0}^{n} c_{n-m}^{(2m+1)} A_{2m+1}$$

for $n = 0, 1, 2, \ldots$; these equations give the coefficients recursively:

$$c_0^{(2n)} A_{2n} = f_{2n} - \sum_{m=0}^{n-1} c_{n-m}^{(2m)} A_{2m},$$

$$c_0^{(2n+1)} A_{2n+1} = f_{2n+1} - \sum_{m=0}^{n-1} c_{n-m}^{(2m+1)} A_{2m+1}$$

for $n \geq 0$, where the sums are absent if $n = 0$.

The problem of expanding an analytic function of w in the form (3.55) is discussed in Watson's book [1298, §16.13]; the series can be called a *Neumann–Gegenbauer series*. It can be shown that the coefficients are given explicitly by

$$A_n = \frac{2n+1}{2^n} \sum_{m=0}^{[n/2]} \frac{(2n-2m)!}{m!\,(n-m)!} f_{n-2m},$$

which leads to the following (unhelpful) quotation:

> *The construction of the Neumann series when the Maclaurin expansion is given is consequently now merely a matter of analytical ingenuity.*
>
> (Watson [1298, p. 523])

The method outlined above is convenient for computations if the coefficients f_n are known explicitly. However, it is a general method in that it does not exploit the fact that our function f is actually a wavefunction. We shall do this in our two proofs of Theorem 3.15.

First proof of Theorem 3.15 Here, we use the differential equation satisfied by spherical Bessel functions. Thus, as z_n satisfies

$$w^2 z_n''(w) + 2w z_n'(w) + [w^2 - n(n+1)]z_n(w) = 0,$$

f satisfies the homogeneous differential equation

$$\left\{(kb - w)\frac{\mathrm{d}^2}{\mathrm{d}w^2} - 2\frac{\mathrm{d}}{\mathrm{d}w} + (kb - w)\right\} f = 0$$

for $|w| < kb$. Enforcing this on the left-hand side of (3.55) yields

$$\sum_{n=0}^{\infty} A_n \left\{kb[j_n''(w) + j_n(w)] - \frac{n(n+1)}{w}\, j_n(w)\right\} = 0. \tag{3.59}$$

Rearrange this equation into the form

$$\sum_{n=0}^{\infty} C_n\, j_n(w) = 0$$

for $|w| < kb$, whence $C_n = 0$ for all $n \geq 0$. To express C_n in terms of A_m, we use

$$(2n+1)\, w^{-1} z_n(w) = z_{n-1}(w) + z_{n+1}(w), \quad n = 1, 2, \ldots, \tag{3.60}$$

$$(2n+1) z_n'(w) = n z_{n-1}(w) - (n+1) z_{n+1}(w), \quad n = 0, 1, 2, \ldots,$$

for j_n, whence

$$(2n+1) j_n''(w) = \frac{n(n-1)}{2n-1}\, j_{n-2} - \left(\frac{n^2}{2n-1} + \frac{(n+1)^2}{2n+3}\right) j_n + \frac{(n+1)(n+2)}{2n+3}\, j_{n+2}$$

for $n \geq 0$. Substituting in (3.59) gives

$$C_n = kb\left\{\frac{(n+1)(n+2)}{2n+3}\frac{A_{n+2}}{2n+5} + \frac{n(n-1)}{2n-1}\frac{A_{n-2}}{2n-3} + \left[2n+1 - \frac{n^2}{2n-1} - \frac{(n+1)^2}{2n+3}\right]\frac{A_n}{2n+1}\right\} - (n+1)(n+2)\frac{A_{n+1}}{2n+3} - n(n-1)\frac{A_{n-1}}{2n-1} = 0$$

for $n \geq 0$. This recurrence relation gives A_{n+2} for $n \geq 0$ in terms of A_0 and A_1, which are known. To simplify it, first define

$$\mathcal{A}_n = (2n+1)^{-1} A_n$$

and then note that the terms inside the square brackets can be rewritten:

$$2n+1 - \frac{n^2}{2n-1} - \frac{(n+1)^2}{2n+3} = \frac{(n+1)(n+2)}{2n+3} + \frac{n(n-1)}{2n-1}.$$

Grouping terms, we obtain

$$kb\left\{\frac{(n+1)(n+2)}{2n+3}(\mathcal{A}_{n+2}+\mathcal{A}_n)+\frac{n(n-1)}{2n-1}(\mathcal{A}_n+\mathcal{A}_{n-2})\right\}$$
$$-(n+1)(n+2)\mathcal{A}_{n+1}-n(n-1)\mathcal{A}_{n-1}=0$$

for $n \geq 0$. So, by inspection, if we define

$$\mathcal{C}_n = \frac{n(n-1)}{2n-1}\left\{kb(\mathcal{A}_n+\mathcal{A}_{n-2})-(2n-1)\,\mathcal{A}_{n-1}\right\},$$

we see that $\mathcal{C}_{n+2}+\mathcal{C}_n=0$ for $n=0,1,2,\ldots$. But $\mathcal{C}_0=\mathcal{C}_1=0$ whence $\mathcal{C}_n=0$ for all $n\geq 0$, that is

$$\mathcal{A}_{n-1}+\mathcal{A}_{n+1}-(2n+1)(kb)^{-1}\,\mathcal{A}_n=0$$

for $n \geq 1$. This is precisely the recurrence relation satisfied by $z_n(kb)$, namely (3.60). Moreover, since

$$\mathcal{A}_0 = A_0 = z_0(kb) \quad \text{and} \quad \mathcal{A}_1 = \frac{1}{3}A_1 = z_1(kb),$$

we have $\mathcal{A}_n = z_n(kb)$, and this completes the proof of Theorem 3.15. □

In two dimensions, we gave a simple proof of Neumann's addition theorem in which we applied the differential operators $\mathcal{D}_m^{\pm}$ and then passed to the limit $r_1 \to 0$. This proof is simple because of the property $\mathcal{D}_m^{\pm}\mathcal{D}_n^{\pm} = \mathcal{D}_{m+n}^{\pm}$. In three dimensions, we do not have this luxury: we have identified the corresponding operators, $\tilde{\mathcal{Y}}_n^m$, but it is not true that $\tilde{\mathcal{Y}}_n^m\tilde{\mathcal{Y}}_N^M$ is simply related to $\tilde{\mathcal{Y}}_{n+N}^k$. (The precise relation will be given below in Theorem 3.21.) The reason for this is also clear: given two harmonic homogeneous polynomials $\mathcal{H}_n$ and $\mathcal{H}_N$ of degree n and N, respectively, their product $\mathcal{H}_n\mathcal{H}_N$ is homogeneous of degree $n+N$, but it is not harmonic, in general (see (3.30)). Nevertheless, the operator $\tilde{\mathcal{Y}}_n^m$ is useful, so let us use it in a second proof of Theorem 3.15.

Second proof of Theorem 3.15 We apply $\tilde{\mathcal{Y}}_m^0$ to (3.54). As the Cartesian axes at O and O_2 are parallel, we obtain

$$\begin{aligned} z_m(kr_2)\,P_m(\cos\theta_2) &= \sum_{n=0}^{\infty} A_n\,\tilde{\mathcal{Y}}_m^0\{j_n(kr)\,P_n(\cos\theta)\} \\ &= \sum_{n=0}^{\infty} A_n\,\tilde{\mathcal{Y}}_m^0\tilde{\mathcal{Y}}_n^0\{j_0(kr)\}, \end{aligned}$$

where we have used Theorem 3.10. Letting $r \to 0$, so that $r_2 \to b$ and $\theta_2 \to \pi$, gives

$$(-1)^m z_m(kb) = \sum_{n=0}^{\infty} A_n\,L_{mn},$$

where

$$L_{mn} = \lim_{r \to 0} \tilde{\mathscr{Y}}_m^0 \tilde{\mathscr{Y}}_n^0 \{j_0(kr)\} = (-1)^n (2n+1)^{-1} \delta_{mn},$$

a result proved later (Lemma 3.18). Hence, $A_n = (2n+1)\, z_n(kb)$, thus completing our second proof. □

The special case of Theorem 3.15 in which $z_0 = j_0$ was published by Clebsch in 1863; see [1298, p. 363]. According to Watson, the special case with $z_0 = y_0$ is due to Gegenbauer: it is formula (3) on p. 365 of [1298] with $\nu = \frac{1}{2}$.

We can generalise Theorem 3.15 slightly. As in Section 2.5, we consider two origins, O_1 and O_2. Let $\boldsymbol{r}_j$ be the position vector of a general point P with respect to O_j, for $j = 1, 2$. Let $\boldsymbol{b}$ be the position vector of O_1 with respect to O_2, so that $\boldsymbol{r}_2 = \boldsymbol{r}_1 + \boldsymbol{b}$. Let $\boldsymbol{r}_1 = r_1 \hat{\boldsymbol{r}}_1$, $\boldsymbol{r}_2 = r_2 \hat{\boldsymbol{r}}_2$ and $\boldsymbol{b} = b\hat{\boldsymbol{b}}$. Then, we have the next theorem.

Theorem 3.16 *Let $\boldsymbol{r}_2 = \boldsymbol{r}_1 + \boldsymbol{b}$. Then*

$$z_0(kr_2) = 4\pi \sum_{n=0}^{\infty} \sum_{m=-n}^{n} (-1)^n\, z_n(kb)\, \overline{Y_n^m(\hat{\boldsymbol{b}})}\, j_n(kr_1)\, Y_n^m(\hat{\boldsymbol{r}}_1) \tag{3.61}$$

for $r_1 < b$, and

$$z_0(kr_2) = 4\pi \sum_{n=0}^{\infty} \sum_{m=-n}^{n} (-1)^n\, j_n(kb)\, \overline{Y_n^m(\hat{\boldsymbol{b}})}\, z_n(kr_1)\, Y_n^m(\hat{\boldsymbol{r}}_1) \tag{3.62}$$

for $r_1 > b$.

Proof In Theorem 3.15, the angle θ satisfies $\cos\theta = -\hat{\boldsymbol{r}}_1 \cdot \hat{\boldsymbol{b}}$, whence $P_n(\cos\theta) = (-1)^n P_n(\hat{\boldsymbol{r}}_1 \cdot \hat{\boldsymbol{b}})$. So, when $r_1 < b$, Theorem 3.15 gives

$$z_0(kr_2) = \sum_{n=0}^{\infty} (2n+1)\,(-1)^n\, z_n(kb)\, j_n(kr_1)\, P_n(\hat{\boldsymbol{r}}_1 \cdot \hat{\boldsymbol{b}}).$$

Then, (3.20) gives (3.61). The result (3.62) follows by interchanging $\boldsymbol{r}_1$ and $\boldsymbol{b}$. □

This theorem is often useful because it gives a convenient expansion of the free-space Green's function G; see Theorem 6.4. A related useful result follows; see [219, eqn (13)], [318, eqn (3.15b)] and [1000, eqn (2.16)].

Theorem 3.17 *Let $\boldsymbol{r}_2 = \boldsymbol{r}_1 + \boldsymbol{b}$, with $r_1 < b$. Then*

$$z_0(kr_2) = \frac{1}{4\pi} \sum_{n=0}^{\infty} (2n+1) \mathrm{i}^n\, z_n(kb) \int_{\Omega} \exp(\mathrm{i}k\boldsymbol{r}_1 \cdot \hat{\boldsymbol{r}})\, P_n(\hat{\boldsymbol{r}} \cdot \hat{\boldsymbol{b}})\, \mathrm{d}\Omega(\hat{\boldsymbol{r}}).$$

Proof Substitute (3.42) in (3.61), and then use (3.20). □

The effect of truncating the sum over n is examined in [1000, 255, 487].

3.9 A useful lemma

Lemma 3.18

$$\lim_{r\to 0} \tilde{\mathcal{Y}}_m^0 \tilde{\mathcal{Y}}_n^0 \{j_0(kr)\} = \frac{(-1)^n}{2n+1}\,\delta_{mn}, \tag{3.63}$$

$$\lim_{r\to 0} \mathcal{Y}_m^0 \mathcal{Y}_n^0 \{j_0(kr)\} = \frac{(-1)^n}{4\pi}\,\delta_{mn}. \tag{3.64}$$

Proof Denote the left-hand side of (3.63) by L_{mn}. Let

$$\widehat{\Omega}_n \equiv \widehat{\Omega}_n^0 = j_n(kr)\,P_n(\cos\theta).$$

Then, from (3.58), we have $\tilde{\mathcal{Y}}_n^0\{j_0(kr)\} = \widehat{\Omega}_n = O(r^n)$ as $r \to 0$. Thus, as $\tilde{\mathcal{Y}}_m^0$ is a differential operator of order m, we must have $L_{mn} = 0$ for $m < n$. But $L_{mn} = L_{nm}$, whence $L_{mn} = 0$ for $n > m$, so that

$$L_{mn} = L_{mm}\,\delta_{mn}. \tag{3.65}$$

As $\tilde{\mathcal{Y}}_0^0 = I$, it is clear that $L_{00} = 1$. In order to calculate L_{nn} for $n \geq 1$, we need an explicit formula for $\tilde{\mathcal{Y}}_n^0 u$, where u is a wavefunction. From (3.39), we have

$$\tilde{\mathcal{Y}}_n^0 u = \sum_{l=0}^{[n/2]} (-1)^l B_l^{(n)} \left(\mathcal{D}_1^0\right)^{n-2l} u, \tag{3.66}$$

where

$$B_l^{(n)} \equiv B_l^{n,0} = \frac{(2n-2l)!}{2^n l!\,(n-l)!\,(n-2l)!}. \tag{3.67}$$

Hence

$$\begin{aligned} L_{nn} &= \lim_{r\to 0} \tilde{\mathcal{Y}}_n^0\,\widehat{\Omega}_n = \lim_{r\to 0} \sum_{l=0}^{[n/2]} (-1)^l B_l^{(n)} \left(\mathcal{D}_1^0\right)^{n-2l} \widehat{\Omega}_n \\ &= \lim_{r\to 0} B_0^{(n)} \left(\mathcal{D}_1^0\right)^n \widehat{\Omega}_n, \end{aligned}$$

as all other terms have zero limits. Now, from Theorem 3.13, we have

$$\mathcal{D}_1^0 \widehat{\Omega}_n = \frac{n+1}{2n+1}\,\widehat{\Omega}_{n+1} - \frac{n}{2n+1}\,\widehat{\Omega}_{n-1};$$

the second term on the right-hand side is $O(r^{n-1})$ as $r \to 0$, whereas the first term is smaller (it is $O(r^{n+1})$). Hence, to leading order

$$\begin{aligned} \left(\mathcal{D}_1^0\right)^n \widehat{\Omega}_n &\sim (-1)^n \frac{n}{2n+1}\,\frac{n-1}{2n-1}\,\frac{n-2}{2n-3} \cdots \frac{2}{3}\,\frac{1}{1}\,\widehat{\Omega}_0 \\ &\to \frac{(-1)^n\,n!}{(2n+1)(2n-1)\cdots 3\cdot 1} = \frac{2^n(-1)^n(n!)^2}{(2n+1)!} = \frac{(-1)^n}{(2n+1)\,B_0^{(n)}} \end{aligned}$$

as $r \to 0$; $B_0^{(n)}$ is defined by (3.67). Hence, $L_{nn} = (-1)^n(2n+1)^{-1}$. The result (3.64) follows by noting that $\mathcal{Y}_n^0 = \sqrt{(2n+1)/(4\pi)}\,\tilde{\mathcal{Y}}_n^0$. □

We can use Lemma 3.18 to prove other useful results. One of these is an expansion of a plane wave in terms of spherical wavefunctions.

Theorem 3.19 *For $0 \le \theta \le \pi$ and $r \ge 0$,*

$$e^{ikr\cos\theta} = \sum_{n=0}^{\infty} (2n+1)\, i^n\, j_n(kr)\, P_n(\cos\theta). \tag{3.68}$$

If r is bounded, the series is uniformly and absolutely convergent.

Proof As the left-hand side is an axisymmetric wavefunction, we have

$$\exp(ikr\cos\theta) = \exp(ik\boldsymbol{r}\cdot\hat{\mathbf{z}}) = \sum_{n=0}^{\infty} A_n\, j_n(kr)\, Y_n^0(\hat{\boldsymbol{r}}),$$

where the coefficients A_n are to be determined and $\hat{\mathbf{z}} = (0, 0, 1)$ is the unit vector along the positive z-axis. Apply $\mathcal{Y}_m^0$ to this equation, using Theorems 3.10 and 3.11, to give

$$(-i)^m Y_m^0(\hat{\mathbf{z}})\exp(ik\boldsymbol{r}\cdot\hat{\mathbf{z}}) = \sum_{n=0}^{\infty} A_n\, \mathcal{Y}_m^0 \mathcal{Y}_n^0\{j_0(kr)\}.$$

Now, letting $r \to 0$, using Lemma 3.18, we obtain

$$(-i)^m Y_m^0(\hat{\mathbf{z}}) = (-1)^m A_m/(4\pi)$$

so that

$$A_n\, Y_n^0(\hat{\boldsymbol{r}}) = 4\pi i^n\, Y_n^0(\hat{\mathbf{z}})\, Y_n^0(\hat{\boldsymbol{r}}) = (2n+1)\, i^n\, P_n(\cos\theta),$$

as required, using $Y_n^0(\hat{\boldsymbol{r}}) = [(2n+1)/(4\pi)]^{1/2} P_n(\cos\theta)$ and $P_n(1) = 1$.

To prove absolute and uniform convergence, use $|P_n(x)| \le 1$ and $|J_\nu(x)| \le |x/2|^\nu/\Gamma(\nu+1)$, whence

$$|(2n+1)\, i^n\, j_n(kr)\, P_n(\cos\theta)| \le \sqrt{\pi}\,(kr/2)^n/\Gamma\left(n+\frac{1}{2}\right).$$

□

The formula (3.68) is sometimes known as *Rayleigh's expansion* ([73, p. 47], [1010, §334]), although, according to Watson [1298, §11.5], it was found by Bauer in 1859. Indeed, solid-state physicists refer to (3.68) as *Bauer's identity* [422, p. 373]. Colton & Kress [225, p. 32] refer to (3.68) as the 'Jacobi–Anger expansion', but this attribution should be reserved for the cylindrical expansion (2.17). Müller [893, §22, Lemma 1] has given a q-dimensional version of Theorem 3.19; it includes both the Jacobi expansion (2.17) and the Bauer expansion (3.68) as special cases.

We can combine Theorem 3.19 with Legendre's addition theorem to give a second proof of the Funk–Hecke formula (Theorem 3.12). Thus, we have

$$\exp(\mathrm{i}k\boldsymbol{r}\cdot\hat{\boldsymbol{r}}_1) = \sum_{p=0}^{\infty}(2p+1)\,\mathrm{i}^p\,j_p(kr)\,P_p(\hat{\boldsymbol{r}}\cdot\hat{\boldsymbol{r}}_1)$$

$$= 4\pi\sum_{p=0}^{\infty}\mathrm{i}^p\,j_p(kr)\sum_{l=-p}^{p}Y_p^l(\hat{\boldsymbol{r}})\,\overline{Y_p^l(\hat{\boldsymbol{r}}_1)}, \tag{3.69}$$

using (3.20). Now, multiply through by $Y_n^m(\hat{\boldsymbol{r}}_1)$ and integrate over Ω using the orthogonality relation (3.9). This gives (3.42), as required.

3.10 Composition formula for the operator $\mathcal{Y}_n^m$

A key ingredient in the efficient derivation of (non-axisymmetric) addition formulae is the following result, which we will refer to as the *linearisation formula* for spherical harmonics. It gives an expression for the product $Y_n^m\,Y_\nu^\mu$ as a finite sum over q of $Y_q^{m+\mu}$. Similar formulae for other special (and elementary) functions are known; see [45], [46, Lecture 5] or [31, §6.8]. A simple familiar example is

$$\cos m\theta\,\cos n\theta = \frac{1}{2}\cos(m+n)\theta + \frac{1}{2}\cos(m-n)\theta.$$

The linearisation formulae for associated Legendre functions and Legendre polynomials are given by (B.8) and (B.10), respectively.

Theorem 3.20 (Linearisation formula for Y_n^m) *Let n and ν be non-negative integers. Then*

$$Y_n^m(\hat{\boldsymbol{r}})\,Y_\nu^\mu(\hat{\boldsymbol{r}}) = \sum_q \mathcal{G}(n,m;\nu,\mu;q)\,Y_q^{m+\mu}(\hat{\boldsymbol{r}}), \tag{3.70}$$

where $\mathcal{G}$ is a Gaunt coefficient. The summation is from $q=|n-\nu|$ to $q=n+\nu$ in steps of 2.

Proof The form of the expansion is clear; evidently, $\mathcal{G}$ is real. From (3.9) and (3.7), we have

$$\mathcal{G}(n,m;\nu,\mu;q) = (-1)^{m+\mu}\int_\Omega Y_n^m\,Y_\nu^\mu\,Y_q^{-m-\mu}\,\mathrm{d}\Omega \tag{3.71}$$

$$= \frac{(-1)^{m+\mu}}{2}\mathcal{S}\sqrt{\frac{(n-m)!\,(\nu-\mu)!\,(q+m+\mu)!}{(n+m)!\,(\nu+\mu)!\,(q-m-\mu)!}}$$

$$\times\int_{-1}^{1}P_n^m(x)\,P_\nu^\mu(x)\,P_q^{-m-\mu}(x)\,\mathrm{d}x \tag{3.72}$$

$$= (-1)^{m+\mu}\mathcal{S}\begin{pmatrix} n & \nu & q\\ 0 & 0 & 0\end{pmatrix}\begin{pmatrix} n & \nu & q\\ m & \mu & -m-\mu\end{pmatrix}, \tag{3.73}$$

where $\mathcal{S} = \sqrt{(2n+1)(2\nu+1)(2q+1)/(4\pi)}$. Here,

$$\begin{pmatrix} n & \nu & N \\ m & \mu & M \end{pmatrix}$$

is a *Wigner 3-j symbol*. For further information and references, see Appendix B. In particular, it is known that

$$\begin{pmatrix} n & \nu & q \\ 0 & 0 & 0 \end{pmatrix} = \frac{(-1)^s\, s!\, \sqrt{(2s-2n)!\,(2s-2\nu)!\,(2s-2q)!}}{(s-n)!\,(s-\nu)!\,(s-q)!\,\sqrt{(2s+1)!}}$$

when $n+\nu+q=2s$ is even, and vanishes otherwise. Moreover, this 3-j symbol also vanishes if $q>(n+\nu)$ or if $q<|n-\nu|$, for then one of the factorials in the denominator has a negative argument. □

Before proceeding, let us be more explicit about the limits on the summation over q in (3.70). The upper limit is $(n+\nu)$. Summing the series backwards, we obtain terms for $q=n+\nu-2, n+\nu-4, \dots,$ the last term being $q=q_0$, where

$$q_0(n,m;\nu,\mu) = \begin{cases} |n-\nu| & \text{if } |n-\nu| \ge |m+\mu|, \\ |m+\mu| & \text{if } |n-\nu| < |m+\mu| \text{ and} \\ & n+\nu+|m+\mu| \text{ is even,} \\ |m+\mu|+1 & \text{if } |n-\nu| < |m+\mu| \text{ and} \\ & n+\nu+|m+\mu| \text{ is odd;} \end{cases} \tag{3.74}$$

this complicated definition of the lower limit q_0 arises due to a combination of the 3-j symbols and the fact that $Y_q^{m+\mu} \equiv 0$ if $q<|m+\mu|$. Note that $q_0(n,m;\nu,\mu) = q_0(\nu,\mu;n,m)$.

It follows from the definition of q_0 that $n+\nu-q_0$ is even, and so we can rewrite (3.70) as

$$Y_n^m(\hat{\boldsymbol{r}})\, Y_\nu^\mu(\hat{\boldsymbol{r}}) = \sum_{q=0}^{Q} \mathcal{G}(n,m;\nu,\mu;q_0+2q)\, Y_{q_0+2q}^{m+\mu}(\hat{\boldsymbol{r}}),$$

where $Q(n,m;\nu,\mu) = Q(\nu,\mu;n,m) = (n+\nu-q_0)/2$; in particular, $q_0+2Q=n+\nu$.

Returning to the Gaunt coefficient, defined by (3.71), we easily obtain the symmetry relations

$$\mathcal{G}(n,m;\nu,\mu;q) = \mathcal{G}(\nu,\mu;n,m;q) = \mathcal{G}(n,-m;\nu,-\mu;q) \tag{3.75}$$

$$= (-1)^m\,\mathcal{G}(n,m;q,-m-\mu;\nu) \tag{3.76}$$

$$= (-1)^\mu\,\mathcal{G}(q,-m-\mu;\nu,\mu;n). \tag{3.77}$$

The linearisation formula for spherical harmonics is given in, for example, [1028, eqn (2.32)], [1029, eqn (4.32)], [846, eqn (C.17)], [416, p. 19], [1326, eqn (5.18)] and [305, eqn (4.6.5)].

Next, we use Theorem 3.20 to obtain a composition formula for the operator $\mathcal{Y}_n^m$.

Theorem 3.21 (Composition formula for $\mathcal{Y}_n^m$) *Let n and ν be non-negative integers. Then*

$$\mathcal{Y}_n^m \, \mathcal{Y}_\nu^\mu = \sum_q \mathcal{G}(n, m; \nu, \mu; q) \, k^{q-n-\nu} \, \mathcal{Y}_q^{m+\mu} \, (\nabla_3^2)^{(n+\nu-q)/2},$$

where the operator $\mathcal{Y}_n^m$ is defined by Definition 3.9. The summation is from $q = q_0(n, m; \nu, \mu)$ to $q = n+\nu$ in steps of 2, where q_0 is defined by (3.74).

Proof Multiplying (3.70) by $r^{n+\nu}$ gives

$$\mathcal{H}_n^m \, \mathcal{H}_\nu^\mu = \sum_q \mathcal{G}(n, m; \nu, \mu; q) \, \mathcal{H}_q^{m+\mu} \, r^{n+\nu-q},$$

where $\mathcal{H}_n^m(x, y, z) = r^n Y_n^m(\theta, \phi)$. Now, multiply by $(-1/k)^{n+\nu}$ and use the definition of $\mathcal{Y}_n^m$. Note that $\frac{1}{2}(n+\nu-q)$ is a non-negative integer. □

Two corollaries follow immediately.

Corollary 3.22 *Let Φ be harmonic, so that $\nabla_3^2 \Phi = 0$. Then*

$$\mathcal{Y}_n^m \, \mathcal{Y}_\nu^\mu \Phi = \mathcal{G}(n, m; \nu, \mu; n+\nu) \, \mathcal{Y}_{n+\nu}^{m+\mu} \Phi;$$

the Gaunt coefficient can be evaluated using (B.6).

Corollary 3.23 *Let u be a wavefunction, so that $\nabla_3^2 u + k^2 u = 0$. Then*

$$\mathcal{Y}_n^m \, \mathcal{Y}_\nu^\mu u = \sum_q \mathcal{G}(n, m; \nu, \mu; q) \, (-1)^{(n+\nu-q)/2} \, \mathcal{Y}_q^{m+\mu} u.$$

We can use this result to give an alternative proof of Lemma 3.18. In fact, we can prove a more general result.

Theorem 3.24

$$\lim_{r \to 0} \mathcal{Y}_n^m \mathcal{Y}_\nu^\mu \{ j_0(kr) \} = \frac{(-1)^{n+m}}{4\pi} \, \delta_{n\nu} \, \delta_{m+\mu,0}.$$

Proof Corollary 3.23 and Theorem 3.10 give

$$\mathcal{Y}_n^m \, \mathcal{Y}_\nu^\mu \{ j_0(kr) \} = \sum_q \mathcal{G}(n, m; \nu, \mu; q) \, (-1)^{(n+\nu-q)/2} \, j_q(kr) \, Y_q^{m+\mu}(\hat{\boldsymbol{r}}).$$

Now, $j_q(0) = \delta_{q0}$ and $Y_0^0 = (4\pi)^{-1/2}$. But the summation starts from $q = |n-\nu|$, so that

$$\lim_{r \to 0} \mathcal{Y}_n^m \mathcal{Y}_\nu^\mu \{ j_0(kr) \} = \delta_{n\nu} \, \delta_{m+\mu,0} (-1)^n (4\pi)^{-1/2} \, \mathcal{G}(n, m; n, -m; 0).$$

From (3.72), the Gaunt coefficient is given by

$$\mathcal{G}(n, m; n, -m; 0) = \frac{2n+1}{2\sqrt{4\pi}} \int_{-1}^{1} P_n^m(x) \, P_n^{-m}(x) \, \mathrm{d}x = \frac{(-1)^m}{\sqrt{4\pi}},$$

using (3.5) and (3.10), and so the result follows. □

A result equivalent to Theorem 3.24 can be found in [1332, eqn (10)].

Next, let us give a unified proof of Theorems 3.13 and 3.14.

Theorem 3.25 *Set* $\mathcal{G}_\mu(n, m; q) = \mathcal{G}(n, m; 1, \mu; q)$. *For* $0 \le |m| \le n$,

$$\mathcal{Y}_1^\mu \Omega_0^0 = \mathcal{G}_\mu(0, 0; 1)\, \Omega_1^\mu,$$

$$\mathcal{Y}_1^\mu \Omega_n^m = \mathcal{G}_\mu(n, m; n+1)\, \Omega_{n+1}^{m+\mu} - \mathcal{G}_\mu(n, m; n-1)\, \Omega_{n-1}^{m+\mu},$$

where $\Omega_n^m = z_n Y_n^m$ *and the Gaunt coefficients are given by*

$$\mathcal{G}_\mu(n, m; n+1) = \sqrt{\frac{3}{4\pi(2n+3)(2n+1)}} \times \sqrt{\frac{(n+m+1+\mu)!\,(n-m+1-\mu)!}{(n+m)!\,(1+\mu)!\,(n-m)!\,(1-\mu)!}},$$

$$\mathcal{G}_\mu(n, m; n-1) = (-1)^\mu \sqrt{\frac{3}{4\pi(4n^2-1)}} \times \sqrt{\frac{(n+m)!\,(n-m)!}{(n+m-1+\mu)!\,(1-\mu)!\,(n-m-1-\mu)!\,(1+\mu)!}}.$$

Proof Corollary 3.23 gives

$$\mathcal{Y}_1^\mu \mathcal{Y}_n^m\{z_0(kr)\} = \sum_q \mathcal{G}(n, m; 1, \mu; q)\, (-1)^{(n+1-q)/2}\, \mathcal{Y}_q^{m+\mu}\{z_0(kr)\},$$

whence Theorem 3.10 gives

$$\mathcal{Y}_1^\mu \Omega_n^m = \sum_q \mathcal{G}_\mu(n, m; q)\, (-1)^{(n+1-q)/2}\, \Omega_q^{m+\mu}.$$

But the summation has at most two terms, namely those for $q = |n-1|$ and for $q = n+1$; the stated results follow. □

Theorem 3.25 can be obtained from eqn (2.57) in Rose's book [1028]; in that equation, put $\Phi(r) = z_0(kr)$. This theorem includes Theorem 3.13 (put $\mu = 0$) and Theorem 3.14 (put $\mu = \pm 1$); $\mathcal{D}_1^0$ and $\mathcal{D}_1^\pm$ are related to $\mathcal{Y}_1^\mu$ by (3.37). Kim [606, eqn (28)] has also obtained the results in Theorem 3.25. The explicit formulae given in Theorem 3.13 were also obtained by Chew [197, eqn (11)] and by Clercx & Schram ([217, p. 352], [218, p. 5298]). (Actually, Clercx & Schram gave their results for modified spherical Bessel functions.) Similarly, in Theorem 3.14, the formulae (3.50) and (3.51) were also obtained by Clercx & Schram [218, p. 5298]; the first of these was also derived by Chew [197, eqn. (18*a*)].

3.11 Addition theorem for $j_n Y_n^m$

In this section, we generalise Theorem 3.16.

Theorem 3.26 *Let $\boldsymbol{r}_2 = \boldsymbol{r}_1 + \boldsymbol{b}$. Then*

$$j_n(kr_2)\, Y_n^m(\hat{\boldsymbol{r}}_2) = \sum_{\nu=0}^{\infty} \sum_{\mu=-\nu}^{\nu} \widehat{S}_{n\nu}^{m\mu}(\boldsymbol{b})\, j_\nu(kr_1)\, Y_\nu^\mu(\hat{\boldsymbol{r}}_1). \tag{3.78}$$

First proof From the Funk–Hecke formula (Theorem 3.12), we have

$$j_n(kr_2)\, Y_n^m(\hat{\boldsymbol{r}}_2) = \frac{\mathrm{i}^{-n}}{4\pi} \int_\Omega \exp(\mathrm{i}k\boldsymbol{r}_1 \cdot \hat{\boldsymbol{r}}) \exp(\mathrm{i}k\boldsymbol{b} \cdot \hat{\boldsymbol{r}})\, Y_n^m(\hat{\boldsymbol{r}})\, \mathrm{d}\Omega(\hat{\boldsymbol{r}}).$$

Use of (3.69) for the first exponential gives the result with

$$\widehat{S}_{n\nu}^{m\mu}(\boldsymbol{b}) = \mathrm{i}^{\nu-n} \int_\Omega \exp(\mathrm{i}k\boldsymbol{b} \cdot \hat{\boldsymbol{r}})\, Y_n^m(\hat{\boldsymbol{r}})\, \overline{Y_\nu^\mu(\hat{\boldsymbol{r}})}\, \mathrm{d}\Omega(\hat{\boldsymbol{r}}). \tag{3.79}$$

If we expand the remaining exponential, using (3.69) again, we obtain

$$\widehat{S}_{n\nu}^{m\mu}(\boldsymbol{b}) = 4\pi \mathrm{i}^{\nu-n} \sum_{q=0}^{\infty} \mathrm{i}^q\, j_q(kb)\, Y_q^{m-\mu}(\hat{\boldsymbol{b}}) \int_\Omega Y_n^m\, \overline{Y_q^{m-\mu}}\, \overline{Y_\nu^\mu}\, \mathrm{d}\Omega,$$

where we have used (2.18). Hence, using (3.71),

$$\begin{aligned} \widehat{S}_{n\nu}^{m\mu}(\boldsymbol{b}) &= 4\pi \mathrm{i}^{\nu-n} \sum_q \mathrm{i}^q\, j_q(kb)\, \overline{Y_q^{\mu-m}(\hat{\boldsymbol{b}})}\, \mathcal{G}(n, m; q, \mu-m; \nu) \\ &= 4\pi \mathrm{i}^{\nu-n} \sum_q \mathrm{i}^q (-1)^m j_q(kb)\, \overline{Y_q^{\mu-m}(\hat{\boldsymbol{b}})}\, \mathcal{G}(n, m; \nu, -\mu; q), \end{aligned} \tag{3.80}$$

where $\mathcal{G}$ is a Gaunt coefficient and we have used (3.7) and (3.76). □

In the above proof, the regular spherical wavefunction $j_n Y_n^m$ was expanded in plane waves, whence the translation is trivial. This idea was used by Friedman & Russek [366], although there are errors in the details of their proof. These were noted by Stein [1143]; see his Appendix II. Stein also made the connection with the earlier studies of angular momentum in the quantum-mechanics literature; see Appendix B for references. The formula (3.78) was also obtained by Kohn & Rostoker [642], using the same method as Friedman & Russek [366], and by Sack [1043].

Second proof of Theorem 3.26 The form of the expansion (3.78) is clear, as the left-hand side can be expanded in terms of regular spherical wavefunctions centred on $r_1 = 0$. It remains to find the coefficients $\widehat{S}_{n\nu}^{m\mu}$. Apply the differential operator $\mathcal{Y}_N^M(\boldsymbol{r}_2)$, meaning that the derivatives are with respect to x_2, y_2 and z_2, where $\boldsymbol{r}_2 = (x_2, y_2, z_2)$. But $\mathcal{Y}_N^M$ is translationally invariant, so that $\mathcal{Y}_N^M(\boldsymbol{r}_2) = \mathcal{Y}_N^M(\boldsymbol{r}_1)$, whence

$$\mathcal{Y}_N^M \mathcal{Y}_n^m\{j_0(kr_2)\} = \sum_{\nu=0}^{\infty} \sum_{\mu=-\nu}^{\nu} \widehat{S}_{n\nu}^{m\mu}(\boldsymbol{b})\, \mathcal{Y}_N^M \mathcal{Y}_\nu^\mu\{j_0(kr_1)\},$$

where we have used Theorem 3.10 twice. Now, let $r_1 \to 0$, using Theorem 3.24, to give

$$\begin{aligned}\widehat{S}_{n\nu}^{m\mu}(\boldsymbol{b}) &= 4\pi(-1)^{\nu+\mu} \lim_{r_1\to 0} \mathcal{Y}_n^m \mathcal{Y}_\nu^{-\mu}\{j_0(kr_2)\} \qquad (3.81)\\ &= 4\pi(-1)^{\nu+\mu} \lim_{r_1\to 0} \sum_q \mathcal{G}(n, m; \nu, -\mu; q)\\ &\qquad\qquad \times (-1)^{(n+\nu-q)/2}\, \mathcal{Y}_q^{m-\mu}\{j_0(kr_2)\}\\ &= 4\pi(-1)^{\mu} \sum_q \mathcal{G}(n, m; \nu, -\mu; q)\,(-1)^{(q+\nu-n)/2}\, j_q(kb)\, Y_q^{m-\mu}(\hat{\boldsymbol{b}})\end{aligned}$$

in agreement with (3.80). □

Notice that (3.81) can also be written as

$$\widehat{S}_{n\nu}^{m\mu}(\boldsymbol{b}) = 4\pi(-1)^{\nu+\mu}\mathcal{Y}_n^m\{j_\nu(kr_2)\, Y_\nu^{-\mu}(\hat{\boldsymbol{r}}_2)\} \quad \text{evaluated at } \boldsymbol{r}_2 = \boldsymbol{b};$$

Ben-Menahem [82] obtained this formula from (3.80).

Before generalising Theorem 3.26, let us consider some special cases; these can serve as simple checks on the calculations, or they lead to simpler formulae.

(i) $\boldsymbol{b} = \boldsymbol{0}$ ($\boldsymbol{r}_2 = \boldsymbol{r}_1$). As $j_q(0) = \delta_{q0}$ and $Y_0^0 = (4\pi)^{-1/2}$, we find that

$$\widehat{S}_{n\nu}^{m\mu}(\boldsymbol{0}) = \mathrm{i}^{\nu-n}\delta_{m\mu}\int_\Omega Y_n^m\, \overline{Y_\nu^m}\, \mathrm{d}\Omega = \delta_{m\mu}\,\delta_{n\nu},$$

as expected.

(ii) $\boldsymbol{r}_1 = \boldsymbol{0}$ ($\boldsymbol{r}_2 = \boldsymbol{b}$). The right-hand side of (3.78) reduces to

$$\widehat{S}_{n0}^{m0}(\boldsymbol{b}) = \mathrm{i}^{-n}\sum_q \mathrm{i}^q\, j_q(kb)\, Y_q^m(\hat{\boldsymbol{b}}) \int_\Omega Y_n^m\, \overline{Y_q^m}\, \mathrm{d}\Omega = j_n(kb)\, Y_n^m(\hat{\boldsymbol{b}}),$$

as expected.

(iii) $n = m = 0$. In this case, we have

$$\begin{aligned}\frac{1}{\sqrt{4\pi}}\widehat{S}_{0\nu}^{0\mu}(\boldsymbol{b}) &= \mathrm{i}^\nu \sum_q \mathrm{i}^q\, j_q(kb)\, Y_q^{-\mu}(\hat{\boldsymbol{b}}) \int_\Omega \overline{Y_q^{-\mu}}\, \overline{Y_\nu^\mu}\, \mathrm{d}\Omega\\ &= (-1)^\nu\, j_\nu(kb)\, \overline{Y_\nu^\mu(\hat{\boldsymbol{b}})},\end{aligned}$$

so that (3.78) reduces to (3.61).

(iv) $\hat{\boldsymbol{b}} = \pm\hat{\mathbf{z}}$, where $\hat{\mathbf{z}} = (0, 0, 1)$ is the unit vector along the positive z-axis. In this case (translation along the z-axis), we obtain

$$\begin{aligned}\widehat{S}_{n\nu}^{m\mu}(\pm b\hat{\mathbf{z}}) &= 2\pi\,\delta_{m\mu}\mathrm{i}^{\nu-n}\int_0^\pi \mathrm{e}^{\pm\mathrm{i}kb\cos\theta}\, Y_n^m(\hat{\boldsymbol{r}})\, \overline{Y_\nu^m(\hat{\boldsymbol{r}})}\, \sin\theta\, \mathrm{d}\theta\\ &= \sqrt{4\pi}\,\delta_{m\mu}\mathrm{i}^{\nu-n}(-1)^m \sum_q \sqrt{2q+1}\,(\pm\mathrm{i})^q\\ &\qquad \times j_q(kb)\,\mathcal{G}(n, m; \nu, -m; q), \qquad (3.82)\end{aligned}$$

where we have used (A.8) and

$$Y_n^m(\pm\hat{\mathbf{z}}) = \sqrt{(2n+1)/(4\pi)}\,(\pm 1)^n\,\delta_{0m}.$$

Observe that (unlike $\widehat{S}_{n\nu}^{m\mu}(\boldsymbol{b})$) $\widehat{S}_{n\nu}^{m\mu}(\pm b\hat{\mathbf{z}})$ is *real*. To see this, we note that the Gaunt coefficient (which is itself real) is non-zero only when $n+\nu+q$ is even (see the proof of Theorem 3.20), whence $\mathrm{i}^{\nu-n+q}$ is real.

(v) $k=0$. This case corresponds to solid harmonics. From (3.58), we have $j_n(w) \sim c_0^{(n)} w^n$ as $w \to 0$. So, we divide (3.78) by $c_0^{(n)} k^n$ and then let $k \to 0$. It turns out that the non-trivial contributions come from $q = n-\nu$ in (3.80) with $0 \le \nu \le n$, whence

$$r_2^n Y_n^m(\hat{\boldsymbol{r}}_2) = \sum_{\nu=0}^{n} \sum_{m=-\mu}^{\mu} \widehat{M}_{n\nu}^{m\mu}\, b^{n-\nu}\, Y_{n-\nu}^{m-\mu}(\hat{\boldsymbol{b}})\, r_1^\nu Y_\nu^\mu(\hat{\boldsymbol{r}}_1) \tag{3.83}$$

for $\boldsymbol{r}_2 = \boldsymbol{r}_1 + \boldsymbol{b}$, where

$$\widehat{M}_{n\nu}^{m\mu} = 4\pi(-1)^\mu \frac{c_0^{(n-\nu)} c_0^{(\nu)}}{c_0^{(n)}}\, \mathcal{G}(n,m;\nu,-\mu;n-\nu)$$

$$= \frac{2\pi^{3/2}(-1)^{m+\mu}\Gamma(n+3/2)}{\Gamma(\nu+3/2)\,\Gamma(n-\nu+3/2)}\, \mathcal{G}(n,m;n-\nu,\mu-m;\nu),$$

using (3.76). The remaining Gaunt coefficient can be evaluated using (3.73), (B.5) and (B.6). According to Sack [1045, p. 783], the formula (3.83) was first derived by Rose [1030, eqn (33)]; see also [887, eqn (46)]. For a direct proof, see [318, Theorem 3].

3.12 Addition theorem for $h_n^{(1)} Y_n^m$

In this problem the net result is that the simple-minded approach which does not probe too deeply gives for most purposes the right answer, and that the doubts raised by a more thorough examination can ultimately be shown to be groundless. This is a not uncommon situation, which Pauli liked to call the "law of conservation of sloppiness".

(Peierls [954, p. 126])

The two proofs of Theorem 3.26 may be adapted to give addition theorems for other spherical wavefunctions, such as $h_n^{(1)} Y_n^m$. For example, Friedman & Russek [366] used a contour-integral representation for $h_n^{(1)} Y_n^m$ in place of the Funk–Hecke formula; see also [133, Appendix A]. However, it is much simpler to combine the operator $\mathcal{Y}_n^m$ with the known addition theorem for $h_0^{(1)}$ (Theorem 3.16).

In our applications, we will require addition theorems for $h_n^{(1)} Y_n^m$, so, henceforth, we restrict ourselves to $z_n = h_n^{(1)}$. Thus, with $\boldsymbol{r}_2 = \boldsymbol{r}_1 + \boldsymbol{b}$ and $r_1 < b$, Theorem 3.16 gives

$$h_0^{(1)}(kr_2) = 4\pi \sum_{q,l} (-1)^q\, h_q^{(1)}(kb)\, \overline{Y_q^l(\hat{\boldsymbol{b}})}\, j_q(kr_1)\, Y_q^l(\hat{\boldsymbol{r}}_1)$$

from (3.61), where, in keeping with 'Pauli's law', we shall worry about the summation limits later: in fact, we can sum over all $q \geq 0$ and over all l as $Y_q^l \equiv 0$ for $|l| > q$. Now, apply $\mathcal{Y}_n^m(\boldsymbol{r}_2)$, giving

$$h_n^{(1)}(kr_2)\, Y_n^m(\hat{\boldsymbol{r}}_2) = 4\pi \sum_{q,l} (-1)^q\, h_q^{(1)}(kb)\, \overline{Y_q^l(\hat{\boldsymbol{b}})}\, \mathcal{Y}_n^m \mathcal{Y}_q^l \{j_0(kr_1)\}$$

$$= 4\pi \sum_{q,l,\nu} (-1)^{(q+\nu-n)/2}\, h_q^{(1)}(kb)\, \overline{Y_q^l(\hat{\boldsymbol{b}})}$$

$$\times\, \mathcal{G}(n, m; q, l; \nu)\, j_\nu(kr_1)\, Y_\nu^{l+m}(\hat{\boldsymbol{r}}_1).$$

Setting $\mu = l + m$ gives the first result in the following theorem. The second result follows similarly, beginning with (3.62).

Theorem 3.27 *Let $\boldsymbol{r}_2 = \boldsymbol{r}_1 + \boldsymbol{b}$. Then*

$$h_n^{(1)}(kr_2)\, Y_n^m(\hat{\boldsymbol{r}}_2) = \sum_{\nu=0}^{\infty} \sum_{\mu=-\nu}^{\nu} S_{n\nu}^{m\mu}(\boldsymbol{b})\, j_\nu(kr_1)\, Y_\nu^\mu(\hat{\boldsymbol{r}}_1) \tag{3.84}$$

for $r_1 < b$, and

$$h_n^{(1)}(kr_2)\, Y_n^m(\hat{\boldsymbol{r}}_2) = \sum_{\nu=0}^{\infty} \sum_{\mu=-\nu}^{\nu} \widehat{S}_{n\nu}^{m\mu}(\boldsymbol{b})\, h_\nu^{(1)}(kr_1)\, Y_\nu^\mu(\hat{\boldsymbol{r}}_1) \tag{3.85}$$

for $r_1 > b$, where $\widehat{S}_{n\nu}^{m\mu}$ is given by (3.80) *and*

$$S_{n\nu}^{m\mu}(\boldsymbol{b}) = 4\pi \mathrm{i}^{\nu-n} \sum_q \mathrm{i}^q\, h_q^{(1)}(kb)\, \overline{Y_q^{\mu-m}(\hat{\boldsymbol{b}})}\, \mathcal{G}(n, m; q, \mu - m; \nu) \tag{3.86}$$

$$= 4\pi \mathrm{i}^{\nu-n} \sum_q \mathrm{i}^q (-1)^m\, h_q^{(1)}(kb)\, \overline{Y_q^{\mu-m}(\hat{\boldsymbol{b}})}\, \mathcal{G}(n, m; \nu, -\mu; q). \tag{3.87}$$

The content of this theorem is well known. We have already cited the papers by Friedman & Russek [366], Stein [1143] and Sack [1043]. Another well-known paper is that of Danos & Maximon [254]; this paper includes a historical survey. More recent references include [133, §5.1], [424, Appendix D] and [285, Chapter VII, §1.2].

The idea of the proof above (apply $\mathcal{Y}_n^m$ to the addition theorem for $h_0^{(1)}$) is explicit in the papers of Nozawa [919] and Weniger & Steinborn [1308]. It can be used to obtain addition theorems for a variety of functions.

As a special case of (3.87), we have

$$S_{n\nu}^{m\mu}(\pm b\hat{\mathbf{z}}) = \sqrt{4\pi}\, \delta_{m\mu} \mathrm{i}^{\nu-n} (-1)^m \sum_q \sqrt{2q+1}\, (\pm \mathrm{i})^q$$

$$\times\, h_q^{(1)}(kb)\, \mathcal{G}(n, m; \nu, -m; q), \tag{3.88}$$

which should be compared with (3.82). Indeed, we have

$$\widehat{S}_{n\nu}^{m\mu}(\pm b\hat{\mathbf{z}}) = \mathrm{Re}\,\{S_{n\nu}^{m\mu}(\pm b\hat{\mathbf{z}})\}; \tag{3.89}$$

the corresponding formula with $\pm b\hat{\mathbf{z}}$ replaced by $\boldsymbol{b}$ *is false*, in general, but see Lemma 3.29 below.

For another special case, we can multiply (3.84) or (3.85) by k^{n+1} and then let $k \to 0$. This will yield addition theorems for the harmonic multipole potential, $r_2^{-n-1} Y_n^m(\hat{\boldsymbol{r}}_2)$. For direct proofs, see [1045, p. 784] and [318, Theorem 2].

3.13 The separation matrices S and $\widehat{\mathbf{S}}$

We have seen that the addition theorems for $j_n Y_n^m$ and $h_n^{(1)} Y_n^m$ involve the separation matrices $\widehat{\mathbf{S}} = (\widehat{S}_{n\nu}^{m\mu})$ and $\mathbf{S} = (S_{n\nu}^{m\mu})$, respectively. The expressions for these matrices are complicated, so we must consider methods for their evaluation.

It is convenient to modify our notation for spherical wavefunctions, so that it is similar to that used in Chapter 2.

Definition 3.28 *Let* $\boldsymbol{r} = r\hat{\boldsymbol{r}}$. *The* outgoing spherical wavefunction ψ_n^m *is defined by*

$$\psi_n^m(\boldsymbol{r}) = h_n^{(1)}(kr)\, Y_n^m(\hat{\boldsymbol{r}}),$$

where $0 \le |m| \le n$ *and* $n \ge 0$. *Similarly, the* regular spherical wavefunction $\hat{\psi}_n^m$ *is defined by*

$$\hat{\psi}_n^m(\boldsymbol{r}) = j_n(kr)\, Y_n^m(\hat{\boldsymbol{r}}).$$

Comparing with Definition 3.8, we see that $\hat{\psi}_n^m \equiv \widehat{\Omega}_n^m$, whereas $\psi_n^m = \Omega_n^m$ when $z_n = h_n^{(1)}$.

The function ψ_n^m is described as 'outgoing' because it corresponds to outgoing waves at infinity. To see this, note that [1, eqn (10.1.16)]

$$h_n^{(1)}(w) \sim \mathrm{i}^{-n-1}\, w^{-1}\, \mathrm{e}^{\mathrm{i}w} \tag{3.90}$$

as $|w| \to \infty$, whence

$$\operatorname{Re}\left\{\psi_n^m(\boldsymbol{r})\, \mathrm{e}^{-\mathrm{i}\omega t}\right\} \sim \frac{A_n^m}{\sqrt{4\pi}} \frac{P_n^m(\cos\theta)}{kr} \sin\left\{(kr - \omega t) + m\phi - \frac{1}{2} n\pi\right\}$$

as $r \to \infty$; here, A_n^m are the real normalisation factors defined by (3.6).

The wavefunctions $\hat{\psi}_n^m$ and ψ_n^m satisfy

$$\hat{\psi}_n^m(\boldsymbol{r}) = (-1)^n\, \hat{\psi}_n^m(-\boldsymbol{r}) = (-1)^m\, \overline{\hat{\psi}_n^{-m}(\boldsymbol{r})} \tag{3.91}$$

and

$$\psi_n^m(\boldsymbol{r}) = (-1)^n\, \psi_n^m(-\boldsymbol{r}),$$

respectively.

The matrix $\widehat{\mathbf{S}}$ is defined by (3.80). Making the summation limits explicit, we obtain

$$\widehat{S}_{n\nu}^{m\mu}(\boldsymbol{b}) = 4\pi(-1)^{\mu+\nu+Q}\sum_{q=0}^{Q}(-1)^q\hat{\psi}_{q_0+2q}^{m-\mu}(\boldsymbol{b})\,\mathcal{G}(n,m;\nu,-\mu;q_0+2q), \tag{3.92}$$

where $Q=(n+\nu-q_0)/2$, $q_0=q_0(n,m;\nu,-\mu)$ and q_0 is defined by (3.74). From this, we obtain

$$\widehat{S}_{\nu n}^{\mu m}(\boldsymbol{b}) = (-1)^{n+\nu}\overline{\widehat{S}_{n\nu}^{m\mu}(\boldsymbol{b})}, \tag{3.93}$$

using (3.7) and (3.75). Similarly,

$$\widehat{S}_{n\nu}^{-m,-\mu}(\boldsymbol{b}) = (-1)^{m+\mu}\overline{\widehat{S}_{n\nu}^{m\mu}(\boldsymbol{b})}. \tag{3.94}$$

If we use (3.8) and $q_0=n+\nu-2Q$, we obtain

$$\widehat{S}_{n\nu}^{m\mu}(-\boldsymbol{b}) = (-1)^{n+\nu}\,\widehat{S}_{n\nu}^{m\mu}(\boldsymbol{b}) = \overline{\widehat{S}_{\nu n}^{\mu m}(\boldsymbol{b})}, \tag{3.95}$$

where we have used (3.93) for the last equality together with the fact that $n+\nu+q_0$ is even.

Next, define vectors $\boldsymbol{\psi}=(\psi_n^m)$ and $\hat{\boldsymbol{\psi}}=(\hat{\psi}_n^m)$. This enables us to write Theorem 3.26 concisely as

$$\hat{\boldsymbol{\psi}}(\boldsymbol{r}_2) = \widehat{\mathbf{S}}(\boldsymbol{b})\,\hat{\boldsymbol{\psi}}(\boldsymbol{r}_1),$$

where $\boldsymbol{r}_2=\boldsymbol{r}_1+\boldsymbol{b}$. Then, exactly as in Section 2.6, we find that

$$\widehat{\mathbf{S}}(\boldsymbol{b})\,\widehat{\mathbf{S}}(\boldsymbol{c}) = \widehat{\mathbf{S}}(\boldsymbol{b}+\boldsymbol{c}) = \widehat{\mathbf{S}}(\boldsymbol{c})\,\widehat{\mathbf{S}}(\boldsymbol{b});$$

this can be used to effect two translations. In particular, taking $\boldsymbol{c}=-\boldsymbol{b}$ gives

$$\widehat{\mathbf{S}}(\boldsymbol{b})\,\widehat{\mathbf{S}}(-\boldsymbol{b}) = \widehat{\mathbf{S}}(-\boldsymbol{b})\,\widehat{\mathbf{S}}(\boldsymbol{b}) = \mathbf{I},$$

the identity, which shows that

$$\left[\widehat{\mathbf{S}}(\boldsymbol{b})\right]^{-1} = \widehat{\mathbf{S}}(-\boldsymbol{b}).$$

Hence, using (3.95), we obtain

$$\widehat{\mathbf{S}}^{-1} = \overline{\widehat{\mathbf{S}}^{\mathrm{T}}},$$

so that $\widehat{\mathbf{S}}$ is unitary. Explicitly, we have

$$\sum_{\nu=0}^{\infty}\sum_{\mu=-\nu}^{\nu}\widehat{S}_{n\nu}^{m\mu}(\boldsymbol{b})\,\overline{\widehat{S}_{N\nu}^{M\mu}(\boldsymbol{b})} = \delta_{nN}\,\delta_{mM}. \tag{3.96}$$

For $\mathbf{S}$, we have a formula similar to (3.92), namely

$$S_{n\nu}^{m\mu}(\boldsymbol{b}) = 4\pi(-1)^{\mu+\nu+Q}\sum_{q=0}^{Q}(-1)^q\psi_{q_0+2q}^{m-\mu}(\boldsymbol{b})\,\mathcal{G}(n,m;\nu,-\mu;q_0+2q). \tag{3.97}$$

Explicit calculation shows that

$$S_{n\nu}^{m\mu}(-\boldsymbol{b}) = (-1)^{n+\nu} S_{n\nu}^{m\mu}(\boldsymbol{b}) \tag{3.98}$$

and

$$S_{\nu n}^{-\mu,-m}(-\boldsymbol{b}) = (-1)^{m+\mu} S_{n\nu}^{m\mu}(\boldsymbol{b}), \tag{3.99}$$

whence

$$S_{n\nu}^{m\mu}(\boldsymbol{b}) = (-1)^{n+m+\nu+\mu} S_{\nu n}^{-\mu,-m}(\boldsymbol{b}).$$

Exactly as in Section 2.6, we find that

$$\mathbf{S}(\boldsymbol{b})\,\widehat{\mathbf{S}}(\boldsymbol{c}) = \mathbf{S}(\boldsymbol{b}+\boldsymbol{c}) = \widehat{\mathbf{S}}(\boldsymbol{c})\,\mathbf{S}(\boldsymbol{b}) \quad \text{for } c < b. \tag{3.100}$$

Another relation between $\mathbf{S}$ and $\widehat{\mathbf{S}}$ is given in the following lemma.

Lemma 3.29

$$S_{n\nu}^{m\mu}(\boldsymbol{b}) + \overline{S_{\nu n}^{\mu m}(-\boldsymbol{b})} = 2\widehat{S}_{n\nu}^{m\mu}(\boldsymbol{b}).$$

Proof From (3.99), (3.97) and (3.75), we obtain

$$\begin{aligned} S_{\nu n}^{\mu m}(-\boldsymbol{b}) &= 4\pi(-1)^{\mu+\nu+Q}(-1)^{m+\mu} \\ &\quad \times \sum_{q=0}^{Q}(-1)^{q}\psi_{q_0+2q}^{\mu-m}(\boldsymbol{b})\,\mathcal{G}(n, m; \nu, -\mu; q_0+2q), \end{aligned}$$

noting that the definitions of Q and q_0 depend only on $|m+\mu|$. But, using (3.7),

$$(-1)^{m+\mu}\overline{\psi_{\ell}^{\mu-m}(\boldsymbol{b})} = \overline{h_{\ell}(kb)}\,Y_{\ell}^{m-\mu}(\hat{\boldsymbol{b}}).$$

The result follows from $2j_n(w) = h_n(w) + \overline{h_n(w)}$ and (3.92). □

3.13.1 Computation of the separation matrices

The expressions for the separation matrices given above, namely (3.92) and (3.97), are complicated because of the presence of the Gaunt coefficient $\mathcal{G}$. These can be defined in terms of Wigner 3-j symbols; see (3.72). Thus, consideration should be given to the efficient computation of these quantities.

One can compute Gaunt coefficients directly. This approach was advocated in [236, §V] because 'it is of great advantage in the organization of a computer program to be able to obtain coefficients in any arbitrary sequence' [236, p. 170]; see also [455]. However, the explicit formulae for $\mathcal{G}$ and the 3-j symbols involve many factorials (Appendix B) and so much care is needed in the computations.

Other methods are based on the use of recursion relations. See, for example, [305, §3.7], [1326, §V], [250, §C.7.3] and [145, 551, 1079, 1160, 1346, 1348]. Direct recurrence relations for the entries in the separation matrices themselves can also be derived, usually by exploiting the first-derivative theorems in Section 3.7; see [400, 763, 197, 1347, 606, 448]. However, it is known that some of these recurrence relations are numerically unstable in certain circumstances [769]. Also, Xu [1347] found that it was more efficient to compute the Gaunt coefficients first (using the method of [1346]) and then to compute the separation matrices, rather than to use direct recurrence relations.

Another approach is to combine rotation matrices (Section 3.15) together with translation along the z-axis, this being effective because the separation matrices simplify when $\boldsymbol{b} = b\hat{\mathbf{z}}$; these simplifications are discussed in detail in Section 3.16.

3.13.2 A wide-spacing approximation

If kb is large, we can obtain a simple approximation to $\mathbf{S}(\boldsymbol{b})$. Thus, using (3.7) and (3.90), (3.87) gives

$$S_{n\nu}^{m\mu}(\boldsymbol{b}) \sim 4\pi \mathrm{i}^{\nu-n}(-1)^{\mu}\frac{\mathrm{e}^{\mathrm{i}kb}}{\mathrm{i}kb}\sum_{q}\mathcal{G}(n, m; \nu, -\mu; q)\, Y_q^{m-\mu}(\hat{\boldsymbol{b}})$$

as $kb \to \infty$. Evaluating the sum, using Theorem 3.20, we obtain

$$S_{n\nu}^{m\mu}(\boldsymbol{b}) \sim 4\pi \mathrm{i}^{\nu-n}(-1)^{\mu}\frac{\mathrm{e}^{\mathrm{i}kb}}{\mathrm{i}kb}\, Y_n^m(\hat{\boldsymbol{b}})\, Y_\nu^{-\mu}(\hat{\boldsymbol{b}}) \tag{3.101}$$

as $kb \to \infty$. Substituting in (3.84), and evaluating the double sum over ν and μ using (3.69), then gives

$$h_n^{(1)}(kr_2)\, Y_n^m(\hat{\boldsymbol{r}}_2) \sim (-\mathrm{i})^n \frac{\mathrm{e}^{\mathrm{i}kb}}{\mathrm{i}kb}\exp\left(\mathrm{i}k\boldsymbol{r}_1 \cdot \hat{\boldsymbol{b}}\right) Y_n^m(\hat{\boldsymbol{b}}) \tag{3.102}$$

as $kb \to \infty$, where $\boldsymbol{r}_2 = \boldsymbol{r}_1 + \boldsymbol{b}$. This approximation can also be obtained directly, using (3.90), $\hat{\boldsymbol{r}}_2 \simeq \hat{\boldsymbol{b}}$ and $r_2 \simeq b + \boldsymbol{r}_1 \cdot \hat{\boldsymbol{b}}$ on the left-hand side.

The approximation (3.102) is known as a *plane-wave approximation*. It is an example of a *wide-spacing approximation*, as the dimensionless spacing between the two origins, O_1 and O_2, is large, $kb \gg 1$. With the given exact formulae for $\mathbf{S}(\boldsymbol{b})$, it is difficult to refine the approximation. This will be done later, using an alternative formula for $\mathbf{S}(\boldsymbol{b})$; see Section 3.16.

3.13.3 Separable formulae for the separation matrices

For some applications, it is convenient if the dependences on (n, m) and (ν, μ) in $S_{n\nu}^{m\mu}$ and $\widehat{S}_{n\nu}^{m\mu}$ are separated. In fact, we have already given a separable formula for $\widehat{S}_{n\nu}^{m\mu}$, namely, (3.79); see [201, eqn (6)]. We now give a similar formula for $S_{n\nu}^{m\mu}$. Thus, from the definitions (3.71) and (3.87), we have

$$S_{n\nu}^{m\mu}(\boldsymbol{b}) = 4\pi \mathrm{i}^{\nu-n}\sum_{q}\mathrm{i}^q\, h_q^{(1)}(kb)\sum_{M}\overline{Y_q^M(\hat{\boldsymbol{b}})}\int_{\Omega} Y_n^m\, \overline{Y_\nu^\mu}\, Y_q^M\, \mathrm{d}\Omega.$$

Notice that the integral vanishes unless $m-\mu+M=0$. Then, using (3.20), we obtain

$$S_{n\nu}^{m\mu}(\boldsymbol{b}) = \mathrm{i}^{\nu-n}\sum_{q}(2q+1)\mathrm{i}^{q}\,h_{q}^{(1)}(kb)\int_{\Omega}Y_{n}^{m}(\hat{\boldsymbol{r}})\,\overline{Y_{\nu}^{\mu}(\hat{\boldsymbol{r}})}\,P_{q}(\hat{\boldsymbol{r}}\cdot\hat{\boldsymbol{b}})\,\mathrm{d}\Omega(\hat{\boldsymbol{r}}). \qquad (3.103)$$

The sum over q is finite; the upper limit is $(n+\nu)$. Thus, we can interchange the order of summation and integration, giving

$$S_{n\nu}^{m\mu}(\boldsymbol{b}) = \mathrm{i}^{\nu-n}\int_{\Omega}Y_{n}^{m}(\hat{\boldsymbol{r}})\,\overline{Y_{\nu}^{\mu}(\hat{\boldsymbol{r}})}\,M_{n+\nu}(\hat{\boldsymbol{r}};\boldsymbol{b})\,\mathrm{d}\Omega(\hat{\boldsymbol{r}}), \qquad (3.104)$$

where

$$M_{N}(\hat{\boldsymbol{r}};\boldsymbol{b}) = \sum_{q}^{N}(2q+1)\mathrm{i}^{q}\,h_{q}^{(1)}(kb)P_{q}(\hat{\boldsymbol{r}}\cdot\hat{\boldsymbol{b}}) \qquad (3.105)$$

is called the *translation function* by Epton & Dembart [318]. However, we cannot use (3.104) in places where $n\to\infty$ or $\nu\to\infty$, because M_N diverges as $N\to\infty$.

One can verify that the addition theorem,

$$\psi_{0}^{0}(\boldsymbol{r}_2) = \sum_{\nu,\mu}S_{0\nu}^{0\mu}(\boldsymbol{b})\,\hat{\psi}_{\nu}^{\mu}(\boldsymbol{r}_1), \quad r_1 < b,$$

combined with (3.103) and (3.69) leads to Theorem 3.17.

The formulae of this section are key ingredients in fast multipole methods; see Section 6.14.

3.14 Two-centre expansions

As in the two-dimensional case, Section 2.8, we consider two-centre expansions, involving two points, P and P', and two origins, O_1 and O_2. We use the same notation which, for convenience, we repeat here. Thus, $\boldsymbol{r}_j$ is the position vector of P with respect to O_j, $j=1,2$, and $\boldsymbol{b}$ is the position vector of O_1 with respect to O_2. Similarly, $\boldsymbol{r}'_j$ is the position vector of P' with respect to O_j, $j=1,2$, and $\boldsymbol{R}$ is the position vector of P with respect to P'. See Fig. 2.3. For most purposes, it is convenient to think of P and P' as being in the vicinity of O_1 and O_2, respectively.

As $\boldsymbol{R}=\boldsymbol{r}_1-\boldsymbol{r}'_1$, Theorem 3.16 gives

$$\frac{\mathrm{e}^{ikR}}{kR} = 4\pi\mathrm{i}\sum_{n,m}(-1)^{m+n}\hat{\psi}_{n}^{-m}(\boldsymbol{r}_1)\,\psi_{n}^{m}(-\boldsymbol{r}'_1) \quad \text{for } r_1 < r'_1,$$

where the summation is over all $n\geq 0$ and over all m with $|m|\leq n$. Similarly, as $-\boldsymbol{r}'_1=\boldsymbol{b}-\boldsymbol{r}'_2$, Theorem 3.27 gives

$$\psi_{n}^{m}(-\boldsymbol{r}'_1) = \sum_{\nu,\mu}S_{n\nu}^{m\mu}(\boldsymbol{b})\,\hat{\psi}_{\nu}^{\mu}(-\boldsymbol{r}'_2) \quad \text{for } r'_2 < b.$$

So, if both inequalities hold, we obtain

$$\frac{e^{ikR}}{kR} = 4\pi i \sum_{n,m} \sum_{\nu,\mu} (-1)^{n+\nu} S_{n\nu}^{m\mu}(\boldsymbol{b}) \, \overline{\hat{\psi}_n^m(\boldsymbol{r}_1)} \, \hat{\psi}_\nu^\mu(\boldsymbol{r}_2').$$

Thus, we have proved the following two-centre expansion.

Theorem 3.30 (Two-centre expansion of e^{ikR}/R**)** *Let* $\boldsymbol{R} = \boldsymbol{a} + \boldsymbol{b} + \boldsymbol{c}$. *Then*

$$\frac{e^{ikR}}{kR} = 4\pi i \sum_{n,m} \sum_{\nu,\mu} (-1)^{n} S_{n\nu}^{m\mu}(\boldsymbol{b}) \, \overline{\hat{\psi}_n^m(\boldsymbol{a})} \, \hat{\psi}_\nu^\mu(\boldsymbol{c})$$

for $a = |\boldsymbol{a}| < |\boldsymbol{b} + \boldsymbol{c}|$ *and* $c = |\boldsymbol{c}| < b = |\boldsymbol{b}|$*; both inequalities hold if* $a + c < b$.

As far as we know, the first proof of this theorem was given by Sack in 1964 [1044, §4]. For an earlier attempt, see [1098]. The corresponding expansion for R^{-1} was obtained by Carlson & Rushbrooke [174]; see also [155] and [505, p. 843].

3.14.1 Second proof of the addition theorem for $h_n^{(1)} Y_n^m$

We can use two-centre expansions to give another proof of Theorem 3.27, making use of Theorems 3.16 and 3.26. This proof has been given by Gonis [422, pp. 374–375].

From $\boldsymbol{R} = \boldsymbol{r}_1 - \boldsymbol{r}_1'$, Theorem 3.16 gives

$$h_0^{(1)}(kR) = 4\pi \sum_{\nu,\mu} (-1)^\mu \hat{\psi}_\nu^\mu(\boldsymbol{r}_1') \, \psi_\nu^{-\mu}(\boldsymbol{r}_1) \quad \text{for } r_1 > r_1'. \tag{3.106}$$

Then, as $\boldsymbol{r}_1' = \boldsymbol{r}_2' - \boldsymbol{b}$, Theorem 3.26 gives

$$\hat{\psi}_\nu^\mu(\boldsymbol{r}_1') = \sum_{n,m} \widehat{S}_{\nu n}^{\mu m}(-\boldsymbol{b}) \, \hat{\psi}_n^m(\boldsymbol{r}_2'),$$

whence (3.106) gives

$$h_0^{(1)}(kR) = 4\pi \sum_{n,m} \hat{\psi}_n^m(\boldsymbol{r}_2') \sum_{\nu,\mu} (-1)^\mu \widehat{S}_{\nu n}^{\mu m}(-\boldsymbol{b}) \, \psi_\nu^{-\mu}(\boldsymbol{r}_1).$$

Alternatively, assume that $r_2 > r_2'$ with $\boldsymbol{R} = \boldsymbol{r}_2 - \boldsymbol{r}_2'$, whence

$$h_0^{(1)}(kR) = 4\pi \sum_{n,m} (-1)^m \hat{\psi}_n^m(\boldsymbol{r}_2') \, \psi_n^{-m}(\boldsymbol{r}_2).$$

Comparing these two expansions, in a neighbourhood of O_2, we obtain

$$(-1)^m \psi_n^{-m}(\boldsymbol{r}_2) = \sum_{\nu,\mu} (-1)^\mu \widehat{S}_{\nu n}^{\mu m}(-\boldsymbol{b}) \, \psi_\nu^{-\mu}(\boldsymbol{r}_1).$$

Changing the signs of m and μ, and noting from (3.94) and (3.95) that

$$(-1)^{m+\mu} \widehat{S}_{\nu n}^{-\mu,-m}(-\boldsymbol{b}) = \widehat{S}_{n\nu}^{m\mu}(\boldsymbol{b}), \tag{3.107}$$

we obtain

$$\psi_n^m(\boldsymbol{r}_2) = \sum_{\nu,\mu} \widehat{S}_{n\nu}^{m\mu}(\boldsymbol{b})\, \psi_\nu^\mu(\boldsymbol{r}_1) \quad \text{for } r_1 > b,$$

which proves (3.85).

For (3.84), we interchange $\boldsymbol{r}_1$ and $\boldsymbol{b}$ to give

$$\begin{aligned}\psi_n^m(\boldsymbol{r}_2) &= \sum_{N,M} \widehat{S}_{nN}^{mM}(\boldsymbol{r}_1)\, \psi_N^M(\boldsymbol{b}) \\ &= 4\pi \sum_{N,M,q} \psi_N^M(\boldsymbol{b})\, \mathrm{i}^{N-n+q}(-1)^M\, \hat{\psi}_q^{m-M}(\boldsymbol{r}_1)\, \mathcal{G}(n,m;N,-M;q),\end{aligned}$$

where we have used (3.80) for $\widehat{S}_{nN}^{mM}$. Now, replace q, N and M by ν, q and $m-\mu$, respectively. This gives (3.84), namely

$$\psi_n^m(\boldsymbol{r}_2) = \sum_{\nu,\mu} S_{n\nu}^{m\mu}(\boldsymbol{b})\, \hat{\psi}_\nu^\mu(\boldsymbol{r}_1) \quad \text{for } r_1 < b,$$

where $S_{n\nu}^{m\mu}$ is given by (3.86).

3.14.2 A formula for $\mathrm{S}(\boldsymbol{a}+\boldsymbol{b})$

From Theorem 3.30, we have

$$h_0(kR) = 4\pi \sum_{\nu,\mu} \sum_{N,M} S_{N\nu}^{M\mu}(\boldsymbol{b})\, \overline{\hat{\psi}_N^M(-\boldsymbol{a})}\, \hat{\psi}_\nu^\mu(\boldsymbol{c}),$$

for $a < |\boldsymbol{b}+\boldsymbol{c}|$ and $c < b$. We also have

$$\psi_n^m(\boldsymbol{R}) = \sum_{\nu,\mu} S_{n\nu}^{m\mu}(\boldsymbol{a}+\boldsymbol{b})\, \hat{\psi}_\nu^\mu(\boldsymbol{c})$$

for $c < |\boldsymbol{a}+\boldsymbol{b}|$. So, assuming that both expansions are valid, we obtain

$$S_{n\nu}^{m\mu}(\boldsymbol{a}+\boldsymbol{b}) = 4\pi \mathcal{Y}_n^m \sum_{N,M} S_{N\nu}^{M\mu}(\boldsymbol{b})\, \overline{\hat{\psi}_N^M(-\boldsymbol{a})},$$

where we have used (3.40). Now, we can regard $\boldsymbol{b}$ as fixed, and then apply the operator $\mathcal{Y}_n^m$ with respect to the Cartesian coordinates of $\boldsymbol{a}$. Thus,

$$\begin{aligned}\mathcal{Y}_n^m \overline{\hat{\psi}_N^M(-\boldsymbol{a})} &= \mathcal{Y}_n^m \frac{\mathrm{i}^N}{4\pi} \int_\Omega \exp(\mathrm{i}k\boldsymbol{a}\cdot\hat{\boldsymbol{r}}_1)\, \overline{Y_N^M(\hat{\boldsymbol{r}}_1)}\, \mathrm{d}\Omega(\hat{\boldsymbol{r}}_1) \\ &= \frac{\mathrm{i}^{N-n}}{4\pi} \int_\Omega \exp(\mathrm{i}k\boldsymbol{a}\cdot\hat{\boldsymbol{r}}_1)\, Y_n^m(\hat{\boldsymbol{r}}_1)\, \overline{Y_N^M(\hat{\boldsymbol{r}}_1)}\, \mathrm{d}\Omega(\hat{\boldsymbol{r}}_1),\end{aligned}$$

where we have used (3.42) and Theorem 3.11. Hence,

$$S_{n\nu}^{m\mu}(\boldsymbol{a}+\boldsymbol{b}) = \sum_{N,M} \mathrm{i}^{N-n} S_{N\nu}^{M\mu}(\boldsymbol{b}) \int_\Omega \exp(\mathrm{i}k\boldsymbol{a}\cdot\hat{\boldsymbol{r}})\, Y_n^m(\hat{\boldsymbol{r}})\, \overline{Y_N^M(\hat{\boldsymbol{r}})}\, \mathrm{d}\Omega(\hat{\boldsymbol{r}}). \tag{3.108}$$

In fact, using (3.79), we recognise this formula as (3.100): it states that

$$\mathbf{S}(\boldsymbol{a}+\boldsymbol{b}) = \widehat{\mathbf{S}}(\boldsymbol{a})\,\mathbf{S}(\boldsymbol{b}) \quad \text{for } a < b.$$

Using (3.103) in (3.108) followed by (3.11) gives

$$S_{n\nu}^{m\mu}(\boldsymbol{a}+\boldsymbol{b}) = \mathrm{i}^{\nu-n}\sum_{q=0}^{\infty}(2q+1)\mathrm{i}^q h_q(kb)$$
$$\times \int_{\Omega} \exp(\mathrm{i}k\boldsymbol{a}\cdot\hat{\boldsymbol{r}})\, Y_n^m(\hat{\boldsymbol{r}})\,\overline{Y_\nu^\mu(\hat{\boldsymbol{r}})}\, P_q(\hat{\boldsymbol{r}}\cdot\hat{\boldsymbol{b}})\,\mathrm{d}\Omega(\hat{\boldsymbol{r}}). \tag{3.109}$$

This formula is derived (differently) in [201] and it is used in [640]. The effect of truncating the sum over q is examined in [641].

3.15 Use of rotation matrices

> [The] *point of view with which we propose to look at the problem* ... [is] *the representation of the translation by an operator. We thus introduce a "standard translation matrix,"* $S(kb)$... [representing] *a translation along the z-axis by a distance b. Any general shift of the coordinate system is then given by* $D(R_2)\,S(kb)\,D(R_1)$ *where D is the usual rotation operator: one first rotates the coordinate system so that the z-axis points in the direction of the displacement (rotation* R_1*), one then shifts along the z-axis, and lastly rotates into the final position (rotation* R_2*).*
>
> (Danos & Maximon [254, p. 767])

As in Section 2.7, we can decompose the translation from O_2 to O_1 into two rotations and a translation in a particular direction. This approach was first exploited by Danos & Maximon [254]. It reduces the general problem to the calculation of $\widehat{\mathbf{S}}(b\hat{\mathbf{z}})$ and $\mathbf{S}(b\hat{\mathbf{z}})$, together with certain rotation matrices, $\mathbf{D}$.

The basic formula is

$$Y_n^m(\hat{\boldsymbol{r}}') = \sum_{l=-n}^{n} D_n^{ml}(\boldsymbol{\omega})\, Y_n^l(\hat{\boldsymbol{r}}), \tag{3.110}$$

where a point P has position vectors $\boldsymbol{r} = r\hat{\boldsymbol{r}}$ and $\boldsymbol{r}' = r\hat{\boldsymbol{r}}'$ with respect to $Oxyz$ and $Ox'y'z'$, respectively. The rotation from $Oxyz$ to $Ox'y'z'$ is described by $\boldsymbol{\omega} = (\alpha, \beta, \gamma)$, where α, β and γ are the three Eulerian angles. The reverse rotation is described by $\boldsymbol{\omega}^{-1} = (-\gamma, -\beta, -\alpha)$; from (C.4), we have

$$D_n^{ml}(\boldsymbol{\omega}^{-1}) = \overline{D_n^{lm}(\boldsymbol{\omega})}. \tag{3.111}$$

Further information and explicit formulae for the rotation matrix $\mathbf{D}$ (with entries D_n^{ml}) can be found in Appendix C.

Let us carry out the three steps for $\hat{\psi}_n^m(\boldsymbol{r}_2)$ in order to obtain the expansion about O_1, and compare the result with Theorem 3.26.

First, we rotate the coordinate axes $O_2x_2y_2z_2$ into O_2xyz, chosen so that O_1 is at $x = y = 0$, $z = b$; this rotation is described by Eulerian angles $\boldsymbol{\omega}$. The result is

$$\hat{\psi}_n^m(\boldsymbol{r}_2) = \sum_{l=-n}^{n} D_n^{ml}(\boldsymbol{\omega})\,\hat{\psi}_n^l(\boldsymbol{r}).$$

Second, we introduce axes $O_1\tilde{x}\tilde{y}\tilde{z}$ at O_1, where $\tilde{x} = x$, $\tilde{y} = y$ and $\tilde{z} = z - b$. Then, translating from O_2 to O_1, we obtain

$$\hat{\psi}_n^l(\boldsymbol{r}) = \sum_{\nu=0}^{\infty}\sum_{M=-\nu}^{\nu} \widehat{S}_{n\nu}^{lM}(b\hat{\mathbf{z}})\,\hat{\psi}_\nu^M(\tilde{\boldsymbol{r}}).$$

Finally, we rotate the axes $O_1\tilde{x}\tilde{y}\tilde{z}$ into $O_1x_1y_1z_1$, using $\boldsymbol{\omega}^{-1}$, giving

$$\hat{\psi}_\nu^M(\tilde{\boldsymbol{r}}) = \sum_{\mu=-\nu}^{\nu} D_\nu^{M\mu}(\boldsymbol{\omega}^{-1})\,\hat{\psi}_\nu^\mu(\boldsymbol{r}_1).$$

Comparing with Theorem 3.26 gives

$$\widehat{S}_{n\nu}^{m\mu}(\boldsymbol{b}) = \sum_{l=-n}^{n}\sum_{M=-\nu}^{\nu} D_n^{ml}(\boldsymbol{\omega})\,\widehat{S}_{n\nu}^{lM}(b\hat{\mathbf{z}})\,\overline{D_\nu^{\mu M}(\boldsymbol{\omega})}, \tag{3.112}$$

where we have used (3.111). There is a similar formula for $\mathbf{S}(\boldsymbol{b})$: merely replace $\widehat{S}$ by S in (3.112).

As rotation does not change the distance from the origin, (3.110) gives

$$\psi_n^m(r\hat{\boldsymbol{r}}') = \sum_{l=-n}^{n} D_n^{ml}(\boldsymbol{\omega})\,\psi_n^l(r\hat{\boldsymbol{r}}) \quad \text{and} \quad \hat{\psi}_n^m(r\hat{\boldsymbol{r}}') = \sum_{l=-n}^{n} D_n^{ml}(\boldsymbol{\omega})\,\hat{\psi}_n^l(r\hat{\boldsymbol{r}}).$$

3.16 Alternative expressions for $\mathbf{S}(b\hat{\mathbf{z}})$

The expressions for $\widehat{\mathbf{S}}(\boldsymbol{b})$ and $\mathbf{S}(\boldsymbol{b})$, given by (3.92) and (3.97), respectively, are complicated. Therefore, the decompositions described in Section 3.15, involving the rotation matrix $\mathbf{D}$, are worth pursuing. They relate $\mathbf{S}(\boldsymbol{b})$ to $\mathbf{S}(b\hat{\mathbf{z}})$ and $\widehat{\mathbf{S}}(\boldsymbol{b})$ to $\widehat{\mathbf{S}}(b\hat{\mathbf{z}})$. In fact, due to (3.89), it is sufficient to consider $\mathbf{S}(b\hat{\mathbf{z}})$.

From (3.88) and (3.98), we have

$$S_{n\nu}^{m\mu}(b\hat{\mathbf{z}}) = (-1)^{n+\nu} S_{n\nu}^{m\mu}(-b\hat{\mathbf{z}}) = (-1)^n\, S_{n\nu}^m(kb)\,\delta_{m\mu}, \tag{3.113}$$

say, where the *z-axis propagator* $S_{n\nu}^m$ satisfies

$$S_{n\nu}^m(kb) = S_{\nu n}^m(kb) = S_{n\nu}^{-m}(kb),$$

which means that we can assume that $m \geq 0$ without loss of generality.

In order to develop alternative expressions for $S_{n\nu}^m$, we begin with an integral representation for e^{ikR}/R.

Theorem 3.31 *For $R = |\boldsymbol{R}| > 0$,*

$$\frac{e^{ikR}}{R} = \frac{1}{4\pi^2} \int_{\mathcal{C}} \frac{\xi^2}{\xi^2 - k^2} \int_{\Omega} \exp(i\xi\hat{\boldsymbol{r}} \cdot \boldsymbol{R})\, d\Omega(\hat{\boldsymbol{r}})\, d\xi. \tag{3.114}$$

The contour $\mathcal{C}$ goes from $\xi = -\infty$ to $\xi = +\infty$, passing above the simple pole at $\xi = -k$ and below the simple pole at $\xi = +k$.

Proof From (3.44), the inner integral is $4\pi\, j_0(\xi R)$, whence the right-hand side of (3.114) is

$$\frac{1}{2\pi i R} \int_{\mathcal{C}} \frac{\xi}{\xi^2 - k^2} \left(e^{i\xi R} - e^{-i\xi R}\right) d\xi = \mathcal{I}_+ - \mathcal{I}_-,$$

say, where

$$\mathcal{I}_\pm = \frac{1}{2\pi i R} \int_{\mathcal{C}} \frac{\xi}{\xi^2 - k^2}\, e^{\pm i\xi R}\, d\xi.$$

These integrals can be evaluated using Cauchy's residue theorem. Thus, for $\mathcal{I}_+$, close the contour using a large semicircle in the upper half of the complex ξ-plane. There is no contribution from this semicircle as it recedes to infinity, by Jordan's lemma. The contour encloses the simple pole at $\xi = +k$ (but not the pole at $\xi = -k$). Hence, evaluating the residue, we see that $\mathcal{I}_+ = \frac{1}{2} e^{ikR}/R$. Similarly, for $\mathcal{I}_-$, close the contour in the lower half of the complex ξ-plane. Taking account of the direction of traversal around the contour, and evaluating the residue of the pole at $\xi = -k$, we find that $\mathcal{I}_- = -\mathcal{I}_+$, and the result follows. □

In the above proof, we split the ξ-integral into two, $\mathcal{I}_+$ and $\mathcal{I}_-$, and then evaluated each separately, one by closing the contour in the upper half-plane and one by closing the contour in the lower half-plane. This is a standard technique. However, care is needed to ensure that $\mathcal{I}_+$ and $\mathcal{I}_-$ exist separately: the splitting is not unique, and some splittings may introduce additional singularities at $\xi = 0$. For example, a common mistake when evaluating related integrals is to split using

$$2j_n(\xi R) = h_n^{(1)}(\xi R) + h_n^{(2)}(\xi R);$$

each of the spherical Hankel functions is $O(\xi^{-n-1})$ as $\xi \to 0$.

The representation (3.114) is a variant of well-known formulae using volume integrals; for example, DeSanto [278, p. 64] shows that

$$\frac{e^{ikR}}{R} = \frac{1}{2\pi^2} \lim_{\varepsilon \to 0+} \iiint \frac{\exp(i\boldsymbol{\xi} \cdot \boldsymbol{R})}{|\boldsymbol{\xi}|^2 - (k + i\varepsilon)^2}\, d\boldsymbol{\xi}. \tag{3.115}$$

However, (3.114) is preferable for at least two reasons: it does not involve the device of adding a small imaginary part to the wavenumber; and it uses an integration along the entire real ξ-axis, so that contour-integral methods are readily available.

Note that one cannot interchange the order of integration in (3.114): the resulting ξ-integral diverges.

We are going to combine Theorem 3.31 with the two-centre expansion, Theorem 3.30, so as to obtain an alternative expression for $\mathbf{S}(b\hat{\mathbf{z}})$. Thus, write $\boldsymbol{R} = \boldsymbol{a}+\boldsymbol{b}+\boldsymbol{c}$, and then use the expansion (3.69) twice,

$$\exp(\mathrm{i}\xi\hat{\boldsymbol{r}}\cdot\boldsymbol{a}) = 4\pi\sum_{n,m}\mathrm{i}^n\, j_n(\xi a)\,\overline{Y_n^m(\hat{\boldsymbol{a}})}\,Y_n^m(\hat{\boldsymbol{r}}),$$

$$\exp(\mathrm{i}\xi\hat{\boldsymbol{r}}\cdot\boldsymbol{c}) = 4\pi\sum_{\nu,\mu}\mathrm{i}^\nu\, j_\nu(\xi c)\,Y_\nu^\mu(\hat{\boldsymbol{c}})\,\overline{Y_\nu^\mu(\hat{\boldsymbol{r}})}.$$

Substituting in (3.114), the inner integral becomes

$$(4\pi)^2\sum_{n,m}\sum_{\nu,\mu}\mathrm{i}^{n+\nu}\, j_n(\xi a)\, j_\nu(\xi c)\,\overline{Y_n^m(\hat{\boldsymbol{a}})}\,Y_\nu^\mu(\hat{\boldsymbol{c}})\,\mathcal{I}, \tag{3.116}$$

where

$$\mathcal{I} = \int_\Omega \exp(\mathrm{i}\xi\hat{\boldsymbol{r}}\cdot\boldsymbol{b})\,Y_n^m(\hat{\boldsymbol{r}})\,\overline{Y_\nu^\mu(\hat{\boldsymbol{r}})}\,\mathrm{d}\Omega(\hat{\boldsymbol{r}}).$$

Now, assume that $\boldsymbol{b} = b\hat{\mathbf{z}}$, whence $\hat{\boldsymbol{r}}\cdot\boldsymbol{b} = b\cos\theta$ and

$$\begin{aligned}\mathcal{I} &= 2\pi\delta_{m\mu}\int_0^\pi \mathrm{e}^{\mathrm{i}\xi b\cos\theta}\,Y_n^m(\hat{\boldsymbol{r}})\,\overline{Y_\nu^m(\hat{\boldsymbol{r}})}\,\sin\theta\,\mathrm{d}\theta\\ &= \frac{1}{2}\,\delta_{m\mu}\int_{-1}^1 \mathrm{e}^{\mathrm{i}\xi bt}\,G_m(t;\,n,\,\nu)\,\mathrm{d}t,\end{aligned} \tag{3.117}$$

where

$$G_m(t;\,n,\,\nu) = A_n^m A_\nu^m\, P_n^m(t)\, P_\nu^m(t) = G_m(t;\,\nu,\,n) \tag{3.118}$$

and A_n^m is a spherical-harmonic normalisation factor; see (3.6). G_m has the following properties:

$$G_m(t;\,n,\,\nu) \text{ is a polynomial of degree } n+\nu; \tag{3.119}$$

$$G_{-m}(t;\,n,\,\nu) = G_m(t;\,n,\,\nu); \tag{3.120}$$

$$G_m(-t;\,n,\,\nu) = (-1)^{n+\nu}G_m(t;\,n,\,\nu); \tag{3.121}$$

$$G_m(t;\,n,\,\nu) \equiv 0 \quad \text{for } |m| > \min(n,\,\nu).$$

The first of these implies that $G_m^{(j)}(t;\,n,\,\nu)\equiv 0$ for $j>n+\nu$, where $G_m^{(j)} = (\mathrm{d}^j/\mathrm{d}t^j)G_m$. Hence, repeated integration by parts gives

$$\int_{-1}^1 \mathrm{e}^{\mathrm{i}Xt}\,G_m(t)\,\mathrm{d}t = \sum_{j=0}^{n+\nu}\left[\frac{\mathrm{e}^{\mathrm{i}Xt}}{\mathrm{i}X}\left(\frac{-1}{\mathrm{i}X}\right)^j G_m^{(j)}(t)\right]_{-1}^1,$$

where we have written $G_m(t)$ for $G_m(t;\,n,\,\nu)$. But, from (3.121),

$$G_m^{(j)}(-t;\,n,\,\nu) = (-1)^{n+\nu+j}G_m^{(j)}(t;\,n,\,\nu),$$

whence

$$\int_{-1}^{1} \mathrm{e}^{\mathrm{i}Xt}\, G_m(t)\,\mathrm{d}t = \sum_{j=0}^{n+\nu} G_m^{(j)}(1)\, \frac{(-1)^j}{(\mathrm{i}X)^{j+1}} \left\{ \mathrm{e}^{\mathrm{i}X} - (-1)^{n+\nu+j}\, \mathrm{e}^{-\mathrm{i}X} \right\}. \tag{3.122}$$

We will evaluate $G_m^{(j)}(1)$ later (Lemma 3.33); it turns out that $G_m^{(j)}(1) = 0$ for $j < |m|$, so that the lower summation limit in (3.122) can be replaced by $j = |m|$.

Having evaluated $\mathcal{J}$, we next consider the outer integral; it is of the form

$$\int_{\mathcal{C}} \frac{\xi^2}{\xi^2 - k^2}\, j_n(\xi a)\, j_\nu(\xi c) \int_{-1}^{1} \mathrm{e}^{\mathrm{i}\xi b t}\, G_m(t; n, \nu)\,\mathrm{d}t\,\mathrm{d}\xi = \mathcal{L},$$

say. From (3.122), we see that the inner integral is $O(\xi^{-n-\nu})$ as $\xi \to 0$. However, $j_n(\xi a)\, j_\nu(\xi c) = O(\xi^{n+\nu})$ as $\xi \to 0$, whence the integrand in the outer integral is analytic at $\xi = 0$. So,

$$\mathcal{L} = \sum_{j=|m|}^{n+\nu} G_m^{(j)}(1)\, \frac{(-1)^j}{(\mathrm{i}b)^{j+1}} \left\{ \mathcal{L}_+ - (-1)^{n+\nu+j}\, \mathcal{L}_- \right\},$$

where

$$\mathcal{L}_\pm = \int_{\mathcal{C}} \frac{\xi^{-j+1}}{\xi^2 - k^2}\, j_n(\xi a)\, j_\nu(\xi c)\, \mathrm{e}^{\pm \mathrm{i}\xi b}\,\mathrm{d}\xi.$$

Assuming that $b > (a + c)$, we can close the contour for $\mathcal{L}_\pm$ as for $\mathcal{J}_\pm$ in the proof of Theorem 3.31, giving

$$\mathcal{L}_+ = \pi \mathrm{i}\, k^{-j}\, j_n(ka)\, j_\nu(kc)\, \mathrm{e}^{\mathrm{i}kb}$$

and $\mathcal{L}_- = (-1)^{n+\nu+j+1} \mathcal{L}_+$, whence

$$\mathcal{L} = 2\pi \mathrm{i} k\, \mathrm{e}^{\mathrm{i}kb}\, j_n(ka)\, j_\nu(kc) \sum_{j=|m|}^{n+\nu} G_m^{(j)}(1)\, \frac{(-1)^j}{(\mathrm{i}kb)^{j+1}}.$$

Combining this formula with (3.114), (3.116) and (3.117), we obtain

$$\frac{\mathrm{e}^{\mathrm{i}kR}}{kR} = 4\pi \mathrm{i}\, \mathrm{e}^{\mathrm{i}kb} \sum_{n,m,\nu} \mathrm{i}^{n+\nu}\, \overline{\hat{\psi}_n^m(\boldsymbol{a})}\, \hat{\psi}_\nu^m(\boldsymbol{c}) \sum_{j-|m|}^{n+\nu} G_m^{(j)}(1)\, \frac{(-1)^j}{(\mathrm{i}kb)^{j+1}}$$

for $b > (a + c)$. Finally, comparison with the two-centre expansion of Theorem 3.30 gives the following result.

Theorem 3.32 (Martin [792]**)**

$$S_{n\nu}^{m\mu}(b\hat{\mathbf{z}}) = \delta_{m\mu}\, \mathrm{e}^{\mathrm{i}kb}\, (-\mathrm{i})^n\, \mathrm{i}^\nu \sum_{j=|m|}^{n+\nu} G_m^{(j)}(1; n, \nu)\, \frac{(-1)^j}{(\mathrm{i}kb)^{j+1}}$$

for all $b > 0$, where $G_m(t; n, \nu)$ is defined by (3.118).

Note that although the argument above gives the result for $b > (a+c)$, $S_{n\nu}^{m\mu}(b\hat{\mathbf{z}})$ itself does not depend on a and c, and so analytic continuation shows that the result must hold for all $b > 0$.

The coefficients $G_m^{(j)}(1; n, \nu)$ are given by the next lemma.

Lemma 3.33 *For* $0 \le |m| \le \min(n, \nu)$ *and* $|m| \le j \le n+\nu$,

$$G_m^{(j)}(1; n, \nu) = (-1)^m \frac{j!}{2^j} \sum_{k=k_0}^{k_1} \mathcal{A}_k(\nu, |m|)\, \mathcal{A}_{j-k}(n, -|m|), \tag{3.123}$$

where $k_0 = \max(0, j-n)$, $k_1 = \min(\nu, j-|m|)$ *and*

$$\mathcal{A}_k(\nu, m) = \sqrt{2\nu+1} \sqrt{\frac{(\nu+m)!}{(\nu-m)!}} \frac{(\nu+k)!}{(m+k)!\, k!\, (\nu-k)!}. \tag{3.124}$$

$G_m^{(j)}(1; n, \nu) = 0$ *for all other values of* j, m, n *and* ν.

Proof From (3.120), it is enough to take $m \ge 0$. From (A.4), we have

$$G_m(t) = A_n^{-m} A_\nu^{-m} (2^m m!)^{-2} (1-t^2)^m F_n F_\nu,$$

where $F_n \equiv F\left(m-n,\, m+n+1;\, m+1;\, \tfrac{1}{2}(1-t)\right)$, F is a hypergeometric function and we have used $[(n+m)!/(n-m)!]\, A_n^m = A_n^{-m}$. (This expression for $G_m(t)$ can be used to show (3.119).) Now, put $z = \tfrac{1}{2}(1-t)$ whence $1-t^2 = 4z(1-z)$. Then, use ([1, eqn (15.3.3)])

$$(1-z)^{\alpha+\beta-\gamma} F(\alpha, \beta; \gamma; z) = F(\gamma-\alpha, \gamma-\beta; \gamma; z)$$

with $\alpha = m-\nu$, $\beta = m+\nu+1$ and $\gamma = m+1$ to give

$$(1-z)^m F_\nu = F(-\nu, \nu+1; m+1; z).$$

Explicitly, we have

$$F(-\nu, \nu+1; m+1; z) = \sum_{k=0}^{\nu} a_k(\nu, m)\, z^k,$$

$$F(m-n, m+n+1; m+1; z) = \sum_{k=0}^{n-m} b_k(n, m)\, z^k,$$

where

$$a_k(\nu, m) = \frac{(-1)^k\, m!\, (\nu+k)!}{(m+k)!\, k!\, (\nu-k)!},$$

$$b_k(n, m) = \frac{(-1)^k\, m!\, (n+m+k)!\, (n-m)!}{(m+k)!\, k!\, (n-m-k)!\, (n+m)!}.$$

Hence, after rearranging the double summation, we find that

$$G_m(t; n, \nu) = z^m \sum_{s=0}^{n+\nu-m} C_s z^s, \tag{3.125}$$

where

$$C_s = A_n^{-m} A_\nu^{-m} \sum_{k=0}^{s} \frac{a_k(\nu, m)}{m!} \frac{b_{s-k}(n, m)}{m!} \quad \text{for } 0 \le s \le n+\nu-m.$$

In fact, as $a_k = 0$ for $k > \nu$ and $b_k = 0$ for $k > n-m$, the summation is actually from $k = \max(0, s-n+m)$ to $k = \min(s, \nu)$. Simplifying further gives

$$C_s = (-1)^s \sum_k \mathcal{A}_k(\nu, m)\, \mathcal{A}_{s+m-k}(n, -m),$$

where $\mathcal{A}_k(\nu, m)$ is defined by (3.124).

Finally, let us compare the expansion (3.125) with the Taylor series for $G_m(t)$ about $t = 1$, which is

$$G_m(t) = \sum_{j=0}^{n+\nu} \frac{(-2)^j}{j!} G_m^{(j)}(1)\, z^j.$$

This comparison shows that $G_m^{(j)}(1) = 0$ for $j < m$, and gives the desired formula (3.123). □

Combining Lemma 3.33 with Theorem 3.32 and (3.113) gives the next result.

Theorem 3.34 (Martin [792]) *For all $kb > 0$,*

$$S_{n\nu}^m(kb) = (-1)^m \mathrm{i}^{n+\nu} \frac{\mathrm{e}^{\mathrm{i}kb}}{\mathrm{i}kb} \sum_{j=|m|}^{n+\nu} j!\, w^j \sum_{s=s_0}^{s_1} \mathcal{A}_s(\nu, |m|)\, \mathcal{A}_{j-s}(n, -|m|), \tag{3.126}$$

where $w = \mathrm{i}/(2kb)$, $s_0 = \max(0, j-n)$, $s_1 = \min(\nu, j-|m|)$ and $\mathcal{A}_s(\nu, m)$ is defined by (3.124).

Before discussing the virtues of the formula (3.126), we give another formula that is used widely by physicists.

Theorem 3.35 (Rehr & Albers [1016], **Fritzsche** [370])

$$S_{n\nu}^m(kb) = (-1)^m \mathrm{i}^{n+\nu} A_n^m A_\nu^{-m} \frac{\mathrm{e}^{\mathrm{i}kb}}{\mathrm{i}kb} \sum_{\ell=0}^{\ell_1} \frac{w^{2\ell+m}}{\ell!\,(\ell+m)!} d_\nu^\ell(w)\, d_n^{\ell+m}(w) \tag{3.127}$$

for $m \ge 0$, where $w = \mathrm{i}/(2kb)$, A_n^m is defined by (3.6), *$\ell_1 = \min(\nu, n-m)$ and*

$$d_n^\ell(w) = \sum_{s=\ell}^{n} \frac{(n+s)!\, w^{s-\ell}}{(n-s)!\,(s-\ell)!}. \tag{3.128}$$

Proof In order to evaluate (3.127), a single sum must be calculated wherein each term consists of the product of two sums. We will show that part of this calculation can be carried out analytically, leaving the double sum that occurs in Theorem 3.34. Thus, explicitly, the sum in (3.127) is

$$\sum_{\ell} \frac{w^{2\ell+m}\, d_\nu^\ell\, d_n^{\ell+m}}{\ell!\,(\ell+m)!} = \sum_{\ell,s,t} \frac{w^{s+t}}{\ell!\,(\ell+m)!} \times \frac{(\nu+s)!}{(\nu-s)!\,(s-\ell)!}\,\frac{(n+t)!}{(n-t)!\,(t-\ell-m)!}, \tag{3.129}$$

where, in accordance with Pauli's 'law of sloppiness' [954, p. 126], we do not worry unduly about the summation limits. Introducing a new summation variable $j = s+t$, the right-hand side becomes

$$\sum_j w^j \sum_s \frac{(\nu+s)!\,(n+j-s)!}{(\nu-s)!\,(n-j+s)!} \sum_\ell \frac{1}{\ell!\,(s-\ell)!\,(\ell+m)!\,(j-s-m-\ell)!},$$

wherein the ℓ-sum can be written in terms of binomial coefficients as

$$\frac{1}{s!\,(j-s)!} \sum_\ell \binom{s}{\ell} \binom{j-s}{j-s-m-\ell} = \frac{1}{s!\,(j-s)!} \binom{j}{j-s-m}.$$

Hence, the left-hand side of (3.129) equals

$$\sum_j w^j\, j! \sum_s \frac{(\nu+s)!}{(m+s)!\,s!\,(\nu-s)!}\,\frac{(n+j-s)!}{(j-s-m)!\,(j-s)!\,(n-j+s)!},$$

and the result follows. □

The formula (3.127) is of the general form

$$S_{n\nu}^{m\mu}(\boldsymbol{b}) = \frac{\mathrm{e}^{\mathrm{i}kb}}{\mathrm{i}kb} \sum_{N,M} \mathcal{S}_{nN}^{mM}(\boldsymbol{b})\, \widetilde{\mathcal{S}}_{\nu N}^{\mu M}(\boldsymbol{b}), \tag{3.130}$$

where the dependence on (n, m) and (ν, μ) is separated. This separation is very convenient from a numerical point of view. Moreover, it is known that many published approximations to $\mathbf{S}$ can be obtained by truncating the series in (3.130); see [370] and [1016]. Numerically, one often retains only a few terms, but this can lead to serious errors, especially if kb is not large [371]. Nevertheless, it has been demonstrated by Sébilleau [1090] that, when properly truncated, (3.130) gives a very efficient and accurate algorithm for computing $\mathbf{S}$; he has also derived recursion relations for calculating the terms in (3.130).

It is interesting to compare the Rehr–Albers–Fritzsche formula (3.127) with the formula (3.126) in Theorem 3.34. Thus, the dependence on n and ν is separated in both, but (3.127) involves the *functions* $d_n^\ell(w)$, whereas (3.126) involves the *coefficients* $\mathcal{A}_s(n, m)$. Formula (3.126) can also be used to obtain *consistent* wide-spacing approximations (large kb). Thus, for the leading-order approximation ($j = 0$), we must take $m = 0$, whence $s_0 = s_1 = 0$ giving

$$S_{n\nu}^m(kb) \sim \delta_{m0}\, \frac{\mathrm{e}^{\mathrm{i}kb}}{\mathrm{i}kb}\, \mathrm{i}^{n+\nu} \sqrt{2n+1}\sqrt{2\nu+1} \quad \text{as } kb \to \infty, \tag{3.131}$$

which is known as the *plane-wave approximation* [699]; it is a special case of (3.101). If we include terms of $O((kb)^{-2})$, we obtain

$$S_{n\nu}^{m}(kb) \sim \delta_{m0}\, S_{n\nu}^{\mathrm{PWA}}\, \{1 + w[n(n+1) + \nu(\nu+1)]\}$$
$$- w\delta_{|m|1}\, S_{n\nu}^{\mathrm{PWA}} \sqrt{n(n+1)}\sqrt{\nu(\nu+1)} \quad \text{as } kb \to \infty,$$

where $\delta_{m0}\, S_{n\nu}^{\mathrm{PWA}}$ is the right-hand side of (3.131) and $w = \mathrm{i}/(2kb)$.

3.17 Vector spherical wavefunctions

Vector spherical harmonics were introduced by Hansen in 1935 [476]. We define them as follows.

Definition 3.36 *For $m = 0, \pm 1, \pm 2, \ldots, n$ and $n = 1, 2, \ldots,$*

$$\mathbf{P}_n^m(\hat{\boldsymbol{r}}) = \hat{\boldsymbol{r}}\, Y_n^m(\hat{\boldsymbol{r}}),$$

$$\mathbf{C}_n^m(\hat{\boldsymbol{r}}) = -\eta_n \hat{\boldsymbol{r}} \times \boldsymbol{D}_a Y_n^m(\hat{\boldsymbol{r}}) = \eta_n \left(\frac{\hat{\boldsymbol{\theta}}}{\sin\theta} \frac{\partial}{\partial\phi} - \hat{\boldsymbol{\phi}} \frac{\partial}{\partial\theta} \right) Y_n^m(\theta, \phi),$$

$$\mathbf{B}_n^m(\hat{\boldsymbol{r}}) = \eta_n \boldsymbol{D}_a Y_n^m(\hat{\boldsymbol{r}}) = \eta_n \left(\hat{\boldsymbol{\theta}} \frac{\partial}{\partial\theta} + \frac{\hat{\boldsymbol{\phi}}}{\sin\theta} \frac{\partial}{\partial\phi} \right) Y_n^m(\theta, \phi),$$

where $\eta_n = [n(n+1)]^{-1/2}$, $\boldsymbol{D}_a$ is the angular differential operator

$$\boldsymbol{D}_a = \hat{\boldsymbol{\theta}} \frac{\partial}{\partial\theta} + \frac{\hat{\boldsymbol{\phi}}}{\sin\theta} \frac{\partial}{\partial\phi},$$

and $\hat{\boldsymbol{r}}$, $\hat{\boldsymbol{\theta}}$ and $\hat{\boldsymbol{\phi}}$ are spherical polar unit vectors. When $n = 0$, we have $\mathbf{C}_0^0 = \mathbf{B}_0^0 = \mathbf{0}$ and $\mathbf{P}_0^0(\hat{\boldsymbol{r}}) = (4\pi)^{-1/2}\hat{\boldsymbol{r}}$.

Our definitions are very similar to those used by Dassios & Rigou [263], except that we use normalised spherical harmonics Y_n^m. See also [250, §B.12]. It is evident that vector spherical harmonics are pointwise perpendicular,

$$\mathbf{P}_n^m \cdot \mathbf{C}_n^m = \mathbf{C}_n^m \cdot \mathbf{B}_n^m = \mathbf{B}_n^m \cdot \mathbf{P}_n^m = 0. \tag{3.132}$$

They are also orthonormal in $L^2(\Omega)$:

$$\int_\Omega \mathbf{P}_n^m \cdot \overline{\mathbf{P}_\nu^\mu}\, \mathrm{d}\Omega = \int_\Omega \mathbf{C}_n^m \cdot \overline{\mathbf{C}_\nu^\mu}\, \mathrm{d}\Omega = \int_\Omega \mathbf{B}_n^m \cdot \overline{\mathbf{B}_\nu^\mu}\, \mathrm{d}\Omega = \delta_{n\nu}\delta_{m\mu}. \tag{3.133}$$

These relations can be proved using (3.9) and (3.18).

Next, we define (dimensionless) *vector spherical wavefunctions*.

Definition 3.37 *For $m = 0, \pm 1, \pm 2, \ldots, n$ and $n = 0, 1, 2, \ldots,$*

$$\begin{aligned}
\mathbf{L}_n^m(\boldsymbol{r}) &= k^{-1}\text{grad}\,\{z_n(kr)\,Y_n^m(\hat{\boldsymbol{r}})\},\\
\mathbf{M}_n^m(\boldsymbol{r}) &= \text{curl}\,\{\boldsymbol{r}\,z_n(Kr)\,Y_n^m(\hat{\boldsymbol{r}})\},\\
\mathbf{N}_n^m(\boldsymbol{r}) &= K^{-1}\text{curl}\,\{\mathbf{M}_n^m(\boldsymbol{r})\},
\end{aligned}$$

where z_n is any spherical Bessel function. In terms of vector spherical harmonics, we have

$$\begin{aligned}
\mathbf{L}_n^m(\boldsymbol{r}) &= z_n'(kr)\,\mathbf{P}_n^m(\hat{\boldsymbol{r}}) + (\eta_n kr)^{-1}\,z_n(kr)\,\mathbf{B}_n^m(\hat{\boldsymbol{r}}),\\
\mathbf{M}_n^m(\boldsymbol{r}) &= \eta_n^{-1}\,z_n(Kr)\,\mathbf{C}_n^m(\hat{\boldsymbol{r}}),\\
\mathbf{N}_n^m(\boldsymbol{r}) &= (\eta_n^2 Kr)^{-1}\,z_n(Kr)\,\mathbf{P}_n^m(\hat{\boldsymbol{r}})\\
&\qquad + \eta_n^{-1}\left\{(Kr)^{-1}\,z_n(Kr) + z_n'(Kr)\right\}\mathbf{B}_n^m(\hat{\boldsymbol{r}})
\end{aligned}$$

where $\eta_n = [n(n+1)]^{-1/2}$.

We define

$$\mathbf{L}_{nm}^{(1)} \equiv \mathbf{L}_n^m \text{ with } z_n = j_n, \text{ and } \mathbf{L}_{nm}^{(3)} \equiv \mathbf{L}_n^m \text{ with } z_n = h_n^{(1)} \equiv h_n,$$

with similar definitions for $\mathbf{M}_{nm}^{(1)}$, $\mathbf{M}_{nm}^{(3)}$, $\mathbf{N}_{nm}^{(1)}$ and $\mathbf{N}_{nm}^{(3)}$.

The functions $\mathbf{L}_{nm}^{(3)}$, $\mathbf{M}_{nm}^{(3)}$ and $\mathbf{N}_{nm}^{(3)}$ are used to represent outgoing elastic waves, such as those scattered by a spherical obstacle. The functions $\mathbf{L}_{nm}^{(1)}$, $\mathbf{M}_{nm}^{(1)}$ and $\mathbf{N}_{nm}^{(1)}$ are used to represent regular elastic waves, such as those inside a spherical inclusion. Similarly, the functions $\mathbf{M}_{nm}^{(3)}$ and $\mathbf{N}_{nm}^{(3)}$ (with $K = k = \omega\sqrt{\varepsilon\mu}$) are used to represent outgoing electromagnetic waves.

Properties of vector spherical wavefunctions can be found in [1149, §§7.11–7.14], [884, pp. 1865–1866], [564, §8.17], [549, §16.2], [83, §2.3], [133, §2.2] and [263]. In particular, vector analogues of the Funk–Hecke formula (Theorem 3.12) have been given in [263].

Addition theorems for vector spherical wavefunctions were first given by Stein [1143] and by Cruzan [247]. For alternative proofs and computational aspects, see [254, 120, 341, 205, 606, 1347, 1348] and [83, §J.3].

3.18 Multipoles for water waves

Consider three-dimensional water-wave problems in deep water. Introduce spherical polar coordinates (r, θ, ϕ), defined by (3.1), so that the free surface is part of $z = 0$ $(\theta = \frac{1}{2}\pi)$. We want to construct multipole solutions; these must satisfy Laplace's equation in the water ($z > 0$) and the free-surface condition on $z = 0$. These are readily found, using the method described in Section 2.11.

Thus, to construct wavefree potentials, we begin by choosing a harmonic function χ that vanishes on $z = 0$; for example, we may take

$$\chi(r, \theta, \phi) = r^{-(n+1)} \tilde{Y}_n^m(\theta, \phi), \quad 0 \le |m| \le n,$$

where $\tilde{Y}_n^m$ is an unnormalised spherical harmonic (Definition 3.2). From (A.7), we have

$$\chi(r, \frac{1}{2}\pi, \phi) = 0 \text{ provided that } n + m \text{ is odd.}$$

Then, $(K - \partial/\partial z)\chi$ is a wavefree potential. In order to calculate $\partial\chi/\partial z$, we can use the following corollary to Theorems 3.13 and 3.14.

Corollary 3.38 *For $0 \le |m| \le n$ and $n = 0, 1, 2, \dots$,*

$$\frac{\partial}{\partial z} \frac{\tilde{Y}_n^m}{r^{n+1}} = -(n - m + 1) \frac{\tilde{Y}_{n+1}^m}{r^{n+2}}. \tag{3.134}$$

For $0 \le |m| \le n$ and $n = 1, 2, \dots$,

$$\frac{\partial}{\partial z} \tilde{\mathcal{H}}_n^m = (n + m) \tilde{\mathcal{H}}_{n-1}^m, \tag{3.135}$$

$$\left(\frac{\partial}{\partial x} + \mathrm{i}\frac{\partial}{\partial y}\right) \tilde{\mathcal{H}}_n^m = -\tilde{\mathcal{H}}_{n-1}^{m+1}, \tag{3.136}$$

$$\left(\frac{\partial}{\partial x} - \mathrm{i}\frac{\partial}{\partial y}\right) \tilde{\mathcal{H}}_n^m = (n + m)(n + m - 1) \tilde{\mathcal{H}}_{n-1}^{m-1}, \tag{3.137}$$

where $\tilde{\mathcal{H}}_n^m(x, y, z) = r^n \tilde{Y}_n^m(\theta, \phi)$.

Proof Let $C_n = 1 \cdot 3 \cdot 5 \cdots (2n - 1)$ with $C_0 = 1$. Then [1, §10.1.2]

$$w^{n+1} h_n^{(1)}(w) \sim -\mathrm{i} C_n \quad \text{and} \quad w^{-n} j_n(w) \sim 1/C_{n+1} \quad \text{as} w \to 0.$$

Choose $z_n = h_n^{(1)}$ in (3.46) and then multiply by k^{n+2}. The result (3.134) follows by letting $k \to 0$. Similarly, we obtain (3.135), (3.136) and (3.137) by choosing $z_n = j_n$ in (3.46), (3.50) and (3.51), respectively, multiplying by k^{1-n}, and then letting $k \to 0$. The formula (3.136) can also be obtained by differentiating (3.26). □

Our specific definitions of wavefree potentials are given next; note that it turns out to be inconvenient to use Y_n^m or $\tilde{Y}_n^m$, because we shall need certain integral representations that only hold for $m \ge 0$.

Definition 3.39 *Let* $\boldsymbol{r} = (r\sin\theta\cos\phi,\ r\sin\theta\sin\phi,\ r\cos\theta)$. *The spherical wavefree potential* Φ_j^m *is defined by*

$$\Phi_j^{\pm m}(\boldsymbol{r}) = \frac{(2j-1)!}{K^{m+2j+1}}\left(K - \frac{\partial}{\partial z}\right)\frac{P_{m+2j-1}^m(\cos\theta)}{r^{m+2j}}\,\mathrm{e}^{\pm \mathrm{i}m\phi} \tag{3.138}$$

$$= \left\{\frac{(2j-1)!}{(Kr)^{m+2j}}P_{m+2j-1}^m(\cos\theta) + \frac{(2j)!}{(Kr)^{m+2j+1}}P_{m+2j}^m(\cos\theta)\right\}\mathrm{e}^{\pm \mathrm{i}m\phi},$$

for $m = 0, 1, 2, \ldots$ *and* $j = 1, 2, \ldots$. *The analogous regular potential* $\widehat{\Phi}_j^m$ *is defined by*

$$\widehat{\Phi}_j^{\pm m}(\boldsymbol{r}) = \frac{K^{m+2j}}{(2m+2j+1)!}\left(K - \frac{\partial}{\partial z}\right)r^{m+2j+1}\,P_{m+2j+1}^m(\cos\theta)\,\mathrm{e}^{\pm \mathrm{i}m\phi}$$

$$= \left\{\frac{(Kr)^{m+2j}}{(2m+2j)!}P_{m+2j}^m(\cos\theta) - \frac{(Kr)^{m+2j+1}}{(2m+2j+1)!}P_{m+2j+1}^m(\cos\theta)\right\}\mathrm{e}^{\pm \mathrm{i}m\phi}, \tag{3.139}$$

for $m = 0, 1, 2, \ldots$ *and* $j = 0, 1, 2, \ldots$.

In addition to wavefree potentials, we also need potentials that generate waves. The simplest of these is an axisymmetric source at the origin; it is given by

$$\Phi_0^0(\boldsymbol{r}) = \frac{1}{K}⨍_0^\infty \mathrm{e}^{-kz}\,J_0(k\sigma)\,\frac{k\,\mathrm{d}k}{k-K} \tag{3.140}$$

$$\sim \pi\mathrm{i}\,\mathrm{e}^{-Kz}\,H_0^{(1)}(K\sigma) \quad \text{as } \sigma = \sqrt{x^2+y^2} \to \infty.$$

To verify directly that Φ_0^0 does satisfy the free-surface condion, we calculate

$$\left(K + \frac{\partial}{\partial z}\right)\Phi_0^0 = -\frac{1}{K}\int_0^\infty k\,\mathrm{e}^{-kz}\,J_0(k\sigma)\,\mathrm{d}k = \frac{1}{K}\frac{\partial}{\partial z}\frac{1}{r} = -\frac{z}{Kr^3},$$

and this vanishes on $z = 0$ for $\sigma > 0$; here, we have used

$$\frac{1}{r} = \frac{1}{\sqrt{x^2+y^2+z^2}} = \int_0^\infty \mathrm{e}^{-kz}\,J_0(k\sigma)\,\mathrm{d}k, \quad \text{for } z > 0, \tag{3.141}$$

with $\sigma = \sqrt{x^2+y^2}$, which is a well-known formula due to Lipschitz (see [31, eqn (4.11.13)] or [427, 6.623(1)]).

Multipole potentials with higher-order singularities at O can be constructed by horizontal differentiation of Φ_0^0. Thus, using Theorem 2.7, we have, for $m = 0, 1, 2, \ldots$,

$$\left(\frac{\partial}{\partial x} \pm \mathrm{i}\frac{\partial}{\partial y}\right)^m J_0(k\sigma) = (-k)^m\,J_m(k\sigma)\,\mathrm{e}^{\pm \mathrm{i}m\phi}, \tag{3.142}$$

where $x=\sigma\cos\phi$ and $y=\sigma\sin\phi$. This suggests defining

$$\Phi_0^{\pm m}(\boldsymbol{r})=\frac{\mathrm{e}^{\pm im\phi}}{K^{m+1}}\,⨍_0^\infty k^{m+1}\,\mathrm{e}^{-kz}\,J_m(k\sigma)\,\frac{\mathrm{d}k}{k-K} \tag{3.143}$$

for $m=0,1,2,\ldots$; in the far field,

$$\Phi_0^{\pm m}(\boldsymbol{r})\sim\pi\mathrm{i}\,\mathrm{e}^{-Kz}\,H_m^{(1)}(K\sigma)\,\mathrm{e}^{\pm im\phi}\quad\text{as }\sigma\to\infty. \tag{3.144}$$

The potentials $\{\Phi_j^m\}$ with $j\geq 0$ and all m are known as *multipole potentials*. They are harmonic, satisfy the free-surface and radiation conditions, and are singular at the origin. Axisymmetric wavefree potentials (Φ_j^0) were found by Havelock [488]. Multipole potentials can also be defined for water of constant finite depth, and with singularities at a submerged point, $(x,y,z)=(\xi,\eta,\zeta)$, instead of at $(x,y,z)=(0,0,0)$. In particular, the wave source at (ξ,η,ζ) is recognised as the fundamental solution $G(x,y,z;\xi,\eta,\zeta)$; see Section 3.18.1. For more information on three-dimensional multipole potentials, see [1176] and [731, Appendix B]. For multipole potentials in prolate spheroidal coordinates (with horizontal axis), see [907].

In order to obtain integral representations for the wavefree potentials, Φ_j^m with $j\geq 1$, we shall use the following result.

Theorem 3.40 *For* $0\leq|m|\leq n$,

$$\frac{P_n^m(\cos\theta)}{r^{n+1}}=\frac{1}{(n-m)!}\int_0^\infty k^n\,\mathrm{e}^{-kz}\,J_m(k\sigma)\,\mathrm{d}k, \tag{3.145}$$

where $z=r\cos\theta$, $\sigma=r\sin\theta$ *and* $z>0$.

Proof For $m=n=0$, (3.145) reduces to (3.141). For $0\leq m\leq n$ and $n>0$, the result (3.145) follows by combining (3.29), (3.141) and (3.142). For $-n\leq m<0$, use (3.5) and $J_{-m}(w)=(-1)^m J_m(w)$. □

Theorem 3.40 is useful because it relates harmonic functions in two coordinate systems, namely spherical and cylindrical polar coordinates. The formula (3.145) itself is discussed in [1298, §13.21] and in [321, §7.8]. There are analogous formulae for wavefunctions; see [133] for details.

Combining (3.138) and (3.145) gives

$$\Phi_j^{\pm m}(\boldsymbol{r})=\frac{\mathrm{e}^{\pm im\phi}}{K^{m+2j+1}}\int_0^\infty (k+K)k^{m+2j-1}\,\mathrm{e}^{-kz}\,J_m(k\sigma)\,\mathrm{d}k, \tag{3.146}$$

for $m=0,1,2,\ldots$ and $j=1,2,\ldots$.

We shall need to calculate various derivatives of Φ_j^m and $\widehat{\Phi}_j^m$, in both vertical and horizontal directions. Formulae for these derivatives of the multipoles Φ_j^m are collected in the next theorem.

Theorem 3.41 *For vertical differentiation, we have*

$$(\partial^2/\partial z^2)\Phi_0^{\pm m} = K^2(\Phi_0^{\pm m} + \Phi_1^{\pm m}), \tag{3.147}$$

$$(\partial^2/\partial z^2)\Phi_j^{\pm m} = K^2\Phi_{j+1}^{\pm m}, \quad j = 1, 2, 3, \ldots. \tag{3.148}$$

For horizontal differentiation, let

$$\tilde{\mathcal{D}}_m^{\pm} = \left(\frac{k}{K}\right)^m \mathcal{D}_m^{\pm} = \left[\frac{-1}{K}\left(\frac{\partial}{\partial x} \pm \mathrm{i}\frac{\partial}{\partial y}\right)\right]^m, \quad m = 0, 1, 2, \ldots; \tag{3.149}$$

see Definition 2.6 for the differential operators $\mathcal{D}_m^{\pm}$. Then

$$\tilde{\mathcal{D}}_m^{+}\Phi_j^{\mu} = \Phi_j^{m+\mu}, \quad \tilde{\mathcal{D}}_m^{-}\Phi_j^{-\mu} = \Phi_j^{-m-\mu}, \quad m \geq 0, \;\mu \geq 0, \; j \geq 0; \tag{3.150}$$

$$\tilde{\mathcal{D}}_m^{+}\Phi_0^{-\mu} = \begin{cases} (-1)^{\mu}\sum_{j=0}^{\mu}\Phi_j^{m-\mu}, & 0 \leq \mu \leq m, \\ (-1)^{m}\sum_{j=0}^{m}\Phi_j^{m-\mu}, & 0 \leq m \leq \mu; \end{cases} \tag{3.151}$$

$$\tilde{\mathcal{D}}_m^{-}\Phi_0^{\mu} = \begin{cases} (-1)^{\mu}\sum_{j=0}^{\mu}\Phi_j^{\mu-m}, & 0 \leq \mu \leq m, \\ (-1)^{m}\sum_{j=0}^{m}\Phi_j^{\mu-m}, & 0 \leq m \leq \mu; \end{cases} \tag{3.152}$$

$$\tilde{\mathcal{D}}_m^{+}\Phi_j^{-\mu} = \begin{cases} (-1)^{\mu}\Phi_{\mu+j}^{m-\mu}, & 0 \leq \mu \leq m, \\ (-1)^{m}\Phi_{m+j}^{m-\mu}, & 0 \leq m \leq \mu, \end{cases} \quad j \geq 1; \tag{3.153}$$

$$\tilde{\mathcal{D}}_m^{-}\Phi_j^{\mu} = \begin{cases} (-1)^{\mu}\Phi_{\mu+j}^{\mu-m}, & 0 \leq \mu \leq m, \\ (-1)^{m}\Phi_{m+j}^{\mu-m}, & 0 \leq m \leq \mu, \end{cases} \quad j \geq 1. \tag{3.154}$$

Proof The properties (3.147) and (3.148) are easily proved using the integral representations (3.143) and (3.146).

Theorem 2.7, (3.140) and (3.143) immediately give

$$\tilde{\mathcal{D}}_m^{\pm}\Phi_0^{0} = \Phi_0^{\pm m}, \quad m = 0, 1, 2, \ldots. \tag{3.155}$$

Similarly, Theorem 2.9 gives

$$\tilde{\mathcal{D}}_m^{+}\Phi_0^{\mu} = \Phi_0^{m+\mu} \quad \text{and} \quad \tilde{\mathcal{D}}_m^{-}\Phi_0^{-\mu} = \Phi_0^{-m-\mu} \tag{3.156}$$

for $m \geq 0$ and $\mu \geq 0$, which are (3.150) with $j = 0$.

Next, consider $\tilde{\mathcal{D}}_m^{+}\Phi_0^{-\mu} = \tilde{\mathcal{D}}_m^{+}\tilde{\mathcal{D}}_\mu^{-}\Phi_0^{0} = \tilde{\mathcal{D}}_{m-\mu}^{+}\tilde{\mathcal{D}}_\mu^{+}\tilde{\mathcal{D}}_\mu^{-}\Phi_0^{0}$, using (3.155) and assuming for the moment that $1 \leq \mu \leq m$. But (cf. Lemma 2.8)

$$\tilde{\mathcal{D}}_\mu^{+}\tilde{\mathcal{D}}_\mu^{-}\{J_0(k\sigma)\} = (-k^2/K^2)^{\mu}J_0(k\sigma),$$

whence Theorem 2.7 and (3.140) give

$$\tilde{\mathcal{D}}_m^{+}\Phi_0^{-\mu} = (-1)^{\mu}\frac{\mathrm{e}^{\mathrm{i}(m-\mu)\phi}}{K^{m+\mu+1}}⨍_0^{\infty} k^{m-\mu+1}\,\mathrm{e}^{-kz}\,J_{m-\mu}(k\sigma)\,\Lambda_\mu(k)\,\mathrm{d}k,$$

where

$$\Lambda_\mu(k) = \frac{k^{2\mu}}{k-K} = \frac{K^{2\mu}}{k-K} + \frac{k^{2\mu}-K^{2\mu}}{k-K}$$
$$= \frac{K^{2\mu}}{k-K} + K^{2\mu}(k+K)\sum_{j=1}^{\mu} K^{-2j}k^{2j-2},$$

and we have used the identities (2.94) and (2.95). Hence, the integral representations (3.143) and (3.146) give the first of (3.151); the second of (3.151) and both of (3.152) are proved similarly.

Moreover, by differentiating (3.151), (3.152) and (3.156) twice with respect to z, we obtain formulae for the action of $\tilde{\mathcal{D}}_m^{\pm}$ on $\Phi_j^{\pm\mu}$. For example, if we apply $\partial^2/\partial z^2$ to the first of (3.156), using (3.147), we obtain

$$\tilde{\mathcal{D}}_m^{+}\{K^2(\Phi_0^{\mu}+\Phi_1^{\mu})\} = K^2(\Phi_0^{m+\mu}+\Phi_1^{m+\mu}).$$

Hence, using $(3.156)_1$ again, we obtain $\tilde{\mathcal{D}}_m^{+}\Phi_1^{\mu} = \Phi_1^{m+\mu}$. Further applications of $\partial^2/\partial z^2$, using (3.148), give the first of (3.150). The second of (3.150), as well as (3.153) and (3.154), are all obtained similarly. □

There are analogous formulae for the regular potentials $\widehat{\Phi}_j^m$; they are collected in the next theorem.

Theorem 3.42 *For vertical differentiation, we have*

$$(\partial^2/\partial z^2)\widehat{\Phi}_j^{\pm m} = K^2\widehat{\Phi}_{j-1}^{\pm m}, \quad j=1,2,3,\ldots. \tag{3.157}$$

For horizontal differentiation, we have

$$\tilde{\mathcal{D}}_m^{+}\widehat{\Phi}_j^{\mu} = \begin{cases} \widehat{\Phi}_{j-m}^{m+\mu}, & 0\le m\le j, \\ 0, & m>j, \end{cases} \quad \mu\ge 0; \tag{3.158}$$

$$\tilde{\mathcal{D}}_m^{-}\widehat{\Phi}_j^{-\mu} = \begin{cases} \widehat{\Phi}_{j-m}^{-m-\mu}, & 0\le m\le j, \\ 0, & m>j, \end{cases} \quad \mu\ge 0; \tag{3.159}$$

$$\mathcal{D}_m^{+}\widehat{\Phi}_j^{-\mu} = \begin{cases} (-1)^m\,\widehat{\Phi}_j^{m-\mu}, & 0\le m\le\mu, \quad j\ge 0, \\ (-1)^{\mu}\,\widehat{\Phi}_{j-m+\mu}^{m-\mu}, & 0<m-\mu\le j, \\ 0, & 0\le j<m-\mu; \end{cases} \tag{3.160}$$

$$\tilde{\mathcal{D}}_m^{-}\widehat{\Phi}_j^{\mu} = \begin{cases} (-1)^m\,\widehat{\Phi}_j^{\mu-m}, & 0\le m\le\mu, \quad j\ge 0, \\ (-1)^{\mu}\,\widehat{\Phi}_{j-m+\mu}^{\mu-m}, & 0<m-\mu\le j, \\ 0, & 0\le j<m-\mu; \end{cases} \tag{3.161}$$

where $\tilde{\mathcal{D}}_m^{\pm}$ is defined by (3.149).

Proof The property (3.157) follows by using (3.135). From (3.136) and (3.139), we easily obtain $\tilde{\mathscr{D}}_1^+\widehat{\Phi}_j^\mu = \widehat{\Phi}_{j-1}^{\mu+1}$; repeated application of $\tilde{\mathscr{D}}_1^+$ then yields the first of (3.158). In particular,

$$\tilde{\mathscr{D}}_\mu^+\widehat{\Phi}_j^\mu = \widehat{\Phi}_0^{2\mu} \quad \text{for } j \geq 0 \text{ and } \mu \geq 0.$$

But, from (3.139) and (3.26), we have

$$\widehat{\Phi}_0^m = (\tfrac{1}{2}K)^m(1-Kz)(x+\mathrm{i}y)^m/m!,$$

whence $\tilde{\mathscr{D}}_1^+\widehat{\Phi}_0^m = 0$; this proves the second of (3.158). The formulae (3.159) are obtained by complex conjugation of (3.158).

Next, consider $\tilde{\mathscr{D}}_m^-\widehat{\Phi}_j^\mu$. Using (3.137) and (3.139), we obtain $\tilde{\mathscr{D}}_1^-\widehat{\Phi}_j^\mu = -\widehat{\Phi}_j^{\mu-1}$; repeated application of $\tilde{\mathscr{D}}_1^-$ then yields the first of (3.161). In particular, $\tilde{\mathscr{D}}_\mu^-\widehat{\Phi}_j^\mu = (-1)^\mu\widehat{\Phi}_j^0$ for $j \geq 0$ and $\mu \geq 0$, where, from (3.139),

$$\widehat{\Phi}_j^0 = \frac{K^{2j}}{(2j)!}\tilde{\mathscr{H}}_{2j}^0 - \frac{K^{2j+1}}{(2j+1)!}\tilde{\mathscr{H}}_{2j+1}^0.$$

Repeated use of (3.137) gives

$$\tilde{\mathscr{D}}_N^-\widehat{\Phi}_j^0 = (-1)^N\left\{\frac{(Kr)^{2j-N}}{(2j-2N)!}\tilde{Y}_{2j-N}^{-N} - \frac{(Kr)^{2j-N+1}}{(2j-2N+1)!}\tilde{Y}_{2j-N+1}^{-N}\right\}.$$

But, using (3.5), we have

$$\tilde{Y}_n^{-N}(\theta,\phi) = (-1)^N\frac{(n-N)!}{(n+N)!}P_n^N(\cos\theta)\,\mathrm{e}^{-\mathrm{i}N\phi},$$

whence $\tilde{\mathscr{D}}_N^-\widehat{\Phi}_j^0 = \widehat{\Phi}_{j-N}^{-N}$ for $0 \leq N \leq j$. In particular, $\tilde{\mathscr{D}}_j^-\widehat{\Phi}_j^0 = \widehat{\Phi}_0^{-j}$, whence the second of (3.159) shows that $\tilde{\mathscr{D}}_N^-\widehat{\Phi}_j^0 = 0$ for $N > j \geq 0$. So, if $m > \mu \geq 0$, we have

$$\tilde{\mathscr{D}}_m^-\widehat{\Phi}_j^\mu = \tilde{\mathscr{D}}_{m-\mu}^-\tilde{\mathscr{D}}_\mu^-\widehat{\Phi}_j^\mu = (-1)^\mu\tilde{\mathscr{D}}_{m-\mu}^-\widehat{\Phi}_j^0,$$

and the second and third of (3.161) follow. The formulae (3.160) follow by complex conjugation. □

3.18.1 A fundamental solution for deep water

The function Φ_0^0 corresponds to a wave source at the origin. More generally, we can place the source at an arbitrary submerged point; the corresponding potential is a fundamental solution, denoted by G. For deep water, we have

$$G(P,Q) \equiv G(x,y,z;\xi,\eta,\zeta) = -\frac{1}{2\pi}\left(\frac{1}{R}+\frac{1}{R'}\right) - \frac{K}{\pi}⨍_0^\infty \mathrm{e}^{-k(z+\zeta)}\,J_0(k\rho)\,\frac{\mathrm{d}k}{k-K} \tag{3.162}$$

$$= -\frac{1}{2\pi R} - \frac{1}{2\pi} \fint_0^\infty \mathrm{e}^{-k(z+\zeta)} J_0(k\rho) \frac{k+K}{k-K} \, \mathrm{d}k \tag{3.163}$$

$$= -\frac{1}{2\pi} \fint_0^\infty \left\{ \mathrm{e}^{-k|z-\zeta|} + \frac{k+K}{k-K} \mathrm{e}^{-k(z+\zeta)} \right\} J_0(k\rho) \, \mathrm{d}k, \tag{3.164}$$

where $R = |\boldsymbol{r}_P - \boldsymbol{r}_Q| = \{\rho^2 + (z-\zeta)^2\}^{1/2}$, $R' = \{\rho^2 + (z+\zeta)^2\}^{1/2}$ and $\rho^2 = (x-\xi)^2 + (y-\eta)^2$. Equations (3.163) and (3.164) follow from (3.162) using (3.141).

Equation (3.163) shows that

$$G \sim -(2\pi R)^{-1} \quad \text{as } R \to 0. \tag{3.165}$$

In the far field, we have

$$G(x, y, z; \xi, \eta, \zeta) \sim -\mathrm{i}\, K \,\mathrm{e}^{-K(z+\zeta)} H_0^{(1)}(K\rho) \quad \text{as } \rho \to \infty. \tag{3.166}$$

Notice that $G(P, O) = -(K/\pi)\, \Phi_0^0(\boldsymbol{r})$, where Φ_0^0 is defined by (3.140).

The construction of G can be found in [1303, pp. 475–478], together with the analogous formula for constant finite depth. See also the Appendix to John's famous paper [559] and [675, §1.1].

Theorem 3.43 (Martin [780]**)** *Suppose that $R = |\boldsymbol{r}_P - \boldsymbol{r}_Q|$ with $|\boldsymbol{r}_P| < |\boldsymbol{r}_Q|$. Then the three-dimensional deep-water fundamental solution $G(P, Q)$, defined by* (3.162)*, has the expansion*

$$G(\boldsymbol{r}_P, \boldsymbol{r}_Q) = \frac{K}{\pi} \sum_{n=0}^{\infty} \sum_{m=-\infty}^{\infty} (-1)^{m+1} \alpha_n^{-m}(\boldsymbol{r}_P)\, \Phi_n^m(\boldsymbol{r}_Q), \tag{3.167}$$

where

$$\alpha_n^{\pm m}(\boldsymbol{r}) = \sum_{l=0}^{\infty} \widehat{\Phi}_{n+l}^{\pm m}(\boldsymbol{r}), \quad m = 0, 1, 2, \ldots. \tag{3.168}$$

Proof Fix the point $P = (x, y, z)$. We choose $\alpha_0^m(x, y, z)$ so that

$$G(x, y, z; \xi, \eta, \zeta) + \frac{K}{\pi} \sum_{m=-\infty}^{\infty} (-1)^m \alpha_0^{-m}(x, y, z)\, \Phi_0^m(\xi, \eta, \zeta) \equiv H, \tag{3.169}$$

say, is wavefree as $\varpi = \sqrt{\xi^2 + \eta^2} \to \infty$. From (2.30), we have

$$H_0^{(1)}(K\rho) = \sum_{m=-\infty}^{\infty} (-1)^m H_m^{(1)}(K\varpi)\, J_m(K\sigma)\, \mathrm{e}^{\mathrm{i}m(\varphi-\phi)} \quad \text{for } \varpi > \sigma,$$

where $\xi = \varpi \cos\varphi$ and $\eta = \varpi \sin\varphi$. Hence, making use of (3.166) and (3.144), we obtain

$$\alpha_0^{\pm m}(x, y, z) = \mathrm{e}^{-Kz} J_m(K\sigma)\, \mathrm{e}^{\pm \mathrm{i}m\phi}, \quad m = 0, 1, 2, \ldots. \tag{3.170}$$

As H is wavefree, harmonic and satisfies the free-surface condition, it must have an expansion in terms of wavefree potentials. To find this expansion, we first use (3.143) and (3.170) to obtain

$$\sum_{m=-\infty}^{\infty} (-1)^m \alpha_0^{-m}\, \Phi_0^m = \mathrm{e}^{-Kz} \sum_{m=0}^{\infty} \epsilon_m J_m(K\sigma)\cos m(\phi-\varphi)$$
$$\times \frac{(-1)^m}{K^{m+1}} ⨍_0^{\infty} \mathrm{e}^{-k\zeta} J_m(k\varpi)\, \frac{k^{m+1}}{k-K}\, \mathrm{d}k,$$

where $\epsilon_0 = 1$ and $\epsilon_m = 2$ for $m > 0$. Next, assuming for now that $z < \zeta$, (3.164) and (2.15) give

$$G = -\frac{1}{\pi} \sum_{m=0}^{\infty} \epsilon_m\, (-1)^m \cos m(\phi-\varphi)$$
$$\times ⨍_0^{\infty} \left(\cosh kz + \frac{K\,\mathrm{e}^{-kz}}{k-K}\right) \mathrm{e}^{-k\zeta}\, J_m(k\sigma)\, J_m(k\varpi)\, \mathrm{d}k.$$

Hence,

$$H = -\frac{1}{\pi} \sum_{m=0}^{\infty} \epsilon_m\, (-1)^m \cos m(\phi-\varphi) ⨍_0^{\infty} \mathrm{e}^{-k\zeta}\, J_m(k\varpi)\, G_m\, \mathrm{d}k, \tag{3.171}$$

where

$$G_m = J_m(k\sigma)\cosh kz + \frac{1}{k-K}\left\{K\mathrm{e}^{-kz} J_m(k\sigma) - \frac{k^{m+1}}{K^m}\mathrm{e}^{-Kz} J_m(K\sigma)\right\}$$
$$= \sum_{q=0}^{\infty} Q_q^m\, \frac{P_{m+q}^m(\cos\theta)}{(2m+q)!}, \tag{3.172}$$

where

$$Q_q^m = \frac{1}{2}[1+(-1)^q](kr)^{m+q} + (-r)^q (kr)^m\, \frac{Kk^q - kK^q}{k-K}$$

and we have used the following result.

Lemma 3.44 (Thorne [1176]**)** *For* $m = 0, 1, 2, \ldots,$

$$\mathrm{e}^{\pm kz} J_m(k\sigma) = \sum_{q=0}^{\infty} \frac{(\pm 1)^q}{(2m+q)!}\, (kr)^{m+q}\, P_{m+q}^m(\cos\theta). \tag{3.173}$$

Proof For $m = 0$, (3.173) reduces to

$$\mathrm{e}^{\pm kz} J_0(k\sigma) = \sum_{q=0}^{\infty} \frac{(\pm 1)^q}{q!}\, (kr)^q\, P_q(\cos\theta). \tag{3.174}$$

For $\sigma = 0$, this formula reduces further to the known power series for $\mathrm{e}^{\pm kz}$, and so (3.174) is valid as both sides are axisymmetric harmonic functions; see, for example,

[597, p. 255] or [549, §3.3]. For $m > 0$, we differentiate (3.174) using (3.142) and (3.136) to give

$$\mathrm{e}^{\pm kz}\,(-k)^m\,J_m(k\sigma)\,\mathrm{e}^{\mathrm{i}m\phi} = \sum_{q=0}^{\infty} \frac{(\pm k)^q}{q!}\left(\frac{\partial}{\partial x} + \mathrm{i}\frac{\partial}{\partial y}\right)^m \tilde{Y}_q^0$$
$$= (-1)^m\,\mathrm{e}^{\mathrm{i}m\phi} \sum_{q=0}^{\infty} \frac{(\pm k)^q}{q!}\, r^{q-m}\, P_{q-m}^m(\cos\theta).$$

But $P_n^m \equiv 0$ for $m > n$ and so the first $2m$ terms in the summation are zero; replacing q by $q+2m$ then gives the result. □

For another proof of (3.173), see [1045, p. 780].

Returning to (3.172), we can see that the $q = 0$ and $q = 1$ terms are both zero. For $q > 1$, we have

$$Kk^q - kK^q = (k-K)\sum_{s=1}^{q-1} k^s K^{q-s},$$

and so, if we define

$$F_{qs} = \frac{(-r)^q (kr)^m}{(2m+q)!}\, k^s K^{q-s}\, P_{m+q}^m(\cos\theta),$$

we obtain

$$G_m = \sum_{j=1}^{\infty} F_{2j,2j} + \sum_{q=2}^{\infty}\sum_{s=1}^{q-1} F_{qs} = \sum_{j=1}^{\infty} F_{2j,2j} + \sum_{s=1}^{\infty}\sum_{q=s+1}^{\infty} F_{qs}$$
$$= \sum_{j=1}^{\infty} F_{2j,2j} + \sum_{j=1}^{\infty}\left(\sum_{q=2j+1}^{\infty} F_{q,2j} + \sum_{q=2j}^{\infty} F_{q,2j-1}\right)$$
$$= \sum_{j=1}^{\infty}\sum_{q=2j}^{\infty}(F_{q,2j} + F_{q,2j-1}) = \sum_{j=1}^{\infty}\sum_{s=0}^{\infty}(F_{2j+s,2j} + F_{2j+s,2j-1})$$
$$= \sum_{j=1}^{\infty}\sum_{s=0}^{\infty} \frac{(-1)^s (Kr)^{m+2j+s}}{(2m+2j+s)!}\, P_{m+2j+s}^m(\cos\theta)\, \frac{(k+K)}{K^{m+2j}}\, k^{m+2j-1}.$$

Substituting in (3.171), using (3.146) and $2\cos m(\phi - \varphi) = \mathrm{e}^{\mathrm{i}m\phi}\mathrm{e}^{-\mathrm{i}m\varphi} + \mathrm{e}^{-\mathrm{i}m\phi}\mathrm{e}^{\mathrm{i}m\varphi}$, gives

$$H = \frac{K}{\pi}\sum_{j=1}^{\infty}\sum_{m=-\infty}^{\infty} (-1)^{m+1}\alpha_j^{-m}(\boldsymbol{r}_P)\,\Phi_j^m(\boldsymbol{r}_Q), \tag{3.175}$$

where

$$\alpha_j^{\pm m} = \mathrm{e}^{\pm\mathrm{i}m\phi}\sum_{s=0}^{\infty}(-1)^s \frac{(Kr)^{m+2j+s}}{(2m+2j+s)!}\, P_{m+2j+s}^m(\cos\theta) = \sum_{l=0}^{\infty}\widehat{\Phi}_{j+l}^{\pm m}. \tag{3.176}$$

The result (3.167) follows by combining (3.169) and (3.175), assuming that $z < \zeta$. However, both sides of (3.167) are regular harmonic functions for $|\boldsymbol{r}_P| < |\boldsymbol{r}_Q|$, and so analytic continuation implies that the expansion (3.167) is valid for all points P and Q satisfying $|\boldsymbol{r}_P| < |\boldsymbol{r}_Q|$. Finally, note that Lemma 3.44 shows that (3.176) reduces to (3.170) when $j = 0$, so that the definition (3.168) holds for $j = 0, 1, 2, \ldots$ □

As a special case of Theorem 3.43, put P at a point O_2 in the free surface; let $\boldsymbol{c}$ be the position vector of O_2 with respect to O. Then, as $G(O_2, Q) = -(K/\pi)\Phi_0^0(\boldsymbol{r}_Q - \boldsymbol{c})$, we obtain

$$\Phi_0^0(\boldsymbol{r}_Q - \boldsymbol{c}) = \sum_{n=0}^{\infty} \sum_{m=-\infty}^{\infty} (-1)^m \alpha_n^{-m}(\boldsymbol{c})\, \Phi_n^m(\boldsymbol{r}_Q), \quad |\boldsymbol{c}| < |\boldsymbol{r}_Q|. \tag{3.177}$$

This formula will be used in the proof of an addition theorem for three-dimensional multipole potentials; see Theorem 3.46.

3.18.2 Addition theorems

Let O_1 and O_2 be two origins in the free surface. Let $\boldsymbol{r}_j$ be the position vector of a general point P with respect to O_j, for $j = 1, 2$. Let $\boldsymbol{b}$ be the position vector of O_1 with respect to O_2, so that $\boldsymbol{r}_2 = \boldsymbol{r}_1 + \boldsymbol{b}$.

Our addition theorems for $\Phi_n^{\pm m}$ involve the functions $\alpha_n^{\pm m}$, defined by (3.168); in particular, $\alpha_0^{\pm m}$ is given by (3.170). The functions $\alpha_n^{\pm m}$ are harmonic and regular, and they satisfy the free-surface boundary condition. The definition (3.168) yields

$$\alpha_n^{\pm m} - \alpha_{n+1}^{\pm m} = \widehat{\Phi}_n^{\pm m}, \quad n = 0, 1, 2, \ldots. \tag{3.178}$$

Various derivatives of α_n^m are collected in the next theorem.

Theorem 3.45 *For vertical differentiation, we have*

$$(\partial^2/\partial z^2)\alpha_0^{\pm m} = K^2 \alpha_0^{\pm m} \quad \text{and} \quad (\partial^2/\partial z^2)\alpha_n^{\pm m} = K^2 \alpha_{n-1}^{\pm m}, \tag{3.179}$$

for $n = 1, 2, 3, \ldots$. *For horizontal differentiation, using the operator* $\tilde{\mathcal{D}}_m^{\pm}$ *defined by* (3.149), *we have*

$$\tilde{\mathcal{D}}_m^{+}\alpha_n^{\mu} = \begin{cases} \alpha_{n-m}^{m+\mu}, & 0 \le m \le n, \\ \alpha_0^{m+\mu}, & 0 \le n \le m, \end{cases} \quad \mu \ge 0; \tag{3.180}$$

$$\tilde{\mathcal{D}}_m^{-}\alpha_n^{-\mu} = \begin{cases} \alpha_{n-m}^{-m-\mu}, & 0 \le m \le n, \\ \alpha_0^{-m-\mu}, & 0 \le n \le m, \end{cases} \quad \mu \ge 0; \tag{3.181}$$

$$\tilde{\mathcal{D}}_m^{+}\alpha_n^{-\mu} = \begin{cases} (-1)^m \alpha_n^{m-\mu}, & 0 \le m \le \mu, \\ (-1)^{\mu} \alpha_{n-m+\mu}^{m-\mu}, & 0 \le m-\mu \le n, \quad \mu \ge 0, \\ (-1)^{\mu} \alpha_0^{m-\mu}, & 0 \le n \le m-\mu, \quad \mu \ge 0; \end{cases} \tag{3.182}$$

$$\tilde{\mathcal{D}}_m^{-}\alpha_n^{\mu} = \begin{cases} (-1)^m \alpha_n^{\mu-m}, & 0 \le m \le \mu, \\ (-1)^{\mu} \alpha_{n-m+\mu}^{\mu-m}, & 0 \le m-\mu \le n, \quad \mu \ge 0, \\ (-1)^{\mu} \alpha_0^{\mu-m}, & 0 \le n \le m-\mu, \quad \mu \ge 0. \end{cases} \tag{3.183}$$

Proof The formulae (3.179) are a consequence of (3.157) and (3.170). The formulae (3.180)–(3.183) follow by applying Theorem 3.42 to the definition (3.168). □

Now, we can state the addition theorem for multipole potentials.

Theorem 3.46 *Let $\boldsymbol{r}_2 = \boldsymbol{r}_1 + \boldsymbol{b}$. For $n = 0, 1, 2, \ldots$, and for all m,*

$$\Phi_n^m(\boldsymbol{r}_2) = \sum_{\nu=0}^{\infty} \sum_{\mu=-\infty}^{\infty} \widehat{S}_{n\nu}^{m\mu}(\boldsymbol{b})\, \Phi_\nu^\mu(\boldsymbol{r}_1) \quad \textit{for } r_1 > b, \tag{3.184}$$

$$\Phi_n^m(\boldsymbol{r}_2) = \sum_{\nu=0}^{\infty} \sum_{\mu=-\infty}^{\infty} S_{n\nu}^{m\mu}(\boldsymbol{b})\, \alpha_\nu^\mu(\boldsymbol{r}_1) \quad \textit{for } r_1 < b. \tag{3.185}$$

The quantities $\widehat{S}_{n\nu}^{m\mu}$ and $S_{n\nu}^{m\mu}$ are defined as follows:

$$\widehat{S}_{0\nu}^{0\mu}(\boldsymbol{b}) = (-1)^\mu \alpha_\nu^{-\mu}(-\boldsymbol{b}); \quad \widehat{S}_{n\nu}^{0\mu}(\boldsymbol{b}) = (-1)^\mu \widehat{\Phi}_{\nu-n}^{-\mu}(-\boldsymbol{b}), \quad \nu \geq n;$$

$$\widehat{S}_{n\nu}^{0\mu}(\boldsymbol{b}) = 0, \quad 0 \leq \nu < n, \quad n \geq 1;$$

$$\widehat{S}_{0\nu}^{m\mu}(\boldsymbol{b}) = \begin{cases} (-1)^\mu \alpha_0^{m-\mu}, & \mu \leq -1, \quad 0 \leq \nu \leq m, \\ (-1)^\mu \alpha_{\nu-m}^{m-\mu}, & \mu \leq -1, \quad \nu \geq m+1, \\ \alpha_0^{m-\mu}, & 0 \leq \mu \leq m-1,\ 0 \leq \nu \leq m-\mu, \\ \alpha_{\nu+\mu-m}^{m-\mu}, & 0 \leq \mu \leq m-1,\ \nu \geq m-\mu+1, \\ (-1)^{m+\mu} \alpha_\nu^{m-\mu}, & \mu \geq m, \quad \nu \geq 0; \end{cases}$$

$$\widehat{S}_{0\nu}^{-m,\mu}(\boldsymbol{b}) = \begin{cases} (-1)^{m+\mu} \alpha_\nu^{-m-\mu}, & \mu \leq -m, \quad \nu \geq 0, \\ \alpha_0^{-m-\mu}, & -m+1 \leq \mu \leq 0,\ 0 \leq \nu \leq m+\mu, \\ \alpha_{\nu-\mu-m}^{-m-\mu}, & -m+1 \leq \mu \leq 0,\ \nu \geq m+\mu+1, \\ (-1)^\mu \alpha_0^{-m-\mu}, & \mu \geq 1, \quad 0 \leq \nu \leq m, \\ (-1)^\mu \alpha_{\nu-m}^{-m-\mu}, & \mu \geq 1, \quad \nu \geq m+1; \end{cases}$$

$$\widehat{S}_{1\nu}^{m\mu}(\boldsymbol{b}) = \widehat{S}_{0,\nu-1}^{m\mu}(\boldsymbol{b}) - \widehat{S}_{0\nu}^{m\mu}(\boldsymbol{b}), \quad \nu \geq 1; \quad \widehat{S}_{10}^{m\mu}(\boldsymbol{b}) = 0;$$

$$\widehat{S}_{j\nu}^{m\mu}(\boldsymbol{b}) = \widehat{S}_{1,\nu-j+1}^{m\mu}(\boldsymbol{b}), \quad \nu \geq j \geq 1;$$

$$\widehat{S}_{j\nu}^{m\mu}(\boldsymbol{b}) = 0, \quad 0 \leq \nu \leq j-1, \quad j \geq 1;$$

$$S_{0\nu}^{0\mu}(\boldsymbol{b}) = (-1)^\mu \Phi_\nu^{-\mu}(-\boldsymbol{b}); \quad S_{n0}^{0\mu}(\boldsymbol{b}) = (-1)^\mu \Phi_n^{-\mu}(-\boldsymbol{b});$$

$$S_{n\nu}^{0\mu}(\boldsymbol{b}) = (-1)^\mu \left(\Phi_{\nu+n}^{-\mu} - \Phi_{\nu+n-1}^{-\mu}\right), \quad \nu \geq 1, \quad n \geq 1;$$

$$S_{0\nu}^{m\mu}(\boldsymbol{b}) = \begin{cases} (-1)^\mu \Phi_\nu^{m-\mu}, & \mu \leq -1, \quad \nu \geq 0, \\ \Phi_{\nu+\mu}^{m-\mu}, & 0 \leq \mu \leq m-1,\ \nu \geq 1, \\ (-1)^{m+\mu} \Phi_{\nu+m}^{m-\mu}, & \mu \geq m, \quad \nu \geq 1; \end{cases}$$

$$S_{00}^{m\mu}(\boldsymbol{b}) = \begin{cases} \sum_{j=0}^{\mu} \Phi_j^{m-\mu}, & 0 \le \mu \le m-1, \\ (-1)^{m+\mu} \sum_{j=0}^{m} \Phi_j^{m-\mu}, & \mu \ge m; \end{cases}$$

$$S_{0\nu}^{-m,\mu}(\boldsymbol{b}) = \begin{cases} (-1)^{m+\mu} \Phi_{\nu+m}^{-m-\mu}, & \mu \le -m, \quad \nu \ge 1, \\ \Phi_{\nu+\mu}^{-m-\mu}, & -m+1 \le \mu \le 0, \ \nu \ge 1, \\ (-1)^{\mu} \Phi_{\nu}^{-m-\mu}, & \mu \ge 1, \quad \nu \ge 0; \end{cases}$$

$$S_{00}^{-m,\mu}(\boldsymbol{b}) = \begin{cases} (-1)^{m+\mu} \sum_{j=0}^{m} \Phi_j^{-m-\mu}, & \mu \le -m, \\ \sum_{j=0}^{-\mu} \Phi_j^{-m-\mu}, & -m+1 \le \mu \le 0; \end{cases}$$

$$S_{1\nu}^{m\mu}(\boldsymbol{b}) = S_{0,\nu+1}^{m\mu}(\boldsymbol{b}) - S_{0\nu}^{m\mu}(\boldsymbol{b}), \quad \nu \ge 1; \quad S_{10}^{m\mu}(\boldsymbol{b}) = S_{01}^{m\mu}(\boldsymbol{b});$$

$$S_{j+1,0}^{m\mu}(\boldsymbol{b}) = S_{j0}^{m\mu}(\boldsymbol{b}) + S_{j1}^{m\mu}(\boldsymbol{b}), \quad j \ge 1;$$

$$S_{j+1,\nu}^{m\mu}(\boldsymbol{b}) = S_{j,\nu+1}^{m\mu}(\boldsymbol{b}), \quad j \ge 1, \quad \nu \ge 1.$$

In the formulae for $\widehat{S}_{0\nu}^{\pm m,\mu}$ and $S_{0\nu}^{\pm m,\mu}$, the integer m satisfies $m \ge 0$.

Proof When $n = m = 0$, (3.184) reduces to

$$\Phi_0^0(\boldsymbol{r}_2) = \sum_{\nu=0}^{\infty} \sum_{\mu=-\infty}^{\infty} (-1)^{\mu} \alpha_{\nu}^{-\mu}(-\boldsymbol{b})\, \Phi_{\nu}^{\mu}(\boldsymbol{r}_1), \quad r_1 > b, \tag{3.186}$$

which is (3.177): it is a special case of the bilinear expansion of G (Theorem 3.43). To prove (3.184) for $n = 1$ and $m = 0$, we differentiate (3.186) twice with respect to z, using (3.147) and (3.148), giving

$$\begin{aligned} \Phi_0^0(\boldsymbol{r}_2) + \Phi_1^0(\boldsymbol{r}_2) &= \sum_{\mu=-\infty}^{\infty} (-1)^{\mu} \alpha_0^{-\mu} (\Phi_0^{\mu} + \Phi_1^{\mu}) \\ &\quad + \sum_{\nu=1}^{\infty} \sum_{\mu=-\infty}^{\infty} (-1)^{\mu} \alpha_{\nu}^{-\mu} \Phi_{\nu+1}^{\mu} \\ &= \sum_{\mu=-\infty}^{\infty} (-1)^{\mu} \alpha_0^{-\mu} \Phi_0^{\mu} + \sum_{\nu=1}^{\infty} \sum_{\mu=-\infty}^{\infty} (-1)^{\mu} \alpha_{\nu-1}^{-\mu} \Phi_{\nu}^{\mu}. \end{aligned}$$

Making use of (3.186), we obtain

$$\Phi_1^0(\boldsymbol{r}_2) = \sum_{\nu=1}^{\infty} \sum_{\mu=-\infty}^{\infty} (-1)^{\mu} (\alpha_{\nu-1}^{-\mu} - \alpha_{\nu}^{-\mu}) \Phi_{\nu}^{\mu}(\boldsymbol{r}_1), \tag{3.187}$$

which gives the stated result for $n = 1$ and $m = 0$, once (3.178) has been used. For $n > 1$ and $m = 0$, repeatedly differentiate (3.187) twice with respect to z, using (3.148).

Next, we apply $\tilde{\mathscr{D}}_m^{+}$ to (3.186), which we write as

$$\Phi_0^0(\boldsymbol{r}_2) = \sum_{s=0}^{\infty} \sum_{\nu=0}^{\infty} (-1)^s \alpha_{\nu}^{-s}(-\boldsymbol{b})\, \Phi_{\nu}^{s}(\boldsymbol{r}_1) + \sum_{s=1}^{\infty} \sum_{\nu=0}^{\infty} (-1)^s \alpha_{\nu}^{s}(-\boldsymbol{b})\, \Phi_{\nu}^{-s}(\boldsymbol{r}_1).$$

Then, applying $\tilde{\mathscr{D}}_m^+$, using (3.151), $(3.150)_1$ and (3.153), gives

$$\begin{aligned}
\Phi_0^m(\boldsymbol{r}_2) &= \sum_{s=0}^{\infty}\sum_{\nu=0}^{\infty}(-1)^s\alpha_\nu^{-s}\Phi_\nu^{m+s} \\
&\quad + \sum_{s=1}^{m}\left\{\alpha_0^s\sum_{n=0}^{s}\Phi_n^{m-s}+\sum_{n=1}^{\infty}\alpha_n^s\Phi_{n+s}^{m-s}\right\} \\
&\quad + \sum_{s=m+1}^{\infty}(-1)^{m+s}\left\{\alpha_0^s\sum_{n=0}^{m}\Phi_n^{m-s}+\sum_{n=1}^{\infty}\alpha_n^s\Phi_{n+m}^{m-s}\right\} \\
&= \sum_{\mu=m}^{\infty}\sum_{\nu=0}^{\infty}(-1)^{m+\mu}\alpha_\nu^{m-\mu}\Phi_\nu^\mu \\
&\quad + \sum_{\mu=0}^{m-1}\left\{\alpha_0^{m-\mu}\sum_{\nu=0}^{m-\mu}\Phi_\nu^\mu+\sum_{\nu=m+1-\mu}^{\infty}\alpha_{\nu-m+\mu}^{m-\mu}\Phi_\nu^\mu\right\} \\
&\quad + \sum_{\mu=-\infty}^{-1}(-1)^\mu\left\{\alpha_0^{m-\mu}\sum_{\nu=0}^{m}\Phi_\nu^\mu+\sum_{\nu=m+1}^{\infty}\alpha_{\nu-m}^{m-\mu}\Phi_\nu^\mu\right\}.
\end{aligned}$$

The result for Φ_0^{-m} is obtained similarly, using $\tilde{\mathscr{D}}_m^-$.

At this stage, we have obtained

$$\Phi_0^m(\boldsymbol{r}_2) = \sum_{\nu=0}^{\infty}\sum_{\mu=-\infty}^{\infty}\widehat{S}_{0\nu}^{m\mu}(\boldsymbol{b})\,\Phi_\nu^\mu(\boldsymbol{r}_1), \tag{3.188}$$

valid for $r_1 > b$ and for all integers m. Further results are obtained by repeated applications of $\partial^2/\partial z^2$. One application gives

$$\Phi_0^m(\boldsymbol{r}_2)+\Phi_1^m(\boldsymbol{r}_2) = \sum_{\mu=-\infty}^{\infty}\left\{\widehat{S}_{00}^{m\mu}\,(\Phi_0^\mu+\Phi_1^\mu)+\sum_{\nu=1}^{\infty}\widehat{S}_{0\nu}^{m\mu}\,\Phi_{\nu+1}^\mu\right\},$$

using (3.147) and (3.148). Hence, using (3.188), we obtain

$$\Phi_1^m(\boldsymbol{r}_2) = \sum_{\nu=1}^{\infty}\sum_{\mu=-\infty}^{\infty}\left(\widehat{S}_{0,\nu-1}^{m\mu}-\widehat{S}_{0\nu}^{m\mu}\right)\Phi_\nu^\mu(\boldsymbol{r}_1),$$

and this defines $\widehat{S}_{1\nu}^{m\mu}$. Repeated differentiation then gives

$$\Phi_j^m(\boldsymbol{r}_2) = \sum_{\nu=j}^{\infty}\sum_{\mu=-\infty}^{\infty}\widehat{S}_{1,\nu-j+1}^{m\mu}(\boldsymbol{b})\,\Phi_\nu^\mu(\boldsymbol{r}_1), \quad \text{for } j \ge 1.$$

This completes the proof of (3.184).

To prove (3.185), we start from

$$\Phi_0^0(\boldsymbol{r}_2) = \sum_{\nu=0}^{\infty}\sum_{\mu=-\infty}^{\infty}(-1)^\mu\Phi_\nu^{-\mu}(-\boldsymbol{b})\,\alpha_\nu^\mu(\boldsymbol{r}_1), \quad r_1 < b, \tag{3.189}$$

which also follows from Theorem 3.43. Repeated applications of $\partial^2/\partial z^2$, using (3.147), (3.148) and (3.179), give

$$\Phi_n^0(\boldsymbol{r}_2) = \sum_{\mu=-\infty}^{\infty} (-1)^\mu \left\{ \Phi_n^{-\mu}\, \alpha_0^\mu + \sum_{\nu=1}^{\infty} \left(\Phi_{\nu+n}^{-\mu} - \Phi_{\nu+n-1}^{-\mu} \right) \alpha_\nu^\mu(\boldsymbol{r}_1) \right\},$$

and this defines $S_{n\nu}^{0\mu}$. Next, we apply $\tilde{\mathcal{D}}_m^+$ to (3.189), using (3.180) and (3.182); this gives $S_{0\nu}^{m\mu}$ for $m > 0$. Similarly, use of $\tilde{\mathcal{D}}_m^-$, (3.181) and (3.183) gives $S_{0\nu}^{-m,\mu}$ for $m > 0$. Finally, repeated applications of $\partial^2/\partial z^2$ to the addition formulae for $\Phi_0^{\pm m}$, using (3.147), (3.148) and (3.179), give $S_{n\nu}^{m\mu}$ for $n > 0$. □

4

Methods based on separation of variables

Today, the separation of variables derivations . . . are only of academic interest.
(Burke & Twersky [163, p. 501])

4.1 Introduction

The method of separation of variables can be used to study acoustic scattering by a single obstacle, provided its surface coincides with a coordinate surface. In fact, the Helmholtz equation separates in eleven three-dimensional coordinate systems; see [41] for a review. Of these eleven, only six are useful for bounded obstacles: circles and ellipses in two dimensions; spheres, prolate spheroids, oblate spheroids and ellipsoids in three dimensions.

For two, or more, obstacles, we can proceed by combining separable solutions appropriate to each obstacle with an appropriate addition theorem. This method was used by Záviška in 1913 [1372] for two-dimensional scattering by circular cylinders. It is exact, and leads to an infinite system of simultaneous algebraic equations.

Despite the opening quotation (from 1964), the method is widely used, probably because it is both conceptually simple and numerically effective. Consequently, we give a detailed derivation for acoustic scattering by several circular cylinders and by several spheres, and then we discuss various extensions of the method.

4.2 Separation of variables for one circular cylinder

Consider a circular cylinder of radius a. Choose Cartesian coordinates (x, y), with the origin O at the centre of a typical cross-section, and plane polar coordinates (r, θ), so that $x = r\cos\theta$ and $y = r\sin\theta$.

We suppose that a plane wave is incident upon the cylinder, so that

$$u_{\text{inc}} = \mathrm{e}^{ik(x\cos\alpha + y\sin\alpha)} = \mathrm{e}^{ikr\cos(\theta-\alpha)},$$

where α is the angle of incidence. From the Jacobi expansion (2.17), we have

$$\mathrm{e}^{\mathrm{i}w\cos\varphi} = \sum_{n=-\infty}^{\infty} \mathrm{i}^n J_n(w)\,\mathrm{e}^{\mathrm{i}n\varphi} = \sum_{n=0}^{\infty} \epsilon_n \mathrm{i}^n J_n(w) \cos n\varphi, \tag{4.1}$$

where ϵ_n is the Neumann factor: $\epsilon_0 = 1$ and $\epsilon_n = 2$ for $n \geq 1$. Hence

$$u_{\mathrm{inc}} = \sum_{n=-\infty}^{\infty} \mathrm{i}^n J_n(kr)\,\mathrm{e}^{\mathrm{i}n(\theta-\alpha)} = \sum_{n=0}^{\infty} \epsilon_n \mathrm{i}^n J_n(kr) \cos n(\theta-\alpha). \tag{4.2}$$

Now, in plane polar coordinates, the Helmholtz equation has separated solutions of the form

$$J_n(kr)\,\mathrm{e}^{\pm\mathrm{i}n\theta} \quad \text{and} \quad Y_n(kr)\,\mathrm{e}^{\pm\mathrm{i}n\theta}.$$

In order to satisfy the radiation condition at infinity, we take the combination

$$\{J_n(kr) + \mathrm{i}Y_n(kr)\}\mathrm{e}^{\pm\mathrm{i}n\theta} \equiv H_n^{(1)}(kr)\,\mathrm{e}^{\pm\mathrm{i}n\theta},$$

since the Hankel function $H_n^{(1)}$ satisfies (2.28) giving rise to a cylindrical wave that propagates outwards (this is where we take note of the assumed time dependence, $\mathrm{e}^{-\mathrm{i}\omega t}$). Furthermore, in order to have the same symmetries as the incident wave, we take the combination

$$u_{\mathrm{sc}} = \sum_{n=0}^{\infty} \epsilon_n \mathrm{i}^n B_n H_n(kr) \cos n(\theta-\alpha), \tag{4.3}$$

where, from now on, we write H_n for $H_n^{(1)}$. So, for any choice of the coefficients B_n, the expression on the right-hand side of (4.3) satisfies the two-dimensional Helmholtz equation and the radiation condition (subject only to the convergence of the series in (4.3)).

To complete the problem, we require a boundary condition on the cylinder. This is usually given in terms of the total field u, defined by

$$u = u_{\mathrm{inc}} + u_{\mathrm{sc}}.$$

In the sequel, we are mainly concerned with sound-hard bodies, for which

$$\frac{\partial u}{\partial r} = 0 \quad \text{on } r = a. \tag{4.4}$$

Differentiating (4.2) and (4.3), and setting r equal to a gives

$$\sum_{n=0}^{\infty} \epsilon_n k\, \mathrm{i}^n \{J_n'(ka) + B_n H_n'(ka)\} \cos n(\theta-\alpha) = 0, \qquad -\pi \leq \theta < \pi,$$

where $J_m'(w) = (\mathrm{d}/\mathrm{d}w)J_m(w)$ and $H_m'(w) = (\mathrm{d}/\mathrm{d}w)H_m(w)$. Then, orthogonality of the trigonometric functions implies that the expression inside the braces must be zero for each n, whence

$$B_n = -\frac{J_n'(ka)}{H_n'(ka)} \tag{4.5}$$

and then the scattered field is given everywhere in $r \geq a$ by (4.3). In particular, the expression for the total field on the cylinder simplifies to

$$u(a, \theta) = \frac{2\mathrm{i}}{\pi ka} \sum_{n=0}^{\infty} \epsilon_n \frac{\mathrm{i}^n}{H_n'(ka)} \cos n(\theta - \alpha),$$

having made use of the Wronskian for Bessel functions,

$$J_m(z)Y_m'(z) - J_m'(z)Y_m(z) = 2/(\pi z). \tag{4.6}$$

For sound-soft bodies, the boundary condition is

$$u = 0 \quad \text{on } r = a. \tag{4.7}$$

The corresponding solution for the scattered field is also given by (4.3), but with

$$B_n = -\frac{J_n(ka)}{H_n(ka)}. \tag{4.8}$$

4.2.1 The far field

The scattered field far from a cylinder is characterised by its far-field pattern.

Definition 4.1 *Let* $\boldsymbol{r}$ *be the position vector of the point at* (x, y) *with respect to the origin O. Let* $r = |\boldsymbol{r}|$ *and* $\hat{\boldsymbol{r}} = \boldsymbol{r}/r$. *For two-dimensional problems, the* far-field pattern, f, *is defined by*

$$u_{\mathrm{sc}}(\boldsymbol{r}) \sim \sqrt{\frac{2}{\pi}}\, \mathrm{e}^{-\mathrm{i}\pi/4} \frac{\mathrm{e}^{\mathrm{i}kr}}{\sqrt{kr}} f(\hat{\boldsymbol{r}}) \quad \text{as } r \to \infty. \tag{4.9}$$

The factor $(2/\pi)^{1/2}\mathrm{e}^{-\mathrm{i}\pi/4}$ is convenient as it leads to simpler formulae for f; it is sometimes absorbed into f, as is the factor $k^{-1/2}$.

The far-field pattern $f(\hat{\boldsymbol{r}}) \equiv f(\theta)$ gives the field scattered in the direction of the unit vector $\hat{\boldsymbol{r}} = (\cos\theta, \sin\theta)$. If the incident field is a plane wave, propagating in the direction α, we write $f(\theta; \alpha)$. Notice that Definition 4.1 holds for any incident field, for cylinders of any cross-section, and for any finite number of cylinders.

With our definition of f, we obtain

$$f(\theta; \alpha) = \sum_{n=0}^{\infty} \epsilon_n B_n \cos n(\theta - \alpha), \tag{4.10}$$

where we have used (2.28) in (4.3). The coefficients B_n are given by (4.5) and (4.8) for hard and soft circular cylinders, respectively.

4.2.2 Literature

It was Fourier who first solved boundary-value problems for second-order partial differential equations by the method of separation of variables, as expounded in his famous essay, *Théorie analytique de la chaleur*, published in 1822 [358]; for an extract and a discussion, see [705, Chapter 13]. Specific applications to the Helmholtz equation exterior to a circular cylinder are given in many places, including [1010, §§341–343], [682, §304], [690], [884, pp. 1376–1379], [564, §8.6], [885, §8.1], [135, Chapter 2] and [944, Chapter II, §1].

4.3 Notation

For scattering by a single cylinder, it is often useful to exploit symmetries. In particular, this motivates the use of trigonometric functions: $\cos n\theta$ is symmetric about $\theta = 0$, whereas $\sin n\theta$ is antisymmetric about $\theta = 0$. However, for multiple-scattering problems, symmetry is less important and many formulae turn out to be simpler when expressed using exponentials. So, it is convenient to introduce the notation of Definition 2.13:

$$\psi_n(\boldsymbol{r}) = H_n(kr)\,\mathrm{e}^{\mathrm{i}n\theta}, \quad \hat{\psi}_n(\boldsymbol{r}) = J_n(kr)\,\mathrm{e}^{\mathrm{i}n\theta}, \quad n = 0, \pm 1, \pm 2, \ldots.$$

The functions $\psi_n(\boldsymbol{r})$ and $\hat{\psi}_n(\boldsymbol{r})$ are separated solutions of the Helmholtz equation in plane polar coordinates. $\psi_n(\boldsymbol{r})$ satisfies the radiation condition at infinity and is singular at $r = |\boldsymbol{r}| = 0$: we call ψ_n an *outgoing cylindrical wavefunction*. $\hat{\psi}_n(\boldsymbol{r})$ is regular at $r = 0$: we call $\hat{\psi}_n$ a *regular cylindrical wavefunction*. Properties of ψ_n and $\hat{\psi}_n$ are given in Section 2.6.

With the above notation, we can reconsider the problem of scattering by one circular cylinder. Thus, the incident wave can be expanded as

$$u_{\mathrm{inc}} = \sum_m d_m \hat{\psi}_m(\boldsymbol{r}), \tag{4.11}$$

where the notation implies that the summation is from $m = -\infty$ to $m = \infty$, and

$$d_m = \mathrm{i}^m \mathrm{e}^{-\mathrm{i}m\alpha}. \tag{4.12}$$

Similarly, the scattered field can be expanded as

$$u_{\mathrm{sc}} = \sum_m c_m \psi_m(\boldsymbol{r}). \tag{4.13}$$

Application of the boundary condition (4.4) on $r = a$ shows that c_m and d_m are related by

$$c_m H_m'(ka) + d_m J_m'(ka) = 0, \quad m = 0, \pm 1, \pm 2, \ldots; \tag{4.14}$$

this relation determines the coefficients c_m in terms of the known coefficients d_m:

$$c_m = \sum_n T_{mn} d_n, \tag{4.15}$$

where

$$T_{mn} = -\,[J_m'(ka)/H_m'(ka)]\,\delta_{mn} \tag{4.16}$$

and δ_{ij} is the Kronecker delta ($\delta_{ij} = 0$ if $i \neq j$ and $\delta_{ij} = 1$ if $i = j$). Although the matrix $\mathbf{T}$ with entries T_{mn} used to be called the *transition matrix*, it is now invariably called the *T-matrix*! We will discuss its properties in detail in Chapter 7; here, we simply note that, for this simple example, $\mathbf{T}$ is a diagonal matrix.

For a sound-soft cylinder, the corresponding T-matrix has entries

$$T_{mn} = -\,[J_m(ka)/H_m(ka)]\,\delta_{mn}. \tag{4.17}$$

In terms of the T-matrix, the far-field pattern is given by

$$f(\theta) = \sum_n \sum_m (-\mathrm{i})^n \, \mathrm{e}^{\mathrm{i}n\theta} \, T_{nm} \, d_m, \tag{4.18}$$

where we have used (4.10), (4.13) and (2.28). For circular cylinders, the T-matrix is diagonal, and so (4.18) simplifies. However, the formula (4.18) is general: it holds for cylinders of any cross-section once the corresponding T-matrix is known.

4.4 Multipole method for two circular cylinders

Let O be the origin of two-dimensional Cartesian coordinates, so that a typical point has position vector $\boldsymbol{r} = (x, y)$ with respect to O. Consider two circular cylinders, S_j, $j = 1, 2$. The circle S_j has radius a_j and centre O_j at $\boldsymbol{r} = \boldsymbol{b}_j$. Define plane polar coordinates (r_j, θ_j) at O_j, so that $\boldsymbol{r} = \boldsymbol{r}_j + \boldsymbol{b}_j$ with $\boldsymbol{r}_j = r_j(\cos\theta_j, \sin\theta_j)$. See Fig. 4.1.

The given incident field u_{inc} is scattered by the cylinders. We assume that, in the neighbourhood of each cylinder (including the interior of each cylinder), u_{inc} is a regular solution of the Helmholtz equation, so that

$$u_{\text{inc}} = \sum_m d_m^j \hat{\psi}_m(\boldsymbol{r}_j), \quad j = 1, 2. \tag{4.19}$$

The coefficients d_m^j are known in terms of u_{inc}.

Generalising (4.13), we express the scattered field as an infinite sum of multipoles at the centre of each circle,

$$u_{\text{sc}} = \sum_m c_m^1 \psi_m(\boldsymbol{r}_1) + \sum_m c_m^2 \psi_m(\boldsymbol{r}_2). \tag{4.20}$$

This representation will be justified later; clearly it provides a radiating solution of the Helmholtz equation for any reasonable choice of the coefficients c_m^1 and c_m^2. These coefficients will now be determined by applying the boundary condition on each cylinder.

Consider the cylinder S_1. In order to use the boundary condition on $r_1 = a_1$, we must express each term as a function of θ_1. Thus, we must express $\psi_m(\boldsymbol{r}_2)$ in terms of functions of $\boldsymbol{r}_1$. Now, in the neighbourhood of O_1, $\psi_m(\boldsymbol{r}_2)$ is a regular solution of the

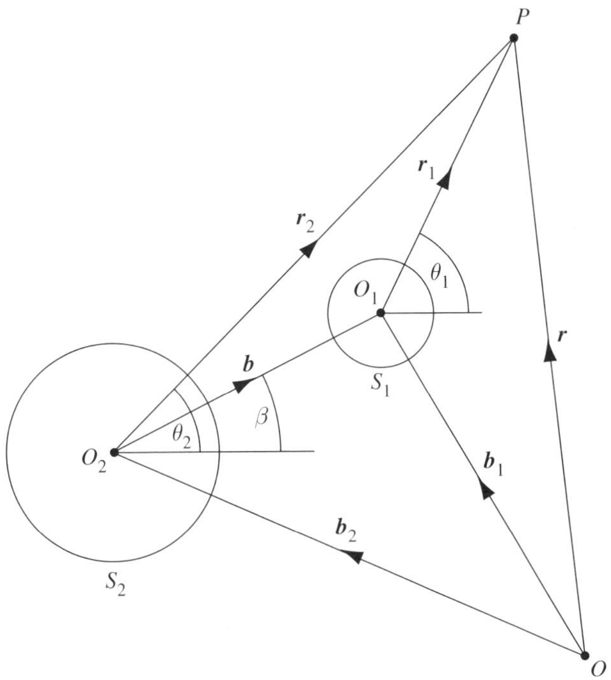

Fig. 4.1. Notation for scattering by two circular cylinders.

Helmholtz equation; hence, it can be expanded in terms of $\hat{\psi}_n(\boldsymbol{r}_1)$. Let $\boldsymbol{r}_2 = \boldsymbol{r}_1 + \boldsymbol{b}$, where $\boldsymbol{b} = \boldsymbol{b}_1 - \boldsymbol{b}_2$ is the position vector of O_1 with respect to O_2. Then, we have

$$\psi_m(\boldsymbol{r}_2) = \sum_n S_{mn}(\boldsymbol{b})\,\hat{\psi}_n(\boldsymbol{r}_1) \quad \text{for } r_1 < b = |\boldsymbol{b}|, \tag{4.21}$$

this inequality coming from the fact that the left-hand side of (4.21) is singular when $\boldsymbol{r}_1 = -\boldsymbol{b}$.

The matrix $\mathbf{S}(\boldsymbol{b})$ is known explicitly; its entries are given by

$$S_{mn}(\boldsymbol{b}) = \psi_{m-n}(\boldsymbol{b}) = H_{m-n}(kb)\,\mathrm{e}^{\mathrm{i}(m-n)\beta},$$

where $\boldsymbol{b} = (b\cos\beta, b\sin\beta)$. The formula (4.21) is Graf's addition theorem in disguise [1298, §11.3]; see (2.34) and Theorem 2.12.

Returning to (4.20), we use the addition theorem (4.21) to expand the scattered field near S_1. When combined with (4.19), we obtain

$$\begin{aligned} u &= \sum_m \left\{ d_m^1 \hat{\psi}_m(\boldsymbol{r}_1) + c_m^1 \psi_m(\boldsymbol{r}_1) \right\} + \sum_m c_m^2 \sum_n S_{mn}(\boldsymbol{b})\,\hat{\psi}_n(\boldsymbol{r}_1) \\ &= \sum_m \left\{ d_m^1 \hat{\psi}_m(\boldsymbol{r}_1) + c_m^1 \psi_m(\boldsymbol{r}_1) + \hat{\psi}_m(\boldsymbol{r}_1) \sum_n S_{nm}(\boldsymbol{b})\, c_n^2 \right\} \end{aligned}$$

for $r_1 < b$. Next, we apply the boundary condition on S_1. Assuming that S_1 is sound-hard, we differentiate with respect to r_1, set r_1 equal to a_1 and use orthogonality of $\{e^{im\theta_1}\}$ to give, for $m = 0, \pm 1, \pm 2, \ldots$,

$$c_m^1 H_m'(ka_1) + J_m'(ka_1) \sum_n S_{nm}(\boldsymbol{b})\, c_n^2 = -d_m^1 J_m'(ka_1). \tag{4.22}$$

If we assume that the other cylinder, S_2, is also sound-hard, a similar argument gives, for $m = 0, \pm 1, \pm 2, \ldots$,

$$c_m^2 H_m'(ka_2) + J_m'(ka_2) \sum_n S_{nm}(-\boldsymbol{b})\, c_n^1 = -d_m^2 J_m'(ka_2). \tag{4.23}$$

Equations (4.22) and (4.23) form a coupled infinite system of simultaneous linear algebraic equations for the coefficients c_m^1 and c_m^2, $m = 0, \pm 1, \pm 2, \ldots$. The literature on this system is discussed in Section 4.5.1.

4.5 Multipole method for N circular cylinders

The method described in Section 4.4 extends easily to N circular cylinders, S_j, $j = 1, 2, \ldots, N$. The circle S_j has radius a_j and centre O_j at $(x, y) = (\xi_j, \eta_j)$. As before, plane polar coordinates (r_j, θ_j) are defined at O_j, with $\theta_j = 0$ being in the x-direction.

The scattered field is expressed as

$$u_{\mathrm{sc}} = \sum_{j=1}^{N} \sum_m c_m^j \psi_m(\boldsymbol{r}_j), \tag{4.24}$$

which is an infinite number of outgoing cylindrical wavefunctions at the centre of each circle, with unknown coefficients c_m^j.

Using the addition theorem (4.21), we find that the total field in the vicinity of S_l is given by

$$u = \sum_m \left\{ d_m^l \hat{\psi}_m(\boldsymbol{r}_l) + c_m^l \psi_m(\boldsymbol{r}_l) + \hat{\psi}_m(\boldsymbol{r}_l) \sum_{\substack{j=1 \\ j \neq l}}^{N} \sum_n S_{nm}(\boldsymbol{b}_{lj})\, c_n^j \right\} \tag{4.25}$$

for $r_l < b_l$, where

$$b_l = \min_{\substack{1 \leq j \leq N \\ j \neq l}} |\boldsymbol{b}_{lj}| \tag{4.26}$$

and $\boldsymbol{b}_{lj}$ is the position vector of O_l relative to O_j (so that $\boldsymbol{b}_{jl} = -\boldsymbol{b}_{lj}$).

Let us assume that all the cylinders are sound-hard. Then, application of the boundary condition on S_l gives

$$c_m^l H_m'(ka_l) + J_m'(ka_l) \sum_{\substack{j=1 \\ j \neq l}}^{N} \sum_n S_{nm}(\boldsymbol{b}_{lj})\, c_n^j = -d_m^l J_m'(ka_l), \tag{4.27}$$

for $m = 0, \pm 1, \pm 2, \ldots$ and $l = 1, 2, \ldots, N$. This is an algebraic system of equations for the coefficients c_m^l. It reduces to (4.22) and (4.23) when $N = 2$, and to (4.14) when $N = 1$. If the system (4.27) can be solved, its solution will yield an exact solution for multiple scattering by N circular cylinders – no approximations have been made.

Similar systems can be derived for sound-soft cylinders, and for any combination of soft and hard cylinders. Indeed, the method works for impedance boundary conditions and also for penetrable cylinders. The latter situation leads to a transmission problem, in which acoustic fields also exist inside each cylinder and are coupled to the exterior field through appropriate transmission conditions on the surfaces of each cylinder; see Section 1.3.3.

4.5.1 Literature

The multipole method, leading to the algebraic system (4.27), was apparently devised by Záviška [1372]. In fact, Záviška treated the corresponding transmission problem.

Row [1035] derived the system (4.27) (for sound-soft cylinders) by specialising an integral equation for cylinders of arbitrary cross-section to circular cylinders; the integral equation is (5.74).

Záviška's method has been rediscovered and used by many subsequent authors for a variety of scalar problems. For two cylinders, see [932, 934, 513, 748, 1136, 1365, 864, 1074, 1195, 1367, 108, 1034]. For more than two cylinders, see [933, 995, 1335, 180] from the 1970s, [456, 996, 997, 387] from the 1980s, [700, 726, 97, 607, 311, 1329, 330, 273, 274, 1328] from the 1990s, and [1171, 1322, 190, 1101] since 2000. There is also a description of the method in [731, §6.1]. For example, several authors have used Záviška's method to study scattering by identical circles centred at the vertices of a regular polygon: for three circles, see [387, 274, 1328]; for four circles, see [726]; for five and six circles, see [330]. Radlinski & Meyers [995] studied similar problems with an additional (larger) circle at the centre of the polygon; see also [410, 721]. Duclos & Clément [298] have considered scattering by regular arrangements of 16 and 38 identical circular cylinders, and then examined the effect of small perturbations in the geometrical arrangement. In [722], the authors examine scattering by clusters of identical circular cylinders, where the cylinders are free to move in response to the incident wave. For an infinite grating of circular cylinders, see [727] and references therein.

Grote & Kirsch [444] have developed a numerical method for scattering by N cylinders, in which each cylinder is enclosed by a circular cylinder. (The circular cylinders are assumed to be disjoint.) The region exterior to the circular cylinders is treated using Záviška's method. The region between each pair of cylinders is treated using finite elements or finite differences. One can regard Záviška's method as providing a boundary condition for the numerical method employed in each of the N bounded interior regions. Numerical results for scattering by two soft elliptic cylinders are given.

For transmission problems, a variety of interface conditions on $r_j = a_j$ can be imposed; see, for example, the paper by Shindo & Togawa [1102], where certain imperfect interfaces are modelled. By using interface conditions on $r_j = a_j$ that vary with θ_j, scattering by an array of thin curved screens can be analysed [1362]. A similar method can be used for scattering by two (or more) circular cylinders in the flat interface between two acoustic half-planes; one of the half-planes may be empty [1368].

The orders-of-scattering approach to the problem of scattering by N circular cylinders was described in detail by Twersky [1196]. He specialised his results to a line of cylinders with collinear centres and to a pair of identical cylinders [1197]. However, Radlinski & Meyers [995] found difficulties with this approach: 'An attempt to obtain the unknown series coefficients from an iterative [orders-of-scattering] technique [based on [1196]] was abandoned because of the difficulty in obtaining convergence for problems involving many closely spaced scatterers' [995, p. 842].

4.5.2 Properties of the system (4.27)

From (4.27), we see that c^l_m must be proportional to $J'_m(ka_l)$; in particular, c^l_m and c^l_{-m} both vanish whenever ka_l is such that $J'_m(ka_l) = 0$. This suggests defining modified coefficients $\tilde{c}^l_m$ by

$$c^l_m = \tilde{c}^l_m \, J'_m(ka_l),$$

whence (4.27) becomes

$$\tilde{c}^l_m H'_m(ka_l) + \sum_{\substack{j=1 \\ j\neq l}}^{N} \sum_n S_{nm}(\boldsymbol{b}_{lj}) \, J'_n(ka_j) \, \tilde{c}^j_n = -d^l_m, \tag{4.28}$$

for $m = 0, \pm 1, \pm 2, \ldots$ and $l = 1, 2, \ldots, N$. Similarly, from (4.25), the total field in the vicinity of S_l is given by

$$u = \sum_m \left\{ d^l_m \hat{\psi}_m(\boldsymbol{r}_l) + \tilde{c}^l_m J'_m(ka_l) \, \psi_m(\boldsymbol{r}_l) \right.$$
$$\left. + \hat{\psi}_m(\boldsymbol{r}_l) \sum_{\substack{j=1 \\ j\neq l}}^{N} \sum_n S_{nm}(\boldsymbol{b}_{lj}) \, J'_n(ka_j) \, \tilde{c}^j_n \right\} \tag{4.29}$$

for $r_l < b_l$. It was observed by Linton & Evans [726] that the double sum inside the braces in (4.29) can be eliminated, using the system (4.28); the result is

$$u = \sum_m \tilde{c}^l_m \left\{ J'_m(ka_l) \, \psi_m(\boldsymbol{r}_l) - H'_m(ka_l) \, \hat{\psi}_m(\boldsymbol{r}_l) \right\} \tag{4.30}$$

for $r_l < b_l$. This formula makes it much simpler to evaluate the field near S_l, once the coefficients $\tilde{c}^l_m$ have been found. In particular, when $r_l = a_l$, we have

$$u(a_l, \theta_l) = -\frac{2\mathrm{i}}{\pi k a_l} \sum_m \tilde{c}^l_m \, \mathrm{e}^{\mathrm{i}m\theta_l}$$

for $0 \le \theta_l < 2\pi$, after using the Wronskian relation (4.6).

4.5.3 Numerical solution of the system (4.27)

Without the use of large-scale automatic computing machinery, it would be impractical to compute the solutions to the system (4.27) [for two cylinders] *for any appreciable range of radii and spacings.*

(Row [1035, p. 674])

Despite this pessimistic quotation, it turns out that Záviška's method is very efficient for numerical calculations. Of course, the infinite system (4.27), or (4.28), must be truncated: for example, (4.28) should be truncated to the system

$$\tilde{c}^l_m H'_m(ka_l) + \sum_{\substack{j=1\\ j\ne l}}^{N} \sum_{n=-M}^{M} S_{nm}(\boldsymbol{b}_{lj})\, J'_n(ka_j)\, \tilde{c}^j_n = -d^l_m,$$

for $m = 0, \pm1, \pm2, \ldots, \pm M$ and $l = 1, 2, \ldots, N$. This is a system of $N(2M+1)$ equations in $N(2M+1)$ unknowns. However, simplifications are possible if symmetry can be exploited.

Row [1035] was the first to solve (4.27) numerically. He considered two identical cylinders, of radius a, centred at $(x, y) = (0, \pm\frac{1}{2}b)$; the incident field was generated by a line-source at $(x, y) = (x_0, 0)$, so that

$$u_{\mathrm{inc}}(x, y) = H^{(1)}_0\left(k\sqrt{(x-x_0)^2 + y^2}\right).$$

For this symmetric configuration, we have

$$c^1_n = c^2_{-n}.$$

Thus, the truncated form of (4.22) appropriate to sound-soft cylinders reduces to

$$c^1_m H_m(ka) + J_m(ka) \sum_{n=-M}^{M} S_{nm}(\boldsymbol{b})\, c^1_{-n} = -d^1_m J_m(ka), \tag{4.31}$$

for $m = 0, \pm1, \pm2, \ldots, \pm M$. This is a system of $(2M+1)$ equations in $(2M+1)$ unknowns. Row [1035] solved it for $ka = 2$ and $b/a = \pi$ with $M = 6$ (his Fig. 6). As the quotation above suggests, this was a formidable calculation (using desk calculators and three assistants) in 1953. Moreover, the results were shown to be in excellent agreement with experiments.

Linton & Evans [726] found that taking $M = 6$ gave results accurate to four significant figures, except when the cylinders were very close together. See also [731, §6.1], where some of the numerical results in [726] are corrected.

Instead of solving the system (4.27) directly, one could try to solve it iteratively [1360]. This is equivalent to an orders-of-scattering approach.

4.6 Separation of variables for one sphere

Consider acoustic scattering by a sphere of radius a. Let (r, θ, ϕ) be spherical polar coordinates with origin at the sphere's centre. We suppose that a plane wave is incident on the sphere, so that

$$u_{\text{inc}}(\boldsymbol{r}) = \exp(\mathrm{i}k\boldsymbol{r}\cdot\hat{\boldsymbol{\alpha}}), \tag{4.32}$$

where $\boldsymbol{r} = (x, y, z)$ and $\hat{\boldsymbol{\alpha}}$ is a unit vector in the direction of propagation. (As the problem is axisymmetric, we could simplify the following calculations by taking $\hat{\boldsymbol{\alpha}}$ along the z-axis. However, we want to illustrate how we might proceed in more general situations.) From (3.69), we have

$$u_{\text{inc}}(\boldsymbol{r}) = 4\pi \sum_{n=0}^{\infty} \sum_{m=-n}^{n} \mathrm{i}^n j_n(kr)\, Y_n^m(\hat{\boldsymbol{r}})\, \overline{Y_n^m(\hat{\boldsymbol{\alpha}})}. \tag{4.33}$$

Similarly, for the scattered field, we can write

$$u_{\text{sc}}(\boldsymbol{r}) = 4\pi \sum_{n=0}^{\infty} \sum_{m=-n}^{n} \mathrm{i}^n B_n^m h_n(kr)\, Y_n^m(\hat{\boldsymbol{r}})\, \overline{Y_n^m(\hat{\boldsymbol{\alpha}})}, \tag{4.34}$$

where $h_n(w) \equiv h_n^{(1)}(w)$ and the coefficients B_n^m are to be found. Evidently, for any reasonable choice of B_n^m, the expression on the right-hand side of (4.34) satisfies the three-dimensional Helmholtz equation and the radiation condition.

For a sound-hard sphere, we have $u = u_{\text{inc}} + u_{\text{sc}}$, where u satisfies the boundary condition (4.4). This gives

$$B_n^m = -\frac{j_n'(ka)}{h_n'(ka)}, \tag{4.35}$$

and then the scattered field is given everywhere in $r \geq a$ by (4.34). In particular, the expression for the total field on the sphere simplifies to

$$u(a, \theta, \phi) = \frac{\mathrm{i}}{(ka)^2} \sum_{n=0}^{\infty} \frac{(2n+1)\mathrm{i}^n}{h_n'(ka)} P_n(\hat{\boldsymbol{r}}\cdot\hat{\boldsymbol{\alpha}}), \tag{4.36}$$

after using Legendre's addition theorem (3.20) and

$$j_n(w)\, y_n'(w) - y_n(w)\, j_n'(w) = w^{-2}, \tag{4.37}$$

which is the Wronskian for spherical Bessel functions [1, eqn 10.1.6].

Let us repeat the calculation using the notation of Definition 3.28, namely

$$\psi_n^m(\boldsymbol{r}) = h_n(kr)\, Y_n^m(\hat{\boldsymbol{r}}) \quad \text{and} \quad \hat{\psi}_n^m(\boldsymbol{r}) = j_n(kr)\, Y_n^m(\hat{\boldsymbol{r}}).$$

The incident wave can be expanded using regular wavefunctions,

$$u_{\text{inc}} = \sum_{n,m} d_n^m\, \hat{\psi}_n^m(\boldsymbol{r}), \tag{4.38}$$

where

$$\sum_{n,m} \equiv \sum_{n=0}^{\infty} \sum_{m=-n}^{n} \tag{4.39}$$

and

$$d_n^m = 4\pi \mathrm{i}^n\, \overline{Y_n^m(\hat{\boldsymbol{\alpha}})}. \tag{4.40}$$

Similarly, the scattered field can be expanded using outgoing wavefunctions,

$$u_{\text{sc}} = \sum_{n,m} c_n^m\, \psi_n^m(\boldsymbol{r}). \tag{4.41}$$

The boundary condition (4.4) shows that $c_n^m\, h_n'(ka) + d_n^m\, j_n'(ka) = 0$, which determines c_n^m. Thus,

$$c_n^m = \sum_{\nu,\mu} T_{n\nu}^{m\mu}\, d_\nu^\mu, \tag{4.42}$$

where the entries in the T-matrix are given by

$$T_{n\nu}^{m\mu} = -\left[j_n'(ka)/h_n'(ka) \right] \delta_{n\nu}\, \delta_{m\mu}.$$

In particular, the total field on the sphere is

$$u(a, \theta, \phi) = \sum_{n,m} \left(d_n^m\, \hat{\psi}_n^m + c_n^m\, \psi_n^m \right) = \frac{\mathrm{i}}{(ka)^2} \sum_{n,m} \frac{d_n^m}{h_n'(ka)}\, Y_n^m(\theta, \phi); \tag{4.43}$$

for plane-wave incidence, d_n^m is given by (4.40), and then (4.43) reduces to (4.36).

4.6.1 The far field

Definition 4.2 *For three-dimensional problems,* the far-field pattern, f, *is defined by*

$$u_{\text{sc}}(\boldsymbol{r}) \sim (\mathrm{i}kr)^{-1} \mathrm{e}^{\mathrm{i}kr} f(\hat{\boldsymbol{r}}) \quad \textit{as } r \to \infty. \tag{4.44}$$

For an incident plane wave, propagating in the direction of the unit vector $\hat{\boldsymbol{\alpha}}$, we denote the corresponding far-field pattern by $f(\hat{\boldsymbol{r}}; \hat{\boldsymbol{\alpha}})$. It is given by

$$f(\hat{\boldsymbol{r}}; \hat{\boldsymbol{\alpha}}) = \sum_{n=0}^{\infty} (2n+1) B_n^m\, P_n(\hat{\boldsymbol{r}} \cdot \hat{\boldsymbol{\alpha}}), \tag{4.45}$$

where we have used (3.20) and (3.90). The coefficients B_n^m occur in the expansion (4.34); for a sound-hard sphere, they are given explicitly by (4.35).

In terms of the T-matrix, we have

$$f(\hat{\boldsymbol{r}}) = \sum_{n,m} \sum_{\nu,\mu} (-\mathrm{i})^n Y_n^m(\hat{\boldsymbol{r}})\, T_{n\nu}^{m\mu}\, d_\nu^\mu, \tag{4.46}$$

where we have used (4.44), (4.41) and (3.90).

We have calculated the far-field pattern for a sphere at the origin. Suppose now that we keep the same incident wave, (4.32), with the same origin O, but we move (translate) the scatterer: how does this affect f? Let O_1 be the centre of the scatterer in its new position, and suppose that $\boldsymbol{b}_1$ is the position vector of O_1 with respect to O. Let $f_1(\hat{\boldsymbol{r}}; \hat{\boldsymbol{\alpha}})$ be the new far-field pattern, characterising the scattered field with respect to O. Put $\boldsymbol{r}_1 = \boldsymbol{r} - \boldsymbol{b}_1$. Then, the incident field is

$$\exp(\mathrm{i}k\boldsymbol{r} \cdot \hat{\boldsymbol{\alpha}}) = \exp(\mathrm{i}k\boldsymbol{b}_1 \cdot \hat{\boldsymbol{\alpha}}) \exp(\mathrm{i}k\boldsymbol{r}_1 \cdot \hat{\boldsymbol{\alpha}})$$

and so the scattered field is asymptotically equal to

$$\exp(\mathrm{i}k\boldsymbol{b}_1 \cdot \hat{\boldsymbol{\alpha}}) \frac{\mathrm{e}^{\mathrm{i}kr_1}}{\mathrm{i}kr_1} f(\hat{\boldsymbol{r}}_1; \hat{\boldsymbol{\alpha}}) \quad \text{as } r_1 = |\boldsymbol{r}_1| \to \infty,$$

where $\hat{\boldsymbol{r}}_1 = \boldsymbol{r}_1/r_1$. But $r_1 \sim r - \boldsymbol{b}_1 \cdot \hat{\boldsymbol{r}}$ and $\hat{\boldsymbol{r}}_1 \sim \hat{\boldsymbol{r}}$ as $r_1 \to \infty$, for fixed $\boldsymbol{b}_1$, and so

$$f_1(\hat{\boldsymbol{r}}; \hat{\boldsymbol{\alpha}}) = \exp\{\mathrm{i}k\boldsymbol{b}_1 \cdot (\hat{\boldsymbol{\alpha}} - \hat{\boldsymbol{r}})\}\, f(\hat{\boldsymbol{r}}; \hat{\boldsymbol{\alpha}}). \tag{4.47}$$

Therefore, translating a scatterer (which need not be spherical) merely changes the phase of the far-field pattern. Exactly the same formula holds in two dimensions.

Alternatively, suppose we keep the same scattering problem but we ask for the far-field pattern with respect to distance from O_1, defined by

$$u_{\mathrm{sc}} \sim f(\hat{\boldsymbol{r}}_1 \,|\, \boldsymbol{b}_1) \frac{\mathrm{e}^{\mathrm{i}kr_1}}{\mathrm{i}kr_1} \quad \text{as } r_1 \to \infty.$$

Then, we easily obtain

$$f(\hat{\boldsymbol{r}}_1 \,|\, \boldsymbol{b}_1) = f(\hat{\boldsymbol{r}}_1) \exp(\mathrm{i}k\boldsymbol{b}_1 \cdot \hat{\boldsymbol{r}}_1) \tag{4.48}$$

and, more generally,

$$f(\hat{\boldsymbol{r}} \,|\, \boldsymbol{b}_2) = \exp\{\mathrm{i}k(\boldsymbol{b}_2 - \boldsymbol{b}_1) \cdot \hat{\boldsymbol{r}}\}\, f(\hat{\boldsymbol{r}} \,|\, \boldsymbol{b}_1). \tag{4.49}$$

4.6.2 Literature

Lord Rayleigh considered acoustic scattering of a plane wave by a fixed, hard sphere in 1872; see [1010, §334] and [1205, §(2)]. For other discussions, see, for example, [682, §297], [610], [690], [884, pp. 1483–1486], [885, §8.2] and [135, Chapter 10]. Movable hard spheres are considered in [499].

Numerical results for the transmission problem of acoustic scattering by a fluid sphere were presented by Anderson in 1950: the computations ‘required the full time service of two computers [i.e. people] for a period of about two months' [29, p. 429]. For further results, see [561].

Inhomogeneous spheres with spherical symmetry have been considered by several authors. For such obstacles, the interior density ρ_0 and speed of sound c_0 are assumed to be given functions of the spherical polar coordinate r (only). Ahner [14] has given low-frequency expansions for problems where ρ_0 is constant and the refractive index n is a smooth function of r, with $n(r) \equiv 1$ for $r > a$ (for some finite a). For problems where n is a smooth function with $n(r) \to 1$ rapidly as $r \to \infty$, Colton [220] has used so-called 'transformation operators', which map solutions of the Helmholtz equation into solutions of (1.28); see also [221]. Colton & Kress [222] have extended this approach to consider related transmission problems. Frisk & DeSanto [369] have exploited the notion of a Jost function from quantum mechanics so as to obtain approximate solutions of (1.28).

Problems where both ρ_0 and c_0 vary have received less attention. The following specific functional forms are considered in [793]:

(i) $\rho_0(r) = \rho_1\,\mathrm{e}^{\beta r}$ and $[k_0(r)]^2 = k_1^2 + \alpha r^{-1}$,
(ii) $\rho_0(r) = \rho_1\,\mathrm{e}^{-\beta r^2}$ and $[k_0(r)]^2 = k_1^2 + \alpha r^2$.

Here, ρ_1, β, k_1^2 and α are adjustable parameters. For both (i) and (ii), explicit solutions of the governing equation, which is Bergmann's equation (1.30), are derived. The radial parts of these solutions are given in terms of known special functions, namely Coulomb wavefunctions and Whittaker functions. These solutions then permit the explicit solution of various scattering problems for inhomogeneous spheres. For spheres in which ρ_0 and c_0 are functions of r, θ and ϕ, see [645, 644], where Dini expansions [1298, Chapter 18] in the radial direction are used.

4.7 Multipole method for two spheres

Let O be the origin of three-dimensional Cartesian coordinates, so that a typical point has position vector $\boldsymbol{r} = (x, y, z)$ with respect to O. Consider two spheres, S_j, $j = 1, 2$. The sphere S_j has radius a_j and centre O_j at $\boldsymbol{r} = \boldsymbol{b}_j$. Define spherical polar coordinates (r_j, θ_j, ϕ_j) at O_j, so that $\boldsymbol{r} = \boldsymbol{r}_j + \boldsymbol{b}_j$ with $\boldsymbol{r}_j = r_j(\sin\theta_j \cos\phi_j, \sin\theta_j \sin\phi_j, \cos\theta_j)$.

The given incident field u_{inc} is scattered by the spheres. We assume that, in the neighbourhood of each sphere, u_{inc} is a regular solution of the Helmholtz equation, so that

$$u_{\mathrm{inc}} = \sum_{n,m} d_{nj}^m \hat{\psi}_n^m(\boldsymbol{r}_j), \quad j = 1, 2. \tag{4.50}$$

The coefficients d_{nj}^m are known in terms of u_{inc}.

The scattered field can be written as an infinite sum of multipoles at the centre of each sphere,

$$u_{\text{sc}} = \sum_{n,m} \{ c_{n1}^m \, \psi_n^m(\boldsymbol{r}_1) + c_{n2}^m \, \psi_n^m(\boldsymbol{r}_2) \} . \tag{4.51}$$

This representation gives a radiating solution of the Helmholtz equation for any reasonable choice of the coefficients c_{n1}^m and c_{n2}^m. We will determine these coefficients by applying the boundary condition on each sphere.

Consider the sphere S_1. As in Section 4.4, we expand $\psi_n^m(\boldsymbol{r}_2)$ near O_1 in terms of regular wavefunctions, $\hat{\psi}_\nu^\mu(\boldsymbol{r}_1)$. Let $\boldsymbol{r}_2 = \boldsymbol{r}_1 + \boldsymbol{b}$, where $\boldsymbol{b} = \boldsymbol{b}_1 - \boldsymbol{b}_2$ is the position vector of O_1 with respect to O_2. Then,

$$\psi_n^m(\boldsymbol{r}_2) = \sum_{\nu,\mu} S_{n\nu}^{m\mu}(\boldsymbol{b}) \, \hat{\psi}_\nu^\mu(\boldsymbol{r}_1) \quad \text{for } r_1 < b. \tag{4.52}$$

The matrix $\mathbf{S}(\boldsymbol{b})$, with entries $S_{n\nu}^{m\mu}(\boldsymbol{b})$, is known explicitly; see Theorem 3.27.

Making use of the addition theorem (4.52) in (4.51), we obtain an expansion of u_{sc} that is valid near S_1. When this is combined with (4.50), we obtain

$$u = \sum_{n,m} \left\{ d_{n1}^m \, \hat{\psi}_n^m(\boldsymbol{r}_1) + c_{n1}^m \, \psi_n^m(\boldsymbol{r}_1) + \hat{\psi}_n^m(\boldsymbol{r}_1) \sum_{\nu,\mu} S_{\nu n}^{\mu m}(\boldsymbol{b}) \, c_{\nu 2}^\mu \right\}$$

for $r_1 < b$. Next, we apply the boundary condition on S_1. Assuming that S_1 is sound-hard, we differentiate with respect to r_1, set r_1 equal to a_1 and use orthogonality of $\{Y_n^m(\theta_1, \phi_1)\}$ to give

$$c_{n1}^m \, h_n'(ka_1) + j_n'(ka_1) \sum_{\nu,\mu} S_{\nu n}^{\mu m}(\boldsymbol{b}) \, c_{\nu 2}^\mu = -d_{n1}^m \, j_n'(ka_1), \tag{4.53}$$

for $n = 0, 1, 2, \ldots$ and $m = -n, \ldots, n$.

If we assume that the other sphere, S_2, is also sound-hard, a similar argument gives

$$c_{n2}^m \, h_n'(ka_2) + j_n'(ka_2) \sum_{\nu,\mu} S_{\nu n}^{\mu m}(-\boldsymbol{b}) \, c_{\nu 1}^\mu = -d_{n2}^m \, j_n'(ka_2), \tag{4.54}$$

for $n = 0, 1, 2, \ldots$ and $m = -n, \ldots, n$.

Equations (4.53) and (4.54) form a coupled infinite system of simultaneous linear algebraic equations for the coefficients c_{nj}^m, where $j = 1, 2$, $n = 0, 1, 2, \ldots$ and $m = -n, \ldots, n$.

In fact, the system can be simplified by choosing the coordinates so that $\boldsymbol{b} = b\hat{\mathbf{z}}$, for then we can write

$$S_{n\nu}^{m\mu}(b\hat{\mathbf{z}}) = (-1)^n S_{n\nu}^m(kb) \, \delta_{m\mu} = (-1)^{n+\nu} S_{n\nu}^{m\mu}(-b\hat{\mathbf{z}}), \tag{4.55}$$

where $S_{n\nu}^m = S_{\nu n}^m$ and is much easier to calculate; see, for example, Theorem 3.34 and Theorem 3.35. Moreover, the Kronecker delta in (4.55) implies that (4.53) and

(4.54) decouple with respect to m, giving

$$c_{n1}^m\, h_n'(ka_1) + j_n'(ka_1) \sum_{\nu=0}^{\infty} (-1)^\nu S_{\nu n}^m(kb)\, c_{\nu 2}^m = -d_{n1}^m\, j_n'(ka_1), \tag{4.56}$$

$$c_{n2}^m\, h_n'(ka_2) + (-1)^n\, j_n'(ka_2) \sum_{\nu=0}^{\infty} S_{\nu n}^m(kb)\, c_{\nu 1}^m = -d_{n2}^m\, j_n'(ka_2), \tag{4.57}$$

where $n = 0, 1, 2, \ldots,$ and $m = -n, \ldots, n$.

4.7.1 Literature

The axisymmetric problem of scattering of plane acoustic waves by two identical soft spheres was considered by Eyges in 1957 [332]. He outlined the method described above, except that he was not aware of the addition theorem; see also the review [1072]. Embleton considered a similar axisymmetric problem in 1962 [313]. He used an orders-of-scattering approach, up to and including the effect on the first sphere of waves that had been scattered twice, once by the first sphere and once by the second sphere. For more recent applications of the same approach, see [1377, 584].

The exact method described above, leading to equations such as (4.56) and (4.57), was used by Marnevskaya for two sound-soft [776] and two sound-hard [777] spheres. A similar method was used by Golovchan [420] for a sphere near a plane wall, and by New & Eisler [904] for a point source near a pair of radiating spheres. For two identical radiating spheres (exterior Neumann problem), see [1174, 1014]. Extensive numerical results for plane-wave scattering by two identical sound-hard spheres have been presented in [392]; similar results have also been obtained for single spheres near plane boundaries, using the method of images [388, 389, 390, 391]. For scattering of waves from a point source by two touching spheres, see [447] and [449, §4.3.3]. For similar problems involving acoustic interactions between spherical bubbles, see [1053, Appendix A] for two bubbles and [1054] for N bubbles inside a cube.

For two identical spheres, and also for certain symmetric configurations of N spheres, it is possible to use group-theoretic arguments so as to obtain simplified algebraic systems; essentially, one uses symmetrised sets of spherical wavefunctions. This method has been developed in [118] and [377] for spheres, and also in [273, 274, 275] for identical parallel circular cylinders.

In [880], the authors began with a truncated form of (4.51) and then applied the boundary condition at a finite number of points on the spheres. Let M be the total number of coefficients c_{nj}^m (actually, the authors consider an axisymmetric problem, so that $m = 0$) and let P be the total number of points on the two spheres. An overdetermined system ($P > M$) was solved in a least-squares sense; the results were found to be sensitive to the values of M and P. The method used in [880] is usually known as 'point-matching'; there is an extensive literature on its implementation and limitations [76, 625, 213]. Note that, for multiple-scattering problems, point-matching yields a numerical method that avoids addition theorems but one whose conditioning should be monitored carefully. For some generalisations of this method, see Section 7.3.2. For the analogous method in two dimensions, starting from (4.24), see [103], where results for 162 cylinders are given.

4.8 Multipole method for N spheres

Let us generalise the method described in Section 4.7 to N spheres, S_j, $j = 1, 2, \ldots, N$. The sphere S_j has radius a_j and centre O_j at $\boldsymbol{r} = \boldsymbol{b}_j$. As before, spherical polar coordinates (r_j, θ_j, ϕ_j) are defined at O_j, with $\theta_j = 0$ being in the z-direction.

The scattered field is expressed as

$$u_{\mathrm{sc}} = \sum_{j=1}^{N} \sum_{n,m} c_{nj}^m \, \psi_n^m(\boldsymbol{r}_j). \tag{4.58}$$

Using the addition theorem (4.52), the total field in the vicinity of S_l is given by

$$u = \sum_{n,m} \left\{ d_{nl}^m \, \hat{\psi}_n^m(\boldsymbol{r}_l) + c_{nl}^m \, \psi_n^m(\boldsymbol{r}_l) + \hat{\psi}_n^m(\boldsymbol{r}_l) \sum_{\substack{j=1 \\ j\neq l}}^{N} \sum_{\nu,\mu} S_{\nu n}^{\mu m}(\boldsymbol{b}_{lj}) \, c_{\nu j}^{\mu} \right\} \tag{4.59}$$

for $r_l < b_l$, where b_l is defined by (4.26) and $\boldsymbol{b}_{lj}$ is the position vector of O_l relative to O_j.

If we assume that all the spheres are sound-hard, application of the boundary condition on S_l gives

$$c_{nl}^m \, h_n'(ka_l) + j_n'(ka_l) \sum_{\substack{j=1 \\ j\neq l}}^{N} \sum_{\nu,\mu} S_{\nu n}^{\mu m}(\boldsymbol{b}_{lj}) \, c_{\nu j}^{\mu} = -d_{nl}^m \, j_n'(ka_l), \tag{4.60}$$

for $n = 0, 1, 2, \ldots,\ m = -n, \ldots, n$ and $l = 1, 2, \ldots, N$. This is an algebraic system of equations for the coefficients c_{nl}^m. If the system can be solved, its solution will give an exact solution for multiple scattering by N spheres – again, no approximations have been made.

Sleeman [1116] has shown how the theory of this section can be modified to account for spheres with a spherically symmetric refractive index, $n(r)$.

4.8.1 Properties of the system (4.60)

As in Section 4.5.2, (4.60) shows that the coefficient c_{nl}^m must be proportional to $j_n'(ka_l)$, which suggests defining modified coefficients $\tilde{c}_{nl}^m$ by

$$c_{nl}^m = \tilde{c}_{nl}^m \, j_n'(ka_l). \tag{4.61}$$

Thus, the system (4.60) becomes

$$\tilde{c}_{nl}^m \, h_n'(ka_l) + \sum_{\substack{j=1 \\ j \neq l}}^{N} \sum_{\nu,\mu} S_{\nu n}^{\mu m}(\boldsymbol{b}_{lj}) \, j_\nu'(ka_j) \, \tilde{c}_{\nu j}^{\mu} = -d_{nl}^m, \quad \begin{array}{l} n = 0, 1, 2, \ldots, \\ m = -n, \ldots, n, \\ l = 1, 2, \ldots, N. \end{array} \tag{4.62}$$

Then, the total field in the vicinity of S_l is given by (4.59) and (4.61); making use of (4.62), we obtain

$$u = \sum_{n,m} \tilde{c}_{nl}^m \left\{ j_n'(ka_l) \, \psi_n^m(\boldsymbol{r}_l) - h_n'(ka_l) \, \hat{\psi}_n^m(\boldsymbol{r}_l) \right\}$$

for $r_l < b_l$. In particular, when $r_l = a_l$, we can use (4.37), giving

$$u(a_l, \theta_l, \phi_l) = -\frac{\mathrm{i}}{(ka_l)^2} \sum_{n,m} \tilde{c}_{nl}^m \, Y_n^m(\theta_l, \phi_l).$$

4.8.2 The far field

The far field of the N-sphere cluster is given by

$$u_{\mathrm{sc}}(\boldsymbol{r}) \sim (\mathrm{i}kr)^{-1} \mathrm{e}^{\mathrm{i}kr} F(\hat{\boldsymbol{r}}) \quad \text{as } r \to \infty, \tag{4.63}$$

where F is the far-field pattern. The scattered field is given everywhere by (4.58). When $r_j > b_j$, we have

$$\psi_n^m(\boldsymbol{r}_j) = \sum_{\nu,\mu} (-1)^{n+\nu} \widehat{S}_{n\nu}^{m\mu}(\boldsymbol{b}_j) \, \psi_\nu^\mu(\boldsymbol{r}),$$

where we have used (3.85) and (3.95), and then (4.58), (3.90) and (4.63) give

$$F(\hat{\boldsymbol{r}}) = \sum_{j=1}^{N} \sum_{n,m} (-1)^n c_{nj}^m \sum_{\nu,\mu} \mathrm{i}^\nu \, \widehat{S}_{n\nu}^{m\mu}(\boldsymbol{b}_j) \, Y_\nu^\mu(\hat{\boldsymbol{r}}).$$

4.8.3 Literature

The literature on scattering by two spheres is described in Section 4.7.1; this includes some references to scattering by N spheres. Gumerov & Duraiswami [447] have given results for scattering of waves from a point source by three spheres, by a random arrangement of 16 spheres, and by a square planar array of 121 identical spheres. Skaropoulos *et al.* [1111] have given results for plane-wave scattering by two and three bubbles of air in water (transmission problem); some results for the same problem but with 147 bubbles are given in [574]. Henseler *et al.* [491] are interested in the computation of scattering frequencies for symmetric configurations of two, three and four identical soft spheres; these correspond to complex frequencies for which the determinant of the system matrix (given by (4.60) for sound-hard spheres) is zero. For the scattering of thermal waves by coated spheres, see [382].

4.9 Electromagnetic waves

The classic scattering problem in electromagnetics is the transmission problem for a homogeneous sphere; the solution using the method of separation of variables and vector spherical wavefunctions is often known as *Mie theory* because of the famous 1908 paper by Mie [849, 715].

In fact, this problem has a long history which is described by Logan [743]. Derivations are given in many books, for example, [73, §17], [1149, §9.25], [1233, Chapter 9], [564, §8.19], [549, §16.9], [111, Chapter 4], [121, §13.5], [598], [431, Chapter 3], [1190, Chapter 1, §6.1] and [270, §3.6]; see also [1279] for computational aspects. The simpler problem of scattering by a perfectly-conducting sphere is discussed in [884, pp. 1882–1885], [940, §12-9] and [135, §10.4].

There are many papers on electromagnetic scattering by inhomogeneous spheres. For an early treatment, see the paper by Wyatt [1343]; see also [117, §4.2]. One popular technique is to replace the inhomogeneous sphere by many concentric spherical layers, and then to use a simple approximation to the refractive index within each layer; see, for example, [955]. Kai & Massoli [579] have reported the results of computations with as many as 10 000 layers. For more information on inhomogeneous scatterers, see [866].

The scattering of plane electromagnetic waves by a pair of dielectric spheres was first considered by Trinks in 1935 [1184]. He used an orders-of-scattering approach; thus, he scattered a plane wave from one sphere (using Mie theory) and then attempted to calculate how this scattered field was scattered by the second sphere. Furthermore, he approximated the scattered fields using an electric dipole only, arguing that this was reasonable for $ka \ll 1$, where both spheres have radius a. Higher-order terms were subsequently included by Germogenova [403]. Corrections to both papers were made by Levine & Olaofe [707]. All three papers mentioned in this paragraph used scalar potentials and ad hoc methods in order to effect the re-expansion of fields about two different origins.

Meanwhile, addition theorems for vector spherical wavefunctions had appeared; see Section 3.17 for references. This paved the way for a more direct treatment, using vector spherical wavefunctions, and this was given by Liang & Lo in 1967 [712] for two perfectly-conducting spheres. They obtained the electromagnetic equivalent of (4.56) and (4.57), but they did not solve this system directly, they iterated it once. It was Bruning & Lo [145, 146] who first solved the linear system 'exactly': they used the Gauss–Seidel iterative method. They considered two spheres with possibly different radii and different internal properties, and presented many numerical results, using recursion formulae for the Gaunt coefficients. They also gave numerical results for two perfectly-conducting spheres. Summarising, the

> *landmark investigations into dependent scattering by spheres clearly were made by Liang and Lo* [712] *and especially by Bruning and Lo* [145, 146] *in the late 1960's. This constituted the first comprehensive, computationally viable*

solution to the problem, complete with experimental verification. These authors assimilated the relevant theorems into a workable theory, which was then applied to the calculation of radar or bistatic cross sections....

(Fuller [374, p. 4718])

Many papers on electromagnetic scattering by clusters of spheres have appeared since those of Bruning & Lo [145, 146]. Examples are [119, 400, 375, 376, 473, 374, 763, 990, 764, 536, 868, 1345, 1386, 1033, 1348]. The paper by Fuller [374] contains a good review of the methods described above; see also [870] and [117]. Mishchenko [865] used these methods to demonstrate that the phenomenon of 'enhanced backscattering' was present with just two spheres; previously, it was known that multiple scattering was involved but it was thought that many scatterers were needed. For some experimental results on scattering by two spheres, see [1304].

4.10 Elastic waves

The scattering of elastic waves by a sphere was first considered by Clebsch in 1863. He supposed that the sphere was rigid ($\boldsymbol{u} = \mathbf{0}$ on $r = a$) and that the incident waves were generated by a point source. His paper is described in some detail in Logan's review [743]; see also [944, p. 18]. It is interesting to note that the elastic (vector) problem was solved before Rayleigh published his solution of the corresponding scalar problem!

There is also much related early work on the vibrations of elastic spheres; see, for example, [745, Chapter XII]. This topic continues to be of interest to global seismologists [83, Chapter 6], [250, Chapter 8]. Finite-element results for vibrations of anisotropic spheres can be found in [153], together with further references.

In 1927, Sezawa [1096] published a paper in which formal solutions to several problems were given, including the scattering of elastic waves by circular (and elliptic) cylinders and spheres; the cylinders and spheres could be cavities, rigid or movable. At about the same time, Herzfeld [494] considered scattering by an elastic sphere in a fluid of small viscosity; he neglected the scattered shear waves.

Elastic waves exterior to single circular cylinders are considered in [1315, 1217] and in [944, Chapter III]. For movable rigid circular cylinders, see [854]. Early studies of acoustic scattering by an elastic cylinder and by an elastic sphere were made by Faran [339]. These solutions are described in Hackman's review [462], in [1219] and in the book by Veksler [1261].

The problem of elastic-wave scattering by single spheres was revived in the late 1950s. Sivukhin [1110] and Ying & Truell [1363] considered P-wave scattering by a spherical cavity. Ying & Truell [1363] also considered fixed rigid spheres and homogeneous elastic spheres (inclusions); numerical results are given in [636, 560], and fluid spheres are treated in [307]. (See also [1185, §32] and [429, §7.1.2].) The analogous problems for incident shear waves are treated in [637, 308, 650, 819].

Several errors in [1363, 308, 560] are corrected in [1291, 446]. Movable rigid spheres are considered in [1333, 943]. The scattering of elastic waves from a point source by a spherical cavity is analysed in [899, 34, 35]. Radiation from a spherical cavity is discussed in [83, §4.6].

The 1973 book by Pao & Mow [944, Chapter IV] contains a detailed description of the use of the method of separation of variables for elastodynamic problems involving one spherical scatterer. Specific results are given for P-wave scattering by a fluid sphere, a movable rigid sphere, a cavity and an elastic inclusion. Numerical results are given for P-wave scattering by a spherical cavity, and when the incident waves are generated by a compressional point source.

Norris [918] has given a systematic treatment of scattering by a homogeneous spherical inclusion, covering both incident P-waves and incident S-waves. Korneev & Johnson [647] also examined this problem for incident P-waves. They noted that much 'work on the scattering of elastic waves by a spherical inclusion has been associated with the names of G.I. Petrashen in the USSR and R. Truell in the USA' [647, p. 231], and gave a good description of the work of Petrashen and his students. For another lengthy review, see [139]. For transient problems involving solid spheres, see the book by Gorshkov & Tarlakovsky [426].

The scattering of P-waves by N identical circular cylinders has been considered by several authors. For example, Glazanov [409] extended Záviška's method to a row of N cylinders, with equally-spaced collinear centres, and then let $N \to \infty$, obtaining results for a grating of cylindrical cavities; see also [421]. More recently, Robert *et al.* [1024] have given numerical results for scattering by two identical cavities, and by two and three closely-spaced fluid-filled cavities. Cheng used an orders-of-scattering approach and gave numerical results for two rigid circular cylinders [192] and for two circular cavities [193]. For applications to fibre-reinforced materials, see [1247, 1060]. For doubly-periodic rectangular arrays of circular cavities and inclusions, see [984] and [843], respectively.

There are several Russian papers by Guz' and Golovchan on multiple scattering of elastic waves by circular cylinders and spheres; an example is [421]. Much of this work is presented in their book, published in Kiev in 1972; an English translation is available [458]. There is also a later review [459].

For a more numerical approach, one could surround a group of scatterers by a large circle (or sphere), and then use separation of variables outside combined with finite elements in the bounded region between the circle and the scatterers. Such hybrid methods have been used for the scattering of elastic waves by two cylinders [1097]. The use of separation of variables ensures that the governing differential equations and radiation condition are satisfied. The use of finite elements permits geometric flexibility, although remeshing is necessary if the geometry is altered.

The problem of scattering by a pipe in an elastic half-space has been approximated by the problem of scattering by two cylindrical cavities in a full-space, where one circular cylinder has a radius that is much larger than that of the other; see [269] and references therein. In fact, the two-dimensional problem of scattering by one circular cavity in an elastic half-plane can be solved exactly using the half-plane multipoles defined in Section 2.10. This method was developed by Gregory in 1970 [439]. Twenty years later, Gregory & Austin [441] developed a similar method for a semi-circular groove in a half-plane. They gave numerical results and, for an incident Rayleigh wave, they found good agreement with the experimental results in [1265, Fig. 26b].

The scattering of elastic waves by N identical spherical inclusions was considered by Mal & Bose [767]. They obtained the algebraic systems analogous to (4.60); layered inclusions ('onions') are treated in [1061]. Mal & Bose [767] went on to derive approximate solutions valid for small spheres, and then used their approximations to estimate the effective properties of a medium containing a random, dilute distribution of spherical inclusions. (For more on this topic, see Section 8.6.) Doyle [292] has used an orders-of-scattering approach, and given numerical results for 2, 649 and 736 spherical inclusions.

In [490], the authors give experimental results for scattering by a periodic row of 20 spherical inclusions embedded in a viscoelastic solid; the incident field is a P-wave propagating in the direction of the row.

Scattering by N rigid spheres is considered in [406]; numerical results are given for two spheres. Acoustic scattering by spherical elastic shells is considered in [517, 518].

In [741] and [988], the authors have considered scattering by a stack of planar layers of spheres, where each layer consists of an infinite periodic arrangement of identical spheres; different layers can have different arrangements. Experimental results for the scattering of a P-wave at normal incidence to a single planar periodic array of spherical inclusions are available [612, 613, 807, 808]; results for random planar arrays are also given in [612, 807].

The techniques used for elastodynamic problems can be extended to thermoviscous fluids. The main difference is that motions are governed by three different Helmholtz equations rather than two. For scattering by one sphere, see [316, 18] and [985, §4.3.4]. For axisymmetric oscillations of two spheres in a thermoviscous fluid, see [484]. For scattering by one poro-elastic sphere, see [93, 1380]; for two poro-elastic spheres, see [485, 486].

4.11 Water waves

> *The theory of ship motions progressed slowly but steadily during the first half of* [the 20th] *century, in the capable hands of such "heroes" of our field as Havelock, Kochin, and Weinblum. ... Lacking contemporary computing facilities, our heroes studied simple geometrical bodies ... Havelock's analysis* [488] *of the floating hemisphere was the ultimate contribution in that era, at the very limits of hand computation.*
>
> (Newman [908, p. 419])

Very few water-wave problems can be solved in closed form. The two main examples are one vertical (circular) cylinder in water of finite depth and one vertical barrier (thin rigid plate) in deep water. The first of these can be solved by separation of variables; see Section 1.6.4, [1303, p. 544], [1058, §6.2], [842, §7.5] and [731, §2.4.1]. For scattering by a thin vertical barrier, see [1303, p. 529], [327] and several chapters in [770].

The radiation of water waves by the forced oscillations of a half-immersed horizontal circular cylinder was studied by Ursell [1222, 1223, 1229], using multipole expansions. For the scattering ('diffraction') problem, suppose that the wetted surface is given by $r = a$, $0 < \theta < \pi$, and that the incident wave is given by

$$\phi_{\rm inc} = {\rm e}^{-Kz+{\rm i}Kx} = \sum_{m=0}^{\infty} \frac{({\rm i}Kr)^m}{m!}\,{\rm e}^{{\rm i}m\theta}.$$

For the scattered wave, we write

$$\phi_{\rm sc} = \sum_{m=0}^{\infty} c_m \Phi_m(r, \theta),$$

where the coefficients c_m are to be found. The multipole potentials Φ_n with $n \geq 2$ are wavefree potentials; see Definition 2.18. The potential Φ_0 is a wave source whereas Φ_1 is a horizontal dipole; see (2.79). By construction, $\phi_{\rm sc}$ satisfies Laplace's equation in the water, the free-surface boundary condition and the radiation condition. The boundary condition on the cylinder, (1.57), gives

$$\sum_{m=0}^{\infty} c_m \Psi_m(\theta) = f(\theta), \quad 0 < \theta < \pi, \tag{4.64}$$

where

$$\Psi_m(\theta) = a\,\frac{\partial \Phi_m}{\partial r} \quad \text{evaluated on } r = a$$

and

$$f(\theta) = -\sum_{m=1}^{\infty} \frac{({\rm i}Ka)^m}{(m-1)!}{\rm e}^{{\rm i}m\theta} = -{\rm i}Ka\,\exp({\rm i}Ka{\rm e}^{{\rm i}\theta} + {\rm i}\theta).$$

It remains to solve (4.64) for c_m.

The functions $\{\Psi_m(\theta)\}$ are not orthogonal over $0 < \theta < \pi$. One way to proceed is as follows. Multiply (4.64) by $\cos n\theta$ and integrate, giving

$$\sum_{m=0}^{\infty} A_{nm} c_m = f_n, \quad n = 0, 1, 2, \dots, \tag{4.65}$$

where

$$A_{nm} = \int_0^{\pi} \Psi_m(\theta) \cos n\theta \,{\rm d}\theta \quad \text{and} \quad f_n = \int_0^{\pi} f(\theta) \cos n\theta \,{\rm d}\theta;$$

the integrals A_{nm} with $m \geq 2$ are elementary. Thus, we have an infinite system of linear algebraic equations to solve for c_m. For a detailed exposition, with numerical results, see [797].

As an alternative, we could truncate the sum in (4.64) at $m = M$, say, and then simply collocate (4.64) at selected values of θ.

Of particular interest are the reflection and transmission coefficients, R and T, defined by

$$\phi_{\text{inc}} + \phi_{\text{sc}} \sim \begin{cases} T\mathrm{e}^{-Kz+\mathrm{i}Kx} & \text{as } x \to \infty, \\ \mathrm{e}^{-Kz}(\mathrm{e}^{\mathrm{i}Kx} + R\mathrm{e}^{-\mathrm{i}Kx}) & \text{as } x \to -\infty. \end{cases}$$

Making use of (2.80), we obtain

$$R = \pi(\mathrm{i}c_0 - c_1) \quad \text{and} \quad T = 1 + \pi(\mathrm{i}c_0 + c_1).$$

These coefficients are tabulated in [797] for $0.01 \le Ka \le 10$.

The basic multipole method can be extended to other problems. For one completely submerged circular cylinder in deep water, see [1224, 1225, 840] and [731, §3.1.1]; the striking result here is that $R \equiv 0$, for all radii and for all depths of submergence.

Multipoles are known for water of constant finite depth; see [1176] and [731, Appendix B]. For radiation from an oscillating half-immersed circular cylinder, see [1369, 1062].

Spherical multipoles have also been used. In deep water, the wavefree potentials are given by Definition 3.39 with source potentials defined by (3.143). For the axisymmetric waves generated by a hemisphere oscillating vertically, see [488, 522], [1303, p. 562] and [731, §3.1.2]. For non-axisymmetric solutions, see [436, 522]. For a submerged sphere in deep water, see [1137, 1281]. For a submerged sphere in water of constant finite depth, see [724].

In the 1970s, an interest in extracting energy from ocean waves developed [241, 824]. Hydrodynamic theory was used to understand the properties of various wave-power devices [328]. In order to account for the interactions within arrays of devices, the notion of *point absorber* was introduced [154, 336]: 'These "point absorbers" can be defined as each possessing a vertical axis of symmetry and being such that the wave field created by the motion of any one of them is not influenced by the presence of the remaining array members' [1173, p. 67]. In other words, the multiple-scattering problem was treated using the single-scattering approximation; see Section 1.1.1. In these analyses, Havelock's solution for one heaving sphere [488] was used; for example, see [1173] for arrays of five and ten identical hemispheres. Kyllingstad [678] also invoked the single-scattering approximation, but he replaced point absorbers with long-wave approximations for small scatterers. (For an overview, with applications to wave-power devices, see the book by Falnes [337].) Greenhow [436] went one step beyond single scattering, for scattering by two floating (movable) hemispheres: he calculated the effect on the first hemisphere of waves that have been scattered twice, once by the first hemisphere and once by the second hemisphere. (Embleton [313] did similar calculations for acoustic scattering by two spheres; see Section 4.7.1.)

The first multi-body problem to be studied extensively was the radiation problem for two half-immersed circular cylinders on deep water; this can be regarded as a simple model of a catamaran. Thus, Ohkusu [925] and Wang & Wahab [1282] extended Ursell's multipole method to analyse the heaving (vertical) motion of a catamaran consisting of two identical, rigidly-connected, half-immersed circular cylinders. Ohkusu [926] has also made similar calculations for the swaying and rolling motions of the same catamaran, whilst Sayer & Spencer [1064] have analysed the motions of two freely-floating identical circular cylinders.

The multipole method has also been used to treat problems involving circular cylinders that are completely submerged. Thus, Wang [1280] has presented extensive numerical results for two identical cylinders, each submerged to the same depth; see also [731, §3.2.1]. O'Leary [935] has shown how the method can be used for an arbitrary number of submerged cylinders; each cylinder can have any radius and any submergence depth. She gave numerical results for several configurations of two and three identical cylinders. Generalisations to trapping problems are made in [983].

Multipole expansions have also been used by Wu [1337] for several submerged spheres. He gives numerical results for N identical spheres, with their centres located at the N vertices of a regular horizontal polygon.

Finally, we note that plane-wave scattering by vertical cylinders in water of constant finite depth leads to a two-dimensional acoustic problem. Thus, for circular cylinders, the method of Záviška can be applied; see Section 4.5. However, if the incident field is not a simple plane wave, then the problem is more complicated: evanescent modes will be excited. As an illustration, we consider a representative problem below.

4.11.1 An exact Green's function

Consider a vertical circular cylinder in water of depth h. The axis of the cylinder is along the z-axis. The incident waves are generated by a point source at $(x, y, z) = (\xi, \eta, \zeta)$. From [559, eqn (A16)], this incident field (which corresponds to a fundamental solution) is given by

$$G(x, y, z; \xi, \eta, \zeta) = C_0 H_0(k\rho) Z_0(z) Z_0(\zeta) + \sum_{n=1}^{\infty} C_n K_0(k_n \rho) Z_n(z) Z_n(\zeta), \tag{4.66}$$

where $K_n(w)$ is a modified Bessel function, $\rho^2 = (x-\xi)^2 + (y-\eta)^2$, k is the positive solution of (1.60), k_n solves (1.65) and the vertical eigenfunction Z_n is defined by (1.64). The coefficients are given by

$$C_0 = \frac{\mathrm{i}(K^2 - k^2)}{K - (K^2 - k^2)h} \quad \text{and} \quad C_n = \frac{2(K^2 + k_n^2)}{\pi[K - (K^2 + k_n^2)h]}$$

for $n = 1, 2, \ldots$. (G has been normalised so that (3.165) is satisfied.)

The scattered field has the representation (1.63), namely

$$\phi_{\mathrm{sc}} = \sum_{n=0}^{\infty} \varphi_n(r,\theta)\, Z_n(z), \tag{4.67}$$

where

$$\varphi_0(r,\theta) = \sum_m c_m^0 H_m(kr)\,\mathrm{e}^{\mathrm{i}m\theta}, \qquad \varphi_n(r,\theta) = \sum_m c_m^n K_m(k_n r)\,\mathrm{e}^{\mathrm{i}m\theta}$$

and the coefficients c_m^n are to be found. Here, $x = r\cos\theta$ and $y = r\sin\theta$. The strategy for determining c_m^n is clear. The boundary condition on the cylinder is

$$\frac{\partial}{\partial r}(G+\phi_{\mathrm{sc}}) = 0 \quad \text{for } r = a,\ -\pi \le \theta < \pi,\ 0 < z < h.$$

Then, the orthogonality relation (1.66) reduces the problem to finding $\varphi_n(r,\theta)$, $n = 0, 1, 2, \ldots$. For φ_0, we expand $H_0(k\rho)$ in (4.66) using the addition theorem, (2.29), whereas for φ_n, we use a similar addition theorem for $K_0(k_n\rho)$; see, for example, [1298, p. 361] or [691, eqn (5.12.15)]. This leads to an expansion of G in terms of functions of r (namely $J_m(kr)$ and $I_m(k_n r)$) and $\mathrm{e}^{\mathrm{i}m\theta}$, so that the boundary condition on $r = a$ can be imposed easily, using the orthogonality of $\mathrm{e}^{\mathrm{i}m\theta}$.

4.11.2 Matched eigenfunction expansions

The expansion (4.67) can always be used for outgoing solutions outside a vertical circular cylinder. The cylinder itself could enclose a floating body, or even several bodies. The potential in the water inside the cylinder could be represented using finite-element approximations (leading to a *hybrid element method* [842, §7.7]) or by another expansion of some kind. For example, consider scattering by a *truncated vertical cylinder*, with its wetted surface given by

$$\{(r,\theta,z) : r = a,\ 0 < z < d,\ -\pi \le \theta < \pi\}$$
$$\cup \{(r,\theta,z) : 0 \le r < a,\ z = d,\ -\pi \le \theta < \pi\},$$

with $d < h$; the first part is the curved part of the cylinder and the second part is the rigid circular base. The expansion (4.67) can be used for $r > a$. In the water below the cylinder, that is, where $r < a$ and $d < z < h$, one can construct a different expansion, using separated solution of Laplace's equation in cylindrical polar coordinates; the expansion must satisfy the boundary conditions at $z = d$ and at $z = h$, and it must be regular at $r = 0$. The boundary condition must be enforced at $r = a$ for $0 < z < d$. The two expansions must be matched across the interface $r = a$ for $d < z < h$, meaning that the potential and its radial derivative must both be continuous across the interface. This leads to a viable numerical method for such problems: it is called the *method of matched eigenfunction expansions*. McIver [828] has used this method to study the interaction between two-dimensional water waves and a pair of immersed rectangular cylinders. For a pair of thin vertical barriers, see [982]. For a vertical stack of aligned horizontal plates, see [1278]. In [927, 815], the authors combined the method of matched eigenfunction expansions with the orders-of-scattering approach in order to calculate interactions between several axisymmetric bodies. For further applications and references, see [381, 1357] and [731, §2.5]. For a related method, in which the horizontal eigenfunctions are represented as integrals (see Chapter 5), see [542]; results for scattering by several truncated vertical cylinders are given.

The method of matched eigenfunction expansions can be used in many problems where the geometrical domain can be decomposed into separate domains in each of which separation of variables can be used. More generally, the following acoustic problem can be solved: a half-plane $x > 0$ has a rigid boundary at $x = 0$ apart from a finite number of gaps, through which sound can propagate into connecting tubes, ducts or bounded cavities (or 'grooves'). For such problems, one can use separated solutions in the cavities combined with Fourier integrals in $x > 0$. As an example, see [643, 1112] for scattering by N rectangular grooves.

4.12 Separation of variables in other coordinate systems

As we remarked in Section 4.1, there are four other coordinate systems in which the Helmholtz equation separates [41]. These are elliptic cylindrical coordinates, prolate spheroidal coordinates, oblate spheroidal coordinates and ellipsoidal coordinates. We shall discuss each of these, briefly, in turn.

4.12.1 Elliptic cylinders

Elliptic cylindrical coordinates, (ξ, η), are defined by (2.49). They can be used to solve the problem of acoustic scattering by one elliptic cylinder, exactly. Thus, consider an ellipse defined by $\xi = \xi_0$, with foci at $(x, y) = (\pm c, 0)$. For an incident plane wave, we have

$$u_{\text{inc}} = \mathrm{e}^{\mathrm{i}k(x\cos\alpha + y\sin\alpha)} = 2\sum_{\sigma,n} \mathrm{i}^n \, \hat{\psi}_{\sigma n}(\xi, \eta, q)\, e_{\sigma n}(\alpha, q),$$

where $e_{\sigma n}(\eta, q)$ is a Mathieu function defined by (2.53), $\hat{\psi}_{\sigma n}$ is a regular elliptical wavefunction (see Definition 2.15), $q = (\frac{1}{2}kc)^2$ and the shorthand notation (2.58) has been used. For the scattered field, we write

$$u_{\text{sc}} = 2\sum_{\sigma,n} \mathrm{i}^n \, B_n^\sigma \, \psi_{\sigma n}(\xi, \eta, q)\, e_{\sigma n}(\alpha, q), \tag{4.68}$$

where $\psi_{\sigma n}$ is an outgoing elliptical wavefunction (see Definition 2.15) and the coefficients B_n^σ are to be determined so as to satisfy the boundary condition on the cylinder. For example, if the cylinder is sound-soft, we have $u = u_{\text{inc}} + u_{\text{sc}} = 0$ on $\xi = \xi_0$, whence

$$\sum_{\sigma,n} \mathrm{i}^n \left\{ \hat{\psi}_{\sigma n}(\xi_0, \eta, q) + B_n^\sigma \, \psi_{\sigma n}(\xi_0, \eta, q) \right\} e_{\sigma n}(\alpha, q) = 0, \quad -\pi \leq \eta < \pi.$$

Then, the orthogonality of $e_{\sigma n}(\eta, q)$ (see (2.54)) gives

$$B_n^\sigma = -\frac{M_{\sigma n}^{(1)}(\xi_0, q)}{M_{\sigma n}^{(3)}(\xi_0, q)}.$$

The corresponding solution for a sound-hard elliptic cylinder has

$$B_n^\sigma = -\frac{M_{\sigma n}^{(1)\prime}(\xi_0, q)}{M_{\sigma n}^{(3)\prime}(\xi_0, q)}.$$

In these expressions, $M_{\sigma n}^{(j)}$ are (radial) Mathieu functions; see Section 2.9 for more information and references.

For scattering of acoustic and electromagnetic waves by one elliptic cylinder, see [886, 61, 162], [835, Chapter XIX], [135, Chapter 3], [944, Chapter IV] and [1139]. For applications to water-wave scattering by vertical elliptic cylinders, and related problems, see [1321, 1375] and references therein.

A similar method can be attempted for two-dimensional in-plane elastodynamic problems, such as scattering by an elliptic cavity. Thus, one constructs vector elliptical wavefunctions (see Section 2.10) and then represents the scattered field with a series analogous to (4.68). However, having imposed the boundary condition on $\xi = \xi_0$, it turns out that the terms in the resulting series are not orthogonal: one obtains a linear system of algebraic equations for the coefficients, not an explicit formula. This difficulty was noticed by Sezawa in 1927 [1096]; see also [1078] and [944, Chapter IV, §3.1]. It is typical with vector problems once one departs from circles and spheres.

Acoustic scattering by two elliptic cylinders can be treated by generalising the method of Section 4.4; the appropriate addition theorems for elliptical wavefunctions are given in Section 2.9.1. This method has been used by Sebak [1087, 1089, 1088]. It has also been used for the degenerate case of scattering by two thin strips, of arbitrary width and orientation [998], and also for the transmission problem of scattering by one elliptic cylinder containing another of arbitrary location and eccentricity [999]. For the orders-of-scattering approach for two cylinders, see [402, 474, 472]. Most of the papers mentioned in this paragraph make use of Særmark's form of the addition theorems [1046], which is not the most efficient form; see the remarks below Theorem 2.17.

For scattering by several identical elliptic cylinders, see [701, 702, 703].

4.12.2 Spheroids

A *triaxial ellipsoid* with (positive) semi-axes a_1, a_2 and a_3, is defined by

$$(x/a_1)^2 + (y/a_2)^2 + (z/a_3)^2 = 1, \tag{4.69}$$

using the notation of [262]. If $a_1 = a_2 = a$, say, the ellipsoid becomes a *spheroid*, an axially symmetric surface. The spheroid is known as *prolate* if $a_3 > a$ and *oblate* if $a_3 < a$; limiting cases are 'needles' and flat circular discs.

Spheroidal wavefunctions can be constructed by separating variables in prolate or oblate spheroidal coordinates. For prolate spheroids, see [884, pp. 1502–1511] and [135, Chapter 11]. For oblate spheroids, see [884, p. 1512] and [135, Chapter 13]. Standard references on spheroidal wavefunctions are [845], [322, §§16.9–16.13], [350] and [1, Chapter 21].

Early work on acoustic scattering by one spheroid is surveyed in [135, §§11.2, 11.3, 13.2 & 13.3]. For example, Spence & Granger [1135] and Einspruch & Barlow [306] gave results for plane-wave scattering by a hard and by a soft prolate spheroid, respectively.

Substantial applications of vector spheroidal wavefunctions to electromagnetic scattering by homogeneous dielectric (and perfectly conducting) spheroids date from the 1970s. Again, as for elastodynamic problems with elliptic cylindrical coordinates, one has to solve a linear system in order to compute the expansion coefficients. Key references are the papers by Asano & Yamamoto [44] and by Voshchinnikov & Farafonov [1269]; see also [710] and [870, §6.1]. For scattering of elastic waves by a rigid spheroid, see [929].

Addition theorems for scalar spheroidal wavefunctions are available in [845, §3.74] and [609, 1106, 765]. Acoustic radiation from two co-axial spheroids is studied in [1232]. Acoustic scattering by two parallel solid spheroids is studied in [634].

Addition theorems for vector spheroidal wavefunctions are also available [1106, 253]. They have been used extensively for scattering by two perfectly conducting, parallel spheroids [1107, 252], by two dielectric parallel spheroids [234] and by spheroids that are oriented arbitrarily [232, 233, 215]. All of these results are reviewed in Chapter 4 of [866].

4.12.3 Ellipsoids

A general triaxial ellipsoid is defined by (4.69). Ellipsoidal wavefunctions can be constructed by separating variables in ellipsoidal coordinates. However, these wavefunctions have not been studied extensively, and it does not seem worthwhile to do so if the goal is to solve scattering problems.

Ellipsoidal wavefunctions are discussed briefly in [322, §16.14]. A later survey is [41]. Sleeman [1115] solved the problem of plane-wave scattering by one ellipsoid, using ellipsoidal wavefunctions. Low-frequency scattering by ellipsoids is discussed in detail in [262, Chapter 8]. The expansion of radiating wavefunctions outside ellipsoids has been given by Dassios [260],

$$u_{sc}(\rho, \mu, \nu) = \frac{e^{ik\rho}}{\rho} \sum_{n=0}^{\infty} \frac{F_n(\mu, \nu)}{\rho^n},$$

where (ρ, μ, ν) are ellipsoidal coordinates and the functions F_n are given by a six-term recurrence relation, in general.

Although ellipsoidal wavefunctions have not been used extensively, problems involving scattering by ellipsoids have been solved numerically. This has been done using a variety of methods, most of which will be described in subsequent chapters.

5

Integral equation methods, I: basic theory and applications

The first use of integral equations in diffraction theory seems to be due to Rayleigh [1011, 1012], *who deduced approximate solutions for the diffraction of plane waves normally incident on a perfectly reflecting plane having an aperture or slit whose dimensions were small compared with the wave-length.*

(Baker & Copson [53, p. 153])

5.1 Introduction

Multiple-scattering problems can be treated using methods based on integral equations. For acoustic and electromagnetic problems, the theory has been given in detail in the book by Colton & Kress [223]; see also the review article by Kleinman & Roach [628]. Although these authors explicitly consider a single scatterer, many results extend trivially to N scatterers. Here, we try to emphasise those features that are different.

We start with a *fundamental solution*. The simplest is

$$G(P, Q) = -\frac{1}{2}\mathrm{i}\, H_0^{(1)}(kR) \tag{5.1}$$

in two dimensions, where $R = |\boldsymbol{r}_P - \boldsymbol{r}_Q|$ is the distance between the two points P and Q and $H_0^{(1)}$ is a Hankel function. In three dimensions, we take

$$G(P, Q) = -\exp(\mathrm{i}kR)/(2\pi R). \tag{5.2}$$

These fundamental solutions are usually called *free-space Green's functions*. For each fixed Q, they satisfy the Helmholtz equation with respect to P, for $P \neq Q$. They also satisfy the Sommerfeld radiation condition with respect to P, and they are symmetric, $G(P, Q) = G(Q, P)$, so that the roles of P and Q can be interchanged. $G(P, Q)$ is also singular when $P = Q$:

$$G(P, Q) \sim \pi^{-1} \log R \quad \text{as } R \to 0 \text{ in two dimensions} \tag{5.3}$$

and

$$G(P, Q) \sim -(2\pi R)^{-1} \quad \text{as } R \to 0 \text{ in three dimensions.} \tag{5.4}$$

The choice of the factors $-\frac{1}{2}\mathrm{i}$ and $-\frac{1}{2}/\pi$ in (5.1) and (5.2), respectively, is largely a matter of taste. Our choice leads to boundary integral equations such as $(I+K)\mu = f$ (see (5.51)), in both two and three dimensions, without any constant factors multiplying the identity I and without any constant factors in the definition of the integral operator K (see (5.7) below). Other normalisations can be (and frequently are) used.

Other fundamental solutions can be used instead of G, and this may be advantageous in some situations. We will discuss this option later.

5.2 Wave sources

Indeed, wave propagation problems have provided a beautiful training ground for budding applied mathematicians to sharpen their skills in analysis.

(Barnett [67, p. 45])

Physically, $G(P, Q)$ represents a wave source. It is usual to refer to G as a *point source*, even though *line source* may be more accurate in two dimensions. The waves are outgoing: in three dimensions, we have

$$\mathrm{Re}\left\{G(P, Q)\,\mathrm{e}^{-\mathrm{i}\omega t}\right\} = -(2\pi R)^{-1}\cos(kR - \omega t),$$

a wave propagating outwards, away from the source at $P = Q$. It is a spherically-symmetric wave, with an amplitude that decays as R^{-1}.

Similar results obtain in two dimensions, except that one needs the large-argument asymptotic approximation to the Hankel function,

$$H_0^{(1)}(kR) \sim (\pi k R/2)^{-1/2}\exp\{\mathrm{i}(kR - \frac{1}{4}\pi)\} \quad \text{as } R \to \infty,$$

whence

$$\mathrm{Re}\left\{G(P, Q)\,\mathrm{e}^{-\mathrm{i}\omega t}\right\} \sim (2\pi k R)^{-1/2}\sin(kR - \omega t - \frac{1}{4}\pi) \quad \text{as } R \to \infty.$$

The waves are outgoing at infinity, they are cylindrically symmetric and they decay as $R^{-1/2}$ as $R \to \infty$.

Note that, unlike for Laplace's equation, derivatives of G decay at the same rate. For example, in three dimensions, put Q at the origin and P at (x, y, z). Then, elementary differentiation gives

$$\frac{\partial G}{\partial z} = \frac{z}{R}\left(\mathrm{i}k - \frac{1}{R}\right)G,$$

and this decays at the same rate as G if $k \neq 0$ (and $z \neq 0$). This is typical: derivatives of static multipoles decay faster than the multipoles themselves, whereas differentiating dynamic multipoles (such as G) does not change the rate of decay. The computational consequences of this difference between static and dynamic problems have been emphasised by Chew [198].

5.3 Layer potentials

The idea of representing solutions as distributions of singular solutions is classical and extremely useful. The theory was worked out first for Laplace's equation; the standard book is by Kellogg [597]. (Kellogg, an American student of Hilbert, died in 1932 'from an unlooked-for heart attack' [101, p. 171].) For the Helmholtz equation, we use the book by Colton & Kress [223] and the review paper by Kleinman & Roach [628] as our standard references.

Before defining layer potentials, we need some notation. We assume that we have N bounded, simply-connected scatterers, B_i, $i = 1, 2, \ldots, N$. The boundary of B_i is S_i, assumed to be smooth. We define

$$B = \bigcup_{i=1}^{N} B_i \quad \text{and} \quad S = \bigcup_{i=1}^{N} S_i,$$

so that B is the collection of all the interiors of the N scatterers and S is all their boundaries. The unbounded connected exterior is denoted by B_e.

A *single-layer* potential is defined by

$$(S\mu)(P) = \int_S \mu(q)\, G(P, q)\, ds_q, \quad P \in B \cup B_e. \tag{5.5}$$

Here, $\mu(q)$ is called a *density function*; it is defined for $q \in S$. (There should not be any confusion between the boundaries S and the single-layer $S\mu$; the latter will always appear with a density.) Given μ, $(S\mu)(P)$ satisfies the Helmholtz equation and the Sommerfeld radiation condition; these properties are inherited from G. Also, $(S\mu)(P)$ is continuous in P as P crosses S. In particular, $(S\mu)(p)$ is defined as an ordinary improper integral when $p \in S$.

The normal derivative of $S\mu$ is discontinuous across S. Its values on S are given by

$$\frac{\partial}{\partial n_p} S\mu = (\pm I + K)\mu, \tag{5.6}$$

where the upper (lower) sign corresponds to $P \to p \in S$ from the exterior B_e (the interior B). I is the identity operator and K is a boundary integral operator defined by

$$(K\mu)(p) = \int_S \mu(q) \frac{\partial}{\partial n_p} G(p, q)\, ds_q, \quad p \in S. \tag{5.7}$$

In (5.6) and (5.7), $\partial/\partial n_p$ means the normal derivative at $p \in S$; explicitly,

$$\frac{\partial}{\partial n_p} G(p, q) = \boldsymbol{n}(p) \cdot (\operatorname{grad}_P G(P, q))\Big|_{P=p},$$

where grad_P is the gradient operator with respect to the coordinates of P and $\boldsymbol{n}(p)$ is the unit normal vector at $p \in S$ pointing into B_e.

The formulae (5.5)–(5.7) are valid if μ is continuous on S, although this condition can be relaxed. Also, if S is sufficiently smooth (at least twice continuously differentiable), the integrand in (5.7) is continuous in two dimensions (see Section 5.4.2) and weakly singular in three dimensions.

Physically, $(S\mu)(P)$ is the field at P due to a (continuous) distribution of sources over S; the field will vary as μ is varied. Instead of sources, we can also consider *normal dipoles* distributed over S. This gives a *double-layer potential*, defined by

$$(D\nu)(P) = \int_S \nu(q) \frac{\partial}{\partial n_q} G(P, q)\, \mathrm{d}s_q, \quad P \in B \cup B_e, \tag{5.8}$$

where

$$\frac{\partial}{\partial n_q} G(P, q) = \boldsymbol{n}(q) \cdot \left(\text{grad}_Q G(P, Q)\right)\Big|_{Q=q}.$$

$(D\nu)(P)$ is discontinuous across S; its jump behaviour is given by

$$D\nu = (\mp I + \overline{K^*})\nu, \tag{5.9}$$

where, as before, the upper (lower) sign corresponds to $P \to p \in S$ from B_e (B). The boundary integral operator $\overline{K^*}$ is defined by

$$(\overline{K^*}\nu)(p) = \int_S \nu(q) \frac{\partial}{\partial n_q} G(p, q)\, \mathrm{d}s_q, \quad p \in S. \tag{5.10}$$

$\overline{K^*}$ is the Hermitian adjoint of K; the asterisk denotes the adjoint with respect to the inner product in $L^2(S)$, which is itself defined by

$$(u, v) = \int_S u(q)\, \overline{v(q)}\, \mathrm{d}s_q, \tag{5.11}$$

where the overbar denotes complex conjugation. Thus, we see that $(Ku, v) = (u, K^* v)$. Also, we have $S^*\mu = \overline{S}\mu$.

From (5.9), we also have

$$2\nu(p) = \lim_{P_1 \to p} (D\nu)(P_1) - \lim_{P_2 \to p} (D\nu)(P_2), \tag{5.12}$$

where, in this formula, $P_1 \in B$ and $P_2 \in B_e$.

Equations (5.9) and (5.12) are valid if the density ν is continuous, for example. Again, if S is sufficiently smooth, the integrand in (5.10) is continuous in two dimensions and weakly singular in three dimensions.

The operators S, K and $\overline{K^*}$ are well-studied boundary integral operators. We shall also require the normal derivative of the double-layer potential. The treatment of the corresponding operator has generated an enormous literature; the salient points will be summarised next.

5.3.1 The normal derivative of a double-layer potential

We can define the normal derivative of $D\nu$ at $p \in S$ if the density ν is smooth enough. A sufficient condition is that ν has uniformly Hölder-continuous first tangential derivatives on S [223, Theorem 2.23]. Then, we can define

$$(N\nu)(p) = \frac{\partial}{\partial n_p} \int_S \nu(q) \frac{\partial}{\partial n_q} G(p, q)\, \mathrm{d}s_q, \quad p \in S, \tag{5.13}$$

which is often called a *hypersingular operator*.

It is known that the normal derivative of $(D\nu)(P)$ is continuous across S if ν is (merely) continuous on S [223, Theorem 2.21].

The definition (5.13) has caused many difficulties in the literature on diffraction theory. Specifically, consider a flat region $\mathcal{T}$ in the plane $z = 0$, where $Oxyz$ are Cartesian coordinates. Then, (5.13) reduces to

$$(N\nu)(x, y) = -\frac{1}{2\pi} \lim_{z\to 0} \frac{\partial}{\partial z} \int_{\mathcal{T}} \nu(x', y') \left\{ \lim_{z'\to 0} \frac{\partial}{\partial z'} \left(\frac{\mathrm{e}^{\mathrm{i}kR}}{R} \right) \right\} \mathrm{d}x'\, \mathrm{d}y', \tag{5.14}$$

where $R = \{(x-x')^2 + (y-y')^2 + (z-z')^2\}^{1/2}$. In his well-known review, Bouwkamp noted that

> *many authors (e.g. Levine and Schwinger* [706]*) … differentiate with respect to z under the sign of integration in* (5.14), *so as to obtain formally a pure integral equation with the (non-integrable!) kernel*
>
> $$\left[\frac{\partial^2}{\partial z\, \partial z'} \left(\frac{\mathrm{e}^{\mathrm{i}kR}}{R} \right) \right]_{z=z'=0}.$$
>
> *This procedure eventually leads to all kinds of divergent integrals* (*Baker and Copson* [53, p. 186, footnote]) *and should therefore be rejected.*
>
> (Bouwkamp [134, p. 40])

Thus, it is tempting simply to take the normal derivative $\partial/\partial n_p$ in (5.13) under the integral sign, but this leads to a non-integrable integrand: there is a strong singularity at $R = 0$.

It is helpful to consider an explicit example.

Example 5.1 In (5.14), let us take $x = y = 0$ and $\nu = \nu_0$, a constant, and suppose that $\mathcal{T}$ is a circular disc of radius a, centred at the origin. Introducing plane polar coordinates, $x' = r\cos\theta$ and $y' = r\sin\theta$, we obtain

$$(N\nu_0)(0, 0) = -\frac{\nu_0}{2\pi} \lim_{z\to 0} \frac{\partial}{\partial z} \int_0^{2\pi} \int_0^a \left\{ \lim_{z'\to 0} \frac{\partial}{\partial z'} \left(\frac{\mathrm{e}^{\mathrm{i}kR}}{R} \right) \right\} r\, \mathrm{d}r\, \mathrm{d}\theta$$

$$= \nu_0 \lim_{z\to 0} \frac{\partial}{\partial z} \left\{ z \int_0^a \left(\frac{\mathrm{i}k}{R_0^2} - \frac{1}{R_0^3} \right) \mathrm{e}^{\mathrm{i}kR_0}\, r\, \mathrm{d}r \right\},$$

where $R_0^2 = \sqrt{r^2+z^2}$. The integrand is $(\mathrm{d}/\mathrm{d}r)(\mathrm{e}^{\mathrm{i}kR_0}/R_0)$, and so

$$(N\nu_0)(0,0) = \nu_0 \lim_{z\to 0} \frac{\partial}{\partial z}\left\{ z\left(\frac{\mathrm{e}^{\mathrm{i}k\sqrt{a^2+z^2}}}{\sqrt{a^2+z^2}} - \frac{\mathrm{e}^{\mathrm{i}kz}}{z} \right)\right\}$$
$$= (\nu_0/a)\left(\mathrm{e}^{\mathrm{i}ka} - \mathrm{i}ka\right).$$

This shows, for this example at least, that $N\nu$ is well defined, provided the operations are carried out in the correct order.

For flat regions $\mathcal{T}$, we can easily derive alternative expressions for $N\nu$. When $R \neq 0$, we have

$$\frac{\partial^2 G}{\partial z\,\partial z'} = -\frac{\partial^2 G}{\partial z'^2} = \frac{\partial^2 G}{\partial x'^2} + \frac{\partial^2 G}{\partial y'^2} + k^2 G$$
$$= \frac{\partial^2 G}{\partial x^2} + \frac{\partial^2 G}{\partial y^2} + k^2 G. \tag{5.15}$$

The last equality suggests that, *when $\mathcal{T}$ is flat,*

$$(N\nu)(x,y) = \left(\frac{\partial^2}{\partial x^2} + \frac{\partial^2}{\partial y^2} + k^2 \right)(S\nu)(x,y). \tag{5.16}$$

The second equality in (5.15) suggests that, if $\nu = 0$ around the edge of $\mathcal{T}$, we can integrate by parts (actually, we use the two-dimensional divergence theorem in the region $\mathcal{T}$) in order to move one tangential derivative from G onto ν, giving

$$(N\nu)(x,y) = ⨍_{\mathcal{T}} \left(-\frac{\partial\nu}{\partial x'}\frac{\partial G}{\partial x'} - \frac{\partial\nu}{\partial y'}\frac{\partial G}{\partial y'} + k^2\nu G \right) \mathrm{d}x'\,\mathrm{d}y'$$
$$= ⨍_{\mathcal{T}} \left(\frac{\partial\nu}{\partial x'}\frac{\partial G}{\partial x} + \frac{\partial\nu}{\partial y'}\frac{\partial G}{\partial y} + k^2\nu G \right) \mathrm{d}x'\,\mathrm{d}y', \tag{5.17}$$

where the integral must be interpreted as a two-dimensional Cauchy principal-value integral; see Appendix F.

Equations (5.16) and (5.17) are given by Bouwkamp [134, p. 40] and by Hönl *et al.* [512, §§47, 48]. Equation (5.16) is also given as equation (6.16) in [53]. Two-dimensional versions of both formulae can be found in [512, §49], whereas Sommerfeld [1129, §39 A], Bouwkamp [134, p. 43] and Jones [564, p. 622] give the two-dimensional version of (5.16).

Equation (5.17) is an example of a *regularisation*: the singularity in the integrand has been weakened. In fact, (5.17) can be generalised to curved surfaces:

Theorem 5.2

$$(N\nu)(p) = ⨍_S \{ (\boldsymbol{n}(q) \times \operatorname{grad}_q \nu) \cdot (\boldsymbol{n}(p) \times \operatorname{grad}_p G(p,q))$$
$$+ k^2\, \boldsymbol{n}(p)\cdot\boldsymbol{n}(q)\, G(p,q)\, \nu(q)\}\,\mathrm{d}s_q. \tag{5.18}$$

This formula was first given by Maue [813]. A proof is given in Appendix G. Equation (5.18) shows that $N\nu$ can be written as an integral involving ν and its tangential derivatives. It is valid provided that S is a closed surface (or a collection of closed surfaces). It is also valid for open surfaces, but only if ν vanishes around the boundary of S, ∂S, otherwise there is an additional line-integral contribution from ∂S; see Appendix G.

Another regularisation, which is useful when solving the hypersingular integral equation $N\nu = f$ by a Galerkin method, can be obtained from Maue's formula.

Theorem 5.3 *Let* $\boldsymbol{a}(q) = \boldsymbol{n}(q) \times \operatorname{grad}_q \nu$ *and* $\boldsymbol{b}(q) = \boldsymbol{n}(q) \times \operatorname{grad}_q \psi$ *for* $q \in S$, *a closed surface. Then*

$$\int_S (N\nu)\,\psi\,\mathrm{d}s = -\int_S \boldsymbol{b}\cdot(S\boldsymbol{a})\,\mathrm{d}s + k^2 \int_S \psi\,\boldsymbol{n}\cdot\{S(\nu\boldsymbol{n})\}\,\mathrm{d}s, \tag{5.19}$$

where, in the integrands, S is the acoustic single-layer operator.

Proof Suppose that $\psi(p)$ is the restriction to S of a function (also called ψ) defined in a three-dimensional neighbourhood of S. If we multiply Maue's formula (5.18) by $\psi(p)$ and integrate over S, we see that the desired result will follow if

$$\int_S \psi(p) \oint_S \boldsymbol{a}(q)\cdot\left\{\boldsymbol{n}(p)\times\operatorname{grad}_p G(p,q)\right\}\mathrm{d}s_q\,\mathrm{d}s_p = -\int_S \boldsymbol{b}\cdot(S\boldsymbol{a})\,\mathrm{d}s. \tag{5.20}$$

Now, $\boldsymbol{a}(q)\cdot\left\{\boldsymbol{n}(p)\times\operatorname{grad}_p G\right\} = \boldsymbol{n}(p)\cdot\operatorname{curl}_p\{G\boldsymbol{a}(q)\}$, so that the left-hand side of (5.20) is

$$\begin{aligned}\int_S \boldsymbol{n}\cdot\{\psi\operatorname{curl}(S\boldsymbol{a})\}\,\mathrm{d}s &= \int_S \boldsymbol{n}\cdot\{\operatorname{curl}(\psi\, S\boldsymbol{a}) - (\operatorname{grad}\psi)\times(S\boldsymbol{a})\}\,\mathrm{d}s \\ &= \int_S \boldsymbol{n}\cdot\operatorname{curl}(\psi\, S\boldsymbol{a})\,\mathrm{d}s - \int_S (S\boldsymbol{a})\cdot\boldsymbol{b}\,\mathrm{d}s,\end{aligned}$$

which reduces to the right-hand side of (5.20) when we use Stokes's theorem (recalling that S is closed). □

The formula (5.19) is due to Nedelec [902]; see also [836, Theorem 9.15] and [903, Theorem 3.4.2]. From (5.19), we can also see that

$$\int_S (N\nu)\,\psi\,\mathrm{d}s = \int_S (N\psi)\,\nu\,\mathrm{d}s,$$

whence $N^* = \overline{N}$; see (5.11).

An alternative to regularisation is provided by the following formula:

$$\frac{\partial}{\partial n_p}\int_S \nu(q)\,\frac{\partial}{\partial n_q}G(p,q)\,\mathrm{d}s_q = \int\!\!\!\!\times_S \nu(q)\,\frac{\partial^2}{\partial n_p\,\partial n_q}G(p,q)\,\mathrm{d}s_q, \quad p\in S. \tag{5.21}$$

Here, the '×' on the integral sign means that the integral must be interpreted as a *Hadamard finite-part integral* [466]. (Our notation follows Mangler [772]; other notations are common.) Finite-part integrals are discussed in Appendix D.

The formula (5.21) is valid provided ν is smooth enough ($\nu \in C^{1,\alpha}$). Note that the left-hand side is defined as a limit as $P \to p$; first one integrates over S with P not on S, and then one computes the normal derivative. On the right-hand side, the integrand is not integrable in the ordinary sense (or in the Cauchy principal-value sense), but it is integrable in the finite-part sense. In other words, we can apply the normal derivative at p under the integral sign, provided we interpret the integral as a finite-part integral.

A proof of (5.21) in two dimensions is given in Appendix E; see Theorem 5.4. For three dimensions, see [836, pp. 223 & 249].

Observe that the two sides of (5.21) are defined independently of each other: the left-hand side could be defined by (5.18), for example, whereas the right-hand side is defined by Definition D.5 when S is a curve and by Definition F.4 when S is a surface. Actually, some authors (for example, Filippi [349, p. 480]) take (5.21) as *defining* the finite-part integral on the right-hand side.

5.4 Explicit formulae in two dimensions

For several purposes, including analytical clarity and computational treatments, it is worthwhile to give explicit formulae for the various integral operators defined above. We do this in two dimensions by parametrising the curve S; recall that S has N disjoint components. Thus, we suppose that

$$S = \{(x, y) : x = x(t),\ \ y = y(t),\ \ 0 \le t \le 1\}.$$

We assume that the point (x, y) moves around (each component of) S in the anti-clockwise sense as t increases. We suppose that

$$v(t) = \left\{[x'(t)]^2 + [y'(t)]^2\right\}^{1/2} > 0 \quad \text{for } 0 \le t \le 1.$$

We put $v \equiv v(s)$, $0 \le s \le 1$, and we locate points p and q on S using $p = (x(s),\ y(s))$ and $q = (x(t),\ y(t))$.

5.4.1 The single-layer operator

Define

$$\mathcal{R}(s, t) = |\boldsymbol{r}_p - \boldsymbol{r}_q| = \left\{[x(t) - x(s)]^2 + [y(t) - y(s)]^2\right\}^{1/2} \ge 0.$$

Then, as $\mathrm{d}s_q = v(t)\,\mathrm{d}t$, we obtain

$$(S\mu)(p) = \int_0^1 M(s, t)\,\varphi(t)\,\mathrm{d}t,$$

where $\varphi(t) = \mu(q(t))$ and

$$M(s, t) = -\frac{1}{2}\mathrm{i}\,H_0^{(1)}(k\mathcal{R}(s, t))\,v(t)$$

is logarithmically singular as $|s-t| \to 0$:

$$M(s,t) \sim \pi^{-1} v \log(k\mathcal{R}) \sim \pi^{-1} v \log|s-t|$$

as $|s-t| \to 0$, using (5.3) and

$$\mathcal{R}(s,t) \sim |s-t|\, v \quad \text{as } |s-t| \to 0. \tag{5.22}$$

5.4.2 The operators K and $\overline{K^*}$

For these operators, we use

$$\boldsymbol{n}(q) = (y'(t), -x'(t))/v(t);$$

this unit normal vector at $q \in S$ points into B_e. There is a similar expression for $\boldsymbol{n}(p)$. Then, we find that

$$(K\mu)(p) = \int_0^1 L(s,t)\, \varphi(t)\, dt,$$

where

$$L(s,t) = \frac{ik}{2\mathcal{R}} \frac{v(t)}{v(s)} H_1^{(1)}(k\mathcal{R}) \{y'(s)[x(s)-x(t)] - x'(s)[y(s)-y(t)]\}$$

and we have used $(d/dw)H_0^{(1)}(w) = -H_1^{(1)}(w)$. Provided S is at least twice continuously differentiable, $L(s,t)$ is continuous: it is bounded as $|s-t| \to 0$. Taylor-series expansions show that

$$y'(s)[x(s)-x(t)] - x'(s)[y(s)-y(t)] \sim \frac{1}{2}(s-t)^2[x'(s)\,y''(s) - x''(s)\,y'(s)]$$

as $|s-t| \to 0$. Then, $H_1^{(1)}(w) \sim 2/(i\pi w)$ as $w \to 0$ and (5.22) show that

$$L(s,t) \sim \frac{x'(s)\,y''(s) - x''(s)\,y'(s)}{2\pi v^2} = \frac{\kappa(s)\, v}{2\pi} \tag{5.23}$$

as $|s-t| \to 0$, where $\kappa(s)$ is the curvature at p [1146, eqn (2.20)]. Thus, $L(s,s)$ is defined: 'The anticipated difficulty is one of appearance only' [115, p. 77]. This is one of those results 'that is well known to those to whom it is well known'. It can be found, for example, in the books by Lovitt [746, §41], Riesz & Szőkefalvi-Nagy [1020, §81], Smirnov [1118, §199], Kress [655, Problem 6.1] and Chen & Zhou [185, §6.6]. (Interestingly, Lovitt [746, p. III] remarks that, in preparing his book, he made extensive use of his notes on lectures given by Professor Oskar Bolza at the University of Chicago in 1913. Many copies of these handwritten notes were distributed [115].)

Similar results obtain for $\overline{K^*}$. We have

$$(\overline{K^*}\mu)(p) = \int_0^1 L^*(s,t)\, \varphi(t)\, dt,$$

where

$$L^*(s,t) = \frac{ik}{2\mathcal{R}} H_1^{(1)}(k\mathcal{R}) \{y'(t)[x(t)-x(s)] - x'(t)[y(t)-y(s)]\}.$$

We note that the limiting behaviour of $L^*(s,t)$ as $|s-t| \to 0$ is given precisely by the right-hand side of (5.23).

All the formulae given above can be found in a paper by Kress [654].

5.4.3 The hypersingular operator N

In two dimensions, Maue's formula (5.18) reduces to

$$(N\mu)(p) = -\frac{\mathrm{i}}{2v} \oint_0^1 \frac{\mathrm{d}\varphi}{\mathrm{d}t} \frac{\mathrm{d}}{\mathrm{d}s} H_0^{(1)}(k\mathcal{R})\,\mathrm{d}t$$
$$- \frac{\mathrm{i}k^2}{2v} \int_0^1 \{x'(t)x'(s) + y'(t)y'(s)\}\, H_0^{(1)}(k\mathcal{R})\,\varphi(t)\,\mathrm{d}t; \qquad (5.24)$$

compare this with (5.17) when $x(t) = t$ and $y(t)$ is constant. For a direct proof of (5.24), see [1055, Theorem 2.5.4].

The first term on the right-hand side of (5.24) involves the tangential derivative of $\mu(q(t)) = \varphi(t)$; it can also be written as

$$+\frac{\mathrm{i}}{2v} \frac{\mathrm{d}}{\mathrm{d}s} \int_0^1 \varphi(t) \frac{\mathrm{d}}{\mathrm{d}t} H_0^{(1)}(k\mathcal{R})\,\mathrm{d}t.$$

In two-dimensional diffraction theory, equations with $N\mu$ expressed in terms of tangential derivatives have been used in [883, 363, 1370].

$N\mu$ can also be expressed as a finite-part integral, as follows.

Theorem 5.4 (Martin & Rizzo [800]**)** *Let $\mu \in C^{1,\alpha}(S)$, $0 < \alpha \le 1$. Then, the normal derivative of a double-layer potential, defined by* (5.13), *is given by*

$$(N\mu)(p) = \oint_0^1 Q(s,t)\,\varphi(t)\,\mathrm{d}t,$$

where

$$Q(s,t) = \frac{\mathrm{i}k}{2} \frac{v(t)}{\mathcal{R}} \Big\{ (\boldsymbol{n}(p)\cdot\boldsymbol{R})(\boldsymbol{n}(q)\cdot\boldsymbol{R}) \frac{k}{\mathcal{R}} H_2^{(1)}(k\mathcal{R})$$
$$- (\boldsymbol{n}(p)\cdot\boldsymbol{n}(q))\, H_1^{(1)}(k\mathcal{R}) \Big\}, \qquad (5.25)$$

$\boldsymbol{R} = \boldsymbol{r}_p - \boldsymbol{r}_q$ *and* $\mathcal{R} = |\boldsymbol{R}|$.

A proof of this theorem is given in Appendix E. Also, Taylor-series expansions show that

$$Q(s,t) \sim -\frac{1}{\pi v (s-t)^2} + \frac{k^2 v}{2\pi} \log|s-t| \quad \text{as } |s-t| \to 0.$$

5.5 Explicit formulae in three dimensions

In two dimensions, we can usually construct a global parametrisation of the boundary curve S without difficulty. In three dimensions, this is seldom true, and it may even be impossible if S has a complicated shape. If S is reasonably simple (for example, if S is star-shaped), global parametrisations will exist and their use may lead to excellent numerical methods; see Section 5.5.3, [49, §5.5] and [430, 378]. More generally, we suppose that S has been covered by a set of patches $\mathcal{P}_\kappa$, $\kappa = 1, 2, \dots, K$, where each patch is the image of a two-dimensional parameter domain, $\mathcal{T}_\kappa$. Let $\mathcal{P} \subset S$ be a typical patch, defined by

$$\mathcal{P} = \{(x, y, z) : x = x(t_1, t_2),\ \ y = y(t_1, t_2),\ \ z = z(t_1, t_2),\ \ (t_1, t_2) \in \mathcal{T}\}.$$

With $\boldsymbol{r} = (x, y, z)$, we locate points p and q on $\mathcal{P}$ using

$$\boldsymbol{r}_p = \boldsymbol{r}(s_1, s_2) \quad \text{and} \quad \boldsymbol{r}_q = \boldsymbol{r}(t_1, t_2),$$

where $(s_1, s_2) \in \mathcal{T}$ and $(t_1, t_2) \in \mathcal{T}$; in general, p and q need not lie on the same patch, but we are mainly interested here in the behaviour of various kernels when $|\boldsymbol{r}_p - \boldsymbol{r}_q| \to 0$, so we may assume that p and q are both on $\mathcal{P}$.

So, the point $(t_1, t_2) \in \mathcal{T}$ is mapped into a point $q \in \mathcal{P}$; we assume that this mapping is twice continuously differentiable. Note that, having selected a point $p \in \mathcal{P}$, one choice for $\mathcal{T}$ is the tangent plane at p; this is sometimes convenient for analysis. For more details, see [223, §2.1].

Conventional element-based methods for solving boundary integral equations (as described later) use non-overlapping patches $\mathcal{P}_\kappa$, called *boundary elements*. Overlapping elements can also be used, combined with partitions of unity; this approach has been described in detail by Bruno & Kunyansky [149, 150]. (Partitions of unity are defined in, for example, [571, §7.4] and [853, Chapter XIII, §1.3].)

Let

$$\boldsymbol{N}(q) = \frac{\partial \boldsymbol{r}_q}{\partial t_1} \times \frac{\partial \boldsymbol{r}_q}{\partial t_2}. \tag{5.26}$$

Then $\boldsymbol{N}(q)$ is a normal at $q \in \mathcal{P}$; we assume that the local parameters t_1 and t_2 have been chosen so that $\boldsymbol{N}(q)$ points out of B, and that $|\boldsymbol{N}(q)| > 0$ for all $q \in \mathcal{P}$, so that $\boldsymbol{n}(q) = \boldsymbol{N}(q)/|\boldsymbol{N}(q)|$. Also, we have $\mathrm{d}s_q = |\boldsymbol{N}(q)|\,\mathrm{d}t_1\,\mathrm{d}t_2$.

5.5.1 The single-layer operator

Define

$$\boldsymbol{R} = \boldsymbol{r}_p - \boldsymbol{r}_q \quad \text{and} \quad \mathcal{R}(s_1, s_2; t_1, t_2) = |\boldsymbol{R}| = |\boldsymbol{r}(s_1, s_2) - \boldsymbol{r}(t_1, t_2)| \geq 0.$$

Then, we obtain

$$\int_{\mathcal{P}} \mu(q)\, G(p, q)\, \mathrm{d}s_q = \int_{\mathcal{T}} M(s_1, s_2; t_1, t_2)\, \varphi(t_1, t_2)\, \mathrm{d}t_1\, \mathrm{d}t_2, \tag{5.27}$$

where $\varphi(t_1, t_2) = \mu(q(t_1, t_2))$ and

$$M(s_1, s_2; t_1, t_2) = -|\boldsymbol{N}(q)| \exp(\mathrm{i}k\mathcal{R})/(2\pi\mathcal{R}).$$

To examine the singular behaviour as $\mathcal{R} \to 0$, we introduce plane polar coordinates (r, θ) at (s_1, s_2), so that

$$t_1 = s_1 + r\cos\theta \quad \text{and} \quad t_2 = s_2 + r\sin\theta.$$

Then, we have $\mathcal{R}^2 = r^2\ell_2(\theta) + O(r^3)$ as $r \to 0$, where $\ell_2(\theta)$ is defined by (F.15) in terms of the local geometry near p. Hence

$$M(s_1, s_2; t_1, t_2) \simeq -|\boldsymbol{N}(p)|\, [\ell_2(\theta)]^{-1/2} /(2\pi r)$$

for small r. As $\mathrm{d}t_1\,\mathrm{d}t_2 = r\,\mathrm{d}r\,\mathrm{d}\theta$, the integrand in (5.27) is seen to be bounded at $r = 0$.

5.5.2 The operators K and $\overline{K^*}$

Elementary calculations show that

$$\int_{\mathcal{P}} \mu(q) \frac{\partial}{\partial n_p} G(p, q)\,\mathrm{d}s_q = \int_{\mathcal{T}} L(s_1, s_2; t_1, t_2)\,\varphi(t_1, t_2)\,\mathrm{d}t_1\,\mathrm{d}t_2,$$

where

$$L(s_1, s_2; t_1, t_2) = \frac{|\boldsymbol{N}(q)|}{2\pi}(\boldsymbol{n}(p)\cdot\boldsymbol{R})\left(\frac{\mathrm{i}k}{\mathcal{R}^2} - \frac{1}{\mathcal{R}^3}\right)\mathrm{e}^{\mathrm{i}k\mathcal{R}}.$$

Similarly,

$$\int_{\mathcal{P}} \mu(q) \frac{\partial}{\partial n_q} G(p, q)\,\mathrm{d}s_q = \int_{\mathcal{T}} L^*(s_1, s_2; t_1, t_2)\,\varphi(t_1, t_2)\,\mathrm{d}t_1\,\mathrm{d}t_2,$$

where

$$L^*(s_1, s_2; t_1, t_2) = \frac{1}{2\pi}(\boldsymbol{N}(q)\cdot\boldsymbol{R})\left(\frac{\mathrm{i}k}{\mathcal{R}^2} - \frac{1}{\mathcal{R}^3}\right)\mathrm{e}^{\mathrm{i}k\mathcal{R}}.$$

Let $\boldsymbol{X}_i = \partial\boldsymbol{r}/\partial t_i$ evaluated at (s_1, s_2). Then, Taylor's theorem gives

$$\boldsymbol{R} = -r(\boldsymbol{X}_1\cos\theta + \boldsymbol{X}_2\sin\theta) + O(r^2) \quad \text{as } r \to 0;$$

see (F.14). As $\boldsymbol{N}(p) = \boldsymbol{X}_1 \times \boldsymbol{X}_2$ is orthogonal to $\boldsymbol{X}_1$ and $\boldsymbol{X}_2$, we deduce that $\boldsymbol{N}(p)\cdot\boldsymbol{R} = O(r^2)$ as $r \to 0$; see [223, Theorem 2.2]. Hence, both L and L^* are $O(r^{-1})$ as $r \to 0$.

The formulae giving the singular behaviour of M, L and L^* are well known because they are identical to the classical results for Laplace's equation ($k = 0$) [597].

5.5.3 Global parametrisations

Suppose that S is diffeomorphic to the unit sphere Ω. As a special case of these surfaces, we consider *star-shaped* surfaces; these can be defined in terms of spherical polar coordinates, (r, θ, ϕ). Thus, we suppose that S is defined by

$$r = \rho(\hat{\boldsymbol{r}}) = \rho(\theta, \phi), \quad \hat{\boldsymbol{r}} \in \Omega, \tag{5.28}$$

where ρ is a specified (smooth) function of the two polar angles. For surfaces defined by (5.28), integrals over S can be transformed into integrals over Ω:

$$\int_S f(q)\,\mathrm{d}s_q = \int_\Omega f\left(\rho(\hat{\boldsymbol{r}})\hat{\boldsymbol{r}}\right) |\boldsymbol{N}(q)|\,\rho^2\,\mathrm{d}\Omega(\hat{\boldsymbol{r}}). \tag{5.29}$$

Here, $|\boldsymbol{N}(q)| = \left\{1 + (\rho_\theta/\rho)^2 + (\rho_\phi/[\rho\sin\theta])^2\right\}^{1/2}$, where $\rho_\theta = \partial\rho/\partial\theta$ and $\rho_\phi = \partial\rho/\partial\phi$. We also have

$$\frac{\partial u}{\partial n_q} = \frac{1}{|\boldsymbol{N}(q)|}\left\{\frac{\partial u}{\partial r} - \frac{\rho_\theta}{\rho^2}\frac{\partial u}{\partial\theta} - \frac{\rho_\phi}{(\rho\sin\theta)^2}\frac{\partial u}{\partial\phi}\right\}, \tag{5.30}$$

where all the derivatives are evaluated on $r = \rho(\theta, \phi)$.

5.6 Green's theorem

We were being addressed on Green's theorem by the old curmudgeon Joseph Larmor. ... It was in fact a poor show, and Larmor didn't even understand what I said was the "point" of Green: to make an infinity of a function to do positive work instead of being a disaster.

(Littlewood [734, p. 135])

The divergence theorem is

$$\int_{\mathcal{B}} \operatorname{div}\boldsymbol{w}\,\mathrm{d}V = \int_{\mathcal{S}} \boldsymbol{w}\cdot\boldsymbol{\nu}\,\mathrm{d}S, \tag{5.31}$$

where $\mathcal{B}$ is a bounded region with boundary $\mathcal{S}$, $\boldsymbol{w}$ is a continuously differentiable vector field, defined in $\mathcal{B}\cup\mathcal{S}$, and $\boldsymbol{\nu}(q)$ is the unit normal vector at $q \in \mathcal{S}$ pointing *out* of $\mathcal{B}$. If we set $\boldsymbol{w} = \phi\,\operatorname{grad}\psi$, where ϕ and ψ are scalar fields, we obtain

$$\int_{\mathcal{B}} \left\{\phi\,\nabla^2\psi + (\operatorname{grad}\phi)\cdot(\operatorname{grad}\psi)\right\}\mathrm{d}V = \int_{\mathcal{S}} \phi\,\frac{\partial\psi}{\partial\nu}\,\mathrm{d}S, \tag{5.32}$$

which is *Green's first theorem*. If we interchange ϕ and ψ, and subtract the result from (5.32), we obtain *Green's second theorem*,

$$\int_{\mathcal{B}} (\phi\,\nabla^2\psi - \psi\,\nabla^2\phi)\,\mathrm{d}V = \int_{\mathcal{S}} \left(\phi\,\frac{\partial\psi}{\partial\nu} - \psi\,\frac{\partial\phi}{\partial\nu}\right)\mathrm{d}S. \tag{5.33}$$

We use Green's (second) theorem to derive integral representations for solutions of the Helmholtz equation. We start by considering the interior region B with boundary S. Choose a point $P \in B$, and then surround P by a small ball (circle in two dimensions) B_ε of radius ε. Apply Green's theorem in $B\backslash B_\varepsilon$ to $w(Q)$ and $G(P, Q)$, where w is a regular wavefunction. As both w and G satisfy the Helmholtz equation in $B\backslash B_\varepsilon$, the left-hand side of (5.33) is identically zero. The singularity in $G(P, Q)$ at $P = Q$ leads to a non-zero contribution on the right-hand side of (5.33) as $\varepsilon \to 0$. Noting that $\boldsymbol{\nu} = \boldsymbol{n}$, the final result is

$$\begin{aligned} -2w(P) &= \int_S \left\{ G(P, q) \frac{\partial w}{\partial n_q} - w(q) \frac{\partial}{\partial n_q} G(P, q) \right\} \mathrm{d}s_q \\ &= \left(S \frac{\partial w}{\partial n} \right)(P) - (Dw)(P), \quad P \in B. \end{aligned} \tag{5.34}$$

This representation for $w(P)$ at interior points P not on the boundary in terms of the boundary values of w and $\partial w/\partial n$ dates back to Helmholtz himself in 1859 [682, §290], [564, §1.26], [53, §4.2].

If $P \in B_e$, $G(P, Q)$ is non-singular for $Q \in B$, whence

$$0 = \int_S \left\{ G(P, q) \frac{\partial w}{\partial n_q} - w(q) \frac{\partial}{\partial n_q} G(P, q) \right\} \mathrm{d}s_q, \quad P \in B_e. \tag{5.35}$$

For exterior regions, we proceed similarly. Consider a region B_r bounded internally by S and externally by a large sphere S_r of radius r. Choose a point $P \in B_r$, and then surround P by a small ball B_ε of radius ε. Apply Green's theorem in $B_r\backslash B_\varepsilon$ to $w_e(Q)$ and $G(P, Q)$, where w_e is a radiating wavefunction. As both w_e and G satisfy the Helmholtz equation in $B_r\backslash B_\varepsilon$, the left-hand side of (5.33) is identically zero. As both w_e and G satisfy the Sommerfeld radiation condition, there is no contribution from integrating over S_r as $r \to \infty$. The singularity in $G(P, Q)$ at $P = Q$ leads to a non-zero contribution as $\varepsilon \to 0$. Noting that $\boldsymbol{\nu} = -\boldsymbol{n}$, the final result is

$$\begin{aligned} 2w_e(P) &= \int_S \left\{ G(P, q) \frac{\partial w_e}{\partial n_q} - w_e(q) \frac{\partial}{\partial n_q} G(P, q) \right\} \mathrm{d}s_q \\ &= \left(S \frac{\partial w_e}{\partial n} \right)(P) - (Dw_e)(P), \quad P \in B_e. \end{aligned} \tag{5.36}$$

This representation for radiating wavefunctions is due to Sommerfeld in 1912; see, for example, [1129, §34 B], [564, §1.31] or [53, §4.2].

If $P \in B$, $G(P, Q)$ is non-singular for $Q \in B_e$, whence

$$0 = \int_S \left\{ G(P, q) \frac{\partial w_e}{\partial n_q} - w_e(q) \frac{\partial}{\partial n_q} G(P, q) \right\} \mathrm{d}s_q, \quad P \in B. \tag{5.37}$$

We call this the *interior integral relation*. It will be used later in certain methods for removing irregular frequencies (Section 6.11) and it is at the heart of null-field methods, to be described in Chapter 7.

5.6.1 Limiting values on S

I remember being delighted [in 1963] *when I realized that this very formula is a constraint between surface displacement and surface traction pertaining to one and the same elastic field in a body occupying D.*

(Rizzo [1022, p. 7])

The formulae (5.34)–(5.37) hold for points P not on the boundary S. We can calculate the limiting values as $P \to p \in S$, using the jump relations given in Section 5.3. We find that both (5.34) and (5.35) give

$$(I - \overline{K^*})w + S(\partial w/\partial n) = 0; \tag{5.38}$$

in this formula, w gives the boundary values of the regular wavefunction $w(P)$ defined for $P \in B$, and $\partial w/\partial n$ is the corresponding normal derivative of $w(P)$ on S. Thus, paraphrasing the quotation from Rizzo's paper [1022] above (where he outlines the development of integral-equation methods for elastostatics), the formula (5.38) is a constraint between surface values and the surface normal derivative pertaining to one and the same wavefunction in a body occupying B.

In the same way, both (5.36) and (5.37) give

$$(I + \overline{K^*})w_e - S(\partial w_e/\partial n) = 0, \tag{5.39}$$

where $w_e(P)$ is a radiating wavefunction, defined for $P \in B_e$; this equation can be found in [244, p. 318].

We can also calculate the normal derivative of (5.34)–(5.37) on S; this will involve the hypersingular operator N. We obtain

$$(I + K)(\partial w/\partial n) - Nw = 0 \tag{5.40}$$

from (5.34) or (5.35), and

$$(I - K)(\partial w_e/\partial n) + Nw_e = 0 \tag{5.41}$$

from (5.36) or (5.37).

5.7 Scattering and radiation problems

Let us formulate a few boundary-value problems, beginning with a specific radiation problem.

Exterior Neumann problem *Find a function $u(P)$ for $P \in B_e$, where*

$$(\nabla^2 + k^2)u = 0 \quad \text{in } B_e, \tag{5.42}$$

$$u \text{ satisfies the radiation condition at infinity} \tag{5.43}$$

and

$$\frac{\partial u}{\partial n} = f \quad on\ S. \tag{5.44}$$

Here, $f(q)$ is a given function, defined for $q \in S$.

A standard argument [223, Theorem 3.13] using Rellich's lemma [223, Lemma 3.11] shows that the exterior Neumann problem has at most one solution: $f \equiv 0$ on S implies that $u \equiv 0$ in B_e.

Exterior Dirichlet problem *This is formulated in the same way as the Neumann problem, except that* (5.44) *is replaced by*

$$u = g \quad on\ S, \tag{5.45}$$

where $g(q)$ is a given function, defined for $q \in S$.

Again, a standard argument shows that the exterior Dirichlet problem has at most one solution.

Scattering problems were discussed in Section 1.3. For sound-hard scatterers, we can formulate the following problem.

Sound-hard scattering problem *Find a function $u_{sc}(P)$ for $P \in B_e$, where*

$$(\nabla^2 + k^2)u_{sc} = 0 \quad in\ B_e, \tag{5.46}$$

$$u_{sc}\ \text{satisfies the radiation condition at infinity} \tag{5.47}$$

and the boundary condition

$$\frac{\partial u}{\partial n} = 0 \quad on\ S, \tag{5.48}$$

where the total field u is defined by

$$u = u_{sc} + u_{inc} \tag{5.49}$$

and u_{inc} is the given incident field.

We can assume that u_{inc} is a regular solution of the Helmholtz equation everywhere, except perhaps at some isolated points in B_e.

Evidently, we can combine (5.48) and (5.49) to give

$$\frac{\partial u_{sc}}{\partial n} = -\frac{\partial u_{inc}}{\partial n} \quad \text{on } S,$$

which shows that the sound-hard scattering problem can be expressed as an exterior Neumann problem for u_{sc} with a particular forcing f. However, sometimes the properties of u_{inc} can be exploited to good effect.

Sound-soft scattering problem *This is formulated in the same way as the sound-hard problem, except that* (5.48) *is replaced by $u = 0$ on S.*

5.8 Integral equations: general remarks

To continue about my failures, I shall mention one which I particularly regret. It concerns the celebrated Dirichlet problem which I, for years, tried to solve ... [but] *physical interpretation, which is in general a very sure guide and had been most often such for me, misled me in that case. It suggested an attempt to solve the problem by a "potential of simple layer" – in that question, a blind alley – while the solution was to be looked for in the introduction of a "potential of a double layer".*

(Hadamard [467, p. 52])

There are two main methods for reducing boundary-value problems to boundary integral equations. These are usually called the *indirect* and the *direct* methods. The indirect method is the oldest, and dates back to Fredholm in 1900. (McLean [836, Chapter 1] gives a nice historical summary; see also [744, 212, 191].) It begins by writing the solution u as a single-layer or a double-layer potential with an unknown density function μ. The boundary condition on S then leads to an integral equation for μ. The choice of integral representation (single-layer or double-layer) determines what type of integral equation is obtained. Traditionally, the goal is to obtain a Fredholm integral equation of the second kind, with a weakly-singular kernel. This approach is described in detail in [597] for Laplace's equation and in [223] for the Helmholtz equation. Much later, it was recognised that other types of integral equations, such as Fredholm integral equations of the first kind or hypersingular equations, could be used, both analytically and computationally.

The direct method begins with an integral representation for $u(P)$, obtained by applying Green's theorem to u and a fundamental solution; for radiating wavefunctions, an example is (5.36). By letting $P \to p$, a point on the boundary S, one obtains a relation between the boundary values of u and its normal derivative; for the Neumann problem, this yields an integral equation for u on S. The direct method was not used extensively until numerical methods for solving boundary integral equations were developed in the early 1960s. Nowadays, direct methods are the methods of choice, especially when combined with numerical methods such as the Boundary Element Method.

There are many books and reviews covering the theory and application of integral equations. The theory of Fredholm integral equations of the second kind is given in many textbooks. Good examples are [1183, 655, 981, 978]; for their numerical treatment, see [49, 415]. The books by Constanda [230] and by Saranen & Vainikko [1055] are devoted entirely to the analysis of boundary integral equations arising from two-dimensional boundary-value problems.

Boundary integral equations of various types can be viewed and analysed as pseudo-differential equations. Presentations of this point of view are given in several reviews by Wendland (such as [1306, 1307]) and by McLean [836]. For pseudo-differential equations on curves, see [1055].

There are several books in which the use of boundary elements is described in the context of acoustics; examples are [216, 1015, 1340, 1267].

In this book, we assume that the (closed) surface S is smooth, so that corners and edges are not permitted. If S is not smooth, the theory must be changed substantially. For two-dimensional problems, we refer the reader to [248, §9.7], [655, §6.5], [653, 654] and [49, Chapter 8]. For three-dimensional problems, see [1305, 633, 238, 836] and [49, §9.1.5].

5.9 Integral equations: indirect method

The area rule involves calculating the [potential] *flow about a bumpy body of revolution... K. E. Van Every, my boss, talked to me about looking over the available methods and seeing what was best. I did and discovered an entirely new method that was far more powerful than any of the existing ones. I got an OK to try to work out the details of a computer programming method and got it to work quite successfully. The method is now known as the Panel Method and is universally used for low speed flow analysis... Our first successful calculations were in early 1955.*

(A.M.O. Smith [179, p. 36])

The main idea with this method is to look for a solution in the form of single-layer or double-layer potentials, or some combination of both. The particular choice made will determine the kind of integral equation obtained.

Let us look for a solution of the exterior Neumann problem in the form of a single-layer potential; other options will be discussed later. Thus, we write

$$u(P) = (S\mu)(P), \quad P \in B_e, \tag{5.50}$$

where the density μ is to be found. For any reasonable μ, (5.42) and (5.43) are satisfied. It remains to satisfy the boundary condition (5.44). Imposing this, using (5.6), we obtain

$$(I + K)\mu = f, \tag{5.51}$$

which is a *Fredholm integral equation of the second kind* for μ. If we can solve (5.51), we will have solved the exterior Neumann problem for u.

It turns out that (5.51) is uniquely solvable for μ, for any f, except when k^2 coincides with an eigenvalue of the interior Dirichlet problem. At these *irregular values* of k^2, or *irregular frequencies*, the following boundary-value problem has a non-trivial solution.

Interior Dirichlet problem *Find a function $\psi(P)$ for $P \in B$, where*

$$(\nabla^2 + k^2)\psi = 0 \textit{ in } B, \textit{ and } \psi = 0 \textit{ on } S.$$

Let us denote the set of irregular values by $\mathrm{IV}(S)$. It is clear that

$$\mathrm{IV}(S) = \bigcup_{i=1}^{N} \mathrm{IV}(S_i), \tag{5.52}$$

because we can obtain a non-trivial solution of the interior Dirichlet problem for S by taking $\psi(P)$ to be an eigenfunction of the interior Dirichlet problem for S_j, say, with $\psi(P) \equiv 0$ for $P \in B_i$, $i = 1, 2, \ldots, N$, $i \neq j$; see [243, Chapter VI, §1.3].

The fact (5.52) is unfortunate, because it means that, in general, there are N times as many irregular frequencies as there are for a single scatterer. Methods for eliminating irregular frequencies are described in Chapter 6; see Section 6.8 for an overview of these methods.

Example 5.5 Let S_i be a sphere of radius a_i, $i = 1, 2, \ldots, N$. Then

$$\mathrm{IV}(S_i) = \{k^2 : j_n(ka_i) = 0, \quad n = 0, 1, 2, \ldots\}.$$

For each n, the spherical Bessel function $j_n(x)$ has an infinite number of positive zeros; denote them by $x_{n,m}$, $m = 1, 2, \ldots$. Then

$$\mathrm{IV}(S_i) = \{k^2 : k = x_{n,m}/a_i, \quad n = 0, 1, 2, \ldots, \quad m = 1, 2, \ldots\}.$$

Note that the irregular values do not depend on the relative location or orientation of the scatterers, merely on their shape.

As an alternative to using a single-layer potential, we may look for a solution of the exterior Neumann problem in the form of a double-layer potential. Thus, we write

$$u(P) = (D\nu)(P), \quad P \in B_e, \tag{5.53}$$

where the density ν is to be found. Application of the boundary condition (5.44) yields

$$N\nu = f, \tag{5.54}$$

which is a hypersingular integral equation for ν. Again, if we can solve (5.54), we will have solved the exterior Neumann problem for u. It turns out that (5.54) is uniquely solvable for ν, for any f, except when k^2 coincides with an eigenvalue of the interior *Neumann* problem. See Corollary 3.3.1 in [628] for an identification of these irregular values and Theorem 3.3.2 in [223] for a proof of solvability.

For the exterior Dirichlet problem, use of the single-layer representation (5.50) leads to

$$S\mu = g \tag{5.55}$$

whereas the double-layer representation (5.53) leads to

$$(I - \overline{K^*})\nu = -g. \tag{5.56}$$

This last equation is a Fredholm integral equation of the second kind for ν, whereas (5.55) is a Fredholm integral equation of the first kind for μ. Equations (5.55) and (5.56) have the same irregular frequencies as (5.51) and (5.54), respectively.

Kleinman & Roach [628] give a complete catalogue of all the available boundary integral equations for the Dirichlet and Neumann problems.

The integral equations (5.51), (5.54), (5.55) and (5.56) are examples of *indirect boundary integral equations*, so called because the unknown density function does not have a clear physical interpretation.

For scattering problems, the indirect method can be used by starting with an integral representation for the scattered field u_{sc}.

Mixture problems can be treated in a similar way. Thus, write

$$S_N = \bigcup_{i=1}^{N_1} S_i \quad \text{and} \quad S_D = \bigcup_{i=N_1+1}^{N} S_i,$$

where the Neumann condition is to be imposed on S_N and the Dirichlet condition on S_D. Look for a solution in the form

$$u(P) = \int_{S_N} \mu(q)\, G(P,q)\, \mathrm{d}s_q + \int_{S_D} \nu(q) \frac{\partial}{\partial n_q} G(P,q)\, \mathrm{d}s_q, \qquad P \in B_{\mathrm{e}}.$$

Then, apply the Neumann condition on S_N, leading to one integral equation for $p \in S_N$. Apply the Dirichlet condition on S_D, giving an integral equation for $p \in S_D$. This coupled pair of integral equations is to be solved for $\mu(q)$ with $q \in S_N$ and for $\nu(q)$ with $q \in S_D$. For more details, see [657]. David [267] has considered a more general problem, where a Robin (impedance) condition is imposed on some of the obstacles.

5.9.1 Literature

> *While I cannot be absolutely certain, it appears that Kupradze in the period 1934–1936 was the first to apply the Fredholm integral equation method to an arbitrary exterior problem* [for the Helmholtz equation] *under Dirichlet boundary conditions.*
>
> (Dolph [288, p. 205])

> *Kupradze was born on 2 November 1903.... From 1930 to 1933 Kupradze completed the graduate curriculum in Leningrad under the supervision of Krylov and Smirnov.*
>
> *From 1933 to 1935 Kupradze was a scientific secretary of the Steklov Institute of Mathematics of the Soviet Academy of Sciences. Here in 1935 he defended his doctoral dissertation on the topic "Some fundamental problems of diffraction of electromagnetic waves".*
>
> (Burchuladze *et al.* [159, p. 177])

Kupradze's work on the use of layer potentials to solve boundary-value problems for the Helmholtz equation is described in [670, Chapter II]. He obtained (5.51) for the exterior Neumann problem and (5.56) for the exterior Dirichlet problem. He also showed how the analysis could be modified at the irregular values of k^2. See also [850, §57] and [1118, §231].

Perhaps the earliest numerical application of the indirect method (for the Helmholtz equation) is due to Chen & Schweikert [188] in 1963. They solved (5.51) for a piston vibrating in the surface of a sound-hard sphere of radius a. They used a *panel method*: the sphere's surface was approximated by 320 flat triangular panels, μ was taken as constant over each panel, and then these source strengths were determined by collocation; the computation for $ka = 2$ took 1.75 hours. Panel methods of this kind were developed in the late 1950s by Smith, Hess and others at the Douglas Aircraft Company for potential-flow calculations; see, for example, [498] and [179, Chapter 4]. (Hess has written several reviews of panel methods and their historical development; examples are [495, 496, 497].)

Other early computations were made by Brundrit [143] (although some of his numerical results are incorrect [135, p. 469]). He solved (5.51) for scattering of a plane wave by axisymmetric bodies. Sound radiation from elastic structures was calculated using an indirect method in [68, 823]. For multiple-scattering problems, the indirect method has also been used in [538, 539, 1131, 994, 312, 1008, 735]. For scattering by two square cylinders, see [1058, §6.7].

The classical approach is to reduce boundary-value problems to Fredholm integral equations of the second kind. This is the approach emphasised by Colton & Kress [223] and by Kleinman & Roach [628]. However, these references also cover the use of (5.54) for the exterior Neumann problem and (5.55) for the exterior Dirichlet problem. For more information on the use of single-layer potentials to solve Dirichlet problems, leading to first-kind integral equations, see [516] and [407]. For the use of double-layer potentials to solve Neumann problems, see [408, 1066, 902]. See also [836, Chapter 9] and [903, Chapter 3].

5.10 Integral equations: direct method

We apply Green's theorem to the free-space Green's function and to the required unknown solution in the region external to all scatterers, and thereby represent the scattered field as an integral over some surface inclosing the region. Contracting the surface and breaking it up into individual portions inclosing a single object, leads to a representation of the total scattered field as a sum of surface integrals... Then imposing the boundary condition at each object leads to a determinate set of coupled integral equations for the fields on all scatterers, and could these values be obtained explicitly, the total field in space would follow on integration.

(Twersky [1198, p. 716])

The basic method of this section was described succinctly by Twersky in his 1960 review [1198]. We shall give a more detailed description. Thus, consider the integral representation obtained from Green's theorem, (5.36). When used for the exterior Neumann problem, it gives

$$2u(P) = (Sf)(P) - (Du)(P), \quad P \in B_e, \tag{5.57}$$

where we have used (5.44). This formula gives u at P in terms of $u(q)$, $q \in S$. To find $u(q)$, we let $P \to p \in S$, using (5.9), and obtain

$$(I+\overline{K^*})u = Sf. \tag{5.58}$$

This is another Fredholm integral equation of the second kind. The unknown is $u(q)$, $q \in S$. As (5.58) is the Hermitian adjoint of (5.51), it has the same irregular values, namely IV(S).

There are some differences between (5.51) and (5.58). First, the unknown μ does not have any physical significance, whereas the solution of (5.58) may be the desired physical quantity. For this reason, integral equations based on Green's theorem are often known as *direct boundary integral equations*.

Second, the right-hand side of (5.58) involves an integration over S.

Third, at irregular frequencies, (5.51) has no solutions in general, whereas (5.58) has infinitely many solutions. For, to deduce existence, f must be such that $(Sf, \overline{\mu_0}) = 0$ for every solution μ_0 of $(I+K)\mu_0 = 0$. Define $U(P) = (S\mu_0)(P)$ for $P \in B_e$. Clearly, U satisfies the Helmholtz equation in B_e and the radiation condition at infinity, and (5.6) shows that $\partial U/\partial n = 0$ on S. Thus, U solves the homogeneous exterior Neumann problem (with $f \equiv 0$) and so, by the uniqueness theorem, $U \equiv 0$ in B_e. In particular, $U = 0$ on S, whence $S\mu_0 = 0$. Hence, $(Sf, \overline{\mu_0}) = (f, \overline{S\mu_0}) = 0$, and so the solvability condition is satisfied automatically.

Equation (5.58) can be obtained directly from (5.39). Similarly, if we use (5.41), we obtain

$$Nu = (K-I)f, \tag{5.59}$$

which is a hypersingular integral equation for $u(q)$.

For the exterior Dirichlet problem, we are given $u = g$ on S, and so Green's theorem gives the representation

$$2u(P) = (Sv)(P) - (Dg)(P), \quad P \in B_e, \tag{5.60}$$

where $v(q) \equiv \partial u/\partial n$. We may find v by solving (5.39) or (5.41), which become

$$Sv = (I+\overline{K^*})g \tag{5.61}$$

and

$$(I-K)v = -Ng, \tag{5.62}$$

respectively. Equation (5.61) is a Fredholm integral equation of the first kind. Equation (5.62) is a Fredholm integral equation of the second kind; it has the same irregular frequencies as (5.54) and (5.56).

5.10.1 Scattering problems

For the sound-hard scattering problem, we can replace u and f by u_{sc} and $-\partial u_{inc}/\partial n$, respectively, in (5.57), (5.58) and (5.59). Thus, we obtain

$$2u_{sc}(P) = -(Sv_{inc})(P) - (Du_{sc})(P), \quad P \in B_e, \tag{5.63}$$

where $v_{inc} \equiv \partial u_{inc}/\partial n$ and $u_{sc}(q)$ solves

$$(I + \overline{K^*})u_{sc} = -Sv_{inc} \tag{5.64}$$

or

$$Nu_{sc} = (I - K)v_{inc}. \tag{5.65}$$

However, as $u_{inc}(P)$ is a regular wavefunction in B, (5.35) (with $w = u_{inc}$) gives

$$0 = (Sv_{inc})(P) - (Du_{inc})(P), \quad P \in B_e. \tag{5.66}$$

When this is added to (5.63), we obtain

$$2u_{sc}(P) = -(Du)(P), \quad P \in B_e, \tag{5.67}$$

where $u = u_{sc} + u_{inc}$ is the total field. Thus, the field scattered by sound-hard obstacles can *always* be represented as a double-layer potential, with density $u(q)$.

Letting $P \to p \in S$ in (5.67), we obtain

$$(I + \overline{K^*})u = 2u_{inc}, \tag{5.68}$$

which is a Fredholm integral equation of the second kind for the boundary values of the total field u. It is often known as the *Helmholtz integral equation*. It is similar to (5.64) except that the right-hand side is simpler. Of course, changing the right-hand side does not change the irregular frequencies: both equations have the same irregular frequencies, namely IV(S).

Evaluating the normal derivative of (5.67) on S, we obtain an alternative integral equation for u, namely

$$Nu = 2v_{inc}. \tag{5.69}$$

For the sound-soft scattering problem, we can replace v and g by $v_{sc} \equiv \partial u_{sc}/\partial n$ and $-u_{inc}$, respectively, in (5.60), (5.61) and (5.62). This gives

$$2u_{sc}(P) = (Sv_{sc})(P) + (Du_{inc})(P), \quad P \in B_e, \tag{5.70}$$

where $v_{sc}(q)$ solves

$$Sv_{sc} = -(I + \overline{K^*})u_{inc} \tag{5.71}$$

or

$$(I - K)v_{sc} = Nu_{inc}. \tag{5.72}$$

Adding (5.66) to (5.70) gives

$$2u_{\mathrm{sc}}(P) = (Sv)(P), \quad P \in B_{\mathrm{e}}, \tag{5.73}$$

where $v = \partial u/\partial n$ is the normal derivative of total field on S. Thus, the field scattered by sound-soft obstacles can *always* be represented as a single-layer potential, with density $v(q)$.

Letting $P \to p \in S$ in (5.73), we obtain

$$Sv = -2u_{\mathrm{inc}}, \tag{5.74}$$

which is a Fredholm integral equation of the first kind for v. For a Fredholm integral equation of the second kind, we calculate the normal derivative of (5.73) on S, and obtain

$$(I - K)v = 2v_{\mathrm{inc}}, \tag{5.75}$$

which should be compared with (5.72).

5.10.2 The far field: two dimensions

For sound-hard obstacles in two dimensions, the scattered field is given by (5.67) as

$$\begin{aligned} u_{\mathrm{sc}}(\boldsymbol{r}) &= \frac{\mathrm{i}}{4} \int_S u(q) \frac{\partial}{\partial n_q} \left\{ H_0^{(1)}(kR) \right\} \mathrm{d}s_q \\ &= -\frac{\mathrm{i}k}{4} \int_S u(q) \frac{\partial R}{\partial n_q} H_1^{(1)}(kR) \,\mathrm{d}s_q, \end{aligned} \tag{5.76}$$

where $R = |\boldsymbol{r} - \boldsymbol{r}_q|$. When kR is large, we can use (2.28). We also have

$$R = \left\{ (\boldsymbol{r} - \boldsymbol{r}_q) \cdot (\boldsymbol{r} - \boldsymbol{r}_q) \right\}^{1/2} \sim r - \hat{\boldsymbol{r}} \cdot \boldsymbol{r}_q \quad \text{as } r \to \infty,$$

where $\boldsymbol{r} = r\hat{\boldsymbol{r}}$. Hence, the far field is given by Definition 4.1, where the far-field pattern is given by

$$f(\hat{\boldsymbol{r}}) = \frac{k}{4} \int_S u(q) \, [\hat{\boldsymbol{r}} \cdot \boldsymbol{n}(q)] \exp\left(-\mathrm{i}k\hat{\boldsymbol{r}} \cdot \boldsymbol{r}_q\right) \mathrm{d}s_q. \tag{5.77}$$

For sound-soft obstacles, (5.73) gives

$$u_{\mathrm{sc}}(\boldsymbol{r}) = -\frac{\mathrm{i}}{4} \int_S \frac{\partial u}{\partial n_q} H_0^{(1)}(kR) \,\mathrm{d}s_q, \tag{5.78}$$

and then the far-field pattern is given by

$$f(\hat{\boldsymbol{r}}) = -\frac{\mathrm{i}}{4} \int_S \frac{\partial u}{\partial n_q} \exp\left(-\mathrm{i}k\hat{\boldsymbol{r}} \cdot \boldsymbol{r}_q\right) \mathrm{d}s_q. \tag{5.79}$$

5.10.3 The far field: three dimensions

For sound-hard obstacles in three dimensions, (5.67) gives

$$u_{sc}(\boldsymbol{r}) = \frac{1}{4\pi}\int_S u(q)\,\frac{\partial}{\partial n_q}\left(\frac{e^{ikR}}{R}\right) ds_q. \tag{5.80}$$

When R is large, we have

$$\frac{\partial}{\partial n_q}\left(\frac{e^{ikR}}{R}\right) \sim \frac{ik}{R}\,e^{ikR}\,\frac{\partial R}{\partial n_q}.$$

Hence, the far field is given by Definition 4.2, where the far-field pattern is given by

$$f(\hat{\boldsymbol{r}}) = \frac{k^2}{4\pi}\int_S u(q)\,[\hat{\boldsymbol{r}}\cdot\boldsymbol{n}(q)]\exp\left(-ik\hat{\boldsymbol{r}}\cdot\boldsymbol{r}_q\right) ds_q. \tag{5.81}$$

In the special case of plane-wave scattering by a hard sphere of radius a, centred at the origin, we have $\boldsymbol{r}_q = a\hat{\boldsymbol{r}}_q$, $\boldsymbol{n}(q) = \hat{\boldsymbol{r}}_q$, $ds_q = a^2\,d\Omega(\hat{\boldsymbol{r}}_q)$ and

$$u(q) = \sum_{n,m} u_n^m\,Y_n^m(\hat{\boldsymbol{r}}_q),$$

where, from Section 4.6,

$$u_n^m = \frac{4\pi i}{(ka)^2}\,\frac{i^n}{h_n'(ka)}\,\overline{Y_n^m(\hat{\boldsymbol{\alpha}})}, \tag{5.82}$$

$\hat{\boldsymbol{\alpha}}$ is the propagation direction of the incident wave, and we have used (4.39). Hence

$$f(\hat{\boldsymbol{r}};\hat{\boldsymbol{\alpha}}) = \frac{(ka)^2}{4\pi}\sum_{n,m} u_n^m \int_\Omega Y_n^m(\hat{\boldsymbol{r}}_q)\,(\hat{\boldsymbol{r}}\cdot\hat{\boldsymbol{r}}_q)\exp\left(-ika\hat{\boldsymbol{r}}\cdot\hat{\boldsymbol{r}}_q\right) d\Omega(\hat{\boldsymbol{r}}_q). \tag{5.83}$$

Now, from (3.69), we have

$$\exp\left(-ika\hat{\boldsymbol{r}}\cdot\hat{\boldsymbol{r}}_q\right) = 4\pi\sum_{\nu,\mu}(-i)^\nu j_\nu(ka)\,Y_\nu^\mu(\hat{\boldsymbol{r}})\,\overline{Y_\nu^\mu(\hat{\boldsymbol{r}}_q)},$$

whence differentiating once with respect to ka gives

$$(\hat{\boldsymbol{r}}\cdot\hat{\boldsymbol{r}}_q)\exp\left(-ika\hat{\boldsymbol{r}}\cdot\hat{\boldsymbol{r}}_q\right) = 4\pi i\sum_{\nu,\mu}(-i)^\nu j_\nu'(ka)\,Y_\nu^\mu(\hat{\boldsymbol{r}})\,\overline{Y_\nu^\mu(\hat{\boldsymbol{r}}_q)}.$$

Substituting in (5.83), we can integrate over Ω using the orthogonality relation (3.9), obtaining

$$f(\hat{\boldsymbol{r}};\hat{\boldsymbol{\alpha}}) = i(ka)^2\sum_{n,m}(-i)^n u_n^m\,j_n'(ka)\,Y_n^m(\hat{\boldsymbol{r}}).$$

This reduces to (4.45), as obtained by the method of separation of variables, after use of (5.82) and (3.20).

For sound-soft obstacles, (5.73) gives

$$u_{sc}(\boldsymbol{r}) = -\frac{1}{4\pi}\int_S \frac{\partial u}{\partial n_q}\,\frac{e^{ikR}}{R}\,ds_q. \tag{5.84}$$

Then, the far field is given by (4.44) wherein

$$f(\hat{\boldsymbol{r}}) = -\frac{\mathrm{i}k}{4\pi}\int_S \frac{\partial u}{\partial n_q}\exp\left(-\mathrm{i}k\hat{\boldsymbol{r}}\cdot\boldsymbol{r}_q\right)\mathrm{d}s_q. \tag{5.85}$$

Similar formulae can be derived for other scatterers.

5.10.4 Literature

This paper makes a short study of Fredholm integral equations related to potential theory and elasticity, with a view to preparing the ground for their exploitation in the numerical solution of difficult boundary-value problems. Attention is drawn to the advantages of Fredholm's first equation and of Green's boundary formula.

(Jaswon [552, Abstract])

According to Wendland [1306, p. 513], direct boundary integral equations were first used by Sobolev in 1936 [1125]. However, integral equations are not derived in this paper; rather, Sobolev shows how the Kirchhoff formula for the wave equation ([1149, §8.1], [564, §1.17], [1127, Lecture 14]) can be generalised to other linear hyperbolic equations. See also [1126, Chapter III]. In fact, Prager [986] used a direct method for two-dimensional potential flow in 1928. He used a Nyström method ([655, Chapter 12], [49, Chapter 4]) to solve (5.75) (with $k = 0$), numerically, for an elliptic cylinder. Prager's paper is also described in [212].

For the three-dimensional Helmholtz equation, Maue [813] obtained both (5.68) and (5.75) in 1949. He also obtained (5.69) and (5.74), but he regarded the second-kind equations as preferable. Papas [946] has derived the two-dimensional version of the first-kind equation (5.74) for a sound-soft cylinder.

In 1960, Millar [856] studied the multiple-scattering problem for N sound-hard parallel cylinders, using (5.64), and he used (5.74) for N sound-soft cylinders. In both cases, he found asymptotic solutions for small cylinders. Row [1035] also began with (5.74), which he then specialised to circular cylinders; see Section 4.5. Noble [916] has obtained (5.68), (5.69), (5.71) and (5.72).

To solve a boundary integral equation analytically is, generally speaking, out of the question. A straightforward numerical approach is to replace the equation by a system of simultaneous linear equations, referring to a set of nodal points spaced along the boundary L *of the relevant domain* D, *these equations being assembled and solved by digital computer techniques.*

(Jaswon [552, p. 24])

The direct method began producing numerical results in the early 1960s. Jaswon [552] clearly saw the advantages of the direct method over the indirect method, and he did not flinch from using first-kind equations; he also considered mixed boundary-value problems. Symm [1162], Jaswon's student, gave some numerical solutions for Laplace's equation in two dimensions. See also the book by Jaswon & Symm [553].

Numerical results for the Helmholtz equation were obtained by Banaugh & Goldsmith [58] and Mei & Van Bladel [844], in two dimensions. Banaugh & Goldsmith [58] solved (5.64). Mei & Van Bladel [844] solved both (5.68) and (5.74). Chertock [194] solved (5.58) for radiation from three-dimensional axisymmetric bodies. Andreasen [30] also solved both (5.68) and (5.74) for a two-dimensional multiple-scattering problem with two identical elliptic cylinders with collinear major axes. Zhu & Moule [1378] solved (5.64) for four identical circular cylinders. Perrey-Debain *et al.* solved (5.68) for three [957] and four [956] identical circular cylinders and obtained excellent agreement with the results of Linton & Evans [726]. For applications of direct methods to the scattering of a plane wave by two identical sound-hard spheres, see [1094, 1095, 573, 880, 1353]; for analogous radiation problems, see [287, 1353].

Plane-wave scattering by an infinite periodic row of identical parallel cylinders can be reduced to a problem for one cylinder. This problem has been solved in [7] for oblique incidence and sound-hard circular cylinders, beginning from (5.68).

A non-standard integral equation method has been used by Hron & Razavy [515]. For the problem of scattering by two identical circular cylinders, they begin by introducing bipolar coordinates. This maps the problem into a new problem with a simpler geometry (the interior of a rectangle) but with a more complicated partial differential equation, namely $[\nabla^2 + \kappa(\boldsymbol{r})]v = 0$, where κ is a known function of position. Green's function techniques then lead to a complicated Volterra integral equation of the second kind. Scattering by two spheres is treated similarly, using bispherical coordinates.

6

Integral equation methods, II: further results and applications

6.1 Introduction

In Chapter 5, we described basic integral-equation methods, and we showed how they can be used to solve the basic problems of acoustic scattering by sound-hard and sound-soft obstacles. In this chapter, we consider many other applications, including transmission problems, inhomogeneous obstacles, electromagnetic problems, elastodynamic problems and hydrodynamic problems.

We also discuss modified integral equations that do not suffer from irregular frequencies: an overview of these modifications is given in Section 6.8. Some applications of exact Green's functions are considered in Section 6.12. Another approach, due to Twersky, is described in Section 6.13. His method makes use of certain integral equations but they are not *boundary* integral equations. The chapter ends with a discussion of 'fast multipole methods' in Section 6.14.

6.2 Transmission problems

Acoustic transmission problems were formulated in Section 1.3.3. Physically, they correspond to scattering of a sound wave in a homogeneous fluid by blobs of another homogeneous fluid. Such problems can be reduced to boundary integral equations over the interface S.

The transmission problem is usually reduced to a pair of coupled boundary integral equations for a pair of unknowns. For example, we have the following pair [299, 624]:

$$\left.\begin{aligned}(1+\gamma)u+(\overline{K^*}-\gamma\overline{K_0^*})u-(S-S_0)v &= 2u_{\mathrm{inc}},\\ (1+\gamma)v+\gamma(N-N_0)u-(\gamma K-K_0)v &= 2\gamma v_{\mathrm{inc}}.\end{aligned}\right\}\tag{6.1}$$

Here, γ is the parameter (density ratio) occurring in the second transmission condition $(1.26)_2$, $v=\partial u/\partial n$ and $v_{\mathrm{inc}}=\partial u_{\mathrm{inc}}/\partial n$ on S, and the operators with the zero subscript are the operators of Section 5.3 with k replaced by the interior wavenumber k_0.

It turns out that the pair (6.1) is always uniquely solvable (subject to reasonable assumptions on the parameters of the problem); see [627, Theorem 4.2].

The pair (6.1) gives a direct method. Examples of other direct methods, with applications, can be found in [882, 239, 1082, 1091, 825, 1262, 470]. For indirect methods, see, for example, [658, 1025] and [223, §3.8].

It is also possible to reduce the transmission problem to a single boundary integral equation for a single unknown function. In order to obtain such equations, one uses a layer *ansatz* in one region (B_e, say) and Green's theorem in the other; different combinations lead to different integral equations with different properties. For example, suppose we write

$$u_{sc}(P) = (S\mu)(P), \quad P \in B_e,$$

whence $u_{sc} = S\mu$ and $\partial u_{sc}/\partial n = (I+K)\mu$ on S. On the other hand, the normal derivative of the interior Helmholtz integral representation, (5.40), gives

$$(I+K_0)(\partial u_0/\partial n) - N_0 u_0 = 0.$$

Hence, the transmission conditions (1.26) give

$$\{(I+K_0)(I+K) - \gamma N_0 S\}\mu = f(p), \quad p \in S, \tag{6.2}$$

where $f(p) = \gamma N_0 u_{inc} - (I+K_0)v_{inc}$. Equation (6.2) is a Fredholm integral equation of the second kind for $\mu(p)$. It is uniquely solvable except when $k^2 \in \mathrm{IV}(S)$; see (5.52).

Many other single integral equations have been derived and analysed in [627]. These include Fredholm integral equations of the first kind (as found by Maystre & Vincent [817], Marx [805, 806] and Knockaert & De Zutter [635]), hypersingular integral equations, and equations that do not suffer from irregular frequencies. Obstacles with internal interfaces are considered in [1161].

6.3 Inhomogeneous obstacles

The problem of acoustic scattering by bounded inhomogeneities embedded in an unbounded homogeneous medium was formulated in Section 1.3.4. The density and sound-speed are assumed to be functions of position within the inhomogeneities, and they can be discontinuous across the interface between the inhomogeneities and the surrounding fluid, S. The transmission conditions are

$$p_e = p_0 \quad \text{and} \quad \frac{1}{\rho_e}\frac{\partial p_e}{\partial n} = \frac{1}{\rho_0}\frac{\partial p_0}{\partial n} \quad \text{on } S, \tag{6.3}$$

where ρ_e is the (constant) density of the fluid in B_e. Within the inhomogeneous obstacles, B, the governing equation is Bergmann's equation, (1.27), which we may write as

$$(\nabla^2 + k^2)p_0 = Vp_0 \quad \text{in } B,$$

where k is the exterior wavenumber and

$$Vp_0 = (k^2 - k_0^2)p_0 + \rho_0^{-1}(\operatorname{grad}\rho_0)\cdot\operatorname{grad}p_0; \tag{6.4}$$

here, ρ_0 and $k_0 = \omega/c_0$ are functions of position in B. An equivalent form of Bergmann's equation is

$$(\nabla^2 + k^2)p_0 = (k^2 - k_0^2\rho_e/\rho_0)p_0 + \operatorname{div}\{(1-\rho_e/\rho_0)\operatorname{grad}p_0\}; \tag{6.5}$$

see, for example, [885, eqn (8.1.12)] or [974, eqn (9-1.19)].

Werner wrote an important paper on the problem of scattering by a bounded inhomogeneous obstacle in 1963 [1312]. He reduced the problem to a system of coupled integral equations, using single-layer, double-layer and volume potentials; this is an example of an *indirect method*. He proved that the scattering problem has exactly one solution; his uniqueness result is in [1311]. (Werner's system of integral equations and his uniqueness proof are summarised in [796].) However, as far as we know, his system of integral equations has not been used in computations. Later, we shall use a *direct method*, and we shall only use volume potentials. (Properties of acoustic volume potentials are discussed in Appendix H.)

Wall [1272] has given a uniqueness theorem for the scattering problem in which the density and sound-speed are allowed to be piecewise-analytic functions of position within B.

6.3.1 A formal method

There is a formal method for solving scattering problems involving inhomogeneous media. To explain it, let us define a function $U(P)$ by

$$U(P) = \begin{cases} p_e(P) - p_{inc}(P), & P \in B_e, \\ p_0(P) - p_{inc}(P), & P \in B, \end{cases} \tag{6.6}$$

so that U is a radiating wavefunction (as $p_e - p_{inc} = p_{sc}$). Evidently,

$$(\nabla^2 + k^2)U(P) = 2f(P), \quad P \in B \cup B_e, \tag{6.7}$$

where

$$f(P) = \begin{cases} 0, & P \in B_e, \\ \frac{1}{2}(Vp_0)(P), & P \in B, \end{cases}$$

and we have used the facts that $(\nabla^2 + k^2)p_{sc} = 0$ in B_e and $(\nabla^2 + k^2)p_{inc} = 0$ in B. Note also that U is continuous across S, whereas $\partial U/\partial n$ is not (unless $\rho_e = \rho_0(q)$, $q \in S$).

Now, regard the right-hand side of (6.7) as a source term, and then note that

$$u(P) = \int G(P,Q)\, f(Q)\, \mathrm{d}V_Q \quad \text{solves} \quad (\nabla^2 + k^2)u = 2f. \tag{6.8}$$

Application of (6.8) to (6.7) gives

$$p_0(P) - \frac{1}{2}\int_B G(P,Q)\,(Vp_0)(Q)\,\mathrm{d}V_Q = p_{\mathrm{inc}}(P), \quad P \in B, \tag{6.9}$$

which is an equation for p_0 in B. A variant of this equation for the total pressure due to point-source insonification is given in the review paper [1182]. In fact, as we shall see, (6.9) is not valid when the density is not continuous across S. (This is unfortunate, because most of the applications in [1182] are to arrays of discrete scatterers.)

Formal derivations of this kind are often found in textbooks; see, for example, [196, §8.9.1]. Indeed, the result (6.8) can be justified readily if one assumes that f is Hölder continuous. However, for discrete scatterers, there will be interfaces across which the wavenumber, the density and the normal derivative of the pressure will be discontinuous.

The formal derivation given above can be repaired [796]. Instead, in Section 6.3.2, we shall derive an equation, similar to (6.9), that respects the proper transmission conditions across interfaces [796]. We prove that solving this equation is equivalent to solving the transmission problem for the acoustic pressure (Theorem 6.1). It reduces to the well-known Lippmann–Schwinger equation when the density in the inhomogeneity is constant and equal to the density of the surrounding homogeneous fluid. It also reduces to the equation derived formally above, namely (6.9), but only when there is no discontinuity in the density across the boundary of the inhomogeneity. If there is such a discontinuity (as is typical in applications), an extra term is needed; see (6.28) below.

6.3.2 An integral representation and an integro-differential equation

We shall consider integral representations obtained using the free-space Green's function for the exterior fluid, $G(P,Q) = -\exp(\mathrm{i}kR)/(2\pi R)$. An application of Green's second theorem in B_e to p_{sc} and G gives

$$\int_S \left\{ G(P,q)\frac{\partial p_{\mathrm{sc}}}{\partial n_q} - p_{\mathrm{sc}}(q)\frac{\partial}{\partial n_q}G(P,q)\right\}\mathrm{d}s_q = \begin{cases} 2p_{\mathrm{sc}}(P), & P \in B_\mathrm{e}, \\ 0, & P \in B. \end{cases}$$

A similar application in B to p_{inc} and G gives

$$\int_S \left\{ G(P,q)\frac{\partial p_{\mathrm{inc}}}{\partial n_q} - p_{\mathrm{inc}}(q)\frac{\partial}{\partial n_q}G(P,q)\right\}\mathrm{d}s_q = \begin{cases} 0, & P \in B_\mathrm{e}, \\ -2p_{\mathrm{inc}}(P), & P \in B. \end{cases}$$

Adding these gives

$$\int_S \left\{ G(P,q)\frac{\rho_\mathrm{e}}{\rho_0}\frac{\partial p_0}{\partial n_q} - p_0(q)\frac{\partial}{\partial n_q}G(P,q)\right\}\mathrm{d}s_q = \begin{cases} 2p_{\mathrm{sc}}(P), & P \in B_\mathrm{e}, \\ -2p_{\mathrm{inc}}(P), & P \in B, \end{cases} \tag{6.10}$$

where we have used $p_e = p_{inc} + p_{sc}$ and the transmission conditions (6.3). The first of (6.10) gives an integral representation for $p_{sc}(P)$ in terms of a distribution of sources and dipoles over S. This representation is not very convenient here because we do not know p_0 or $\partial p_0/\partial n$ on S.

To make progress, recall Green's first theorem,

$$\int_B \{\phi \nabla^2 \psi + (\operatorname{grad}\phi)\cdot(\operatorname{grad}\psi)\}\,\mathrm{d}V = \int_S \phi \frac{\partial \psi}{\partial n}\,\mathrm{d}s,$$

where ϕ and ψ are sufficiently smooth in B. Choose $\phi(Q) = p_0(Q)$ and $\psi(Q) = G(P, Q)$ with $P \in B_e$, whence

$$\int_S p_0(q) \frac{\partial}{\partial n_q} G(P, q)\,\mathrm{d}s_q = \int_B \{(\operatorname{grad} p_0)\cdot(\operatorname{grad}_Q G) - k^2 p_0 G\}\,\mathrm{d}V_Q, \tag{6.11}$$

where we have used $(\nabla^2 + k^2)G(P, Q) = 0$ for $P \neq Q$. Similarly, if we choose $\psi(Q) = p_0(Q)$ and $\phi(Q) = (\rho_e/\rho_0)G(P, Q)$ with $P \in B_e$, we obtain

$$\int_S \frac{\rho_e}{\rho_0}\frac{\partial p_0}{\partial n_q} G(P, q)\,\mathrm{d}s_q = \int_B \frac{\rho_e}{\rho_0}\{(\operatorname{grad} p_0)\cdot(\operatorname{grad}_Q G) - k_0^2(Q)\,p_0(Q)\,G(P, Q)\}\,\mathrm{d}V_Q, \tag{6.12}$$

where we have used Bergmann's equation, (1.27). Subtracting (6.11) from (6.12) gives the left-hand side of (6.10) for $P \in B_e$, whence

$$p_{sc}(P) = (\mathcal{L}p_0)(P), \quad P \in B_e, \tag{6.13}$$

where

$$(\mathcal{L}v)(P) = \frac{1}{2}\int_B \Big\{\Big(\frac{\rho_e}{\rho_0} - 1\Big)(\operatorname{grad} v)\cdot(\operatorname{grad}_Q G(P, Q)) + \Big(k^2 - k_0^2\frac{\rho_e}{\rho_0}\Big)v(Q)\,G(P, Q)\Big\}\,\mathrm{d}V_Q. \tag{6.14}$$

We repeat the calculations for $P \in B$, having excised a small sphere centred at P. The singularity at $P = Q$ has no effect on (6.12) but it causes $-2p_0(P)$ to be added to the left-hand side of (6.11). Then, (6.10) for $P \in B$ becomes

$$-p_{inc}(P) = -p_0(P) + (\mathcal{L}p_0)(P), \quad P \in B.$$

At this stage, we have proved one half of the following theorem.

Theorem 6.1 *Let the pair $\{p_e, p_0\}$ solve the scattering problem formulated in Section* 1.3.4. *Then $v(P) \equiv p_0(P)$ solves*

$$v(P) - (\mathcal{L}v)(P) = p_{inc}(P), \quad P \in B, \tag{6.15}$$

where $\mathcal{L}v$ *is defined by* (6.14). *Conversely, let* v *solve* (6.15). *Then the pair* $\{p_e, p_0\}$, *defined by*

$$p_e(P) = p_{inc}(P) + (\mathcal{L}v)(P) \quad \text{for } P \in B_e \tag{6.16}$$

and $p_0(P) = v(P)$ *for* $P \in B$, *solves the scattering problem.*

Proof We have to prove the second half of the theorem. From (6.16), we define p_{sc} using

$$p_{sc}(P) = (\mathcal{L}v)(P), \qquad P \in B_e; \tag{6.17}$$

evidently, p_{sc} satisfies $(\nabla^2 + k^2)p_{sc} = 0$ in B_e and the Sommerfeld radiation condition, as it inherits these properties from G.

Next, let us show that $p_0 \equiv v$ satisfies Bergmann's equation. As p_{inc} satisfies $(\nabla^2 + k^2)p_{inc} = 0$ in B, (6.15) gives

$$(\nabla^2 + k^2)(v - \mathcal{L}v) = 0 \quad \text{in } B. \tag{6.18}$$

Now, from the definition (6.14), we have

$$\begin{aligned} 2(\mathcal{L}v)(P) = &-\frac{\partial}{\partial x_j^P}\int_B \left(\frac{\rho_e}{\rho_0} - 1\right)\frac{\partial v}{\partial x_j^Q}\, G(P,Q)\,\mathrm{d}V_Q \\ &+ \int_B \left(k^2 - k_0^2\frac{\rho_e}{\rho_0}\right) v(Q)\, G(P,Q)\,\mathrm{d}V_Q, \end{aligned} \tag{6.19}$$

where $P \equiv (x_1^P, x_2^P, x_3^P)$, $Q \equiv (x_1^Q, x_2^Q, x_3^Q)$ and summation over j is implied. The second integral in (6.19) is an acoustic volume potential and the first term is the sum of three first derivatives of volume potentials. The properties of volume potentials are summarised in Appendix H. In particular, the result of applying $(\nabla^2 + k^2)$ is given by (H.1), so that we obtain

$$\begin{aligned} (\nabla^2 + k^2)(\mathcal{L}v) &= -\frac{\partial}{\partial x_j^P}\left\{\left(\frac{\rho_e}{\rho_0} - 1\right)\frac{\partial v}{\partial x_j^P}\right\} + \left(k^2 - k_0^2\frac{\rho_e}{\rho_0}\right)v(P) \\ &= (\nabla^2 + k^2)v - \rho_e \operatorname{div}\left(\rho_0^{-1}\operatorname{grad} v\right) - k_0^2(\rho_e/\rho_0)v, \end{aligned}$$

for $P \in B$, whence (6.18) gives the desired result.

To verify the transmission conditions, observe that (6.15) gives

$$p_0(P) = p_{inc}(P) + (\mathcal{L}v)(P), \qquad P \in B. \tag{6.20}$$

But, as $\mathcal{L}v$ comprises a volume potential and first derivatives of volume potentials, it follows that $(\mathcal{L}v)(P)$ is continuous as P crosses S (see Appendix H). Thus, (6.17) and (6.20) show that the first transmission condition, $(6.3)_1$, is satisfied.

For the second transmission condition, we take the normal derivative of (6.17) and (6.20) to give

$$\frac{\partial}{\partial n}\{p_{sc} - (p_0 - p_{inc})\} = \left[\frac{\partial}{\partial n}\mathcal{L}v\right] \quad \text{on } S, \tag{6.21}$$

where $[f]$ is the discontinuity in f across S, defined by

$$[f(p)] = \lim_{P_e \to p} f(P_e) - \lim_{P \to p} f(P), \quad P_e \in B_e, \quad P \in B, \quad p \in S. \tag{6.22}$$

It is shown in Appendix H that

$$\left[\frac{\partial}{\partial n} \mathcal{L} v \right] = \left(\frac{\rho_e}{\rho_0} - 1 \right) \frac{\partial v}{\partial n}, \tag{6.23}$$

and then (6.21) and $v \equiv p_0$ imply that $(6.3)_2$ is satisfied. This completes the proof of Theorem 6.1. □

We have seen that solving the scattering problem for inhomogeneous obstacles is equivalent to solving equation (6.15), which is an integro-differential equation for $v(P)$, $P \in B$. This equation is uniquely solvable. To see this, we appeal to Werner's existence and uniqueness results [1312]. Thus, the solution $\{p_e, p_0\}$ of the scattering problem exists and, by the first half of Theorem 6.1, p_0 solves (6.15). For uniqueness, suppose that $v_0(P)$ solves (6.15) with $p_{\text{inc}} \equiv 0$. Construct $p_e = (\mathcal{L} v_0)(P)$ for $P \in B_e$ and $p_0 = v_0(P)$ for $P \in B$. By the second half of Theorem 6.1, these fields solve the homogeneous scattering problem; they must vanish identically by the uniqueness theorem for the scattering problem. In particular, $v_0(P) \equiv 0$ for $P \in B$, as required.

We note that an integro-differential equation equivalent to (6.15) was derived by Gerjuoy & Saxon [401] in 1954. In fact, they derived a coupled system, involving the pressure and the velocity, which they regarded as preferable to a single equation for the pressure as they were motivated by the construction of variational principles. Equation (6.15) was also obtained fortuitously by Morse & Ingard [885, eqn (8.1.13)]. They applied the formal method of Section 6.3.1 to (6.5), used the identity

$$\begin{aligned} G \operatorname{div} \left\{ \left(\frac{\rho_e}{\rho_0} - 1 \right) \operatorname{grad} v \right\} &= \left(1 - \frac{\rho_e}{\rho_0} \right) (\operatorname{grad} v) \cdot (\operatorname{grad}_Q G) \\ &\quad + \operatorname{div} \left\{ \left(\frac{\rho_e}{\rho_0} - 1 \right) G \operatorname{grad} v \right\} \end{aligned} \tag{6.24}$$

and then discarded the contribution from the last term.

6.3.3 The Lippmann–Schwinger equation

Consider the special case of the scattering problem where $\rho_0(Q) = \rho_e$ for all $Q \in B$, so that the density of all the scatterers is the same as that of the surrounding homogeneous fluid. Then, the integro-differential equation (6.15) reduces to the integral equation

$$v(P) - \frac{1}{2} k^2 \int_B \{1 - n(Q)\} v(Q) \, G(P, Q) \, \mathrm{d}V_Q = p_{\text{inc}}(P), \quad P \in B, \tag{6.25}$$

where $n(Q) = (k_0/k)^2 = \{c/c_0(Q)\}^2$. This integral equation and its numerical treatment have been discussed in [202, 151, 1344, 189, 152, 148, 525] and [196, §8.9.1].

For example, Hyde & Bruno [525] gave results for scattering by a regular $5 \times 5 \times 5$ array of identical spherical inhomogeneities.

Let us define $n(P) = 1$ for $P \in B_{\rm e}$ and

$$w(P) = \begin{cases} p_{\rm e}(P), & P \in B_{\rm e}, \\ p_0(P), & P \in B. \end{cases}$$

Then, we can combine (6.25) with the representation (6.16) to obtain

$$w(P) - \frac{1}{2} k^2 \int \{1 - n(Q)\} w(Q)\, G(P, Q)\, {\rm d}V_Q = p_{\rm inc}(P) \tag{6.26}$$

for all $P \in B \cup B_{\rm e}$, where the integration is over all Q. We recognise this equation as the *Lippmann–Schwinger equation* [733]; see, for example, [14], [225, §8.2], [911, §10.3] and [953, Theorem 9.4]. Notice that our derivation shows that the Lippmann–Schwinger equation is valid even when $n(Q)$ is discontinuous as Q crosses S. This fact is implicit in [949] and explicit in [1313, 1314].

6.3.4 An alternative equation

As we know that $v \equiv p_0$ solves Bergmann's equation in B, we can use this fact to rewrite the expression for $\mathcal{L}v$. Thus, using (6.24) and

$$\begin{aligned} \operatorname{div}\left\{\left(\frac{\rho_{\rm e}}{\rho_0} - 1\right) \operatorname{grad} v\right\} &= \left(\frac{\rho_{\rm e}}{\rho_0} - 1\right) \nabla^2 v + \rho_{\rm e} \left(\operatorname{grad} \rho_0^{-1}\right) \cdot \operatorname{grad} v \\ &= \left(\frac{\rho_{\rm e}}{\rho_0} - 1\right) \left\{\nabla^2 v + \rho_0 \left(\operatorname{grad} \rho_0^{-1}\right) \cdot \operatorname{grad} v\right\} \\ &\quad + \rho_0 \left(\operatorname{grad} \rho_0^{-1}\right) \cdot \operatorname{grad} v \\ &= \left(1 - \frac{\rho_{\rm e}}{\rho_0}\right) k_0^2 v - \rho_0^{-1} (\operatorname{grad} \rho_0) \cdot \operatorname{grad} v \end{aligned}$$

in (6.14), we obtain

$$(\mathcal{L}v)(P) = \frac{1}{2} \int_B G(P, Q)\, (Vv)(Q)\, {\rm d}V_Q + (\mathcal{L}_E v)(P),$$

where Vv is defined by (6.4) and

$$\begin{aligned} (\mathcal{L}_E v)(P) &= \frac{1}{2} \int_B \operatorname{div}\left\{\left(\frac{\rho_{\rm e}}{\rho_0} - 1\right) G \operatorname{grad} v\right\} {\rm d}V_Q \\ &= \frac{1}{2} \int_S \left(\frac{\rho_{\rm e}}{\rho_0(q)} - 1\right) G(P, q) \frac{\partial v}{\partial n}\, {\rm d}s_q, \end{aligned} \tag{6.27}$$

by the divergence theorem. Thus, the scattering problem can be reduced to solving

$$p_0(P) = p_{\rm inc}(P) + \frac{1}{2} \int_B G(P, Q)\, (Vp_0)(Q)\, {\rm d}V_Q + p_E(P) \tag{6.28}$$

for $P \in B$, where

$$p_E(P) = (\mathcal{L}_E p_0)(P) = \frac{1}{2}\int_S \left(\frac{\partial p_e}{\partial n} - \frac{\partial p_0}{\partial n} \right) G(P, q)\, \mathrm{d}s_q$$

and we have used $(6.3)_2$ in (6.27).

Equation (6.28) reduces to an equation obtained by Ramm [1003] when both ρ_0 and k_0 are constants. However, in this situation, the scatterer is homogeneous and the problem reduces to the transmission problem discussed in Section 6.2: it can be solved using boundary integral equations over S.

If we had attempted to solve the scattering problem using the formal method described in Section 6.3.1, we would have obtained (6.9), which is precisely (6.28) but with $p_E(P) \equiv 0$. In general, this extra term is not zero, and its magnitude is difficult to estimate. Observe that, from (6.27), p_E does vanish if $\rho_0(q) = \rho_e$ for all $q \in S$, which means that the density is continuous across S. Otherwise, the single-layer potential $p_E(P)$ should be retained.

6.3.5 The far field

We have seen that the scattered field is given by (6.13). The far field is given by Definition 4.2. Standard calculations then show that the far-field pattern is given by

$$f(\hat{\boldsymbol{r}}) = \frac{\mathrm{i}k}{4\pi}\int_B \left\{ \mathrm{i}k\left(\frac{\rho_e}{\rho_0} - 1\right)\hat{\boldsymbol{r}}\cdot \operatorname{grad} p_0 - \left(k^2 - k_0^2\frac{\rho_e}{\rho_0}\right)p_0 \right\} \exp\left(-\mathrm{i}k\hat{\boldsymbol{r}}\cdot\boldsymbol{r}_Q\right) \mathrm{d}V_Q. \qquad (6.29)$$

An equivalent formula can be found in [885, eqn (8.1.14)].

The formula (6.29) leads to various approximations. For example, replacing p_0 by p_{inc} on the right-hand side of (6.29) gives the *Born approximation* for f.

6.4 Electromagnetic waves

Electromagnetic scattering problems were formulated in Section 1.4, and the results obtained by the method of separation of variables were reviewed in Section 4.9. Here, we summarise how integral equations can be used to solve electromagnetic problems.

6.4.1 Layer potentials and the indirect method

Recall the acoustic single-layer potential, $(S\nu)(P)$, defined by (5.5). In electromagnetic theory, we usually apply the operator S to a vector-valued function of position, $\boldsymbol{a}(q)$, say; we define

$$(\mathcal{C}\boldsymbol{a})(P) = \operatorname{curl}\{S\boldsymbol{a}\} \quad \text{and} \quad (\mathcal{F}\boldsymbol{a})(P) = \operatorname{curl}\{\mathcal{C}\boldsymbol{a}\}. \qquad (6.30)$$

We are interested in the tangential components of these vector fields evaluated on S when $\boldsymbol{a}(q)$ itself is a tangential density (so that $\boldsymbol{a}(q)\cdot\boldsymbol{n}(q)=0$ for all $q\in S$). For continuous tangential densities, we have

$$\boldsymbol{n}\times\mathcal{C}\boldsymbol{a}=\mp\boldsymbol{a}+\mathcal{M}\boldsymbol{a}, \tag{6.31}$$

where the upper (lower) sign corresponds to $P\to p\in S$ from B_e (B) and $\mathcal{M}$ is a boundary integral operator defined by

$$(\mathcal{M}\boldsymbol{a})(p)=\boldsymbol{n}(p)\times\operatorname{curl}\{S\boldsymbol{a}\},\quad p\in S.$$

For sufficiently smooth tangential densities $\boldsymbol{a}$, we also have

$$\boldsymbol{n}\times\mathcal{F}\boldsymbol{a}=\mathcal{P}\boldsymbol{a} \tag{6.32}$$

on S, where

$$(\mathcal{P}\boldsymbol{a})(p)=\boldsymbol{n}(p)\times\operatorname{curl}\operatorname{curl}\{S\boldsymbol{a}\},\quad p\in S.$$

Further properties of the operators $\mathcal{M}$ and $\mathcal{P}$ can be found in [223, §2.7] and in [799].

For scattering by perfect conductors, we may seek a solution in the form

$$\boldsymbol{E}_{\mathrm{sc}}(P)=(\mathcal{C}\boldsymbol{j})(P),\quad \boldsymbol{H}_{\mathrm{sc}}(P)=(\mathrm{i}\omega\mu)^{-1}\operatorname{curl}\boldsymbol{E}_{\mathrm{sc}}(P),\quad P\in B_e, \tag{6.33}$$

where $\boldsymbol{j}(q)$ is a Hölder-continuous tangential vector density. The boundary condition $(1.41)_2$ and the jump relation (6.31) then yield

$$(I-\mathcal{M})\boldsymbol{j}=\boldsymbol{n}\times\boldsymbol{E}_{\mathrm{inc}}, \tag{6.34}$$

which is a Fredholm integral equation of the second kind for $\boldsymbol{j}(q)$, $q\in S$. It turns out (see [223, §4.4]) that this equation has exactly one solution $\boldsymbol{j}$ except when the following interior problem has a non-trivial solution.

Interior Maxwell problem *Find a field $\{\boldsymbol{E},\boldsymbol{H}\}$ that satisfies Maxwell's equations* (1.37) *in the bounded domain B together with the boundary condition $\boldsymbol{n}\times\boldsymbol{E}=\boldsymbol{0}$ on S.*

If this problem has a non-trivial solution, we say that $k^2=\omega^2\mu\varepsilon$ is an eigenvalue of the interior Maxwell problem. All such eigenvalues are known to be real [223, p. 125]. Physically, the interior Maxwell problem corresponds to a perfectly-conducting cavity resonator.

As an alternative to (6.33), we can look for a solution in the form

$$\boldsymbol{E}_{\mathrm{sc}}(P)=(\mathcal{F}\boldsymbol{j})(P)=(\operatorname{grad}\operatorname{div}+k^2)(S\boldsymbol{j})(P),\quad P\in B_e; \tag{6.35}$$

it follows that $\boldsymbol{j}$ solves $\mathcal{P}\boldsymbol{j}=-\boldsymbol{n}\times\boldsymbol{E}_{\mathrm{inc}}$ on S.

6.4.2 Thin wires

The representation (6.35) has been used to solve problems involving scattering by thin wires. Thus, for a straight wire of unit length, lying on the z-axis, with $0 \le z \le 1$, the governing integro-differential equation can be reduced to

$$\left(\frac{\mathrm{d}^2}{\mathrm{d}z^2}+k^2\right)\int_0^1 I(\zeta)\,K(z-\zeta)\,\mathrm{d}\zeta = f(z), \quad 0<z<1, \tag{6.36}$$

where I is the unknown current in the wire and f is known in terms of the incident field. The kernel is given by

$$K(w) = \frac{1}{\pi}\int_0^\pi \frac{\exp\left\{\mathrm{i}k\sqrt{w^2+4a^2\sin^2(\theta/2)}\right\}}{\sqrt{w^2+4a^2\sin^2(\theta/2)}}\,\mathrm{d}\theta,$$

where a is the (small) radius of the wire's circular cross-section. Equation (6.36) is to be solved subject to $I(0)=I(1)=0$. For a derivation of (6.36), and some discussion of its properties, see [564, §3.14] and [567]; in particular, $K(w) \sim -(a\pi)^{-1}\log|w|$ as $w \to 0$. Numerical methods for solving (6.36) are developed in [268]. Many authors replace K by a 'reduced kernel'

$$K_\mathrm{r}(w) = \frac{\exp\{\mathrm{i}k\sqrt{w^2+a^2}\}}{\sqrt{w^2+a^2}},$$

but this replacement makes (6.36) ill-posed [268]; evidently, $K_\mathrm{r}(0)$ is defined whereas $K(w)$ is logarithmically singular as $w \to 0$.

For multiple scattering by several parallel thin wires, one can derive systems of one-dimensional integro-differential equations. Such equations, using K_r, have been studied at length in the book by King *et al.* [611]; see also [1114].

6.4.3 The Stratton–Chu formula and the direct method

The vector analogues of Green's theorem are obtained by applying the divergence theorem to $\boldsymbol{w} = \boldsymbol{P} \times \operatorname{curl}\boldsymbol{Q}$, giving

$$\int_\mathcal{B}\{\boldsymbol{P}\cdot\operatorname{curl}\operatorname{curl}\boldsymbol{Q} - (\operatorname{curl}\boldsymbol{P})\cdot(\operatorname{curl}\boldsymbol{Q})\}\,\mathrm{d}V = \int_\mathcal{S}\boldsymbol{P}\cdot(\boldsymbol{\nu}\times\operatorname{curl}\boldsymbol{Q})\,\mathrm{d}S,$$

where $\mathcal{B}$ is a bounded region with boundary $\mathcal{S}$ and $\boldsymbol{\nu}(q)$ is the unit normal vector at $q \in \mathcal{S}$ pointing out of $\mathcal{B}$. Interchanging $\boldsymbol{P}$ and $\boldsymbol{Q}$, and then subtracting the two results gives [1150]

$$\int_\mathcal{B}\{\boldsymbol{Q}\cdot\operatorname{curl}\operatorname{curl}\boldsymbol{P} - \boldsymbol{P}\cdot\operatorname{curl}\operatorname{curl}\boldsymbol{Q}\}\,\mathrm{d}V$$
$$= \int_\mathcal{S}\{\boldsymbol{P}\times\operatorname{curl}\boldsymbol{Q} - \boldsymbol{Q}\times\operatorname{curl}\boldsymbol{P}\}\cdot\boldsymbol{\nu}\,\mathrm{d}S. \tag{6.37}$$

Following Stratton & Chu [1150] (see also [1149, §8.14], [1152] and [223, Theorem 4.1]), apply (6.37) in B to $\boldsymbol{P}=\boldsymbol{E}(Q)$ and $\boldsymbol{Q}=\boldsymbol{c}G(P,Q)$, where $\mathcal{B}=B$, $\mathcal{S}=S$, $\boldsymbol{\nu}=\boldsymbol{n}$, $\boldsymbol{c}$ is an arbitrary constant vector and $P\in B$. The result is

$$2\boldsymbol{E}(P)=\int_S\{\mathrm{i}\omega\mu(\boldsymbol{n}\times\boldsymbol{H})G+(\boldsymbol{n}\times\boldsymbol{E})\times\operatorname{grad}_q G+(\boldsymbol{n}\cdot\boldsymbol{E})\operatorname{grad}_q G\}\,\mathrm{d}s_q; \tag{6.38}$$

this is [1149, §8.14, eqn (19)] and [979, eqn (4.8a)]. An equivalent but more concise formula, making use of (6.30), is

$$2\boldsymbol{E}(P)=\mathcal{C}\{\boldsymbol{n}\times\boldsymbol{E}\}-(\mathrm{i}\omega\varepsilon)^{-1}\mathcal{F}\{\boldsymbol{n}\times\boldsymbol{H}\},\quad P\in B. \tag{6.39}$$

To see that these are indeed equivalent, note first that

$$(\mathcal{C}\boldsymbol{a})(P)=\int_S(\operatorname{grad}_P G)\times\boldsymbol{a}(q)\,\mathrm{d}s_q,$$

using $\operatorname{curl}(\psi\boldsymbol{c})=(\operatorname{grad}\psi)\times\boldsymbol{c}$ for constant $\boldsymbol{c}$. Thus, as $G(P,Q)$ is a function of the distance between P and Q, we see that

$$\mathcal{C}\{\boldsymbol{n}\times\boldsymbol{E}\}=\int_S(\boldsymbol{n}\times\boldsymbol{E})\times\operatorname{grad}_q G\,\mathrm{d}s_q. \tag{6.40}$$

Second, using the vector identity $\operatorname{curl}\operatorname{curl}=\operatorname{grad}\operatorname{div}-\nabla^2$ and the equation $(\nabla^2+k^2)G=0$, we obtain

$$(\mathcal{F}\boldsymbol{a})(P)=\operatorname{grad}_P\int_S(\operatorname{Div}\boldsymbol{a})\,G(P,q)\,\mathrm{d}s_q+k^2\int_S\boldsymbol{a}(q)\,G(P,q)\,\mathrm{d}s_q,$$

where $\operatorname{Div}\boldsymbol{a}$ is the surface divergence and we have used the fact that

$$\operatorname{div}\{S\boldsymbol{a}\}=S\{\operatorname{Div}\boldsymbol{a}\};$$

see [223, Theorem 2.29]. Also, as

$$\operatorname{Div}(\boldsymbol{n}\times\boldsymbol{H})=-\boldsymbol{n}\cdot\operatorname{curl}\boldsymbol{H}=\mathrm{i}\omega\varepsilon\,\boldsymbol{n}\cdot\boldsymbol{E}$$

(see [223, eqn (2.75)]), we obtain

$$\mathcal{F}\{\boldsymbol{n}\times\boldsymbol{H}\}=-\mathrm{i}\omega\varepsilon\int_S(\boldsymbol{n}\cdot\boldsymbol{E})\operatorname{grad}_q G\,\mathrm{d}s_q+\omega^2\mu\varepsilon\int_S(\boldsymbol{n}\times\boldsymbol{H})G\,\mathrm{d}s_q. \tag{6.41}$$

Thus, (6.38) and (6.39) are equivalent.

Equation (6.38) or (6.39) is an example of what are usually known as the *Stratton–Chu representations*. There are similar formulae for the exterior region B_e, which make use of the Silver–Müller radiation conditions (see Section 1.4.1). Thus, we obtain

$$\mathcal{C}\{\boldsymbol{n}\times\boldsymbol{E}_{\mathrm{sc}}\}-(\mathrm{i}\omega\varepsilon)^{-1}\mathcal{F}\{\boldsymbol{n}\times\boldsymbol{H}_{\mathrm{sc}}\}=\begin{cases}-2\boldsymbol{E}_{\mathrm{sc}}(P), & P\in B_e,\\ \boldsymbol{0}, & P\in B,\end{cases}$$

and

$$\mathcal{C}\{\boldsymbol{n}\times\boldsymbol{H}_{\mathrm{sc}}\}+(\mathrm{i}\omega\mu)^{-1}\mathcal{F}\{\boldsymbol{n}\times\boldsymbol{E}_{\mathrm{sc}}\}=\begin{cases}-2\boldsymbol{H}_{\mathrm{sc}}(P), & P\in B_e,\\ \boldsymbol{0}, & P\in B.\end{cases}$$

Similarly, we have (cf. (6.39))

$$\mathcal{C}\{\boldsymbol{n}\times\boldsymbol{E}_{\mathrm{inc}}\}-(\mathrm{i}\omega\varepsilon)^{-1}\mathcal{F}\{\boldsymbol{n}\times\boldsymbol{H}_{\mathrm{inc}}\}=\begin{cases}\mathbf{0}, & P\in B_{\mathrm{e}},\\ 2\boldsymbol{E}_{\mathrm{inc}}(P), & P\in B,\end{cases}$$

and

$$\mathcal{C}\{\boldsymbol{n}\times\boldsymbol{H}_{\mathrm{inc}}\}+(\mathrm{i}\omega\mu)^{-1}\mathcal{F}\{\boldsymbol{n}\times\boldsymbol{E}_{\mathrm{inc}}\}=\begin{cases}\mathbf{0}, & P\in B_{\mathrm{e}},\\ 2\boldsymbol{H}_{\mathrm{inc}}(P), & P\in B.\end{cases}$$

Adding these equations, using (1.38), gives

$$\mathcal{C}\{\boldsymbol{n}\times\boldsymbol{E}\}-(\mathrm{i}\omega\varepsilon)^{-1}\mathcal{F}\{\boldsymbol{n}\times\boldsymbol{H}\}=\begin{cases}-2\boldsymbol{E}_{\mathrm{sc}}(P), & P\in B_{\mathrm{e}},\\ 2\boldsymbol{E}_{\mathrm{inc}}(P), & P\in B,\end{cases} \tag{6.42}$$

and

$$\mathcal{C}\{\boldsymbol{n}\times\boldsymbol{H}\}+(\mathrm{i}\omega\mu)^{-1}\mathcal{F}\{\boldsymbol{n}\times\boldsymbol{E}\}=\begin{cases}-2\boldsymbol{H}_{\mathrm{sc}}(P), & P\in B_{\mathrm{e}},\\ 2\boldsymbol{H}_{\mathrm{inc}}(P), & P\in B,\end{cases} \tag{6.43}$$

where $\{\boldsymbol{E},\boldsymbol{H}\}$ are the total electromagnetic fields.

Computing the tangential components of (6.42) and (6.43) on S, using (6.31) and (6.32), we obtain

$$(I+\mathcal{M})(\boldsymbol{n}\times\boldsymbol{E})-(\mathrm{i}\omega\varepsilon)^{-1}\mathcal{P}(\boldsymbol{n}\times\boldsymbol{H})=2\boldsymbol{n}\times\boldsymbol{E}_{\mathrm{inc}} \tag{6.44}$$

and

$$(I+\mathcal{M})(\boldsymbol{n}\times\boldsymbol{H})+(\mathrm{i}\omega\mu)^{-1}\mathcal{P}(\boldsymbol{n}\times\boldsymbol{E})=2\boldsymbol{n}\times\boldsymbol{H}_{\mathrm{inc}}. \tag{6.45}$$

For scattering by perfect conductors, we have $\boldsymbol{n}\times\boldsymbol{E}=\mathbf{0}$ on S, and then the first of (6.42) gives the representation

$$2\boldsymbol{E}_{\mathrm{sc}}(P)=(\mathrm{i}\omega\varepsilon)^{-1}(\mathcal{F}\boldsymbol{J})(P),\quad P\in B_{\mathrm{e}}, \tag{6.46}$$

where $\boldsymbol{J}(q)=\boldsymbol{n}\times\boldsymbol{H}$, $q\in S$. Moreover, (6.44) and (6.45) reduce to

$$\mathcal{P}\boldsymbol{J}=-2\mathrm{i}\omega\varepsilon\,\boldsymbol{n}\times\boldsymbol{E}_{\mathrm{inc}} \tag{6.47}$$

and

$$(I+\mathcal{M})\boldsymbol{J}=2\boldsymbol{n}\times\boldsymbol{H}_{\mathrm{inc}}, \tag{6.48}$$

respectively. Equations (6.47) and (6.48) are known as the *electric field integral equation* (EFIE) and the *magnetic field integral equation* (MFIE), respectively. The EFIE is a singular integral equation. The MFIE is a Fredholm integral of the second kind; it suffers from the same irregular frequencies as (6.34). For an alternative discussion of irregular frequencies in the electromagnetic context, see [683].

6.4.4 Transmission problems

If the region B is filled with a different homogeneous material, one can reduce the corresponding transmission problem (formulated in Section 1.4.3) to pairs of coupled boundary integral equations over the interface S, using G and G_0, where G_0 is G with k replaced by $k_0 = \omega\sqrt{\varepsilon_0\mu_0}$.

It is also possible to reduce electromagnetic transmission problems to single integral equations over S.

6.4.5 Inhomogeneous obstacles

Suppose that μ_0 and ε_0 are functions of position in B; see Section 1.4.4 for the equations governing $\boldsymbol{E}_0$ and $\boldsymbol{H}_0$ within B. Then, from [1152, eqn (2.10)], we have

$$\int_B \boldsymbol{v}_0(Q)\, G(P,Q)\,\mathrm{d}V_Q + \int_S \boldsymbol{F}_0(P,q)\,\mathrm{d}s_q = \begin{cases} \boldsymbol{0}, & P \in B_\mathrm{e}, \\ 2\boldsymbol{E}_0(P), & P \in B, \end{cases} \tag{6.49}$$

where

$$\boldsymbol{v}_0 = (k^2 - k_0^2)\boldsymbol{E}_0 - (\mu_0^{-1}\operatorname{grad}\mu_0) \times \operatorname{curl}\boldsymbol{E}_0 + \operatorname{grad}\operatorname{div}\boldsymbol{E}_0, \tag{6.50}$$

$k_0^2 = \omega^2\mu_0\varepsilon_0$ and

$$\boldsymbol{F}_0 = (\boldsymbol{n} \times \operatorname{curl}\boldsymbol{E}_0 - \boldsymbol{n}\operatorname{Div}\boldsymbol{E}_0)G + (\boldsymbol{n} \times \boldsymbol{E}_0) \times \operatorname{grad}_q G + (\boldsymbol{n}\cdot\boldsymbol{E}_0)\operatorname{grad}_q G.$$

Maxwell's equations in B, (1.42), give

$$\begin{aligned} \boldsymbol{F}_0 = {} & (\boldsymbol{n} \times \boldsymbol{E}_0) \times \operatorname{grad}_q G + (\boldsymbol{n}\cdot\boldsymbol{E}_0)\operatorname{grad}_q G \\ & + \{\mathrm{i}\omega\mu_0(\boldsymbol{n} \times \boldsymbol{H}_0) + \boldsymbol{n}\,(\varepsilon_0^{-1}\operatorname{grad}\varepsilon_0)\cdot\boldsymbol{E}_0\}G. \end{aligned}$$

On the other hand, (6.42) and the transmission conditions (1.43) give

$$\mathcal{C}\{\boldsymbol{n} \times \boldsymbol{E}_0\} - (\mathrm{i}\omega\varepsilon)^{-1}\mathcal{F}\{\boldsymbol{n} \times \boldsymbol{H}_0\} = \begin{cases} -2\boldsymbol{E}_\mathrm{sc}(P), & P \in B_\mathrm{e}, \\ 2\boldsymbol{E}_\mathrm{inc}(P), & P \in B. \end{cases}$$

Making use of (6.40) and (6.41), we can write these as

$$\int_S \boldsymbol{F}_1(P,q)\,\mathrm{d}s_q = \begin{cases} -2\boldsymbol{E}_\mathrm{sc}(P), & P \in B_\mathrm{e}, \\ 2\boldsymbol{E}_\mathrm{inc}(P), & P \in B, \end{cases} \tag{6.51}$$

where

$$\boldsymbol{F}_1 = (\boldsymbol{n} \times \boldsymbol{E}_0) \times \operatorname{grad}_q G + (\varepsilon_0/\varepsilon)(\boldsymbol{n}\cdot\boldsymbol{E}_0)\operatorname{grad}_q G + \mathrm{i}\omega\mu(\boldsymbol{n} \times \boldsymbol{H}_0)G.$$

Subtracting (6.51) from (6.49), we obtain

$$\int_B \boldsymbol{v}_0 G(P,Q)\,\mathrm{d}V_Q + \int_S \boldsymbol{F}(P,q)\,\mathrm{d}s_q = \begin{cases} 2\boldsymbol{E}_\mathrm{sc}(P), & P \in B_\mathrm{e}, \\ 2\boldsymbol{E}_0(P) - 2\boldsymbol{E}_\mathrm{inc}(P), & P \in B, \end{cases} \tag{6.52}$$

where

$$\boldsymbol{F} = \left(1 - \frac{\varepsilon_0}{\varepsilon}\right)(\boldsymbol{n} \cdot \boldsymbol{E}_0)\operatorname{grad}_q G + \left\{\left(1 - \frac{\mu}{\mu_0}\right)(\boldsymbol{n} \times \operatorname{curl} \boldsymbol{E}_0) + \boldsymbol{n}\,(\varepsilon_0^{-1} \operatorname{grad} \varepsilon_0) \cdot \boldsymbol{E}_0\right\} G.$$

The first of (6.52) gives an integral representation for the scattered electric field in terms of the field $\boldsymbol{E}_0$ inside B and on S, whereas the second of (6.52) can be solved, in principle, for $\boldsymbol{E}_0$. The combination of volume and surface integrals is inconvenient, computationally, but notice that $\boldsymbol{F}$ vanishes if $\varepsilon_0(q) = \varepsilon$, $\mu_0(q) = \mu$ and $\varepsilon_0(P)$ is constant near S.

When $\mu_0(Q) \equiv \mu$, the second of (6.52) reduces to [979, eqn (4.18)].

Before eliminating the surface integral in (6.52), let us recall a formal derivation of an equation that is very similar to (6.52).

6.4.5.1 A formal derivation

We can write (1.44) as

$$(\nabla^2 + k^2)\boldsymbol{E}_0 = \boldsymbol{v}_0 \quad \text{in } B,$$

where $\boldsymbol{v}_0$ is defined by (6.50). So, if we define a vector field $\boldsymbol{U}$ by (cf. (6.6))

$$\boldsymbol{U}(P) = \begin{cases} \boldsymbol{E}(P) - \boldsymbol{E}_{\text{inc}}(P), & P \in B_{\text{e}}, \\ \boldsymbol{E}_0(P) - \boldsymbol{E}_{\text{inc}}(P), & P \in B, \end{cases}$$

we see that $\boldsymbol{U}$ satisfies the Silver–Müller radiation conditions, that $\boldsymbol{n} \times \boldsymbol{U}$ is continuous across S, and that $(\nabla^2 + k^2)\boldsymbol{U} = 2\boldsymbol{f}$, where $\boldsymbol{f}(P) = \boldsymbol{0}$ for $P \in B_{\text{e}}$ and $\boldsymbol{f}(P) = \frac{1}{2}\boldsymbol{v}_0(P)$ for $P \in B$. Hence, as each Cartesian component of $\boldsymbol{U}$ satisfies the Sommerfeld radiation condition ([223, Corollary 4.7]), (6.8) gives

$$2\boldsymbol{U}(P) = \int_B \boldsymbol{v}_0(Q)\, G(P, Q)\, \mathrm{d}V_q. \tag{6.53}$$

This formula should be compared with (6.52); we see that (6.53) is incorrect unless the surface integral in (6.52) vanishes.

6.4.5.2 An alternative formulation

Consider (6.51). We can write $\boldsymbol{c} \cdot \boldsymbol{F}_1 = \boldsymbol{n} \cdot \boldsymbol{X}$, where $\boldsymbol{c}$ is an arbitrary constant vector and

$$\boldsymbol{X}(Q) = \boldsymbol{E}_0 \times \operatorname{curl}_Q (G\boldsymbol{c}) + (\varepsilon_0/\varepsilon)(\boldsymbol{c} \cdot \operatorname{grad}_Q G)\boldsymbol{E}_0 - (\mu/\mu_0)G\boldsymbol{c} \times \operatorname{curl} \boldsymbol{E}_0. \tag{6.54}$$

Then, the divergence theorem gives

$$-2\boldsymbol{c} \cdot \boldsymbol{E}_{\text{sc}}(P) = \int_B \operatorname{div} \boldsymbol{X}\, \mathrm{d}V_Q, \quad P \in B_{\text{e}},$$

whence

$$2\boldsymbol{E}_{\mathrm{sc}}(P) = \int_B \boldsymbol{W}(P,Q)\,\mathrm{d}V_Q, \quad P \in B_\mathrm{e}, \tag{6.55}$$

where $\operatorname{div}\boldsymbol{X} = -\boldsymbol{c}\cdot\boldsymbol{W}$ and

$$\begin{aligned}\boldsymbol{W}(P,Q) = {} & \Big(1-\frac{\mu}{\mu_0}\Big)(\operatorname{grad}_Q G)\times\operatorname{curl}\boldsymbol{E}_0(Q) \\ & + \Big(1-\frac{\varepsilon_0}{\varepsilon}\Big)\big\{k^2 G\boldsymbol{E}_0 - \operatorname{grad}_P(\boldsymbol{E}_0\cdot\operatorname{grad}_Q G)\big\}.\end{aligned} \tag{6.56}$$

For $P \in B$, this calculation must be modified because of the singularity of $G(P,Q)$ at $P = Q$. So, we remove a ball B_δ from B, where B_δ has radius δ, centre P and spherical boundary S_δ. Thus, from (6.51), we have

$$2\boldsymbol{c}\cdot\boldsymbol{E}_{\mathrm{inc}}(P) = \int_S \boldsymbol{n}\cdot\boldsymbol{X}\,\mathrm{d}s_q = \int_{B\setminus B_\delta} \operatorname{div}\boldsymbol{X}\,\mathrm{d}V_Q + \int_{S_\delta} \boldsymbol{n}\cdot\boldsymbol{X}\,\mathrm{d}s_q. \tag{6.57}$$

(In this formula, we could replace B_δ by any other volume containing P with corresponding boundary S_δ, and we do not have to assume that δ is small.)

We are now going to let $\delta \to 0$ in (6.57). Inspection of (6.54) shows that $\boldsymbol{X}(q) = O(\delta^{-2})$ as $\delta \to 0$, and so the surface integral over S_δ in (6.57) exists and has a limit as $\delta \to 0$; specifically, we find that

$$\lim_{\delta\to 0}\int_{S_\delta} \boldsymbol{n}\cdot\boldsymbol{X}\,\mathrm{d}s_q = \left\{2-\frac{2}{3}\left(1-\frac{\varepsilon_0(P)}{\varepsilon}\right)\right\}\boldsymbol{c}\cdot\boldsymbol{E}_0(P).$$

It follows that the volume integral over $B\setminus B_\delta$ in (6.57) must also have a limit as $\delta \to 0$, even though (6.56) shows that $\operatorname{div}\boldsymbol{X}$ is $O(\delta^{-3})$ as $\delta \to 0$. In fact, the integral exists as a Cauchy principal-value integral; see Appendix H.2. Thus, we obtain

$$\left\{2-\frac{2}{3}\left(1-\frac{\varepsilon_0(P)}{\varepsilon}\right)\right\}\boldsymbol{E}_0(P) - ⨍_B \boldsymbol{W}(P,Q)\,\mathrm{d}V_Q = 2\boldsymbol{E}_{\mathrm{inc}}(P) \tag{6.58}$$

for $P \in B$, which is a singular integral equation for $\boldsymbol{E}_0$.

As a special case, suppose that $\mu_0(Q) \equiv \mu$; this is a common assumption. Then

$$\boldsymbol{W}(P,Q) = k^2\left(1-\frac{\varepsilon_0(Q)}{\varepsilon}\right)\mathbf{G}(P,Q)\cdot\boldsymbol{E}_0(Q), \tag{6.59}$$

where

$$(\mathbf{G}(P,Q))_{ij} = G\delta_{ij} + \frac{1}{k^2}\frac{\partial^2 G}{\partial x_i^Q\,\partial x_j^Q}. \tag{6.60}$$

When (6.59) is used in (6.58), we obtain [1191, eqn. (2.1.41)], whereas (6.55) gives

$$2\boldsymbol{E}_{\mathrm{sc}}(P) = k^2\int_B \left(1-\frac{\varepsilon_0(Q)}{\varepsilon}\right)\mathbf{G}(P,Q)\cdot\boldsymbol{E}_0(Q)\,\mathrm{d}V_Q, \quad P \in B_\mathrm{e}. \tag{6.61}$$

This representation is the starting point for developing various approximations; for example, the Born or Rayleigh–Gans approximation is obtained by replacing $\boldsymbol{E}_0$ by $\boldsymbol{E}_{\mathrm{inc}}$ on the right-hand side of (6.61).

$\mathbf{G}$, defined by (6.60), is known as the *free-space dyadic Green's function*. It can be used instead of G so as to obtain integral representations and integral equations; for example, for perfect conductors, the representation (6.46) can be written as

$$2\boldsymbol{E}_{\mathrm{sc}}(P) = -\mathrm{i}\omega\mu \int_S \mathbf{G}(P, q) \cdot \boldsymbol{J}(q) \, \mathrm{d}s_q, \quad P \in B_{\mathrm{e}}. \tag{6.62}$$

However, the strong singularity ($O(R^{-3})$ as $R \to 0$) means that much care is needed whenever $\mathbf{G}$ is used. For lengthy discussions, see [196, §7.1], [1231, Chapter 3], [1277, Chapter 7], [1190, Chapter 2, §1.1] and [1191, Chapter 2, §1].

6.4.6 Literature

From a mathematical point of view, we recommend the book by Colton & Kress [223] for the basic theory of integral-equation methods for electromagnetic scattering problems. See also [207] and [262]. For non-smooth S, see [875, 874, 156] and references therein.

The famous paper of Stratton & Chu [1150] was published in 1939. According to Sommerfeld [1129, p. 325], the equivalent formula (6.39) was obtained by W. Franz in 1948. In fact, both (6.38) and (6.39) can be found in a 1931 paper by Murray [897]. This paper also contains the MFIE, (6.48). Both the MFIE and the EFIE were derived by Maue [813] in 1949.

In 1954, Storer & Sevick [1147] used (6.62) for multiple-scattering problems; they obtained approximate solutions for scattering by two parallel wires of finite length.

Perhaps the most influential book on computational electromagnetics is Harrington's *Field Computation by Moment Methods*, published in 1968 [481]; see also [480]. Many examples are given, including Galerkin-type methods for solving boundary integral equations. Since then, many books and reviews have appeared: we mention the long review by Poggio & Miller [979], the reviews by Ström [1152] and Wilton [1325], and the books by Chew [196], Wang [1277], Peterson *et al.* [959] and Tsang *et al.* [1191].

According to Poggio & Miller [979, p. 212], 'the first three-dimensional body to be treated using a numerical integral equation approach was the sphere' by F.K. Oshiro in 1965, using the MFIE.

For transmission problems, using pairs of boundary integral equations, see [482, 799] for reviews. In particular, Arvas & Mautz [42] have considered multiple-scattering problems; their paper includes numerical results for two-dimensional problems, as does [43]. Results for clusters of two, three and four identical touching spheres are presented in [838]; see also [617, 471]. Methods based on the use of single integral equations have been developed in [805, 411, 806, 814, 799, 1356].

Several numerical methods have been devised for solving the volume integral equation (6.58). For details and many references, see [1191, Chapter 2] and [870, §6.5]. As an example, the authors of [351] solved the problem of scattering by two touching spheres using two methods, the 'discrete-dipole approximation' (which may be viewed as a method for solving volume integral equations) and the multipole method of Chapter 4; good agreement was reported.

6.5 Elastic waves

We remember very well the time when almost every author dealing with a non-trivial elasticity problem considered it very nearly a matter of his honour to reduce it by all means to a Fredholm equation of the second kind. After this, he was prone at least to think that his investigation was completed theoretically without concerning himself with the implementation of the solution. (People wonder at it now.)

(D.I. Sherman, foreword to [948, p. 10])

Elastodynamic scattering problems were formulated in Section 1.5, and the results obtained by the method of separation of variables were reviewed in Section 4.10. Here, we summarise how integral equations can be used to solve elastodynamic problems.

6.5.1 Layer potentials and the indirect method

The basic fundamental solution is an $n \times n$ matrix $\mathbf{G}$ (sometimes known as the *Kupradze matrix*), where $n = 2$ for plane-strain problems and $n = 3$ for three-dimensional problems. Explicitly,

$$(\mathbf{G}(P; Q))_{ij} = \frac{1}{\mu}\left\{\Psi\delta_{ij} + \frac{1}{K^2}\frac{\partial^2}{\partial x_i \partial x_j}(\Psi - \Phi)\right\},$$

where μ is the shear modulus,

$$\Phi = -\frac{1}{2}\mathrm{i}\, H_0^{(1)}(kR) \quad \text{and} \quad \Psi = -\frac{1}{2}\mathrm{i}\, H_0^{(1)}(KR) \quad \text{when } n = 2,$$

$$\Phi = -\exp(\mathrm{i}kR)/(2\pi R) \quad \text{and} \quad \Psi = -\exp(\mathrm{i}KR)/(2\pi R) \quad \text{when } n = 3,$$

K is the shear wavenumber, k is the compressional wavenumber, $R = |\boldsymbol{r}_P - \boldsymbol{r}_Q|$ and x_i $(1 \le i \le n)$ are the Cartesian components of P. By construction, for each Q, $\mathbf{G}(P; Q)$ satisfies the equation of motion (1.50) with respect to P for all $P \neq Q$ and the Kupradze radiation conditions (Section 1.5.1) with respect to P as $r_P = |\boldsymbol{r}_P| \to \infty$. The matrix $\mathbf{G}(P; Q)$ also has these properties with respect to Q for each P.

Next, we define elastic single-layer and double-layer potentials by

$$(\mathbf{S}f)(P) = \int_S f(q) \cdot \mathbf{G}(q; P)\, \mathrm{d}s_q$$

and

$$(\mathbf{D}f)(P) = \int_S f(q) \cdot T_q \mathbf{G}(q; P)\, \mathrm{d}s_q,$$

respectively, where T_q means that the traction operator T (defined by (1.53)) is applied at $q \in S$. These elastic layer potentials have well-known properties [673]. Thus, $(\mathbf{S}f)(P)$ is continuous as P crosses S, whereas both $\mathbf{D}$ and $T_p\mathbf{S}$ exhibit jumps given by

$$\mathbf{D}f = (\mp\mathbf{I} + \overline{\mathbf{K}^*})f \tag{6.63}$$

and

$$T_p\mathbf{S}f = (\pm\mathbf{I} + \mathbf{K})f \tag{6.64}$$

(cf. (5.9) and (5.6), respectively), where the upper (lower) sign corresponds to $P \to p \in S$ from the exterior B_e (the interior B) and $\mathbf{I}$ is the $n \times n$ identity matrix. $\mathbf{K}$ and $\overline{\mathbf{K}^*}$ are *singular* integral operators, defined, for $p \in S$, by

$$(\mathbf{K}f)(p) = \int_S f(q) \cdot T_p\mathbf{G}(q;\, p)\,\mathrm{d}s_q$$

and

$$(\overline{\mathbf{K}^*}f)(p) = \int_S f(q) \cdot T_q\mathbf{G}(q;\, p)\,\mathrm{d}s_q.$$

In all of the above formulae, it is sufficient that the density f be Hölder continuous on S [673, Chapter 5]. However, we sometimes require the tractions corresponding to the elastic double-layer potential, defined by

$$\mathbf{N}f = T_p\mathbf{D}f. \tag{6.65}$$

The existence of $\mathbf{N}f$ requires that f be smoother: a sufficient condition is that the tangential derivative of $f(q)$ be Hölder continuous for $q \in S$ ($f \in C^{1,\beta}(S)$), and then the right-hand side of (6.65) is continuous across S [673, p. 320].

$\mathbf{S}$ has a weakly singular kernel, but $\mathbf{K}$ and $\overline{\mathbf{K}^*}$ have singular kernels. More precisely, if we write

$$(\mathbf{K}f)_i = \int_S f_j(q)\, K_{ji}(q;\, p)\,\mathrm{d}s_q,$$

then the diagonal entries K_{ii} (no sum) are weakly singular (continuous when $n = 2$), whereas the off-diagonal entries K_{ji} $(i \neq j)$ have Cauchy singularities: the corresponding integrals must be interpreted as Cauchy principal-value integrals; see Appendix D when $n = 2$ and Appendix F when $n = 3$. Thus, for the existence of $\mathbf{S}f$, it is sufficient that f be continuous on S. For the existence of $\mathbf{K}f$ and $\overline{\mathbf{K}^*}f$, f must belong to $C^{0,\beta}(S)$, with $0 < \beta < 1$; however, $\mathbf{K}$ and $\overline{\mathbf{K}^*}$ are not compact on this space. $\mathbf{S}$ and $\mathbf{N}$ are self-adjoint when the inner product is taken as

$$\langle f, g\rangle = \int_S f \cdot \overline{g}\,\mathrm{d}s;$$

$\mathbf{K}$ and $\overline{\mathbf{K}^*}$ are mutually adjoint with respect to this inner product. $\mathbf{N}$ is a hypersingular operator. For further information on all these operators, see [673].

For scattering by cavities, we may seek a solution in the form

$$\boldsymbol{u}_{\mathrm{sc}}(P) = (\mathbf{S}f)(P), \quad P \in B_{\mathrm{e}}.$$

The boundary condition $(1.52)_2$ and the jump relation (6.64) then yield

$$(\mathbf{I}+\mathbf{K})f = -T\boldsymbol{u}_{\mathrm{inc}}, \tag{6.66}$$

which is a singular integral equation for $\boldsymbol{f}(q)$, $q \in S$. It turns out that this equation has exactly one solution $\boldsymbol{f}$ except when the following interior Dirichlet problem has a non-trivial solution.

Vibration problem *Find a displacement* $\boldsymbol{u}$ *that satisfies* (1.50) *in the bounded domain* B *together with the boundary condition* $\boldsymbol{u} = \mathbf{0}$ *on* S.

This eigenvalue problem only has non-trivial solutions for certain values of the frequency ω. It is known that these eigenfrequencies form an infinite, discrete set, and that each eigenfrequency has a finite multiplicity. For proofs of these results, see [453, §§75–78], [1031, §III.3] or [1050, Chapter 2, §7].

6.5.2 Betti's theorem and the direct method

The elastic analogue of Green's theorem is Betti's reciprocal theorem. To formulate this, let

$$\tau_{ij} = c_{ijkl}(\partial/\partial x_k)u_l \quad \text{and} \quad \tau'_{ij} = c'_{ijkl}(\partial/\partial x_k)u'_l$$

be the stresses corresponding to the displacements u_i and u'_i, respectively. Put $w_i = u'_j\tau_{ji}$ in the divergence theorem (5.31) to give

$$\int_{\mathcal{B}} \left(\frac{\partial u'_j}{\partial x_i}\,\tau_{ji} + u'_j\,\frac{\partial \tau_{ji}}{\partial x_i} \right) \mathrm{d}V = \int_{\mathcal{S}} u'_j\tau_{ji}\nu_i\,\mathrm{d}S, \tag{6.67}$$

where $\mathcal{B}$ is a bounded region with boundary $\mathcal{S}$ and $\boldsymbol{\nu}(q)$ is the unit normal vector at $q \in \mathcal{S}$ pointing out of $\mathcal{B}$. Interchange $\boldsymbol{u}$ and $\boldsymbol{u}'$, and subtract the result from (6.67) to give

$$\begin{aligned}\int_{\mathcal{S}} \left(u_i\tau'_{ij} - u'_i\tau_{ij}\right)\nu_i\,\mathrm{d}S = &\int_{\mathcal{B}} \left(u_i\,\frac{\partial \tau'_{ji}}{\partial x_j} - u'_i\,\frac{\partial \tau_{ji}}{\partial x_j} \right) \mathrm{d}V \\ &+ \int_{\mathcal{B}} \left(c'_{ijkl} - c_{ijkl}\right) \frac{\partial u_i}{\partial x_j}\frac{\partial u'_k}{\partial x_l}\,\mathrm{d}V,\end{aligned} \tag{6.68}$$

where we have used (1.47). In particular, if $\boldsymbol{u}$ and $\boldsymbol{u}'$ both satisfy the same equation of motion in $\mathcal{B}$, we obtain

$$\int_{\mathcal{S}} (u_i\tau'_{ij} - u'_i\tau_{ij})\nu_i\,\mathrm{d}S = 0.$$

This is known as Betti's reciprocal theorem.

A careful application of Betti's theorem in B_e to $\boldsymbol{u}_{sc}$ and G gives

$$2\boldsymbol{u}_{sc}(P) = (\mathsf{S}(T\boldsymbol{u}_{sc}))\,(P) - (\mathsf{D}\boldsymbol{u}_{sc})\,(P), \quad P \in B_e, \tag{6.69}$$

where we have made use of the radiation condition satisfied by both $\boldsymbol{u}_{sc}$ and G. Similarly, an application of Betti's theorem in B to $\boldsymbol{u}_{inc}$ and G gives

$$\mathbf{0} = (\mathsf{S}(T\boldsymbol{u}_{inc}))\,(P) - (\mathsf{D}\boldsymbol{u}_{inc})\,(P), \quad P \in B_e.$$

Adding this to (6.69), using $\boldsymbol{u} = \boldsymbol{u}_{sc} + \boldsymbol{u}_{inc}$, gives

$$2\boldsymbol{u}_{sc}(P) = (\mathsf{S}\boldsymbol{t})\,(P) - (\mathsf{D}\boldsymbol{u})\,(P), \quad P \in B_e, \tag{6.70}$$

where $\boldsymbol{t} = T\boldsymbol{u}$. Similarly, we obtain

$$-2\boldsymbol{u}_{inc}(P) = (\mathsf{S}\boldsymbol{t})\,(P) - (\mathsf{D}\boldsymbol{u})\,(P), \quad P \in B. \tag{6.71}$$

Equations (6.70) and (6.71) can be found in [945], for example.

6.5.2.1 Far-field patterns

Asymptotic approximation of (6.70) for large r yields (see, for example, [261])

$$\boldsymbol{u}_{sc}(\boldsymbol{r}) = \boldsymbol{F}^P(\hat{\boldsymbol{r}})\,\frac{e^{ikr}}{r} + \boldsymbol{F}^S(\hat{\boldsymbol{r}})\,\frac{e^{iKr}}{r} + O(r^{-2}) \tag{6.72}$$

as $r = |\boldsymbol{r}| \to \infty$, uniformly with respect to all directions $\hat{\boldsymbol{r}} = \boldsymbol{r}/r$. The vectors $\boldsymbol{F}^P$ and $\boldsymbol{F}^S$ are the elastodynamic far-field patterns. The asymptotic calculation also shows that

$$\boldsymbol{F}^P(\hat{\boldsymbol{r}}) = F^P(\hat{\boldsymbol{r}})\hat{\boldsymbol{r}} \quad \text{and} \quad \hat{\boldsymbol{r}} \cdot \boldsymbol{F}^S(\hat{\boldsymbol{r}}) = 0, \tag{6.73}$$

which reveal that the radiated P-wave propagates in the outward radial direction, whereas the radiated S-wave is polarised in a plane perpendicular to the radial direction.

The far-field patterns can be expressed in terms of boundary integrals of $\boldsymbol{u}$ and $\boldsymbol{t}$. Thus

$$\begin{aligned}\boldsymbol{F}^P(\hat{\boldsymbol{r}}) = -\frac{\hat{\boldsymbol{r}}}{4\pi(\lambda+2\mu)} \int_S \big[& \mathrm{i}k\{\lambda(\boldsymbol{u}\cdot\boldsymbol{n}) + 2\mu(\boldsymbol{n}\cdot\hat{\boldsymbol{r}})(\boldsymbol{u}\cdot\hat{\boldsymbol{r}})\} \\ & + \boldsymbol{t}\cdot\hat{\boldsymbol{r}}\big] \exp(-\mathrm{i}k\hat{\boldsymbol{r}}\cdot\boldsymbol{q})\,\mathrm{d}s_q, \end{aligned} \tag{6.74}$$

$$\begin{aligned}\boldsymbol{F}^S(\hat{\boldsymbol{r}}) = -\frac{1}{4\pi\mu}\hat{\boldsymbol{r}} \times \int_S \big[& \mathrm{i}K\mu\{(\boldsymbol{u}\times\hat{\boldsymbol{r}})(\boldsymbol{n}\cdot\hat{\boldsymbol{r}}) + (\boldsymbol{n}\times\hat{\boldsymbol{r}})(\boldsymbol{u}\cdot\hat{\boldsymbol{r}})\} \\ & + \boldsymbol{t}\times\hat{\boldsymbol{r}}\big] \exp(-\mathrm{i}K\hat{\boldsymbol{r}}\cdot\boldsymbol{q})\,\mathrm{d}s_q, \end{aligned} \tag{6.75}$$

where $\boldsymbol{u}$, $\boldsymbol{t}$ and $\boldsymbol{n}$ are all evaluated at q, the integration point on S with position vector $\boldsymbol{q}$. Wall [1273] has given similar formulae for the far-field patterns, but his involve the scattered field $\boldsymbol{u}_{sc}$ rather than the total field $\boldsymbol{u}$.

6.5.2.2 Boundary integral equations

For scattering by cavities, with $\boldsymbol{t} = \boldsymbol{0}$ on S, we can let $P \to p \in S$ in (6.69) to obtain

$$(\mathsf{I} + \overline{\mathsf{K}^*})\boldsymbol{u}_{\mathrm{sc}} = -\mathsf{S}(T\boldsymbol{u}_{\mathrm{inc}}), \tag{6.76}$$

a boundary integral equation for $\boldsymbol{u}_{\mathrm{sc}}$ on S. Similarly, (6.70) gives

$$(\mathsf{I} + \overline{\mathsf{K}^*})\boldsymbol{u} = 2\boldsymbol{u}_{\mathrm{inc}}, \tag{6.77}$$

a boundary integral equation for $\boldsymbol{u}$ on S. Both of these suffer from the same irregular frequencies as (6.66). They exemplify the direct method for elastodynamic problems.

6.5.3 Transmission problems

If the region B is filled with a different isotropic elastic material, one can reduce the corresponding transmission problem (formulated in Section 1.5.3) to pairs of coupled integral equations over the interface S, using the Kupradze matrix for the material within B as well as G. The derivations are patterned on the corresponding acoustic problems (Section 6.2). The integral equations inevitably involve singular integrals.

6.5.4 Inhomogeneous obstacles

If the region B is filled with inhomogeneous material, we obtain the problem formulated in Section 1.5.4. Such problems can be reduced to integro-differential equations over the volume B. To see this, make the choices

$$u_i(Q) = G_{im}(Q; P), \quad \boldsymbol{u}' = \boldsymbol{u}_0 \quad \text{and} \quad c'_{ijkl} = c^0_{ijkl}$$

in (6.68), with $\mathcal{S} = S$, $\mathcal{B} = B$ and $P \in B_{\mathrm{e}}$, giving

$$(\mathsf{S}\boldsymbol{t}_0)(P) - (\mathsf{D}\boldsymbol{u}_0)(P) = \boldsymbol{v}(P) + \boldsymbol{w}(P), \quad P \in B_{\mathrm{e}}, \tag{6.78}$$

where $\boldsymbol{t}_0 = T_0\boldsymbol{u}_0$,

$$\boldsymbol{v}(P) = \omega^2 \int_B \{\rho - \rho_0(Q)\}\, \boldsymbol{u}_0 \cdot \mathsf{G}(Q; P)\, \mathrm{d}V_Q,$$

$$w_m(P) = \int_B \{c^0_{ijkl}(Q) - c_{ijkl}\} \frac{\partial u^0_i}{\partial x_j} \frac{\partial}{\partial x_l} G_{km}(Q; P)\, \mathrm{d}V_Q.$$

However, we have $\boldsymbol{u} = \boldsymbol{u}_0$ and $\boldsymbol{t} = \boldsymbol{t}_0$ on S, whence (6.70) gives

$$2\boldsymbol{u}_{\mathrm{sc}}(P) = \boldsymbol{v}(P) + \boldsymbol{w}(P), \quad P \in B_{\mathrm{e}},$$

which is a representation for the scattered field as a volume integral. It can be found in [768], for example.

If we repeat the calculation for $P \in B$, we find that $2\boldsymbol{u}_0(P)$ has to be added to the left-hand side of (6.78), due to the singularity of $\mathbf{G}(Q; P)$ at $Q = P$. Then, making use of (6.71), we obtain

$$2\boldsymbol{u}_0(P) - \boldsymbol{v}(P) - \boldsymbol{w}(P) = 2\boldsymbol{u}_{\text{inc}}(P), \quad P \in B, \tag{6.79}$$

which is an integro-differential equation for the interior field, $\boldsymbol{u}_0(P)$. Derivations of (6.79) can be found in, for example, [945, 445, 1122].

6.5.5 Literature

The standard reference on the *theory* of integral-equation methods in the context of elastic waves is the book by Kupradze *et al.* [673]. Kupradze began developing this theory in the 1950s [671, p. 2]; he explicitly permits multiple-scattering problems [671, p. 41]. As he remarks in his 1963 book, the

> *essential circumstance we should bear in mind when using the integral equations* [such as (6.66), (6.76) or (6.77)] ... *consists in the fact that the integrals occurring here are singular and can only be understood in the sense of principal values; therefore, although the equations themselves have the form of Fredholm equations of the second kind, they are, in fact, singular and multidimensional (two-dimensional) and require a special analysis to solve the question whether Fredholm's theory is applicable.*
>
> (Kupradze [671, p. 43])

The relevant theory of multi-dimensional singular integral equations had already been worked out by G. Giraud, S.G. Mikhlin and others: the standard reference is [853]; see also [948] for a brief outline.

In 1963, Banaugh & Goldsmith [59] considered two-dimensional scattering problems, using the representation (1.55) in terms of two acoustic potentials, ϕ and ψ. They used a Sommerfeld representation (5.36) for each potential, and then derived a coupled system of integral equations on S. They considered fixed rigid scatterers, cavities and inclusions. Numerical results were given for P-wave scattering by rigid cylinders with circular, elliptical and three-leafed-clover cross-sections. Banaugh [57] and Sharma [1099] generalised [59] to three dimensions; detailed reductions for axisymmetric obstacles are also given in [1099].

The method of Banaugh & Goldsmith [59] is indirect in the sense that the potentials ϕ and ψ do not have a physical interpretation. Another indirect method, based on the two-dimensional version of (6.66), has been used in [980, 735]: in [735], results for scattering of a P-wave by 30 aligned, identical, randomly placed, elliptical cavities are given.

Direct methods were used to solve some two-dimensional elastostatic problems by Rizzo [1021]; see [1022] for historical remarks. Tan reduced scattering problems to direct boundary integral equations [1166] and then gave numerical results for a variety of two-dimensional problems [1167]. Numerical solutions of (6.77) in three dimensions are reported in [1023, 622, 1019, 740, 1067, 1069]. In particular, [1069] contains numerical results for multiple scattering by two identical spherical cavities and by two identical coaxial spheroidal cavities. In [1067], the effects of irregular frequencies are shown when (6.77) is used for a pair of unequal spherical cavities. A direct method was used in [1164] for the three-dimensional problem of two parallel fluid-filled boreholes, with an acoustic point source inside one of the boreholes; for a point source near two parallel cylindrical cavities (with oval cross-sections), see [38].

The three-dimensional transmission problem for isotropic media has been analysed in [673, Chapter 12, §2] and in [240, 877]. The two-dimensional problem is simpler, and has been studied in [788]. Numerical results for three-dimensional problems have been presented in [623, 1068, 425, 1067].

Numerical solutions of (6.79) for several identical circular cylinders have been described in [697]; each cylinder is homogeneous and isotropic. The authors solve the same problem – a transmission problem – using a pair of coupled boundary integral equations. They conclude [697, p. 153]: 'For multiple inclusions, the volume integral equation method [based on (6.79)] gives very accurate results and is easier and more convenient to apply'. Orthotropic cylinders are considered in [698].

Integral equations for scattering by obstacles in an elastic half-space, $y > 0$, can be obtained using the full-space Green's function $\mathbf{G}$ or using the half-space Green's function $\mathbf{G}^{\mathrm{H}}$, say. (By construction, $\mathbf{G}^{\mathrm{H}}$ satisfies the traction-free boundary condition on $y = 0$.) Using $\mathbf{G}^{\mathrm{H}}$ leads to boundary integral equations over S, whereas using $\mathbf{G}$ leads to boundary integral equations over $S \cup F$, where F is the portion of $y = 0$ not occupied by scatterers. (F is the whole of $y = 0$ if all the scatterers are buried in the half-space.) Thus, there is a trade-off: $\mathbf{G}$ is much simpler to compute than $\mathbf{G}^{\mathrm{H}}$, but using $\mathbf{G}$ means that (part of) the infinite surface F will have to be discretised. Some aspects of this trade-off are discussed in [939], together with numerical comparisons. However, most authors have used $\mathbf{G}$: for the indirect method, see [1052, 752]; for the direct method, see [1018, 593] and the book by Reinoso [1017]. Qian & Beskos [989] have used $\mathbf{G}^{\mathrm{H}}$ in a direct method for scattering by two rigid 'foundations' on $y = 0$; flexible foundations are considered in [1172].

The theory of integral-equation methods for *anisotropic* media has been given by Natroshvili [900]. Two difficulties have to be overcome. First, the far-field behaviour is more complicated than in isotropic media and, second, the fundamental solution analogous to $\mathbf{G}$, which we denote by $\mathbf{G}^{\mathrm{an}}$, is not available explicitly. Nevertheless, methods for computing $\mathbf{G}^{\mathrm{an}}$ have been developed; see [1323, 1275, 1276, 297, 295] and references therein. $\mathbf{G}^{\mathrm{an}}$ can then be used to derive boundary integral equations for scattering problems. Thus, Sáez & Domínguez [1047] have given numerical results for scattering by a pair of identical spherical cavities in a transversely isotropic solid. For arbitrary anisotropy, see [914, 915].

Plane-wave scattering by periodic arrangements of identical obstacles can be reduced to a problem over a reference domain containing a single obstacle. This problem may then be solved using boundary integral equations. For doubly-periodic planar arrays of spherical cavities and inclusions, see [5, 620] and [621], respectively; see [6, 620] for triply-periodic arrays.

The use of boundary element methods to solve elastodynamic problems is described in several books [619, 774, 1015, 116] and in several reviews [95, 638, 96].

6.6 Water waves

Water-wave problems were formulated in Section 1.6, and the results obtained using multipole potentials were reviewed in Section 4.11. Here, we summarise how integral equations can be used.

6.6.1 Layer potentials and the indirect method

The basic fundamental solution, $G(P, Q)$, has already been defined. For deep water in two dimensions, G is defined by (2.87). For deep water in three dimensions, G is defined by (3.164). Both of these have been normalised so that the singular behaviour at $P = Q$ is exactly the same as for our acoustic fundamental solutions; see (5.3) and (5.4). This means that, for example, all the acoustic jump relations for single-layer and double-layer potentials will be valid here, without change. Fundamental solutions for water of constant finite depth are also known. All of the hydrodynamic fundamental solutions mentioned in this paragraph are complicated, and much has been written on effective methods for their numerical evaluation; see [731, §4.6.1] for more information.

The basic water-wave problem is the Neumann problem because most immersed structures are rigid. For this problem, it is natural to write the scattered potential as a hydrodynamic single-layer potential,

$$\phi_{\mathrm{sc}}(P) = \int_S \mu(q)\, G(P, q)\, \mathrm{d}s_q, \quad P \in B_{\mathrm{e}}, \tag{6.80}$$

where $\phi = \phi_{\mathrm{inc}} + \phi_{\mathrm{sc}}$ and μ is an unknown source density. This representation ensures that the free-surface and radiation conditions are satisfied. Then, the boundary condition on the wetted surfaces S, (1.57), together with the jump relation (5.6), give

$$\mu(p) + \int_S \mu(q) \frac{\partial}{\partial n_p} G(p, q)\, \mathrm{d}s_q = -\frac{\partial \phi_{\mathrm{inc}}}{\partial n_p}, \quad p \in S, \tag{6.81}$$

a Fredholm integral equation of the second kind for μ on S. For the corresponding radiation problem, the right-hand side of (6.81) should be replaced by the prescribed normal velocity on S.

For submerged bodies, the integral equation (6.81) is known to be uniquely solvable, for all frequencies. However, there are irregular frequencies for bodies that pierce the free surface. Thus, let B_i denote the interior region bounded by F_i and the wetted surface S_i, where F_i is the portion of the plane $z=0$ inside the ith body. Then, (6.81) is uniquely solvable except when the following interior problem has a non-trivial solution, for at least one value of i, with $i=1, 2, \ldots, N$.

Wave-Dirichlet problem *Find a function ϕ that satisfies $\nabla^2\phi=0$ in B_i, $\phi=0$ on S_i, and the free-surface condition, $K\phi+\partial\phi/\partial z=0$ on F_i.*

Many methods have been developed for eliminating irregular frequencies; some of these will be mentioned later. For a survey, see [675, §3.1.2].

6.6.2 Green's theorem and the direct method

The derivations given in Section 5.10.1 can be used to derive boundary integral equations for water-wave problems. Thus, for scattering problems, (5.63) and (5.64) become

$$2\phi_{sc}(P) = -\int_S \left\{ \phi_{sc}(q)\,\frac{\partial}{\partial n_q}G(P,q) + v_{inc}(q)\,G(P,q)\right\} \mathrm{d}s_q \tag{6.82}$$

for $P \in B_e$ and

$$\phi_{sc}(p) + \int_S \phi_{sc}(q)\,\frac{\partial}{\partial n_q}G(p,q)\,\mathrm{d}s_q = -\int_S v_{inc}(q)\,G(p,q)\,\mathrm{d}s_q, \quad p \in S, \tag{6.83}$$

respectively, where $v_{inc} \equiv \partial\phi_{inc}/\partial n$. Similarly, (5.67) and (5.68) give

$$2\phi_{sc}(P) = -\int_S \phi(q)\,\frac{\partial}{\partial n_q}G(P,q)\,\mathrm{d}s_q, \quad P \in B_e, \tag{6.84}$$

and

$$\phi(p) + \int_S \phi(q)\,\frac{\partial}{\partial n_q}G(p,q)\,\mathrm{d}s_q = 2\phi_{inc}(p), \quad p \in S, \tag{6.85}$$

respectively. Equation (6.83) is an integral equation for the boundary values of ϕ_{sc}, whereas (6.85) is an equation for the boundary values of the total potential. Equations (6.82) and (6.83) can also be used for radiation problems: replace $(-v_{inc})$ by the prescribed normal velocity on S.

Both (6.83) and (6.85) suffer from the same irregular frequencies as (6.81). This fact was known to John [559].

6.6.3 Literature

For submerged bodies, the integral equation (6.81) was used in 1940 by N.E. Kochin; see [1303, p. 543] and [675, §2.1.1.1]. The integral equation (6.85) was used by John [559] for floating bodies; see his eqn (I.10a).

In his 1971 review, Wehausen [1302] made the following remarks on the method of integral equations.

> *Its use as a numerical method for water-wave problems apparently started with MacCamy* [758, 760], *who considered the waves generated by heaving and rolling strips in the free surface....*
>
> (Wehausen [1302, p. 251])

For flat strips (in two dimensions) and discs (in three dimensions), such problems are usually known as *dock problems*. For more on two-dimensional dock problems, see [731, §§2.5.3 & 5.2.2], [725] and references therein. MacCamy [759] also studied scattering by a circular dock; for more references on this problem, see [798].

Kim [608] was the first to use boundary integral equations to obtain numerical solutions for obstacles of non-zero displacement. He solved (6.81) for deep water, in two and three dimensions, and gave numerical results for floating elliptic cylinders, spheroids and ellipsoids.

In the 1970s, several groups developed software for solving three-dimensional water-wave problems in finite-depth water, using (6.81). For example, see papers by Garrison [386, 385, 383, 384], Standing [509, 1141] and their coauthors, van Oortmerssen [1239, 1240] and Faltinsen & Michelsen [338]. Some of these papers treat problems involving two or more scatterers [1141, 1240], as do some later papers [242, 1142, 812, 301, 1063]. For overviews, see [1058, §6.3] and [908].

Duncan & Brown [302] solved (6.81) for N identical floating bodies in an iterative manner, using an approximate orders-of-scattering approach. The approximation arises because it is assumed that each body is in the far field of every other body: interactions are only via cylindrical waves.

The direct method, based on (6.83), was used by Milgram & Halkyard [855] in 1971. Subsequently, this method (together with (6.85)) was used as the basis for *WAMIT*, perhaps the best known software for water-wave problems. For a description and many applications, see [692, 693] and [910]; the last of these includes results for scattering by a linear array of nine floating buoys. For other applications of the direct method to multibody problems, see [303, 510].

Also in the 1970s, several authors used integral-equation methods to treat two-dimensional catamaran problems, typically consisting of two identical, rigidly-connected, half-immersed circular cylinders. For example, Nordenstrøm *et al.* [917], Kim [604], Lee *et al.* [696] and Katory *et al.* [592] all solved (6.81) for deep water, whilst Chung & Coleman [214] did so for water of constant finite depth. Other geometries have also been investigated, including two rectangles [604, 592], two triangles [696] and bulbous-form catamarans [604, 766]. Chakrabarti [181] has solved (6.83) for two submerged circular cylinders. For an application of the indirect method to a real twin-hulled ship, including hydroelastic and forward-speed effects, see [595]; for interactions between two ships, see [837, 591].

For one scatterer with a vertical axis of symmetry, two-dimensional boundary integral equations can be reduced to one-dimensional integral equations by integrating with respect to the azimuthal coordinate. This was noted in [855], and numerical results were given. See [540, 523] and references therein.

For simple geometries, one can write down a boundary integral equation, and then expand the unknown function using a suitable basis. This leads to an algebraic system for the unknown coefficients in the expansion. A form of this method was used by Levine [704] for one submerged circular cylinder: he started with (6.84), inserted a Fourier expansion for $\phi(q)$, and then imposed the boundary condition on the circle. This method was generalised to two submerged circular cylinders by Schnute [1077] and to one submerged sphere by Gray [432]. (Recall that Row [1035] used a similar method for acoustic scattering by two soft circular cylinders; see Section 4.5.1.) Later, Wu & Eatock Taylor [1338, 1339] obtained extensive results for radiation and scattering by one submerged spheroid, beginning with the single-layer representation (6.80).

All the studies mentioned above use a free-surface Green's function, G. Instead, one can obtain integral equations by using the basic fundamental solution for Laplace's equation. Thus, in two dimensions, one applies Green's theorem to ϕ_{sc} and $\log R$, giving an integral equation with a very simple kernel. However, the equation holds over a (much) larger domain, including a piece of the free surface, a piece of the bottom, and two vertical control surfaces. The radiation condition (or an approximation to it) must be imposed over these vertical surfaces, and the bottom is included so as to obtain a finite integration domain. For details of the method, see [841, §3.5] or [1358, pp. 430–434]. (This basic idea is especially useful for *non-linear* free-surface problems, where the linearised radiation condition is inappropriate.) For some two-dimensional applications with two cylinders, see [506, 527, 28].

Another possibility is to use a 'hybrid element method' or 'localised finite-element method', in which a finite-element representation is used in the vicinity of the immersed bodies but an integral representation is used for the fields further away; the two representations are coupled across the interface between the two regions [841, §4]. For some applications of this method to multibody problems, see [304].

6.7 Cracks and other thin scatterers

When the volume of a scatterer reduces to zero, one obtains a simple model of a thin screen or a crack. For a sound-hard screen, the representation (5.67) reduces to

$$2u_{\text{sc}}(P) = -\int_{\Gamma_+} [u(q)] \frac{\partial}{\partial n_q} G(P, q)\, \mathrm{d}s_q, \quad P \in B_{\text{e}}, \tag{6.86}$$

where Γ_+ denotes one side of the screen, which is assumed to be a smooth open surface with boundary (edge) $\partial\Gamma$. (More generally, Γ_+ could be the union of N

disjoint screens.) Let $\boldsymbol{n}(q)$ denote the unit normal vector at $q \in \Gamma_+$, pointing into B_e. Denote the other side of the screen by Γ_-. Then, $[u]$ denotes the discontinuity in u across the screen, defined by

$$[u(p)] = \lim_{P\to p_+} u(P) - \lim_{P\to p_-} u(P),$$

where $p_+ \in \Gamma_+$ and $p_- \in \Gamma_-$ are the same point in space (but on opposite sides of the thin screen). Note that $[u_{\text{inc}}] = 0$ so that $[u] = [u_{\text{sc}}]$.

Applying the boundary condition $\partial u/\partial n = 0$ on Γ_+ to (6.86) gives

$$\frac{\partial}{\partial n_p} \int_{\Gamma_+} [u(q)] \frac{\partial}{\partial n_q} G(p, q)\, \mathrm{d}s_q = 2\frac{\partial u_{\text{inc}}}{\partial n}, \quad p \in \Gamma_+. \tag{6.87}$$

This is a hypersingular boundary integral equation for $[u]$. The same equation is obtained if one applies the boundary condition on Γ_-.

For the scattering of elastic waves by cracks, one obtains a similar equation for $[\boldsymbol{u}]$, the *crack-opening displacement*. The boundary integral equation obtained is sometimes known as the *traction BIE*.

For scattering by sound-soft screens, the basic integral equation is of the form $Sv = f$, where S is the single-layer operator defined on Γ_+.

The reduction of the scattering problem to (6.87) is quite general, in that it does not depend on the shape of the screen. Nevertheless, (6.87) does simplify if the screen is flat. In fact, most reported computations have been for flat screens and cracks. In addition, for flat screens, there are alternative methods, typically involving a combination of dual integral equations and Fourier transforms. For a review of these special methods, see [130]. For scattering by cracks, see [804] for a review of early work and the book by Zhang & Gross [1374] for many applications.

It is convenient now to give separate discussions for two-dimensional and three-dimensional problems.

6.7.1 Two-dimensional problems

In two dimensions, for scattering by one (possibly curved) screen, (6.87) can be reduced to

$$⨎_a^b \frac{u(t)}{(t-x)^2}\, \mathrm{d}t + \int_a^b K(x,t)\, u(t)\, \mathrm{d}t = f(x), \quad a < x < b, \tag{6.88}$$

where f and K are given and u is to be found; often, K has a weak singularity. Early attempts at solving (6.88) numerically, in the context of acoustic or electromagnetic scattering, have been described in [347, 272, 757, 176]. For earlier applications to oscillating flat aerofoils in subsonic flows, see [1234, eqn (3.8)].

Often, (6.88) is to be solved subject to $u(a) = u(b) = 0$. For such problems, one can develop effective numerical algorithms using expansions in terms of Chebyshev polynomials of the second kind. Further details can be found in [363, 413, 414,

594, 326, 809, 947, 412]. For a collocation-based method, see [537]. For treatment using boundary elements, see, for example, [800, 433, 661, 500, 1180, 891] and the references therein.

If there are N screens, one obtains a system of N coupled hypersingular integral equations. The hypersingularity can be treated using Chebyshev expansions. This method has been used for acoustic scattering by several straight rigid strips [443, 1124] and for the scattering of elastic waves by several straight cracks [519, 1373]. For several straight coplanar strips, see [1364]. For two coplanar straight cracks, see [545, 1056]. For three coplanar straight cracks, see [259]. For three parallel straight cracks, see [548]. For the interaction of water waves with thin plates, see [947, 728, 676, 771, 397]. For the scattering of elastic waves by several parallel, fluid-filled cracks, see [245].

6.7.2 Three-dimensional problems

In three dimensions, for scattering by one flat screen, (6.87) can be reduced to a two-dimensional hypersingular integral equation of the form

$$\oint_{\mathcal{T}} \frac{1}{r^3} u(t_1, t_2)\, dt_1\, dt_2 + \int_{\mathcal{T}} K(s_1, s_2; t_1, t_2)\, u(t_1, t_2)\, dt_1\, dt_2 = g(s_1, s_2), \tag{6.89}$$

for $(s_1, s_2) \in \mathcal{T}$, where $\mathcal{T}$ is the screen, K is a given kernel (usually weakly-singular), g is a given function, and u is to be found. In 1969, Filippi & Dumery [347] obtained (6.89) for acoustic scattering by a sound-hard screen; see their eqn (6). For this problem, and for many other similar problems, (6.89) must be solved subject to

$$u(s_1, s_2) = 0 \quad \text{for all } (s_1, s_2) \in \partial\mathcal{T},$$

where $\partial\mathcal{T}$ is the boundary (edge) of $\mathcal{T}$.

In 1975, Cassot [178] solved (6.89) numerically for acoustic scattering by a flat rectangular plate, using a simple method. First, the plate was divided into about 400 smaller rectangles. On each of these, u was assumed to be constant, giving a piecewise-constant approximation to u. Then, a linear algebraic system was obtained by collocating at the centre of each small rectangle. The main complication comes from the evaluation of

$$\oint_{\mathcal{T}_n} \left\{ \frac{1}{r^3} + K(s_1, s_2; t_1, t_2) \right\} dt_1\, dt_2, \tag{6.90}$$

where $\mathcal{T}_n$ is one of the small rectangles and $(s_1, s_2) \in \mathcal{T}_n$. Cassot [178] evaluated (6.90) in two ways: (i) $\mathcal{T}_n$ was replaced by a circular disc of the same area, centred at (s_1, s_2), and then the integral was evaluated analytically (see Example 5.1); (ii) the finite-part integral was evaluated from the definition (F.11), using a small value for ε. Good agreement between the two methods was reported. Kristiansen & Johansen [668] have used similar methods for acoustic radiation from a thin conical horn, this being a plausible model for a loudspeaker. For three non-planar hard screens, see [21].

Itou has given results for two parallel square cracks [547] and for three coplanar rectangular cracks [546]. For scattering by two penny-shaped cracks, see [131, 618, 20]. Thus, Kit *et al.* [618] and Alves [20] solved systems of hypersingular integral equations whereas Boström & Eriksson [131] used a T-matrix formulation; see Chapter 7.

6.8 Modified integral equations: general remarks

There is an extensive literature on devising integral-equation methods that do not suffer from irregular frequencies; for reviews, see [223, §3.6], [958], [85] and [959, Chapter 6]. For a single scatterer, irregular frequencies may be regarded as a nuisance, a small price to pay for having a fairly simple integral equation. However, as we have seen, irregular frequencies are more common for multiple-scattering problems (unless the scatterers are identical). Therefore, it becomes more important that we have uniquely-solvable integral equations for all frequencies.

For a single scatterer, several methods have been devised for eliminating irregular frequencies, and computational experience has been reported. The main ideas are:

- modify the fundamental solution (Section 6.9);
- modify the integral representation (Section 6.10.1);
- combine two different integral equations (Section 6.10.2); and
- augment one integral equation with some constraints (Section 6.11).

We discuss these below. In Section 6.9, we examine modified fundamental solutions. For multiple-scattering problems, a good theoretical result can be proved, showing that all irregular frequencies can be eliminated. However, this method is probably not useful for extensive computations. The same conclusion is reached for the augmentation methods described in Section 6.11; these methods are also less attractive theoretically. The methods described in Section 6.10 seem to be the best available, especially the direct method of Burton & Miller [166] described in Section 6.10.2. When this method was first introduced in 1971, it was not embraced enthusiastically, because it makes essential use of the hypersingular operator N. Subsequently, good methods for computing $N\nu$ have been developed, thus making the method more attractive.

6.9 Modified fundamental solutions

There is no need to use the free-space Green's function G; one may use

$$G_1(P; Q) = G(P, Q) + H(P; Q), \tag{6.91}$$

where H has the following properties: for every $P \in B_e$, $H(P; Q)$ satisfies the Helmholtz equation for *all* $Q \in B_e$, and the radiation condition with respect to Q;

$H(P; Q)$ must have some singularities at $P = Q$ for some $Q \in B$. (If a function v satisfies the Helmholtz equation everywhere in space and the radiation condition, one can prove that v must vanish everywhere; for a proof, see [244, p. 317].)

So, let us modify the fundamental solution with specific choices for H. We do this first for one three-dimensional scatterer ($N = 1$) in Section 6.9.1, so as to review what is known, and then we generalise to N scatterers in Section 6.9.2. Although our analysis is for the three-dimensional exterior Neumann problem using an indirect method, extensions to the Dirichlet problem, to direct methods and to two dimensions can be made.

The idea of replacing the free-space Green's function by a different fundamental solution was developed by Ursell [1226, 1227], Jones [565] and Kleinman & Roach [629, 630], for one scatterer. For multiple-scattering problems, we use a modified fundamental solution which has additional singularities inside every scatterer. We obtain a Fredholm integral equation of the second kind which we prove is always uniquely solvable. Our proof makes essential use of the addition theorems for outgoing and regular spherical wavefunctions, as given in Chapter 3.

A different modification is to simply replace k by $k + \mathrm{i}\varepsilon$ in the selected 'unmodified' integral equation, where ε is small and positive. One then has to extrapolate the solution towards $\varepsilon = 0$. Numerical results using this method applied to (5.56) for the two-dimensional exterior Dirichlet problem are given in [896].

6.9.1 Modified fundamental solutions: one scatterer

Choose the origin O at a point in $B \equiv B_1$, the interior of $S \equiv S_1$. Let B_a denote a ball of radius a, centred at O, with $B_a \subset B$. Let $\boldsymbol{r}_P$ and $\boldsymbol{r}_Q$ denote the position vectors of P and Q, respectively, with respect to O. Then, we replace the free-space Green's function $G(P, Q) \equiv G(\boldsymbol{r}_P, \boldsymbol{r}_Q)$ by

$$G_1(P, Q) = G(\boldsymbol{r}_P, \boldsymbol{r}_Q) - 2\mathrm{i}k \sum_{\ell,m} (-1)^m a_{\ell m} \psi_\ell^m(\boldsymbol{r}_P)\, \psi_\ell^{-m}(\boldsymbol{r}_Q), \tag{6.92}$$

where ψ_ℓ^m is an outgoing spherical wavefunction (Definition 3.28); note that $\psi_\ell^m(\boldsymbol{r})$ is singular at $\boldsymbol{r} = \boldsymbol{0}$. The factors $\{-2\mathrm{i}k(-1)^m\}$ are inserted for later convenience and also render the coefficients $a_{\ell m}$ dimensionless. These coefficients will be chosen later; for now, we merely impose the conditions that the infinite series in (6.92) be uniformly convergent for P and Q outside B_a, and that it can be differentiated twice, term by term.

So, we look for a solution of the exterior Neumann problem in the form

$$u(P) = \int_S \mu(q)\, G_1(P, q)\, \mathrm{d}s_q, \tag{6.93}$$

whence $\mu(q)$ satisfies

$$\mu(p) + \int_S \mu(q) \frac{\partial}{\partial n_p} G_1(p, q)\, \mathrm{d}s_q = f(p), \quad p \in S. \tag{6.94}$$

The solvability of this integral equation is governed by the solvability of the corresponding homogeneous equation, namely

$$\mu(p) + \int_S \mu(q) \frac{\partial}{\partial n_p} G_1(p, q) \, \mathrm{d}s_q = 0, \quad p \in S. \tag{6.95}$$

Theorem 6.2 *Suppose that the homogeneous integral equation* (6.95) *has a non-trivial solution* $\mu(q)$*. Then, the interior wavefunction*

$$U(P) = \int_S \mu(q) \, G_1(P, q) \, \mathrm{d}s_q, \quad P \in B, \tag{6.96}$$

vanishes on S.

Proof ([1226, pp. 120 & 123]) Define $U(P)$ for $P \in B_e$ by (6.96); $\partial U/\partial n$ vanishes on S by (6.95). The uniqueness theorem for the exterior Neumann problem [223, Theorem 3.13] then asserts that $U \equiv 0$ in B_e. The result follows by noting that U is continuous across the source distribution on S [223, Theorem 2.12]. □

If we can show that $U \equiv 0$ in B, it will follow that (6.95) has only the trivial solution (because μ is proportional to the discontinuity in $\partial U/\partial n$ across S) and hence that the inhomogeneous equation (6.94) is uniquely solvable for any f. This can be achieved with some restrictions on the coefficients $a_{\ell m}$.

Theorem 6.3 *Suppose that*

$$\left| a_{\ell m} + \frac{1}{2} \right| > \frac{1}{2} \quad \textit{for } \ell = 0, 1, 2, \ldots \textit{ and } m = -\ell, \ldots, \ell, \tag{6.97}$$

or

$$\left| a_{\ell m} + \frac{1}{2} \right| < \frac{1}{2} \quad \textit{for } \ell = 0, 1, 2, \ldots \textit{ and } m = -\ell, \ldots, \ell. \tag{6.98}$$

Then, every solution of the homogeneous integral equation (6.95) *is a solution of the unmodified homogeneous integral equation*

$$(I + K)\mu = 0, \tag{6.99}$$

which also satisfies

$$A_{\ell m} \equiv -2\mathrm{i}k(-1)^m \int_S \mu(q) \, \psi_\ell^{-m}(\boldsymbol{r}_q) \, \mathrm{d}s_q = 0, \tag{6.100}$$

for $\ell = 0, 1, 2, \ldots$ *and* $m = -\ell, \ldots, \ell$.

Proof For $P \in B$ with $P \neq O$, we have

$$U(P) = \int_S \mu(q) \, G(P, q) \, \mathrm{d}s_q + \sum_{\ell, m} a_{\ell m} A_{\ell m} \psi_\ell^m(\boldsymbol{r}_P), \tag{6.101}$$

where μ is a solution of (6.95). If we restrict P to lie in B_a, we can use the bilinear expansion of G given in the next theorem, which is merely Theorem 3.16 in disguise.

Theorem 6.4 *Suppose that $R = |\boldsymbol{r}_P - \boldsymbol{r}_Q|$ with $|\boldsymbol{r}_P| < |\boldsymbol{r}_Q|$. Then the three-dimensional free-space Green's function, $G(P, Q)$, has the expansion*

$$G(\boldsymbol{r}_P, \boldsymbol{r}_Q) = -\frac{\mathrm{e}^{\mathrm{i}kR}}{2\pi R} = -2\mathrm{i}k \sum_{n=0}^{\infty} \sum_{m=-n}^{n} (-1)^m \hat{\psi}_n^m(\boldsymbol{r}_P)\, \psi_n^{-m}(\boldsymbol{r}_Q). \tag{6.102}$$

Returning to the proof of Theorem 6.3, we use (6.102) in (6.101) to give

$$U(\boldsymbol{r}_P) = \sum_{\ell,m} \left\{ A_{\ell m} \hat{\psi}_\ell^m(\boldsymbol{r}_P) + a_{\ell m} A_{\ell m} \psi_\ell^m(\boldsymbol{r}_P) \right\} \tag{6.103}$$

for $P \in B_a$, $P \neq O$. Here, $\hat{\psi}_\ell^m$ is a regular spherical wavefunction (Definition 3.28). Note that if we can show (6.100), then we can infer from (6.103) that $U \equiv 0$ in B_a and then, by analytic continuation, in B.

Next, following Ursell [1227] and Colton & Kress [223, Theorem 3.35], we consider the integral

$$I \equiv \int_{\Omega_a} \left(U \frac{\partial \overline{U}}{\partial n} - \overline{U} \frac{\partial U}{\partial n} \right) \mathrm{d}s,$$

where Ω_a denotes the spherical boundary of the ball $B_a \subset B$ and the overbar denotes complex conjugation. Using Green's theorem and Theorem 6.2, we see that

$$I = \int_S \left(U \frac{\partial \overline{U}}{\partial n} - \overline{U} \frac{\partial U}{\partial n} \right) \mathrm{d}s = 0.$$

We can also evaluate I directly using a special case of the following lemma.

Lemma 6.5 *Suppose that $U(\boldsymbol{r}_P)$ and $V(\boldsymbol{r}_P)$ have the expansions*

$$U(\boldsymbol{r}_P) = \sum_{n=0}^{\infty} \sum_{m=-n}^{n} \left\{ A_n^m \hat{\psi}_n^m(\boldsymbol{r}_P) + B_n^m \psi_n^m(\boldsymbol{r}_P) \right\} \tag{6.104}$$

and

$$V(\boldsymbol{r}_P) = \sum_{n=0}^{\infty} \sum_{m=-n}^{n} \left\{ C_n^m \hat{\psi}_n^m(\boldsymbol{r}_P) + D_n^m \psi_n^m(\boldsymbol{r}_P) \right\}, \tag{6.105}$$

respectively, for $P \in \Omega_a$, where Ω_a is the sphere of radius a, centred at the origin. Then

$$\int_{\Omega_a} \left(U \frac{\partial V}{\partial r} - V \frac{\partial U}{\partial r} \right) \mathrm{d}s = \frac{\mathrm{i}}{k} \sum_{n=0}^{\infty} \sum_{m=-n}^{n} (-1)^m (A_n^m D_n^{-m} - C_n^m B_n^{-m}), \tag{6.106}$$

$$\int_{\Omega_a} \left(U \frac{\partial \overline{V}}{\partial r} - \overline{V} \frac{\partial U}{\partial r} \right) \mathrm{d}s = -\frac{\mathrm{i}}{k} \sum_{n=0}^{\infty} \sum_{m=-n}^{n} (2B_n^m \overline{D_n^m} + A_n^m \overline{D_n^m} + B_n^m \overline{C_n^m}). \tag{6.107}$$

Proof Let

$$[U, V] = \int_{\Omega_a} \left(U \frac{\partial \overline{V}}{\partial r} - \overline{V} \frac{\partial U}{\partial r} \right) \mathrm{d}s = -[V, U]. \tag{6.108}$$

Substituting for U and V from (6.104) and (6.105) shows that we need the values of similar integrals involving spherical wavefunctions. For example, we have

$$\begin{aligned}\left[\hat{\psi}_n^m, \overline{\psi_\nu^\mu}\right] &= ka^2 \left\{ j_n(ka)\, \overline{h_\nu'(ka)} - j_n'(ka)\, \overline{h_\nu(ka)} \right\} \int_\Omega Y_n^m \overline{Y_\nu^\mu}\, \mathrm{d}\Omega \\ &= -(\mathrm{i}/k)\delta_{n\nu}\delta_{m\mu},\end{aligned}$$

using the orthogonality of the spherical harmonics Y_ℓ^m over the unit sphere Ω, (3.9), and the Wronskian for spherical Bessel functions, (4.37). In a similar way, we obtain

$$\begin{aligned}\left[\hat{\psi}_n^m, \hat{\psi}_\nu^\mu\right] &= \left[\hat{\psi}_n^m, \overline{\hat{\psi}_\nu^\mu}\right] = [\psi_n^m, \psi_\nu^\mu] = 0, \\ \left[\hat{\psi}_n^m, \psi_\nu^\mu\right] &= (\mathrm{i}/k)(-1)^m \delta_{n\nu}\delta_{m+\mu,0}, \\ \left[\psi_n^m, \overline{\psi_\nu^\mu}\right] &= -2(\mathrm{i}/k)\delta_{n\nu}\delta_{m\mu}.\end{aligned} \tag{6.109}$$

The results follow after some straightforward calculation. □

Corollary 6.6 *Suppose that U has the expansion* (6.104). *Then*

$$\int_{\Omega_a} \left(U \frac{\partial \overline{U}}{\partial r} - \overline{U} \frac{\partial U}{\partial r} \right) \mathrm{d}s = -\frac{2\mathrm{i}}{k} \sum_{n=0}^{\infty} \sum_{m=-n}^{n} \left(|B_n^m|^2 + \mathrm{Re}\left\{ A_n^m \overline{B_n^m} \right\} \right). \tag{6.110}$$

Proof Put $U = V$ in (6.107). □

Thus, returning to the proof of Theorem 6.3, we find that

$$0 = I = -\frac{2\mathrm{i}}{k} \sum_{\ell,m} |A_{\ell m}|^2 \left\{ \mathrm{Re}(a_{\ell m}) + |a_{\ell m}|^2 \right\}, \tag{6.111}$$

using Corollary 6.6 with $A_n^m = A_{nm}$ and $B_n^m = a_{nm}A_{nm}$. Since $a_{\ell m}$ satisfies the inequalities (6.97) or (6.98), it follows that (6.111) can only be satisfied if $A_{\ell m} = 0$ for $\ell = 0, 1, 2, \ldots$ and $m = -\ell, \ldots, \ell$. Also, substituting (6.92) in (6.95) shows that μ satisfies (6.99). □

This completes our review of scattering by a single obstacle.

6.9.2 Modified fundamental solutions: several scatterers

Let us now consider the exterior Neumann problem for N three-dimensional scatterers. Choose an origin $O^n \in B_n$, the interior of S_n, and let $\boldsymbol{r}_P^n$ denote the position vector of a point P with respect to O^n. Let B_a^n denote a ball of radius a, centred at O^n; we choose a sufficiently small so that $B_a^n \subset B_n$ for $n = 1, 2, \ldots, N$. Let

$$G_1(P, Q) = G(\boldsymbol{r}_P, \boldsymbol{r}_Q) - 2\mathrm{i}k \sum_{n=1}^{N} \sum_{\ell,m} (-1)^m a_{\ell m}^n \psi_\ell^m(\boldsymbol{r}_P^n)\, \psi_\ell^{-m}(\boldsymbol{r}_Q^n), \tag{6.112}$$

where $a_{\ell m}^n$ are constants. This fundamental solution is singular at every origin O^n, $n = 1, 2, \ldots, N$ (recall that $\psi_\ell^m(\boldsymbol{r}_P^n)$ is singular at O^n). Note that it is essential that our fundamental solution has this property; if we chose a fundamental solution that was not singular inside B_j, say, then we could not eliminate those irregular frequencies associated with S_j.

We look for a solution of our problem in the form (6.93), whence the source density $\mu(q)$ satisfies the integral equation (6.94). Moreover, the same arguments as before show that Theorem 6.2 is true (in the current notation).

Let us now investigate the solvability of the integral equation (6.94) and look for an analogue of Theorem 6.3. Suppose that $\mu(q)$ is any solution of the homogeneous integral equation (6.95). Consider the interior wavefunction $U(P)$, defined by (6.96), for $P \in B^j$ and some j. We restrict P to lie in $B_a^j \subset B_j$, and find that

$$U(\boldsymbol{r}_P^j) = \sum_{\ell,m} A_{\ell m}^j \hat{\psi}_\ell^m(\boldsymbol{r}_P^j) + \sum_{n=1}^{N} \sum_{\ell,m} a_{\ell m}^n A_{\ell m}^n \psi_\ell^m(\boldsymbol{r}_P^n), \tag{6.113}$$

for $P \in B_a^j$, where

$$A_{\ell m}^n = -2\mathrm{i}k(-1)^m \int_S \mu(q)\, \psi_\ell^{-m}(\boldsymbol{r}_q^n)\, \mathrm{d}s_q. \tag{6.114}$$

In order to use Corollary 6.6, we need the expansion of U to be in terms of functions centred on O^j, that is we need the addition theorem for outgoing spherical wavefunctions, Theorem 3.27. This gives the expansion

$$\psi_\ell^m(\boldsymbol{r}_P^n) = \sum_{L,M} S_{\ell L}^{mM}(\boldsymbol{b}^{nj})\, \hat{\psi}_M^L(\boldsymbol{r}_P^j),$$

where $\boldsymbol{r}_P^n = \boldsymbol{b}^{nj} + \boldsymbol{r}_P^j$. This expansion is valid for $|\boldsymbol{r}_P^j| < |\boldsymbol{b}^{nj}|$, which is always true in our application. Note that $\boldsymbol{b}^{nj}$ is the position vector of O^j with respect to O^n, whence $\boldsymbol{b}^{nj} = -\boldsymbol{b}^{jn}$. The separation matrix $S_{\ell L}^{mM}$ is defined by (3.86). Hence, (6.113) becomes

$$U(\boldsymbol{r}_P^j) = \sum_{\ell,m} \hat{\psi}_\ell^m(\boldsymbol{r}_P^j) \left\{ A_{\ell m}^j + \sum_{\substack{n=1 \\ n\neq j}}^{N} \sum_{L,M} a_{LM}^n A_{LM}^n S_{L\ell}^{Mm}(\boldsymbol{b}^{nj}) \right\}$$

$$+ \sum_{\ell,m} \psi_\ell^m(\boldsymbol{r}_P^j)\, a_{\ell m}^j A_{\ell m}^j, \quad P \in B_a^j. \tag{6.115}$$

Using Green's theorem, Corollary 6.6 (for B_a^j) and the fact that $U(\boldsymbol{r}_P^j)$ vanishes on S_j, we obtain

$$0 = \sum_{\ell,m} \left(|\mathcal{A}_{\ell m}^j|^2 + \mathrm{Re}\{a_{\ell m}^j\} \, |A_{\ell m}^j|^2 \right)$$

$$+ \mathrm{Re} \sum_{\ell,m} \overline{\mathcal{A}_{\ell m}^j} \sum_{\substack{n=1 \\ n\neq j}}^{N} \sum_{L,M} \mathcal{A}_{LM}^n S_{L\ell}^{Mm}(\boldsymbol{b}^{nj}), \tag{6.116}$$

where $\mathcal{A}_{\ell m}^n = a_{\ell m}^n A_{\ell m}^n$. Equation (6.116) holds for $j = 1, 2, \ldots, N$.

By comparison with the proof of Theorem 6.3, we expect to be able to deduce from (6.116) that $A_{\ell m}^n = 0$. To do this, we sum over j and obtain

$$0 = \sum_{j=1}^{N} \sum_{\ell,m} \mathrm{Re}(a_{\ell m}^j) \, |A_{\ell m}^j|^2 + K_N, \tag{6.117}$$

where

$$K_N = \sum_{j=1}^{N} \sum_{\ell,m} |\mathcal{A}_{\ell m}^j|^2 + \mathrm{Re} \sum_{j=1}^{N} \sum_{\ell,m} \overline{\mathcal{A}_{\ell m}^j} \sum_{\substack{n=1 \\ n\neq j}}^{N} \sum_{L,M} \mathcal{A}_{LM}^n S_{L\ell}^{Mm}(\boldsymbol{b}^{nj})$$

$$= \frac{1}{2} \sum_{j=1}^{N} \sum_{\ell,m} \overline{\mathcal{A}_{\ell m}^j} \sum_{\substack{n=1 \\ n\neq j}}^{N} \sum_{L,M} \mathcal{A}_{LM}^n \left\{ S_{L\ell}^{Mm}(\boldsymbol{b}^{nj}) + \overline{S_{\ell L}^{mM}(\boldsymbol{b}^{jn})} \right\}$$

$$+ \sum_{j=1}^{N} \sum_{\ell,m} |\mathcal{A}_{\ell m}^j|^2. \tag{6.118}$$

But, from Lemma 3.29, we have

$$S_{L\ell}^{Mm}(\boldsymbol{b}^{nj}) + \overline{S_{\ell L}^{mM}(-\boldsymbol{b}^{nj})} = 2\,\widehat{S}_{L\ell}^{Mm}(\boldsymbol{b}^{nj}),$$

where $\widehat{S}_{L\ell}^{Mm}$ is the separation matrix that occurs both in Theorem 3.27 and in the addition theorem for regular spherical wavefunctions, Theorem 3.26. Thus, (6.118) becomes

$$K_N = \sum_{j=1}^{N} \sum_{\ell,m} |\mathcal{A}_{\ell m}^j|^2 + \sum_{j=1}^{N} \sum_{\ell,m} \overline{\mathcal{A}_{\ell m}^j} \sum_{\substack{n=1 \\ n\neq j}}^{N} \mathcal{C}_{\ell m}^{nj}$$

$$= \sum_{j=1}^{N} \sum_{\ell,m} \left\{ |\mathcal{A}_{\ell m}^j|^2 + \frac{1}{2} \sum_{\substack{n=1 \\ n\neq j}}^{N} \left(\overline{\mathcal{A}_{\ell m}^j} \mathcal{C}_{\ell m}^{nj} + \mathcal{A}_{\ell m}^j \overline{\mathcal{C}_{\ell m}^{nj}} \right) \right\},$$

where

$$\mathcal{C}_{\ell m}^{nj} = \sum_{L,M} \mathcal{A}_{LM}^n \, \widehat{S}_{L\ell}^{Mm}(\boldsymbol{b}^{nj}), \quad n \neq j,$$

and we have noted that K_N is real. Now, for complex quantities A and C, we have

$$\overline{A}C + A\overline{C} = |A+C|^2 - |A|^2 - |C|^2 \tag{6.119}$$

so that

$$K_N = \sum_{j=1}^{N} \sum_{\ell,m} \left\{ |\mathcal{A}_{\ell m}^{j}|^2 - \frac{1}{2} \sum_{\substack{n=1\\ n\neq j}}^{N} \left(|\mathcal{A}_{\ell m}^{j}|^2 + |\mathcal{C}_{\ell m}^{nj}|^2 \right) \right\} + K_N',$$

where

$$K_N' = \frac{1}{2} \sum_{j=1}^{N} \sum_{\ell,m} \sum_{\substack{n=1\\ n\neq j}}^{N} \left| \mathcal{A}_{\ell m}^{j} + \mathcal{C}_{\ell m}^{nj} \right|^2 \geq 0.$$

Further simplifications can be made because the separation matrix $\widehat{S}_{L\ell}^{Mm}$ is a unitary matrix; see (3.96). Thus, we find that

$$\sum_{\ell,m} |\mathcal{C}_{\ell m}^{nj}|^2 = \sum_{\ell,m} |\mathcal{A}_{\ell m}^{n}|^2, \quad n \neq j,$$

whence

$$K_N = (2-N) \sum_{j=1}^{N} \sum_{\ell,m} |\mathcal{A}_{\ell m}^{j}|^2 + K_N'.$$

Thus, (6.117) becomes

$$0 = \sum_{j=1}^{N} \sum_{\ell,m} |A_{\ell m}^{j}|^2 \left\{ \mathrm{Re}(a_{\ell m}^{j}) - (N-2)|a_{\ell m}^{j}|^2 \right\} + K_N'.$$

Hence, we can deduce that $A_{\ell m}^{j} = 0$ provided all the coefficients $a_{\ell m}^{j}$ are such that $\mathrm{Re}(a_{\ell m}^{j}) - (N-2)|a_{\ell m}^{j}|^2 > 0$.

Alternatively, if we use the identity $\overline{A}C + A\overline{C} = -|A-C|^2 + |A|^2 + |C|^2$ instead of (6.119), we obtain

$$0 = \sum_{j=1}^{N} \sum_{\ell,m} |A_{\ell m}^{j}|^2 \left\{ \mathrm{Re}(a_{\ell m}^{j}) + N|a_{\ell m}^{j}|^2 \right\} + K_N'',$$

where $K_N'' \leq 0$. Hence, we deduce that $A_{\ell m}^{j} = 0$ provided that $\mathrm{Re}(a_{\ell m}^{j}) + N|a_{\ell m}^{j}|^2 < 0$.

Summarising, we have proved the following result.

Theorem 6.7 *Suppose that*

$$\mathrm{Re}(a_{\ell m}^{j}) - (N-2)|a_{\ell m}^{j}|^2 > 0 \quad \textit{for } \ell \geq 0,\ |m| \leq \ell \textit{ and } j = 1, 2, \ldots, N$$

or

$$\mathrm{Re}(a_{\ell m}^{j}) + N|a_{\ell m}^{j}|^2 < 0 \quad \textit{for } \ell \geq 0,\ |m| \leq \ell \textit{ and } j = 1, 2, \ldots, N,$$

where N is the number of disjoint scatterers. Then, every solution of the homogeneous integral equation (6.95) *is a solution of* (6.99) *which also satisfies*

$$A^j_{\ell m} = 0 \quad \textit{for } \ell \geq 0,\ |m| \leq \ell \textit{ and } j = 1, 2, \ldots, N,$$

where $A^j_{\ell m}$ *is defined by* (6.114).

If the conditions on $a^j_{\ell m}$ are satisfied, this theorem implies unique solvability of the integral equation (6.94), for all wavenumbers k and for any f. This is an elegant theoretical result, because it yields solvability of the exterior Neumann problem without introducing non-compact operators.

It is noteworthy that Theorem 6.7 reduces formally to Theorem 6.3 when $N = 1$.

When $N = 2$, we obtain uniqueness when $\mathrm{Re}(a^j_{\ell m}) > 0$. This result was obtained previously by Martin [785] in two dimensions. The results for more scatterers, $N > 2$, are in [794]; note that the conditions in Theorem 6.7 do depend on N.

6.9.3 Extensions

Modified fundamental solutions have been used for electromagnetics by Brandt *et al.* [137], Neave [901] and Jost [572]. They have also been used for elastic waves. Two-dimensional analyses have been given in [639, 78], with numerical results in [639, 79]. Three-dimensional analyses have been given in [568, 39].

For two-dimensional water waves and one floating body, the basic paper is by Ursell [1228]. The three-dimensional theory is contained in [780]. For two floating cylinders, see [786, 1342].

6.10 Combination methods

There are two related 'combination methods' that lead to uniquely-solvable boundary integral equations. The methods are dual, in the sense that they give equations that are adjoints with respect to a suitable inner product. Both methods use the free-space Green's function, G. One is an indirect method and one is a direct method. We shall describe them briefly in the context of the exterior Neumann problem. Our description is brief because the analysis of these methods does not depend on the number of scatterers.

6.10.1 A combined-layer method

We use a linear combination of single-layer and double-layer potentials, and write

$$u(P) = (S\mu)(P) + \mathrm{i}\eta(D\mu)(P), \quad P \in B_{\mathrm{e}}, \tag{6.120}$$

where η is a real, non-zero coupling parameter. Thus, for the exterior Neumann problem, the boundary condition (5.44) yields

$$(I + K + \mathrm{i}\eta N)\mu = f. \tag{6.121}$$

This hypersingular integral equation is always uniquely solvable [223, Theorem 3.34].

The idea of using the combination (6.120) emerged in the 1960s; see [223, §3.6] for references. It was first used for the exterior Dirichlet problem; the corresponding integral equation is

$$(I - \overline{K^*} + \mathrm{i}\eta^{-1} S)\mu = \mathrm{i}\eta^{-1} g, \tag{6.122}$$

which is a uniquely-solvable Fredholm integral equation of the second kind for μ [223, Theorem 3.33]. See also [207, §9.3.4], [185, Theorem 7.9.1], [903, p. 120] and [504, 158].

In 1966, Greenspan & Werner [437] gave numerical solutions of (6.122) when S is a circle; see also [348, §6.2.2]. Bolomey & Tabbara [113] solved both (6.122) and (6.121) for circles, ellipses and rectangles, and verified that irregular frequencies were eliminated. Lin [720] solved (6.122) numerically for spheres and ellipsoids, using a Galerkin method; for other three-dimensional computations, see [149, 150].

Theoretically, the real, non-zero parameter η is arbitrary, although its choice can affect numerical computations; this choice has been investigated in [659].

6.10.2 The method of Burton and Miller

This is a direct method. For the exterior Neumann problem, we have two boundary integral equations obtained from the Helmholtz integral formula, namely (5.58) and (5.59). If we add one to a multiple of the other, we obtain

$$(I + \overline{K^*} + \mathrm{i}\eta N)u = \{S + \mathrm{i}\eta(K - I)\}f, \tag{6.123}$$

where η is a real, non-zero coupling parameter. This hypersingular integral equation for $u(q)$ is always uniquely solvable [223, Theorem 3.43]; see also [719]. It was introduced in 1971 by Burton & Miller [166].

For the exterior Dirichlet problem, we also obtain (6.123), except $u = g$ is known and $f = \partial u / \partial n$ is unknown. Also, note that the integral representation for $u(P)$ is not modified: it remains as (5.57) for the exterior Neumann problem and (5.60) for the exterior Dirichlet problem. Buffa & Hiptmair [158] first multiply (5.59) by a regularising operator before combining with (5.58); this simplifies the numerical analysis of the resulting integral equation for the exterior Dirichlet problem.

Similarly, for scattering by sound-hard obstacles, we can combine the Helmholtz integral equation (5.68) with (5.69) to give

$$(I + \overline{K^*} + \mathrm{i}\eta N)u = 2u_{\mathrm{inc}} + 2\mathrm{i}\eta v_{\mathrm{inc}}, \tag{6.124}$$

which is uniquely solvable for the total field $u(q)$.

There are many papers in which the method of Burton & Miller has been implemented, numerically; examples are [165, 848, 847] from the 1970s, [1170, 27, 1178, 22, 1179] from the 1980s and [206, 739, 360, 25, 738] from the 1990s. However, we are not aware of any published numerical results for multiple-scattering problems, based on (6.124). The choice of η has been investigated in [23].

For scattering by sound-soft obstacles, we can combine (5.74) with (5.75) to give

$$(I - K + \mathrm{i}\eta S)v = 2v_{\text{inc}} - 2\mathrm{i}\eta u_{\text{inc}}, \tag{6.125}$$

where $v = \partial u/\partial n$. Numerical solutions of this equation have been presented in [147, 404]; in particular, [404] contains results for scattering by two circular cylinders.

6.10.3 Extensions

Combined-layer representations have been used for electromagnetic scattering; see [223, §4.6] and [157]. The electromagnetic version of the method of Burton & Miller (usually known as the *combined-field method*) has also been used; see [223, Theorem 4.47]. For example, we can combine the EFIE and the MFIE to give

$$(I + \mathcal{M})\boldsymbol{J} + \mathrm{i}\eta\,\boldsymbol{n} \times \mathcal{P}\boldsymbol{J} = 2\boldsymbol{n} \times \boldsymbol{H}_{\text{inc}} + 2\eta\omega\varepsilon\,\boldsymbol{n} \times (\boldsymbol{n} \times \boldsymbol{E}_{\text{inc}}), \tag{6.126}$$

where η is a real, non-zero coupling constant. In fact, K.M. Mitzner used a combined-field method in 1968; his method and some numerical results for scattering by a sphere are presented in [979, pp. 223–225]. Thus, this work was done before the paper of Burton & Miller [166] appeared. Alternative combinations of the EFIE and the MFIE have been considered by Yaghjian [1349]; for numerical comparisons with the combined-field method, see [237]. In [231], the EFIE is first multiplied by an operator before combining with the MFIE, giving an equation that is better conditioned than (6.126) for low frequencies; see also [10].

A combined-layer representation analogous to (6.120) has been used by Ahner [13] for scattering of elastic waves by a rigid obstacle, and by Kiriaki [614] for a cavity. The elastic version of the method of Burton & Miller [166] has been developed by Takakuda [1165], Jones [569] and Liu & Rizzo [740]. The method of Burton & Miller has also been adapted to water-wave problems [695].

6.11 Augmentation methods

The basic idea here is to augment the Helmholtz integral equation with some constraints. Thus, for the exterior Neumann problem, we begin with (5.58), namely $(I + \overline{K^*})u = Sf$. At irregular frequencies, this integral equation is solvable, but not uniquely. This non-uniqueness may be removed by requiring that $u(q)$ also satisfies additional conditions.

In 1968, Schenck [1070] suggested using the interior integral relation (5.37), which becomes

$$\int_S \left\{ u(q) \frac{\partial}{\partial n_q} G(P, q) - G(P, q)\, f(q) \right\} \mathrm{d}s_q = 0, \quad P \in B. \tag{6.127}$$

As the solution of the exterior Neumann problem, $u(P)$, does satisfy this relation, the only issue of concern is uniqueness. This is ensured by the next theorem.

Theorem 6.8 *Suppose that $\nu(q)$ is continuous on S and that ν solves $(D\nu)(P) = 0$ for all $P \in B$. Then $\nu(q) \equiv 0$, $q \in S$.*

Proof Define $U(P) = (D\nu)(P)$ for $P \in B$ and $U_\mathrm{e}(P) = (D\nu)(P)$ for $P \in B_\mathrm{e}$. We are given that $U(P) \equiv 0$ in B, so that, in particular, $\partial U/\partial n = 0$ on S. But the normal derivative of a double-layer potential with a continuous density ν is continuous across S, whence $\partial U_\mathrm{e}/\partial n = 0$ on S. Now, $U_\mathrm{e}(P)$ is an outgoing wavefunction, so that the uniqueness theorem for the exterior Neumann problem gives $U_\mathrm{e}(P) \equiv 0$ in B_e. In particular, $U_\mathrm{e}(p) = 0$, $p \in S$. Hence, (5.12) gives the result. □

We have called (6.127) an integral *relation* because it holds throughout the volume B whereas the unknown $u(q)$ is defined on the boundary S. This means that solving (6.127) directly is unlikely to be efficient. (Note that we do not need (5.58) if we work with (6.127) for all $P \in B$: (5.58) is redundant [468].) However, Schenck's idea [1070] was to use the standard Helmholtz integral equation (5.58) together with (6.127) evaluated at some points $P \in B$:

> *We implement this method in the discrete sense by writing the* [square algebraic] *system of equations that results from the surface Helmholtz integral* [equation (5.58)], *and then overdetermining the solution with additional compatible equations based on the interior Helmholtz integral* [relation (6.127)] *for strategic interior points.*
>
> (Schenck [1070, p. 46])

The main difficulty with the numerical implementation of Schenck's method lies in the word 'strategic'. Where should the interior points be chosen, and how many are needed? When k^2 is an irregular value, we know that there is a non-trivial solution, ν, of $(I + \overline{K^*})\nu = 0$, so that, by (5.9), the function $U(P) = (D\nu)(P)$ is a solution of the interior Dirichlet problem; it is an eigenfunction. Now, such eigenfunctions typically vanish on various surfaces in B, called *nodal surfaces*. Therefore, if we evaluate (6.127) on a nodal surface (when k^2 is an irregular value), we do not obtain a useful constraint. Thus, 'strategic' points cannot lie on nodal surfaces. Of course, in general, we know neither the irregular values nor the associated nodal surfaces. Consequently, all one can do is to choose a few points $P \in B$, and then hope that at least some of them do not lie on a nodal surface. Canning [172] has devised a computational strategy for deciding when enough interior points have been used, but not where to put them.

Schenck [1070] solved his overdetermined system by a least-squares technique. The same method was used in [468, 646, 1093, 1341, 186]. An alternative is to solve a larger square system by introducing Lagrange multipliers [1092]. For further information, see the 1997 review by Benthien & Schenck [85].

Other constraints, which can be derived from (6.127), or directly, can be used. For example, one can choose to solve (5.58) together with

$$\int_S \left(u \frac{\partial}{\partial n_q} \psi_n^m - f \psi_n^m \right) \mathrm{d}s_q = 0$$

for some values of n and m; here, ψ_n^m is an outgoing spherical wavefunction, singular at a point in B. (These are called *null-field equations*; we shall discuss them in detail in Chapter 7.) This method was suggested by Jones [565] and has been implemented numerically [1158]. It was also used by Liapis [713] for water-wave problems.

We know of two numerical implementations of augmentation methods for multiple-scattering problems: Schenck's method has been used for problems involving two identical spheres [287, 1353]. All such methods require imposing at least one constraint for each scatterer. This, together with the unavailability of good criteria for selecting interior points, suggests that augmentation methods are not useful for general multiple-scattering problems.

6.11.1 Extensions

Schenck's method with Lagrange multipliers has been used for elastodynamic problems in [1019, 425]. Its use for electromagnetic problems is mentioned in [566]. For water waves, see [686].

Krutitskii [669] has described another augmentation method. For the exterior Neumann problem, he uses an indirect method, replacing (5.50) by

$$u(P) = \int_{S \cup S'} \mu(q)\, G(P, q)\, \mathrm{d}s_q, \quad P \in B_e,$$

where S' is a set of closed auxiliary surfaces, with one inside each S_i, $i = 1, 2, \dots, N$. To find $\mu(q)$, $q \in S \cup S'$, he applies the Neumann condition on S and a dissipative boundary condition on S'; the resulting integral equations are shown to be uniquely solvable, subject to certain conditions on the choice of S'. For some numerical applications to scattering by two spheres, see [1353]. In fact, this augmentation method was used much earlier for water-wave problems by Ohmatsu [928]. Thus, as well as distributing sources and dipoles over the wetted surface S, he also does the same over pieces of the horizontal plane $z = 0$ *inside* each immersed body. For more details and numerical results, see [694]. It turns out that, computationally, this method is very effective; notice that it is not necessary to discretise the hypersingular operator with this approach.

6.12 Application of exact Green's functions

As it would be difficult to give any idea of this part without employing analytical symbols, we shall content ourselves with remarking, that it contains a number of singular equations of great generality and simplicity, which seem capable of being applied to many departments of the electrical theory besides those considered in the following pages.

(Green [434, p. 13])

These introductory remarks are in Green's famous *Essay on the Application of Mathematical Analysis to the Theories of Electricity and Magnetism*, first published in 1828 [434]. It was here that he introduced what we shall call the *exact Green's function*, for the Dirichlet problem for Laplace's equation in three dimensions [434, article (5.), pp. 31–36]. He also noted explicitly that his results 'may likewise be shown to hold good, for that [region] exterior to a number of closed surfaces, of any forms whatever...' [434, p. 35].

In this section, we introduce the exact Green's function for the exterior Neumann problem for the Helmholtz equation. We then discuss its use in the context of multiple-scattering problems.

6.12.1 The exact Green's function

Recall the integral representation (5.57). It gives the solution of the exterior Neumann problem in terms of certain integrals of the free-space Green's function G, and the unknown boundary values $u(q)$. Moreover, if k^2 is not an eigenvalue of the interior Dirichlet problem, we can find $u(q)$ by solving the integral equation (5.58), so that we can write

$$u = Lf \quad \text{with} \quad L = (I + \overline{K^*})^{-1} S. \tag{6.128}$$

(In this section, we disregard irregular values; we may remove them by modifying L; see Section 6.8.)

Now, we obtain a different integral representation if we use a different fundamental solution, G_1; see (6.91). Explicitly, we obtain

$$2u(P) = \int_S \left\{ G_1(P; q)\, f(q) - u(q) \frac{\partial}{\partial n_q} G_1(P; q) \right\} \mathrm{d}s_q \tag{6.129}$$

for $P \in B_e$. We can simplify this formula by exploiting the flexibility in G_1. Thus, we introduce a special fundamental solution G^E, defined as follows. Fix the point P. Then, write

$$G^E(Q; P) = G(Q, P) + H(Q; P),$$

and choose H so that it satisfies the Helmholtz equation for all $Q \in B_e$, satisfies the radiation condition, and is such that G^E satisfies

$$\frac{\partial}{\partial n_q} G^E(q; P) = 0 \quad \text{for } q \in S. \tag{6.130}$$

We call G^{E} the *exact Green's function*; Bergman & Schiffer [86] call it the *Neumann function*. One can find detailed discussions of (exact) Green's functions in older books on partial differential equations, such as those of Kellogg [597, Chapter 9], Bateman [74, §2.32], Webster [1301, §66] or Garabedian [379, Chapter 7]. For some early applications of exact Green's functions with the Helmholtz equation, see [73, §50] and [356, 355, 856]. For historical remarks, see [1169].

Like G, G^{E} is *symmetric*:

$$G^{\mathrm{E}}(P; Q) = G^{\mathrm{E}}(Q; P) \quad \text{for all } P \text{ and } Q \text{ in } B_{\mathrm{e}} \cup S,\ P \neq Q. \tag{6.131}$$

This is proved in Appendix I.

Since G^{E} is a fundamental solution, we obtain a simplified form of (6.129) when we use (6.130):

$$2u(P) = \int_S f(q)\, G^{\mathrm{E}}(q; P)\, \mathrm{d}s_q, \qquad P \in B_{\mathrm{e}}. \tag{6.132}$$

This is an explicit formula for the solution of the exterior Neumann problem. Of course, we do not have G^{E}! Finding G^{E} is equivalent to solving a particular exterior Neumann problem. However, letting $P \to p \in S$, (6.132) gives

$$u(p) = \frac{1}{2}\int_S f(q)\, G^{\mathrm{E}}(q; p)\, \mathrm{d}s_q \equiv \frac{1}{2} S^{\mathrm{E}} f,$$

say. Comparison of this formula with (6.128) gives

$$Lf = \frac{1}{2} S^{\mathrm{E}} f. \tag{6.133}$$

As this holds for every f, we deduce that

$$L = \frac{1}{2} S^{\mathrm{E}}. \tag{6.134}$$

Moreover, (6.133) and (6.128) give $(I + \overline{K^*}) S^{\mathrm{E}} f = 2Sf$, which implies that G^{E} solves the integral equation

$$G^{\mathrm{E}}(q; p) + \int_S G^{\mathrm{E}}(l; p) \frac{\partial}{\partial n_l} G(l, q)\, \mathrm{d}s_l = 2G(q, p). \tag{6.135}$$

In this equation, the point p occurs as a parameter; indeed, the same equation holds when p is replaced by $P \in B_{\mathrm{e}}$. See Appendix I for a direct derivation of (6.135) and other similar integral equations. The idea of constructing G^{E} by solving a boundary integral equation can be found in a paper by Boley [112].

In summary, if we want to find G^{E} for a particular geometry, we typically have to solve a boundary integral equation such as (6.135): we have shown above that this is equivalent to calculating L.

6.12.2 Multiple scattering by two obstacles

Consider the exterior Neumann problem for two obstacles. Thus, the problem is to solve the Helmholtz equation in B_e, the unbounded region exterior to S_1 and S_2, subject to a radiation condition and the boundary conditions

$$\partial u/\partial n = f_j \quad \text{on } S_j,\ j = 1, 2. \tag{6.136}$$

One way to solve this problem is to solve the standard Helmholtz integral equation (5.58), obtained using G. We can regard this integral equation as a pair of coupled boundary integral equations for two unknown functions, namely u on S_1 and u on S_2. This pair of equations gives equal weight to S_1 and S_2.

In practice, we may already have information on how to scatter by one of the obstacles (S_2, say) in isolation, such as L, G^E or one of their discrete approximations. One way to use this information is to 'partition' the pair of integral equations; another is to replace G by G^E. We can prove that these two approaches lead to exactly the same equations. A third approach is to assume that we have two exact Green's functions, one for each scatterer; this leads naturally to the generalised Born series (Section 6.12.3).

6.12.2.1 Partitioning

The Helmholtz integral equation (5.58) can be written as

$$A_{11}u_1 + A_{12}u_2 = S_{11}f_1 + S_{12}f_2, \tag{6.137}$$

$$A_{21}u_1 + A_{22}u_2 = S_{21}f_1 + S_{22}f_2. \tag{6.138}$$

Here, $u_j = u(p_j)$, where $p_j \in S_j$ for $j = 1, 2$,

$$A_{jk}u_k = \delta_{jk}u(p_k) + \int_{S_k} u(q_k)\,\frac{\partial}{\partial n_q}G(q_k, p_j)\,\mathrm{d}s_q, \tag{6.139}$$

$$S_{jk}f_k = \int_{S_k} f_k(q_k)\,G(q_k, p_j)\,\mathrm{d}s_q \tag{6.140}$$

and δ_{ij} is the Kronecker delta. We note that A_{jj} is simply the operator $(I + \overline{K^*})$ for S_j (both the field point p_j and the source point (integration point) q_j are on S_j), whereas A_{12} and A_{21} give the interactions between S_1 and S_2 (these integral operators have smooth kernels, as the field and source points are on different surfaces).

We know that the pair of integral equations (6.137) and (6.138) suffers from irregular frequencies. Again, we disregard these here.

Now, suppose that we already have (a discrete approximation to) A_{22}^{-1}; for example, we may have solved (5.58) for $S = S_2$ using an accurate boundary-element method. Then, (6.138) gives

$$u_2 = A_{22}^{-1}\{S_{21}f_1 + S_{22}f_2 - A_{21}u_1\}. \tag{6.141}$$

Eliminating u_2 from (6.137), we obtain

$$\mathcal{A}_{11} u_1 = \mathcal{S}_{11} f_1 + \mathcal{S}_{12} f_2, \tag{6.142}$$

where

$$\mathcal{A}_{11} = A_{11} - A_{12} A_{22}^{-1} A_{21}, \tag{6.143}$$

$$\mathcal{S}_{1j} = S_{1j} - A_{12} A_{22}^{-1} S_{2j}, \quad j = 1, 2. \tag{6.144}$$

Equation (6.142) is an integral equation to solve for u on S_1. We shall return to it later.

We could view partitioning as merely a method for solving systems of linear algebraic equations. However, it can be profitable to view partitioning as arising directly from partitions of the boundary.

6.12.2.2 Use of the exact Green's function

Suppose that G^{E} is the exact Green's function for S_2 (in isolation). Proceeding as for the one-obstacle problem, using G^{E} for our chosen fundamental solution, we obtain the following integral representation for the two-obstacle problem:

$$2u(P) = \int_{S_1} \left(f_1(q_1)\, G^{\mathrm{E}}(q_1; P) - u(q_1) \frac{\partial}{\partial n_q} G^{\mathrm{E}}(q_1; P) \right) \mathrm{d}s_q + \int_{S_2} f_2(q_2)\, G^{\mathrm{E}}(q_2; P)\, \mathrm{d}s_q. \tag{6.145}$$

This representation does not involve the unknown boundary values of u on S_2. To find u on S_1, we let $P \to p_1 \in S_1$, as usual; the result can be written as

$$A_{11}^{\mathrm{E}} u_1 = S_{11}^{\mathrm{E}} f_1 + S_{12}^{\mathrm{E}} f_2, \tag{6.146}$$

where

$$A_{11}^{\mathrm{E}} u_1 = u(p_1) + \int_{S_1} u(q_1) \frac{\partial}{\partial n_q} G^{\mathrm{E}}(q_1, p_1)\, \mathrm{d}s_q, \tag{6.147}$$

$$S_{1j}^{\mathrm{E}} f_j = \int_{S_j} f_j(q_j)\, G^{\mathrm{E}}(q_j, p_1)\, \mathrm{d}s_q, \quad j = 1, 2. \tag{6.148}$$

At this point, we have two boundary integral equations for u on S_1, namely (6.142) and (6.146). It turns out that they are identical.

Theorem 6.9

$$\mathcal{A}_{11} = A_{11}^{\mathrm{E}}, \quad \mathcal{S}_{11} = S_{11}^{\mathrm{E}} \quad \textit{and} \quad \mathcal{S}_{12} = S_{12}^{\mathrm{E}}. \tag{6.149}$$

Proof Let us prove that $\mathcal{A}_{11} = A_{11}^{\mathrm{E}}$. We make use of some formulae from Appendix I, but with S replaced by S_2. From (I.3), we can let $P \to p_1 \in S_1$ to give

$$2G^{\mathrm{E}}(Q; p_1) = 2G(Q, p_1) - \int_{S_2} G^{\mathrm{E}}(l_2; Q) \frac{\partial}{\partial n_l} G(l_2, p_1)\, \mathrm{d}s_l. \tag{6.150}$$

Computing the normal derivative at $q_1 \in S_1$ yields

$$\frac{\partial}{\partial n_q} G^{\mathrm{E}}(q_1; p_1) = \frac{\partial}{\partial n_q} G(q_1, p_1) - \frac{1}{2} \int_{S_2} \frac{\partial}{\partial n_q} G^{\mathrm{E}}(l_2; q_1) \frac{\partial}{\partial n_l} G(l_2, p_1)\, \mathrm{d}s_l. \quad (6.151)$$

Multiply this formula by $u(q_1)$ and integrate over S_1. Adding $u(p_1)$ to the result, making use of the definitions (6.139) and (6.147), gives

$$A_{11}^{\mathrm{E}} u_1 = A_{11} u_1 - \frac{1}{2} A_{12} U_2^{\mathrm{E}}, \quad (6.152)$$

where

$$U_2^{\mathrm{E}}(p_2) = \int_{S_1} u(q_1) \frac{\partial}{\partial n_q} G^{\mathrm{E}}(q_1; p_2)\, \mathrm{d}s_q \quad (6.153)$$

for $p_2 \in S_2$. U_2^{E} satisfies a boundary integral equation over S_2: from (I.4), we have

$$G^{\mathrm{E}}(Q; p_2) + \int_{S_2} G^{\mathrm{E}}(Q; l_2) \frac{\partial}{\partial n_l} G(l_2, p_2)\, \mathrm{d}s_l = 2G(Q, p_2); \quad (6.154)$$

multiply the normal derivative of this equation at q_1 by $u(q_1)$, and then integrate over S_1 to give

$$A_{22} U_2^{\mathrm{E}} = 2A_{21} u_1. \quad (6.155)$$

Inverting this equation and then eliminating U_2^{E} from (6.152), we obtain the desired result:

$$A_{11}^{\mathrm{E}} = A_{11} - A_{12} A_{22}^{-1} A_{21} = \mathcal{A}_{11}.$$

The other two identities in (6.149) can be proved similarly; see [801, Appendix II] for details. □

Thus, the two methods for determining u on S_1 are equivalent. If we subsequently want u on S_2, we can calculate it by simply setting $P = p_2$ in (6.145); there is no jump. We can then calculate $u(P)$ anywhere in B_{e}, using the standard Helmholtz integral representation (5.57).

6.12.3 Generalised Born series for two scatterers

The standard integral equations, (6.137) and (6.138), can be written as

$$A_{11} u_1 = F_1 - A_{12} u_2 \quad \text{and} \quad A_{22} u_2 = F_2 - A_{21} u_1,$$

where $F_j = S_{j1} f_1 + S_{j2} f_2$. The generalised Born series [290, 1081, 1069] gives a method for solving this pair iteratively in the context of multiple scattering by two obstacles (actually, it can be recognised as the block Jacobi method for linear algebraic equations): assuming that A_{11} and A_{22} are non-singular, construct $u_j^{(m)}$ according to

$$u_1^{(m+1)} = A_{11}^{-1}\left\{F_1 - A_{12}u_2^{(m)}\right\}, \tag{6.156}$$

$$u_2^{(m+1)} = A_{22}^{-1}\left\{F_2 - A_{21}u_1^{(m)}\right\}, \tag{6.157}$$

with $u_1^{(0)} = u_2^{(0)} = 0$. Eliminating $u_2^{(m)}$, we obtain

$$u_1^{(m+1)} = g_1 + \mathcal{B}u_1^{(m-1)} \tag{6.158}$$

for $m = 1, 2, \ldots$, where

$$g_1 = A_{11}^{-1}\left\{F_1 - A_{12}A_{22}^{-1}F_2\right\} \quad \text{and} \quad \mathcal{B} = A_{11}^{-1}A_{12}A_{22}^{-1}A_{21}.$$

Hence,

$$u_1^{(2M)} = \sum_{m=0}^{M-1} \mathcal{B}^m g_1 \quad \text{for } M = 1, 2, \ldots. \tag{6.159}$$

Also, since $u_1^{(1)} = A_{11}^{-1}F_1$, (6.158) gives

$$u_1^{(2M+1)} = u_1^{(2M)} + \mathcal{B}^M A_{11}^{-1}F_1 \quad \text{for } M = 1, 2, \ldots. \tag{6.160}$$

The (geometric) series in (6.159) and (6.160) converges if

$$\|\mathcal{B}\| < 1, \tag{6.161}$$

with any reasonable norm. This condition will be satisfied if the distance between the two scatterers, b, is sufficiently large; the wide-spacing approximations (6.163) and (6.164), given below, imply that $\|\mathcal{B}\| = O(b^{-2})$ as $b \to \infty$. Then, if (6.161) is true, the last term in (6.160) tends to zero and both sequences ($\{u_1^{(2M)}\}$ and $\{u_1^{(2M+1)}\}$) converge to $u_1^{(\infty)}$, say, where

$$\begin{aligned} u_1^{(\infty)} &= (I - \mathcal{B})^{-1}g_1 = \mathcal{A}_{11}^{-1}A_{11}g_1 = \mathcal{A}_{11}^{-1}\left\{F_1 - A_{12}A_{22}^{-1}F_2\right\} \\ &= \mathcal{A}_{11}^{-1}\left\{\mathcal{S}_{11}f_1 + \mathcal{S}_{12}f_2\right\}, \end{aligned}$$

which is the solution of the partitioning equation (6.142). This latter equation is not subject to the condition (6.161).

In [290], the authors use the method described above for the scattering of elastic waves by two identical spherical cavities. Their numerical results are for $u_1^{(2)}$ and $u_2^{(2)}$. Comparisons with 'exact' numerical solutions, using (6.77), are made in [1069]. A similar method is used in [289] for scattering by a spherical cavity in an elastic half-space.

Geuzaine *et al.* [404] have developed a method for high-frequency problems that begins by combining the generalised Born series with (6.125). They give numerical results for plane-wave scattering by a pair of sound-soft circles of radii a and $3a/2$, with $b/a \simeq 2.8$ and $ka = 1000$.

Note that the generalised Born series requires a knowledge of both A_{11}^{-1} and A_{22}^{-1}; this is equivalent to knowing two exact Green's functions, one for each obstacle. Rudgers [1036] has used the sum of these two exact Green's functions as a fundamental solution, followed by an iterative method. Wirgin [1327] has obtained a single integral equation for the acoustic transmission problem (see Section 6.2), using both the interior and the exterior exact Green's functions, each of which satisfies (6.130).

The iterative method based on (6.156) and (6.157) shows that, at each stage, one solves a scattering problem for S_1 (using A_{11}^{-1} in (6.156)) and a scattering problem for S_2 (using A_{22}^{-1} in (6.157)). This form of the method is called *boundary decomposition* by Balabane [55, 54]. He has extended the method to N scatterers and given convergence theorems.

6.12.4 An application to active sonar

In *active sonar*, a 'source' (perhaps mounted on the surface of one obstacle) radiates sound which is then scattered by a 'target' (a second obstacle); the aim is to detect the target by analysing the scattered waves received back at the source [1220]. This is an example of multiple scattering by two obstacles [1071]. Thus, we take B_2 to be the source and B_1 to be the target. The boundary conditions are (6.136), with $f_1 \equiv 0$ and $f_2 \equiv v$. The problem is to calculate u_2, the boundary values of u on S_2 when the target is present, and then to compare u_2 with u_2^0, which is the corresponding solution when the target is absent. Making use of the exact Green's function for S_2, (6.132) gives

$$2u_2^0 = S^{\mathrm{E}} v = \int_{S_2} v(q_2)\, G^{\mathrm{E}}(q_2; p_2)\, \mathrm{d}s_q, \quad p_2 \in S_2.$$

From (6.145), we have

$$2u(P) = \int_{S_2} v(q_2)\, G^{\mathrm{E}}(q_2; P)\, \mathrm{d}s_q - \int_{S_1} u(q_1)\, \frac{\partial}{\partial n_q} G^{\mathrm{E}}(q_1; P)\, \mathrm{d}s_q$$

for $P \in B_{\mathrm{e}}$. Letting $P \to p_2 \in S_2$ gives

$$2u_2 = S^{\mathrm{E}} v - U_2^{\mathrm{E}},$$

whereas letting $P \to p_1 \in S_1$ gives

$$A_{11}^{\mathrm{E}} u_1 = S_{12}^{\mathrm{E}} v,$$

where U_2^{E}, A_{11}^{E} and S_{12}^{E} are defined by (6.153), (6.147) and (6.148), respectively.

Defining the *echo* at the source by

$$u_2^{\mathrm{e}} = u_2 - u_2^0,$$

we deduce that

$$2u_2^{\mathrm{e}} = -U_2^{\mathrm{E}} = -2A_{22}^{-1} A_{21} u_1,$$

where we have used (6.155). Hence, eliminating u_1,

$$\begin{aligned} u_2^{\mathrm{e}} &= -A_{22}^{-1} A_{21} \left(A_{11}^{\mathrm{E}}\right)^{-1} S_{12}^{\mathrm{E}} v \\ &= -A_{22}^{-1} A_{21} \left(A_{11} - A_{12} A_{22}^{-1} A_{21}\right)^{-1} (S_{12} - A_{12} A_{22}^{-1} S_{22}) v, \end{aligned} \tag{6.162}$$

using Theorem 6.9. This is the formal expression for the echo in terms of many integral operators involving the free-space Green's function; it was first derived by Schenck [1071, eqn (34)], but without using G^{E}.

The expression (6.162) is exact but complicated. In practice, the source and the target are far apart, and so wide-spacing approximations may be used so as to simplify (6.162). Thus, choose origins $O_j \in B_j$, $j = 1, 2$, and let $\boldsymbol{b}$ be the position vector of O_1 with respect to O_2. Let p_j and q_j have position vectors $\boldsymbol{p}_j$ and $\boldsymbol{q}_j$, respectively, with respect to O_j, $j = 1, 2$. Then, the distance between $p_1 \in S_1$ and $q_2 \in S_2$ is

$$\begin{aligned} |\boldsymbol{b} + \boldsymbol{p}_1 - \boldsymbol{q}_2| &= \left\{b^2 + 2\boldsymbol{b} \cdot (\boldsymbol{p}_1 - \boldsymbol{q}_2) + |\boldsymbol{p}_1 - \boldsymbol{q}_2|^2\right\}^{1/2} \\ &= b + \hat{\boldsymbol{b}} \cdot (\boldsymbol{p}_1 - \boldsymbol{q}_2) + O(b^{-1}) \quad \text{as } b \to \infty, \end{aligned}$$

where $b = |\boldsymbol{b}|$ and $\hat{\boldsymbol{b}} = \boldsymbol{b}/b$, so that b is the nominal range of the target from the source. Then, with $E(\boldsymbol{q}) = \exp(\mathrm{i}k\hat{\boldsymbol{b}} \cdot \boldsymbol{q})$, we find that

$$S_{12} f_2 \simeq -\frac{\mathrm{e}^{\mathrm{i}kb}}{2\pi b} E(\boldsymbol{p}_1) \int_{S_2} f_2(q_2)\, E(-\boldsymbol{q}_2)\, \mathrm{d}s_q,$$

$$A_{12} u_2 \simeq \frac{\mathrm{i}k\, \mathrm{e}^{\mathrm{i}kb}}{2\pi b} E(\boldsymbol{p}_1) \int_{S_2} u(q_2)\, \hat{\boldsymbol{b}} \cdot \boldsymbol{n}(q_2)\, E(-\boldsymbol{q}_2)\, \mathrm{d}s_q, \tag{6.163}$$

$$A_{21} u_1 \simeq -\frac{\mathrm{i}k\, \mathrm{e}^{\mathrm{i}kb}}{2\pi b} E(-\boldsymbol{p}_2) \int_{S_1} u(q_1)\, \hat{\boldsymbol{b}} \cdot \boldsymbol{n}(q_1)\, E(\boldsymbol{q}_1)\, \mathrm{d}s_q, \tag{6.164}$$

all with an error of $O(b^{-2})$ as $b \to \infty$. With these approximations, we can see from (6.162) that the echo u_2^{e} is $O(b^{-2})$ as $b \to \infty$, which is what we would expect: the 'outgoing' signal decays as b^{-1} and the 'return' signal is reduced by a further factor of b^{-1}. However, with our wide-spacing approximations, we can simplify (6.162) to

$$u_2^{\mathrm{e}} \simeq -A_{22}^{-1} A_{21} A_{11} (S_{12} - A_{12} A_{22}^{-1} S_{22}) v,$$

and then, within this approximation, we can insert the approximations for A_{21}, S_{12} and A_{12}. The details have been worked out by Schenck [1071]. He has also given two numerical examples, one for two identical spheres and one for two identical coaxial prolate spheroids.

6.13 Twersky's method

Twersky developed a method for two-dimensional multiple-scattering problems that makes clever use of the Sommerfeld contour-integral representation for Hankel functions. We begin with this formula in Section 6.13.1 and then use it to give another proof of the addition theorem for $H_0^{(1)}$ (Section 6.13.2). We then describe Twersky's method; the basic reference is [1203]. Extensions to other multiple-scattering problems are mentioned at the end of Section 6.13.4.

6.13.1 Sommerfeld's integral representation

Sommerfeld's integral for $H_n^{(1)}(w)$ is given by [1298, §6.12], [1128, §19.B], [321, §7.3.5] and [177, eqn (5–156)]. In particular, *if w is real and positive*, we have

$$H_n^{(1)}(w) = \frac{1}{\pi}(-\mathrm{i})^n \int_{\mathcal{C}_1} \mathrm{e}^{\mathrm{i}w\cos\chi}\,\mathrm{e}^{\mathrm{i}n\chi}\,\mathrm{d}\chi,$$

where $\mathcal{C}_1$ is a contour in the complex χ-plane from $-\eta+\mathrm{i}\infty$ to $\eta-\mathrm{i}\infty$, and η is any real number satisfying $0<\eta<\pi$.

Let us put $w=kr>0$ and $\chi=\tau-\theta$, whence

$$H_n^{(1)}(kr) = \frac{1}{\pi}(-\mathrm{i})^n \int_{\mathcal{C}} \mathrm{e}^{\mathrm{i}kr\cos(\tau-\theta)}\,\mathrm{e}^{\mathrm{i}n(\tau-\theta)}\,\mathrm{d}\tau, \tag{6.165}$$

where $\mathcal{C}$ is a contour in the complex τ-plane between $\gamma+\mathrm{i}\infty$ and $\pi+\gamma-\mathrm{i}\infty$, and γ is any number satisfying $0<\theta-\gamma<\pi$; for example, we can take $\gamma=\theta-\pi/2$. In particular, we have

$$H_0^{(1)}(kr) = \frac{1}{\pi}\int_{\mathcal{C}} \mathrm{e}^{\mathrm{i}kr\cos(\tau-\theta)}\,\mathrm{d}\tau. \tag{6.166}$$

Suppose that, in (6.165), θ is allowed to vary, with $\theta_{\min}<\theta<\theta_{\max}$. Then, we can use the *same* contour $\mathcal{C}$ provided we can find a γ with

$$-\pi+\theta_{\max}\le\gamma\le\theta_{\min};$$

this will be possible if $\theta_{\max}-\theta_{\min}\le\pi$.

6.13.2 Another proof of Graf's addition theorem for $H_0^{(1)}$

Let us give another proof of the formula (2.29). Put $\boldsymbol{r}_2=\boldsymbol{r}_1+\boldsymbol{b}$ with $\boldsymbol{r}_j=r_j(\cos\theta_j, \sin\theta_j)$ and $\boldsymbol{b}=b(\cos\beta,\sin\beta)$. Suppose that $r_1<b$, whence $\beta-\frac{1}{2}\pi<\theta_2<\beta+\frac{1}{2}\pi$. Then, with $\gamma=\beta-\frac{1}{2}\pi$, $r=r_2$ and $\theta=\theta_2$ in (6.166), we obtain

$$H_0^{(1)}(kr_2) = \frac{1}{\pi}\int_{\mathcal{C}} \mathrm{e}^{\mathrm{i}kr_1\cos(\tau-\theta_1)}\,\mathrm{e}^{\mathrm{i}kb\cos(\tau-\beta)}\,\mathrm{d}\tau, \quad r_1<b,$$

as $r_2\cos(\tau-\theta_2)=r_1\cos(\tau-\theta_1)+b\cos(\tau-\beta)$.

Now, using the Jacobi expansion (2.17), we have

$$\mathrm{e}^{\mathrm{i}kr_1\cos(\tau-\theta_1)} = \sum_{n=-\infty}^{\infty} \mathrm{i}^n J_n(kr_1)\,\mathrm{e}^{\mathrm{i}n(\tau-\theta_1)},$$

whence

$$H_0^{(1)}(kr_2) = \frac{1}{\pi} \sum_{n=-\infty}^{\infty} \mathrm{i}^n J_n(kr_1) \, \mathrm{e}^{-\mathrm{i}n(\theta_1-\beta)} \int_{\mathcal{C}} \mathrm{e}^{\mathrm{i}kb\cos(\tau-\beta)} \, \mathrm{e}^{\mathrm{i}n(\tau-\beta)} \, \mathrm{d}\tau$$

$$= \sum_{n=-\infty}^{\infty} (-1)^n H_n^{(1)}(kb) \, J_n(kr_1) \, \mathrm{e}^{-\mathrm{i}n(\theta_1-\beta)},$$

which agrees with (2.29) after replacing n by $(-n)$.

6.13.3 Scattering by one cylinder

Consider the scattering of a plane wave by a sound-hard cylinder with cross-section B. Take an origin $O \in B$ and introduce plane polar coordinates so that a typical point P in the exterior B_e has position vector $\boldsymbol{r}$ with respect to O, where $\boldsymbol{r} = r(\cos\theta, \sin\theta)$. Then, the incident wave is

$$u_{\mathrm{inc}} = \exp(\mathrm{i}k\boldsymbol{r}\cdot\hat{\boldsymbol{\alpha}}) = \mathrm{e}^{\mathrm{i}kr\cos(\theta-\alpha)},$$

where $\hat{\boldsymbol{\alpha}} = (\cos\alpha, \sin\alpha)$ is a unit vector in the direction of propagation. From (5.67), the scattered field can be written as

$$u_{\mathrm{sc}}(\boldsymbol{r}; \hat{\boldsymbol{\alpha}}) = \frac{\mathrm{i}}{4} \int_S u(q; \hat{\boldsymbol{\alpha}}) \, \frac{\partial}{\partial n_q} H_0^{(1)}(kR) \, \mathrm{d}s_q, \tag{6.167}$$

where S is the boundary of B, $u = u_{\mathrm{inc}} + u_{\mathrm{sc}}$ is the total field, $R = |\boldsymbol{r} - \boldsymbol{q}|$ and $\boldsymbol{q}$ is the position vector of $q \in S$ with respect to O.

The far field is given by Definition 4.1. The far-field pattern, f, is given by (5.77), which we write as

$$f(\hat{\boldsymbol{r}}; \hat{\boldsymbol{\alpha}}) = \frac{k}{4} \int_S u(q; \hat{\boldsymbol{\alpha}}) \, [\hat{\boldsymbol{r}}\cdot\boldsymbol{n}(q)] \exp(-\mathrm{i}k\hat{\boldsymbol{r}}\cdot\boldsymbol{q}) \, \mathrm{d}s_q. \tag{6.168}$$

Evidently, this formula gives the far-field pattern if we know u on S. Twersky [1203] gives a corresponding inverse, using the representation (6.166). Thus, suppose that S is star-shaped with respect to O, so that $\boldsymbol{q} = \rho(\varphi)(\cos\varphi, \sin\varphi)$. Suppose further that

$$r > \max_{\varphi} \rho(\varphi) \equiv a,$$

say, so that P is outside the escribed circle to S, centred at O. Then, as in Section 6.13.2, we have

$$H_0^{(1)}(kR) = \frac{1}{\pi} \int_{\mathcal{C}} \mathrm{e}^{\mathrm{i}kr\cos(\tau-\theta)} \, \mathrm{e}^{-\mathrm{i}k\rho\cos(\tau-\varphi)} \, \mathrm{d}\tau, \tag{6.169}$$

where the contour $\mathcal{C}$ goes from $\gamma + \mathrm{i}\infty$ to $\pi + \gamma - \mathrm{i}\infty$, and we may take $\gamma = \theta - \pi/2$.

Substituting (6.169) in (6.167) gives

$$u_{\mathrm{sc}}(\boldsymbol{r}; \hat{\boldsymbol{\alpha}}) = \frac{1}{\pi} \int_{\mathcal{C}} f(\hat{\boldsymbol{\tau}}; \hat{\boldsymbol{\alpha}}) \exp(\mathrm{i}k\boldsymbol{r}\cdot\hat{\boldsymbol{\tau}}) \, \mathrm{d}\tau, \quad r > a, \tag{6.170}$$

where $\hat{\boldsymbol{\tau}} = (\cos\tau, \sin\tau)$ is a *complex* vector. The formula (6.170) gives the scattered field outside the circumscribing circle, $r = a$, in terms of an integral of the far-field pattern over certain complex observation directions $\hat{\boldsymbol{\tau}}$.

Let us make two remarks on the formula (6.170). First, it is inefficient, because it is known that u_{sc} can be reconstructed in $r > a$ from a knowledge of $f(\hat{\boldsymbol{r}}; \hat{\boldsymbol{\alpha}})$, where $\hat{\boldsymbol{r}}$ is *real*; see [223, Theorem 3.6 and Corollary 3.8]. Second, functions of the form

$$\int_{-\pi}^{\pi} f(\tau)\exp(\mathrm{i}k\boldsymbol{r}\cdot\hat{\boldsymbol{\tau}})\,\mathrm{d}\tau$$

are known as *Herglotz wavefunctions* [225, Definition 3.14]; such representations have found application in certain methods for solving inverse obstacle problems.

6.13.4 Scattering by N cylinders

Consider N sound-hard cylinders, S_j, $j = 1, 2, \dots, N$. Choose an origin O_j inside S_j, for each j, and define plane polar coordinates (r_j, θ_j) at O_j, with $\theta_j = 0$ being in the x-direction. The position vector of O_j with respect to O is $\boldsymbol{b}_j$. A typical point P has position vectors $\boldsymbol{r}$ and $\boldsymbol{r}_j$ with respect to O and O_j, respectively, whence $\boldsymbol{r} = \boldsymbol{r}_j + \boldsymbol{b}_j$. A typical point $q \in S_j$ has position vector $\boldsymbol{q}_j$ with respect to O_j.

The incident plane wave is

$$u_{\mathrm{inc}} = \exp(\mathrm{i}k\boldsymbol{r}\cdot\hat{\boldsymbol{\alpha}}) = \mathcal{A}_j\exp(\mathrm{i}k\boldsymbol{r}_j\cdot\hat{\boldsymbol{\alpha}})$$

for each $j = 1, 2, \dots, N$, where $\mathcal{A}_j = \exp(\mathrm{i}k\boldsymbol{b}_j\cdot\hat{\boldsymbol{\alpha}})$ is a constant phase factor.

We can write the scattered field as the sum of contributions from each cylinder. Thus, from (6.167), we have

$$u_{sc}(\boldsymbol{r}; \hat{\boldsymbol{\alpha}}) = \sum_{j=1}^{N}\mathcal{A}_j\,U_j(\boldsymbol{r}_j; \hat{\boldsymbol{\alpha}}),$$

where

$$U_j(\boldsymbol{r}_j; \hat{\boldsymbol{\alpha}}) = \frac{\mathrm{i}}{4}\overline{\mathcal{A}_j}\int_{S_j} u(q; \hat{\boldsymbol{\alpha}})\,\frac{\partial}{\partial n_q}H_0^{(1)}(kR)\,\mathrm{d}s_q,$$

$R = |\boldsymbol{r} - \boldsymbol{q}| = |\boldsymbol{r}_j - \boldsymbol{q}_j|$ and u is the total field. If r_j is large, we obtain

$$U_j(\boldsymbol{r}_j; \hat{\boldsymbol{\alpha}}) \sim \sqrt{\frac{2}{\pi}}\,\mathrm{e}^{-\mathrm{i}\pi/4}\,\frac{\mathrm{e}^{\mathrm{i}kr_j}}{\sqrt{kr_j}}\,F_j(\hat{\boldsymbol{r}}_j; \hat{\boldsymbol{\alpha}}) \quad \text{as } r_j \to \infty,$$

where $\boldsymbol{r}_j = r_j\hat{\boldsymbol{r}}_j$ and

$$F_j(\hat{\boldsymbol{r}}_j; \hat{\boldsymbol{\alpha}}) = \frac{k}{4}\overline{\mathcal{A}_j}\int_{S_j} u(q; \hat{\boldsymbol{\alpha}})\,\left[\hat{\boldsymbol{r}}_j\cdot\boldsymbol{n}(q)\right]\exp(-\mathrm{i}k\hat{\boldsymbol{r}}_j\cdot\boldsymbol{q}_j)\,\mathrm{d}s_q.$$

The far-field of the N-cylinder cluster is given by

$$u_{sc}(\boldsymbol{r}; \hat{\boldsymbol{\alpha}}) \sim \sqrt{\frac{2}{\pi}}\,\mathrm{e}^{-\mathrm{i}\pi/4}\,\frac{\mathrm{e}^{\mathrm{i}kr}}{\sqrt{kr}}\,F(\hat{\boldsymbol{r}}; \hat{\boldsymbol{\alpha}}) \quad \text{as } r \to \infty,$$

where

$$F(\hat{\boldsymbol{r}};\, \hat{\boldsymbol{\alpha}}) = \sum_{j=1}^{N} F_j(\hat{\boldsymbol{r}};\, \hat{\boldsymbol{\alpha}}) \exp\{\mathrm{i}k\boldsymbol{b}_j \cdot (\hat{\boldsymbol{\alpha}} - \hat{\boldsymbol{r}})\};$$

note that $\hat{\boldsymbol{r}}_j \sim \hat{\boldsymbol{r}}$ as $r \to \infty$.

Next, we introduce a plane-wave representation for U_j. Put $\boldsymbol{q}_j = \rho_j(\varphi_j)\,(\cos\varphi_j, \sin\varphi_j)$ and suppose that

$$r_j > \max_{\varphi_j} \rho_j(\varphi_j) \equiv a_j,$$

say. Then, we can write

$$H_0^{(1)}(kR) = \frac{1}{\pi} \int_{\mathcal{C}_j} \mathrm{e}^{\mathrm{i}kr_j \cos(\tau - \theta_j)}\, \mathrm{e}^{-\mathrm{i}k\rho_j \cos(\tau - \varphi_j)}\, \mathrm{d}\tau, \tag{6.171}$$

whence

$$U_j(\boldsymbol{r}_j;\, \hat{\boldsymbol{\alpha}}) = \frac{1}{\pi} \int_{\mathcal{C}_j} F_j(\hat{\boldsymbol{\tau}};\, \hat{\boldsymbol{\alpha}}) \exp(\mathrm{i}k\boldsymbol{r}_j \cdot \hat{\boldsymbol{\tau}})\, \mathrm{d}\tau, \qquad r_j > a_j;$$

the contour $\mathcal{C}_j$ is selected as before.

6.13.4.1 Integral equations

Now, suppose we know the far-field pattern f_j for each cylinder S_j in isolation. We want to use this information in order to calculate F_j. To do this, we consider the total field in the vicinity of one of the cylinders, S_l, say. Thus, we have

$$\begin{aligned} u = u_{\text{inc}} + u_{\text{sc}} &= u_{\text{inc}} + \sum_{j=1}^{N} \mathcal{A}_j U_j \\ &= \mathcal{A}_l \exp(\mathrm{i}k\boldsymbol{r}_l \cdot \hat{\boldsymbol{\alpha}}) + \sum_{\substack{j=1 \\ j\neq l}}^{N} \mathcal{A}_j U_j + \mathcal{A}_l U_l \\ &= \mathcal{A}_l \left\{ U_l^{\text{inc}}(\boldsymbol{r}_l;\, \hat{\boldsymbol{\alpha}}) + U_l(\boldsymbol{r}_l;\, \hat{\boldsymbol{\alpha}}) \right\}, \end{aligned} \tag{6.172}$$

say, where

$$\begin{aligned} U_l^{\text{inc}} = {}& \exp(\mathrm{i}k\boldsymbol{r}_l \cdot \hat{\boldsymbol{\alpha}}) \\ & + \frac{1}{\pi} \sum_{\substack{j=1 \\ j\neq l}}^{N} \int_{\mathcal{C}_j} \exp(\mathrm{i}k\boldsymbol{r}_l \cdot \hat{\boldsymbol{\tau}}) \exp\{\mathrm{i}k(\hat{\boldsymbol{\tau}} - \hat{\boldsymbol{\alpha}}) \cdot \boldsymbol{b}_{lj}\}\, F_j(\hat{\boldsymbol{\tau}};\, \hat{\boldsymbol{\alpha}})\, \mathrm{d}\tau, \end{aligned}$$

$\boldsymbol{b}_{lj} = \boldsymbol{b}_l - \boldsymbol{b}_j$ is the position vector of O_l with respect to O_j, and we have used $\boldsymbol{r}_j = \boldsymbol{r}_l + \boldsymbol{b}_{lj}$.

U_l^{inc} can be regarded as a superposition of plane waves, incident upon S_l. According to (6.172), the corresponding scattered field is U_l. However, we already know how to scatter plane waves by S_l, and so this gives a relation that can be used to find F_j.

Specifically, the far field of U_l must be the same as the far field corresponding to the incident field U_l^{inc}; this gives

$$F_l(\hat{\boldsymbol{r}};\,\hat{\boldsymbol{\alpha}}) = f_l(\hat{\boldsymbol{r}};\,\hat{\boldsymbol{\alpha}}) + \frac{1}{\pi}\sum_{\substack{j=1\\ j\neq l}}^{N}\int_{\mathcal{C}_j} f_l(\hat{\boldsymbol{r}};\,\hat{\boldsymbol{\tau}})\exp\{\mathrm{i}k(\hat{\boldsymbol{\tau}}-\hat{\boldsymbol{\alpha}})\cdot\boldsymbol{b}_{lj}\}\,F_j(\hat{\boldsymbol{\tau}};\,\hat{\boldsymbol{\alpha}})\,\mathrm{d}\tau. \qquad (6.173)$$

This equation holds for $l = 1, 2, \ldots, N$, giving what Twersky [1203, p. 371] calls the *self-consistent system of equations*. Millar [857, §4.2] has shown that (6.173) has at most one solution when $N = 2$.

6.13.4.2 Discussion

The system (6.173) is to be solved for F_j, given f_l. However, this is not straightforward, in general. One reason is that the integrals are over infinite contours in the complex τ-plane; the contours $\mathcal{C}_j$ may be different. Notice also that (6.173) has to be evaluated for $\hat{\boldsymbol{r}} \in \mathcal{C}_l$, which means that analytic continuation is implicit: the derivation of (6.173) is for real directions $\hat{\boldsymbol{r}}$, and we may only know $f_l(\hat{\boldsymbol{r}};\,\hat{\boldsymbol{\alpha}})$ for real $\hat{\boldsymbol{r}}$ and $\hat{\boldsymbol{\alpha}}$.

Nevertheless, (6.173) does have some formal advantages. First, it is exact. Second, it can be derived without making any explicit use of the boundary conditions: (6.173) is unchanged if, for example, some (or all) of the cylinders are sound-soft or penetrable. Third, 'no knowledge of addition theorems is required for the present derivation' [1203, p. 361]: this is because plane-wave representations are easily translated. A disadvantage is that, if (6.173) could be solved, it would only yield the scattered field outside the circumscribed circles, $r_j = a_j$, so that surface fields cannot be calculated readily.

We are not aware of any direct *numerical* solutions of the system (6.173). However, it has been used with various simplifying assumptions, such as wide spacings or small cylinders. Twersky [1203, §4.2] also derived algebraic systems from (6.173); these reduce to the equations obtained by Záviška's method for circular cylinders (see Section 4.5). However:

> *In general, for arbitrary configurations and arbitrary spacings, there is little that can be done with these systems of algebraic equations except expand them by Neumann iteration (subject to whatever restrictions on the parameters this entails) to obtain a series of "successive orders of scattering" for the coefficients.*
>
> (Twersky [1203, p. 373])

In order to obtain algebraic equations from (6.173), we follow Twersky and insert Fourier series for F_l and f_j. Thus, proceeding formally, substitute

$$F_l(\hat{\boldsymbol{r}};\,\hat{\boldsymbol{\alpha}}) = \sum_{m,n} A_{mn}^l\,\mathrm{e}^{\mathrm{i}(m\theta+n\alpha)} \qquad (6.174)$$

and

$$f_l(\hat{\boldsymbol{r}};\,\hat{\boldsymbol{\alpha}}) = \sum_{m,n} a_{mn}^l\,\mathrm{e}^{\mathrm{i}(m\theta+n\alpha)} \qquad (6.175)$$

in (6.173), giving

$$A^l_{ms} = a^l_{ms} + \sum_{\substack{j=1 \\ j\neq l}}^{N} \sum_{n,\mu} a^l_{mn} A^j_{\mu s}\, E(l, j, n+\mu),$$

where

$$E(l, j, n) = \frac{1}{\pi} \int_{\mathcal{C}_j} \exp\{ik(\hat{\boldsymbol{\tau}} - \hat{\boldsymbol{\alpha}}) \cdot \boldsymbol{b}_{lj}\}\, \mathrm{e}^{\mathrm{i}n\tau}\, \mathrm{d}\tau = \mathrm{i}^n H_n^{(1)}(kb_{lj})\, \mathrm{e}^{\mathrm{i}n\beta_{lj}} \mathcal{A}_j \overline{\mathcal{A}_l} \qquad (6.176)$$

and $\boldsymbol{b}_{lj} = b_{lj}(\cos\beta_{lj}, \sin\beta_{lj})$. This calculation is formal because we have used the Fourier series (6.174) when $\theta = \tau$ and the Fourier series (6.175) when $\alpha = \tau$: these series may not be convergent for complex τ. Also, the second equality in (6.176) assumes that $\mathcal{C}_j$ (which was selected so that (6.171) is valid) is such that (6.165) can be used.

6.13.5 Extensions

Twersky extended his method to three dimensions [1199] and to electromagnetic problems [1206]. He applied it to gratings of circular [1202] and elliptical cylinders [164], and to periodic lines [1207] and doubly-periodic planar arrays of identical three-dimensional obstacles [1208, 1209, 1210]. Millar used (6.173) as a starting point for studies of infinite [858], semi-infinite [859] and finite [860] gratings of identical cylinders.

For extensions to two-dimensional elastic waves, see [69, 84]. For three-dimensional elastic waves, see [70, 84, 972, 973, 971].

6.14 Fast multipole methods

The fast multipole method (FMM) is one of a number of algorithms designed to accelerate the numerical solution of boundary integral equations. These algorithms have been used extensively since first invented by Rokhlin in 1985, and used later by him for two-dimensional acoustic scattering problems [1026]. Indeed, it is claimed that the FMM is one of the top ten algorithms of the twentieth century [106]. We note that FMM algorithms are not limited to boundary integral equations, but can be used whenever the effects of groups of sources are to be calculated at groups of observation points.

We shall not give a complete exposition of the FMM. We refer the reader to the excellent review by Nishimura [913] for details and references, and we shall use his terminology. See also the book [449] but note that the FMM is currently a thriving research area.

Addition theorems play a key role in FMM algorithms. To illustrate this statement, we shall describe a typical calculation. Thus, consider, for example, a three-dimensional acoustic double-layer potential,

$$u_0(P) = \int_{S_0} \nu(q) \frac{\partial}{\partial n_q} G(P, q) \, ds_q, \tag{6.177}$$

where S_0 is a piece of S; S_0 could be the surface of one scatterer, or an element of such a surface. The point P is not near S_0.

We emphasise that the double-layer potential (6.177) is an example. We could have chosen to consider a single-layer potential, a linear combination of the two potentials or a finite collection of sources and dipoles at discrete points. These choices lead to straightforward changes in the following calculations.

We use the bilinear expansion of G, Theorem 6.4, namely

$$G(P, Q) = -2\mathrm{i}k \sum_{n,m} (-1)^m \hat{\psi}_n^{-m}(\boldsymbol{r}_Q) \, \psi_n^m(\boldsymbol{r}_P), \tag{6.178}$$

where $\boldsymbol{r}_P$ and $\boldsymbol{r}_Q$ are the position vectors of P and Q, respectively, with respect to an origin O, and we have used (4.39). Equation (6.178) is valid for $r_Q = |\boldsymbol{r}_Q| < |\boldsymbol{r}_P| = r_P$.

Consider a second origin, O_1, in the vicinity of S_0; see Fig. 6.1. If $\boldsymbol{b}_1$ is the position vector of O_1 with respect to O, we have

$$G(P, Q) = -2\mathrm{i}k \sum_{n,m} (-1)^m \hat{\psi}_n^{-m}(\boldsymbol{r}_Q - \boldsymbol{b}_1) \, \psi_n^m(\boldsymbol{r}_P - \boldsymbol{b}_1), \tag{6.179}$$

valid for $|\boldsymbol{r}_Q - \boldsymbol{b}_1| < |\boldsymbol{r}_P - \boldsymbol{b}_1|$.

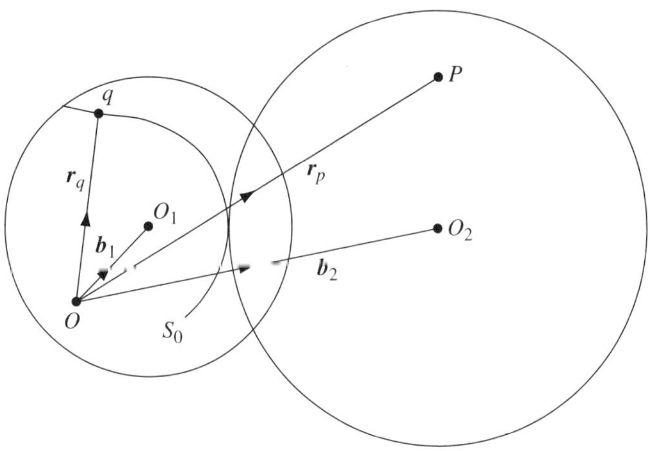

Fig. 6.1. Configuration for translation formulae. The left-hand circle has radius a_1; it encloses S_0. The right-hand circle has radius a_2.

Suppose that a_1 is the radius of the smallest sphere, centred at O_1, that encloses S_0; we suppose that P is outside this sphere. Then, combining (6.179) with (6.177) gives

$$u_0(\boldsymbol{r}_P) = \sum_{n,m} M_n^m(\boldsymbol{b}_1)\, \psi_n^m(\boldsymbol{r}_P - \boldsymbol{b}_1), \quad |\boldsymbol{r}_P - \boldsymbol{b}_1| > a_1, \tag{6.180}$$

where

$$M_n^m(\boldsymbol{b}) = -2\mathrm{i}k(-1)^m \int_{S_0} \nu(q)\, \frac{\partial}{\partial n_q} \hat{\psi}_n^{-m}(\boldsymbol{r}_q - \boldsymbol{b})\, \mathrm{d}s_q, \tag{6.181}$$

and $\boldsymbol{r}_q$ is the position vector of $q \in S_0$ with respect to O. $M_n^m(\boldsymbol{b})$ is called the *multipole moment centred at the point* (with position vector) $\boldsymbol{b}$.

Next, consider a third origin, O_2, with position vector $\boldsymbol{b}_2$ with respect to O; see Fig. 6.1. Using the addition theorem for regular spherical wavefunctions, Theorem 3.26, we obtain

$$\hat{\psi}_n^m(\boldsymbol{r}_q - \boldsymbol{b}_2) = \sum_{\nu,\mu} \widehat{S}_{n\nu}^{m\mu}(\boldsymbol{b}_{12})\, \hat{\psi}_\nu^\mu(\boldsymbol{r}_q - \boldsymbol{b}_1)$$

where $\boldsymbol{b}_{jl} = \boldsymbol{b}_j - \boldsymbol{b}_l = -\boldsymbol{b}_{lj}$ is the position vector of O_j with respect to O_l. Changing the signs of m and μ, and using (3.107) and (6.181), we deduce that

$$M_n^m(\boldsymbol{b}_2) = \sum_{\nu,\mu} \widehat{S}_{\nu n}^{\mu m}(\boldsymbol{b}_{21})\, M_\nu^\mu(\boldsymbol{b}_1). \tag{6.182}$$

This is known as the *M2M formula*.

Equation (6.180) gives an expansion of u_0 in terms of outgoing wavefunctions. However, in a neighbourhood of O_2, u_0 may also be expanded in terms of regular wavefunctions,

$$u_0(\boldsymbol{r}_P) = \sum_{n,m} L_n^m(\boldsymbol{b}_2)\, \hat{\psi}_n^m(\boldsymbol{r}_P - \boldsymbol{b}_2); \tag{6.183}$$

this is a *local expansion*, centred at O_2. To derive (6.183), we first let a_2 be the radius of the largest sphere, centred at O_2, that does not enclose any of S_0; we suppose that P is inside this sphere. Then, we have a formula similar to (6.179),

$$G(P,Q) = -2\mathrm{i}k \sum_{n,m} (-1)^m \psi_n^{-m}(\boldsymbol{r}_Q - \boldsymbol{b}_2)\, \hat{\psi}_n^m(\boldsymbol{r}_P - \boldsymbol{b}_2), \tag{6.184}$$

valid for $|\boldsymbol{r}_P - \boldsymbol{b}_2| < |\boldsymbol{r}_Q - \boldsymbol{b}_2|$. When this is combined with (6.177), we obtain (6.183), wherein

$$L_n^m(\boldsymbol{b}) = -2\mathrm{i}k(-1)^m \int_{S_0} \nu(q)\, \frac{\partial}{\partial n_q} \psi_n^{-m}(\boldsymbol{r}_q - \boldsymbol{b})\, \mathrm{d}s_q; \tag{6.185}$$

we see that the local expansion (6.183) holds for $|\boldsymbol{r}_P - \boldsymbol{b}_2| < a_2$.

From the addition theorem for outgoing spherical wavefunctions, Theorem 3.27, we obtain

$$\psi_n^m(\boldsymbol{r}_q - \boldsymbol{b}_3) = \sum_{\nu,\mu} \widehat{S}_{n\nu}^{m\mu}(\boldsymbol{b}_{23})\, \psi_\nu^\mu(\boldsymbol{r}_q - \boldsymbol{b}_2),$$

valid for $|\boldsymbol{r}_q - \boldsymbol{b}_2| > |\boldsymbol{b}_{23}|$. Hence,

$$L_n^m(\boldsymbol{b}_3) = \sum_{\nu,\mu} \widehat{S}_{\nu n}^{\mu m}(\boldsymbol{b}_{32})\, L_\nu^\mu(\boldsymbol{b}_2), \tag{6.186}$$

provided that $|\boldsymbol{b}_{23}| < a_2$. This is known as the *L2L formula*.

Finally, we relate L_n^m and M_n^m. Using Theorem 3.27 again, we obtain

$$\psi_n^m(\boldsymbol{r}_q - \boldsymbol{b}_2) = \sum_{\nu,\mu} S_{n\nu}^{m\mu}(\boldsymbol{b}_{12})\, \hat{\psi}_\nu^\mu(\boldsymbol{r}_q - \boldsymbol{b}_1)$$

for $|\boldsymbol{r}_q - \boldsymbol{b}_1| < |\boldsymbol{b}_{12}|$. Hence, using (3.99), we obtain the *M2L formula*,

$$L_n^m(\boldsymbol{b}_2) = \sum_{\nu,\mu} S_{\nu n}^{\mu m}(\boldsymbol{b}_{21})\, M_\nu^\mu(\boldsymbol{b}_1), \tag{6.187}$$

which is valid for $|\boldsymbol{b}_{12}| > a_1$.

The M2M, L2L and M2L formulae are all used within the FMM algorithms. See [913, §2.3] for a description of the basic algorithm.

It turns out that the basic M2M, L2L and M2L formulae become more expensive to use as the frequency increases. Thus, diagonalised versions of these formulae are needed; they are derived next.

6.14.1 Diagonal forms

Define a function on the unit sphere by

$$f(\hat{\boldsymbol{r}}\,|\,\boldsymbol{b}) = \sum_{n,m} (-\mathrm{i})^n M_n^m(\boldsymbol{b})\, Y_n^m(\hat{\boldsymbol{r}})$$

for $\hat{\boldsymbol{r}} \in \Omega$, so that

$$M_n^m(\boldsymbol{b}) = \mathrm{i}^n \int_\Omega f(\hat{\boldsymbol{r}}\,|\,\boldsymbol{b})\, \overline{Y_n^m(\hat{\boldsymbol{r}})}\, \mathrm{d}\Omega(\hat{\boldsymbol{r}}). \tag{6.188}$$

Using (3.90), we see from (6.180) that

$$u_0(\boldsymbol{r}_P) \sim \frac{\mathrm{e}^{\mathrm{i}kr_P}}{\mathrm{i}kr_P} f(\hat{\boldsymbol{r}}_P) \quad \text{as } r_P \to \infty,$$

where $\hat{\boldsymbol{r}}_P = \boldsymbol{r}_P / r_P$,

$$f(\hat{\boldsymbol{r}}) = f(\hat{\boldsymbol{r}}\,|\,\boldsymbol{b}_1)\, \exp(-\mathrm{i}k\boldsymbol{b}_1 \cdot \hat{\boldsymbol{r}}) \tag{6.189}$$

is the far-field pattern, and we have used $|\boldsymbol{r}_P - \boldsymbol{b}_1| \sim r_P - \boldsymbol{b}_1 \cdot \hat{\boldsymbol{r}}_P$ as $r_P \to \infty$. Then, from (4.49), we have

$$f(\hat{\boldsymbol{r}}\,|\,\boldsymbol{b}_2) = \exp(\mathrm{i}k\boldsymbol{b}_{21} \cdot \hat{\boldsymbol{r}})\, f(\hat{\boldsymbol{r}}\,|\,\boldsymbol{b}_1). \tag{6.190}$$

This represents a diagonalised version of the M2M formula (6.182).

Let us give a direct proof of (6.190). Multiply (6.182) by $(-\mathrm{i})^n Y_n^m(\hat{\boldsymbol{r}})$ and sum, giving

$$f(\hat{\boldsymbol{r}}\,|\,\boldsymbol{b}_2) = \sum_{n,m}\sum_{\nu,\mu} \widehat{S}_{\nu n}^{\mu m}(\boldsymbol{b}_{21})\,(-\mathrm{i})^n Y_n^m(\hat{\boldsymbol{r}})\, M_\nu^\mu(\boldsymbol{b}_1). \tag{6.191}$$

Now, from (3.79), we have

$$\widehat{S}_{\nu n}^{\mu m}(\boldsymbol{b}) = \mathrm{i}^{n-\nu}\int_\Omega \exp(\mathrm{i}k\boldsymbol{b}\cdot\hat{\boldsymbol{r}}_1)\, Y_\nu^\mu(\hat{\boldsymbol{r}}_1)\,\overline{Y_n^m(\hat{\boldsymbol{r}}_1)}\,\mathrm{d}\Omega(\hat{\boldsymbol{r}}_1), \tag{6.192}$$

whereas (6.188) gives

$$M_\nu^\mu(\boldsymbol{b}) = \mathrm{i}^\nu \int_\Omega f(\hat{\boldsymbol{r}}_2\,|\,\boldsymbol{b})\,\overline{Y_\nu^\mu(\hat{\boldsymbol{r}}_2)}\,\mathrm{d}\Omega(\hat{\boldsymbol{r}}_2). \tag{6.193}$$

Then, substituting (6.192) and (6.193) in (6.191), we obtain

$$f(\hat{\boldsymbol{r}}\,|\,\boldsymbol{b}_2) = \sum_{n,m} Y_n^m(\hat{\boldsymbol{r}}) \int_\Omega \exp(\mathrm{i}k\boldsymbol{b}_{21}\cdot\hat{\boldsymbol{r}}_1)\,\overline{Y_n^m(\hat{\boldsymbol{r}}_1)}\, f(\hat{\boldsymbol{r}}_1\,|\,\boldsymbol{b}_1)\,\mathrm{d}\Omega(\hat{\boldsymbol{r}}_1)$$

after one application of (3.11). The result (6.190) follows after a further application of (3.11).

The virtue of this proof is that it applies to the L2L formula. Thus, if we define

$$g(\hat{\boldsymbol{r}}\,|\,\boldsymbol{b}) = \sum_{n,m}(-\mathrm{i})^n L_n^m(\boldsymbol{b})\, Y_n^m(\hat{\boldsymbol{r}}),$$

then we immediately obtain

$$g(\hat{\boldsymbol{r}}\,|\,\boldsymbol{b}_3) = \exp(\mathrm{i}k\boldsymbol{b}_{32}\cdot\hat{\boldsymbol{r}})\, g(\hat{\boldsymbol{r}}\,|\,\boldsymbol{b}_2), \tag{6.194}$$

even though g does not have an interpretation in terms of far-field patterns.

Finally, we consider the M2L formula, (6.187). From (3.103), we have

$$S_{\nu n}^{\mu m}(\boldsymbol{b}) = \mathrm{i}^{n-\nu}\sum_q (2q+1)\mathrm{i}^q h_q(kb)\int_\Omega Y_\nu^\mu(\hat{\boldsymbol{r}}_1)\,\overline{Y_n^m(\hat{\boldsymbol{r}}_1)}\, P_q(\hat{\boldsymbol{r}}_1\cdot\hat{\boldsymbol{b}})\,\mathrm{d}\Omega(\hat{\boldsymbol{r}}_1).$$

Hence, making use of (6.193) and (3.11), we obtain

$$\begin{aligned} L_n^m(\boldsymbol{b}_2) &= \mathrm{i}^n \sum_{\nu,\mu}\sum_{q=0}^{\infty}(2q+1)\mathrm{i}^q h_q(k|\boldsymbol{b}_{21}|)\int_\Omega\int_\Omega P_q(\hat{\boldsymbol{r}}_1\cdot\hat{\boldsymbol{b}}_{21}) \\ &\qquad \times Y_\nu^\mu(\hat{\boldsymbol{r}}_1)\,\overline{Y_n^m(\hat{\boldsymbol{r}}_1)}\, f(\hat{\boldsymbol{r}}_2\,|\,\boldsymbol{b}_1)\,\overline{Y_\nu^\mu(\hat{\boldsymbol{r}}_2)}\,\mathrm{d}\Omega(\hat{\boldsymbol{r}}_2)\,\mathrm{d}\Omega(\hat{\boldsymbol{r}}_1) \\ &= \mathrm{i}^n \sum_{q=0}^{\infty}(2q+1)\mathrm{i}^q h_q(k|\boldsymbol{b}_{21}|) \\ &\qquad \times \int_\Omega \overline{Y_n^m(\hat{\boldsymbol{r}}_1)}\, P_q(\hat{\boldsymbol{r}}_1\cdot\hat{\boldsymbol{b}}_{21})\, f(\hat{\boldsymbol{r}}_1\,|\,\boldsymbol{b}_1)\,\mathrm{d}\Omega(\hat{\boldsymbol{r}}_1). \end{aligned}$$

When this expression is substituted in (6.183), we find that

$$u_0(\boldsymbol{r}_P) = \frac{1}{4\pi}\sum_{q=0}^{\infty}(2q+1)\mathrm{i}^q h_q(k|\boldsymbol{b}_{21}|)$$
$$\times \int_\Omega P_q(\hat{\boldsymbol{r}}_1\cdot\hat{\boldsymbol{b}}_{21})\, f(\hat{\boldsymbol{r}}_1\,|\,\boldsymbol{b}_1)\exp\{\mathrm{i}k(\boldsymbol{r}_P-\boldsymbol{b}_2)\cdot\hat{\boldsymbol{r}}_1\}\,\mathrm{d}\Omega(\hat{\boldsymbol{r}}_1),$$

having used (3.69) to evaluate the sum over m and n. Note the limits on the remaining sum. Note further that the order of integration and summation cannot be interchanged. Thus, we have

$$u_0(\boldsymbol{r}_P) = \frac{1}{4\pi}\lim_{L\to\infty}\int_\Omega M_L(\hat{\boldsymbol{r}};\boldsymbol{b}_{21})\, f(\hat{\boldsymbol{r}}\,|\,\boldsymbol{b}_1)\exp\{\mathrm{i}k(\boldsymbol{r}_P-\boldsymbol{b}_2)\cdot\hat{\boldsymbol{r}}\}\,\mathrm{d}\Omega(\hat{\boldsymbol{r}}), \tag{6.195}$$

where

$$M_L(\hat{\boldsymbol{r}};\boldsymbol{b}) = \sum_{l=0}^{L}(2l+1)\mathrm{i}^l h_l(kb)\,P_l(\hat{\boldsymbol{r}}\cdot\hat{\boldsymbol{b}})$$

is the translation function; see (3.105).

Equation (6.195) is [318, eqn (3.50)], although our derivation is different. In fact, (6.195) can be obtained directly from (6.177) as follows. Noting that

$$G(P,q) = -\frac{1}{2}\mathrm{i}(k/\pi)\,h_0(k|\boldsymbol{r}_P-\boldsymbol{r}_q|),$$

write $\boldsymbol{r}_P-\boldsymbol{r}_q=\boldsymbol{r}_P+\boldsymbol{b}_1-\boldsymbol{b}_2-\boldsymbol{r}_q+\boldsymbol{b}_{21}$ and then use Theorem 3.17 (with $z_0=h_0$ and $\boldsymbol{b}=\boldsymbol{b}_{21}$). Then, observe that

$$f(\hat{\boldsymbol{r}}) = -\frac{\mathrm{i}k}{2\pi}\int_{S_0}\nu(q)\,\frac{\partial}{\partial n_q}\exp(-\mathrm{i}k\hat{\boldsymbol{r}}\cdot\boldsymbol{r}_q)\,\mathrm{d}s_q$$

(see Section 5.10.3) and use (6.189). This derivation shows clearly that (6.195) is valid if $|\boldsymbol{r}_P-\boldsymbol{b}_2+\boldsymbol{b}_1-\boldsymbol{r}_q|<|\boldsymbol{b}_1-\boldsymbol{b}_2|$ for all $q\in S_0$.

6.14.2 Literature

Rokhlin's 1990 paper considers the two-dimensional acoustic transmission problem [1026]. He uses the indirect method of Kress & Roach [658], and he gives diagonal forms using finite Fourier expansions. Further results and FMM variants in two dimensions are available. Thus, Engheta *et al.* [314] solved the Helmholtz integral equation; they removed irregular frequencies by giving k a small imaginary part [896]. Lu & Chew [747] used a combined-field method. Amini & Profit [26] used the method of Burton & Miller [166].

Three-dimensional diagonal forms (as in Section 6.14.1) were given by Rokhlin [1027]. For numerical results and FMM variants, see, for example, [219, 1130, 1000, 255, 256].

As Nishimura remarks, 'Rokhlin's diagonal form provides a very tricky fast alternative to the original FMM … [that] suffers from instability in Helmholtz' equation for small wave number' [913, p. 310]. Consequently, new FMM algorithms were developed. See, for example, [435, 257, 258].

FMM algorithms have been used by Fujiwara [372, 373] for elastodynamics, and in [310, 708, 709] for multiple scattering of electromagnetic waves. Many other applications are expected as the FMM is developed further.

7
Null-field and T-matrix methods

In our opinion, the full potential of the T *matrix itself has only begun to be realized.... As pointed out by Peterson and Ström, the* T *matrix is an ideal tool for investigating multilayered objects* [964, 965], *as well as interactions (i.e. multiple scattering) between two or more objects* [962, 963]. *It is in these last areas that we believe the richest applications will be found.*

(Waterman [1292, p. 4566])

7.1 Introduction

The null-field and T-matrix methods are used widely for obtaining numerical solutions to various radiation and scattering problems. These methods are related, as we will see. They were first devised by Waterman in 1965 for electromagnetic scattering problems [1287]. Later, they were developed for treating problems in acoustics, elastodynamics and hydrodynamics. For surveys up to 1979, see the book [1248]. Many subsequent applications are reviewed in [1254, 871, 870] and in Chapter 6 of [866]. For electromagnetic problems, see the extensive bibliography [873].

As the opening quotation states, the real power of these methods comes when they are used to solve multiple-scattering problems; the key paper is [963].

7.2 Radiation problems

Suppose that $u(P)$ solves the exterior Neumann problem with $\partial u/\partial n = f$ on S. Then, we have the Helmholtz integral representation (5.57), namely

$$2u(P) = \int_S \left\{ f(q)\, G(P,q) - u(q)\, \frac{\partial}{\partial n_q} G(P,q) \right\} \mathrm{d}s_q, \quad P \in B_\mathrm{e}, \tag{7.1}$$

where $G(P,Q)$ is the free-space Green's function. Similarly, if $P \in B$, we have the interior integral relation (6.127), namely

$$0 = \int_S \left\{ f(q)\, G(P,q) - u(q)\, \frac{\partial}{\partial n_q} G(P,q) \right\} \mathrm{d}s_q, \quad P \in B. \tag{7.2}$$

Similar equations can be derived for the exterior Dirichlet problem, for which $u = g$ on S.

We can regard (7.2) as an equation to be solved for $u(q)$. We know that it is always uniquely solvable (there are no irregular frequencies); see Theorem 6.8. However, (7.2) holds for *all* $P \in B$, and so it is not clear how to solve it. Some methods for doing so are described next.

7.3 Kupradze's method and related methods

Kupradze [672, p. 252] calls (7.2) a 'canonical functional equation'. Having developed methods based on indirect boundary integral equations (see Section 5.9), he chose to abandon them when it came to actual computations: he used (7.2) instead, evaluating it at M points $P_i \in B$, $i = 1, 2, \ldots, M$, and approximating the (non-singular) integral over S with an M-point quadrature rule. Kupradze has developed this method in [671, Chapter VIII] and in [672, Chapter X, §§1–19]. For the Helmholtz equation, see [671, Chapter VIII, §13] and [672, Chapter X, §11], although numerical results are not given.

A continuous version of Kupradze's discrete method is as follows. Consider an auxiliary surface S' in B: for a single obstacle, we may think of S' as being a closed surface similar to but smaller than S, or we may choose S' as a simple surface such as a sphere. See Fig. 7.1. Then, we replace (7.2) by

$$0 = \int_S \left\{ f(q)\, G(p, q) - u(q)\, \frac{\partial}{\partial n_q} G(p, q) \right\} \mathrm{d}s_q, \quad p \in S'. \tag{7.3}$$

Unlike (7.2), this equation does suffer from irregular frequencies: uniqueness is lost whenever k^2 is an eigenvalue of the interior Dirichlet problem for S'.

There are analogous indirect methods. For example, one could seek the solution of the exterior Neumann problem in the form of a source distribution over S',

$$u(P) = \int_{S'} \mu(q)\, G(P, q)\, \mathrm{d}s_q, \quad P \in B_e, \tag{7.4}$$

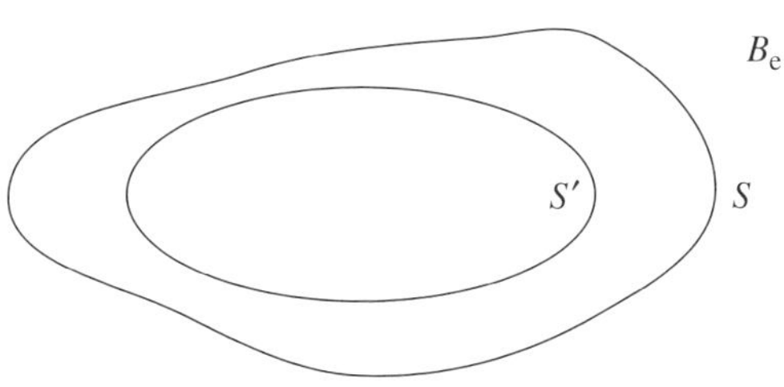

Fig. 7.1. The auxiliary surface S' is inside the scatterer B.

whence the boundary condition gives

$$\int_{S'} \mu(q) \frac{\partial}{\partial n_p} G(p, q) \, \mathrm{d}s_q = f(p), \quad p \in S. \tag{7.5}$$

A difficulty with methods of this type is that $u(P)$ may not be representable as (7.4). We know that when the solution of the exterior Neumann problem is continued analytically from B_e through S into the portion of B between S and S', singularities may be encountered. Indeed, there *must* be singularities within B, but we do not know a priori if they are all inside S'. On the other hand, the representation (7.4) is not singular outside S'. For more information on this difficulty, including computational aspects, see [285, Chapter IV, §2.1].

7.3.1 Literature

Christiansen [211] labels methods based on (7.5) and (7.3) as Method I and Method II, respectively. He gives a thorough review of the literature, and shows that the condition numbers of the corresponding linear $M \times M$ systems worsen as the distance between S and S' increases.

Method I has been used in [718, 950, 656], and reviewed by Ochmann [920]. Electromagnetic applications are reviewed in [580]. Method I has also been used as part of a method for solving inverse problems in acoustics [225, §5.4]. Method II has been used in [987] for acoustics, and in [4] for elastic waves.

For axisymmetric problems, Copley [235] has solved (7.2) using points P along the axis of symmetry. Geer [398] has used an analogous indirect method, with sources (and multipoles) distributed along a piece of the symmetry axis within B. In 1948, Albert & Synge [17] derived the electromagnetic version of (7.2), which they called the 'fundamental integral equation of antenna theory'. They also considered axisymmetric problems and used points P along the symmetry axis; many numerical and asymptotic results are given in [1163]. See also [564, §3.18].

The distinction between the value of the integral (7.2) when $P \in B$ (which is zero) and the analytic continuation of the solution of the exterior boundary-value problem into the interior B (which must have some singularities) has caused some confusion in the literature. For a very clear discussion of the issues and implications, see Millar's paper [862].

7.3.2 Generalised multipole techniques

There are discrete versions of Method I, and generalisations thereof. Thus, let $\{\chi_m(P)\}$ be a set of outgoing wavefunctions, and write

$$u(P) \simeq \sum_{m=1}^{M} c_m \chi_m(P), \quad P \in B_e;$$

the functions $\chi_m(P)$ could be sources, dipoles or multipoles located at points $P_m \in B$. The coefficients c_m can then be determined by applying the boundary condition at M_1 selected points on S. Usually, an overdetermined system ($M_1 > M$) is derived and solved in a least-squares sense; usually, the values found for c_m will depend on both M_1 and M. Methods of this type are sometimes called *generalised multipole techniques* (GMT), although many other names are also used; see, for example, [469, Chapter 8] or [285, Chapter IV].

GMT are essentially numerical. The choices of $\chi_m(P)$ and the collocation points on S are usually made by experience, although some adaptive variants are being developed [469, p. 275]. Thus, at present, GMT are not robust.

A special class of GMT are those with $\chi_m(P) = G(P, P_m)$ and $P_m \in B$; this yields the *method of fundamental solutions* (MFS). In the MFS, the source points P_m may be prescribed or they may be determined as part of a non-linear least-squares scheme. By prescribing an infinite set of points P_m, one can construct a complete set of functions $\{G(P, P_m)\}$, but this set does not form a basis [652]. (The terminology used here is defined in Section 7.7.1.) For extensive reviews of the MFS, see [334, 335].

Another class of GMT uses sources distributed over an internal surface S',

$$\chi_m(P) = \int_{S'} \mu_m(q)\, G(P, q)\, \mathrm{d}s_q,$$

where $\{\mu_m\}$ is a chosen set of densities. For two-dimensional problems, one choice is

$$\mu_m = \exp\{2\pi \mathrm{i} s(m - M' - 1)/L\},$$

where $M = 2M' + 1$, s is arclength and L is the perimeter of S'. This choice was used in [1385] for scattering by two penetrable cylinders, and extended to gratings in [1384].

In the context of multiple-scattering problems, GMT are attractive superficially, at least, because they avoid the use of addition theorems. However, they do not give the desirable separation (discussed in Section 1.1.2) between the scattering properties of individual scatterers and the spacing between them: if the configuration of the scatterers is altered, the whole calculation has to be repeated. Nevertheless, GMT have been used for multiple-scattering problems. For example, Ludwig [749] confirmed the numerical results of Bruning & Lo [145, 146] for electromagnetic scattering by two dielectric spheres. Erez & Leviatan gave results for acoustic scattering by two sound-soft spheres (of different radii) in [324] and for electromagnetic scattering by two perfectly conducting spheres in [325]; they located some of their sources P_m at complex points. Results for electromagnetic scattering by three axisymmetric, coaxial scatterers (raindrops) are given in [323]. Comberg & Wriedt [226] gave results for electromagnetic scattering by two and four spheres, and by two coaxial oblate spheroids. Imhof has presented numerical results for acoustic scattering by two penetrable elliptic cylinders [529]; for extensions to plane elastodynamics and fluid–solid problems, see [530] and [532], respectively. Imhof [531] has also given results for the scattering of elastic waves by two three-dimensional cavities, each cavity being a circular cylinder of finite length.

Benites *et al.* [80] presented many results for scattering of two-dimensional acoustic waves by two sound-hard ellipses and by up to 100 circular cylinders (actually, 50 on one side of a rigid wall). They used a discrete form of (7.4). An extension to two-dimensional elastic waves is described in [81]. Discrete forms of (7.4) have been used extensively for elastodynamic half-space problems, in the seismology context. Thus, Wong [1334] gave results in two dimensions for scattering of body and Rayleigh waves by semi-circular and semi-elliptical canyons; he used line-source solutions that satisfy the traction-free conditions on the boundary of the half-plane. (These are the half-plane multipoles of Section 2.10 with $n=0$.) A similar method was used by Dravinski [293] for multiple scattering by two identical semi-elliptical alluvial valleys (transmission problem). He has extended his approach to anisotropic half-planes [296] and half-spaces [294]. Sánchez-Sesma [1051] gave results for scattering by a hemispherical cavity; he used full-space point sources, and so he had to impose the traction-free condition on (part of) the boundary of the half-space.

7.4 Scattering problems

For scattering problems, (5.36) and (5.37) give

$$\int_S \left\{ G(P,q)\,\frac{\partial u_{\rm sc}}{\partial n_q} - u_{\rm sc}(q)\,\frac{\partial}{\partial n_q} G(P,q) \right\} {\rm d}s_q = \begin{cases} 2u_{\rm sc}(P), & P \in B_{\rm e}, \\ 0, & P \in B, \end{cases}$$

where $u_{\rm sc}$ is the outgoing scattered field. From (5.34) and (5.35), the incident field $u_{\rm inc}$ satisfies

$$\int_S \left\{ G(P,q)\,\frac{\partial u_{\rm inc}}{\partial n_q} - u_{\rm inc}(q)\,\frac{\partial}{\partial n_q} G(P,q) \right\} {\rm d}s_q = \begin{cases} 0, & P \in B_{\rm e}, \\ -2u_{\rm inc}(P), & P \in B, \end{cases}$$

assuming that $u_{\rm inc}$ satisfies the Helmholtz equation in B. Adding these gives

$$\int_S \left\{ G(P,q)\,\frac{\partial u}{\partial n_q} - u(q)\,\frac{\partial}{\partial n_q} G(P,q) \right\} {\rm d}s_q = \begin{cases} 2u_{\rm sc}(P), & P \in B_{\rm e}, \\ -2u_{\rm inc}(P), & P \in B, \end{cases}$$

where $u = u_{\rm sc} + u_{\rm inc}$ is the total field.

So, for sound-hard obstacles, where $\partial u/\partial n = 0$ on S, we obtain the representation (5.67), namely

$$2u_{\rm sc}(P) = -\int_S u(q)\,\frac{\partial}{\partial n_q} G(P,q)\,{\rm d}s_q, \quad P \in B_{\rm e}, \tag{7.6}$$

and the interior relation

$$\int_S u(q)\,\frac{\partial}{\partial n_q} G(P,q)\,{\rm d}s_q = 2u_{\rm inc}(P), \quad P \in B. \tag{7.7}$$

Similar equations can be derived for sound-soft obstacles for which $u=0$ on S.

In a similar way to (7.2), we can regard (7.7) as an equation for $u(q)$. It is always uniquely solvable and it holds for all $P \in B$.

Equation (7.7) has various names. For electromagnetic problems, it is called the 'Ewald–Oseen extinction theorem' because it 'expresses the *extinction* of the incident wave... at any point inside... by interference with... the dipole field' [121, p. 102]. Other terminologies are the 'extended boundary condition' [1287, 16] and the 'extended integral equation' [77].

7.5 Null-field equations for radiation problems: one obstacle

Consider a single obstacle, so that $B \equiv B_1$ is connected and has closed boundary $S \equiv S_1$. We are interested in finding the function $u(q)$ that satisfies the interior integral relation (7.2) for all $P \in B$. Waterman [1287] noted that we only have to satisfy this equation for all P in an open region $B_- \subset B$.

> *Mathematical tractability of the extended integral equation* [i.e. (7.2)] *stems from the observation that one need only make the field vanish in any portion of the interior volume. The field can then be shown to vanish everywhere in the interior in view of the analytic continuability of solutions of differential equations of elliptic type.*
>
> (Waterman [1287, p. 805])

We shall assume here that B_- is a ball in three dimensions. Explicitly, choose an origin $O \in B$. Let S_- be the inscribed sphere to S which is centred at O, so that B_- is the largest ball in B (centred at O). Similarly, let S_+ be the smallest escribed sphere to S, centred at O, so that S_+ encloses the obstacle. The region exterior to S_+ is $B_+ \subset B_e$. See Fig. 7.2. These definitions are convenient because they enable us to use the bilinear expansion of G, Theorem 6.4, namely

$$G(P,Q) = -2ik \sum_{n,m} (-1)^m \hat{\psi}_n^m(\boldsymbol{r}_P)\, \psi_n^{-m}(\boldsymbol{r}_Q), \tag{7.8}$$

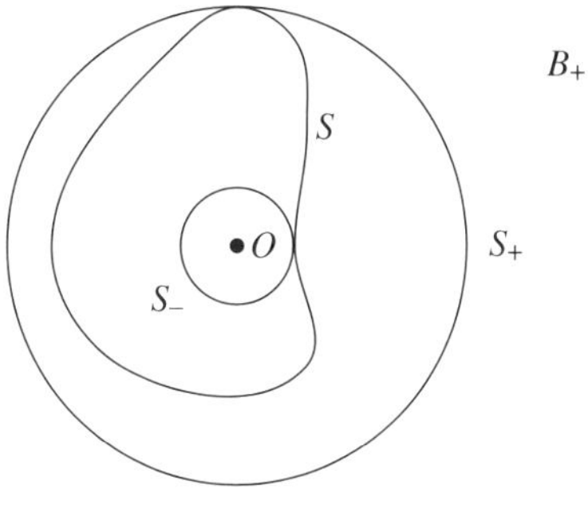

Fig. 7.2. The scatterer has an inscribed sphere S_- (with interior B_-) and a concentric escribed sphere S_+ (with exterior B_+).

where $\boldsymbol{r}_P$ is the position vector of P with respect to O, $r_P = |\boldsymbol{r}_P| < |\boldsymbol{r}_Q| = r_Q$, and we have used (4.39). Thus, when P is restricted to being in B_- (where $r_P < r_q$), we can substitute (7.8) in (7.2) to give

$$\sum_{n,m}(-1)^m \hat{\psi}_n^m(\boldsymbol{r}_P) \int_S \left\{ f(q)\, \psi_n^{-m}(\boldsymbol{r}_q) - u(q) \frac{\partial}{\partial n_q} \psi_n^{-m}(\boldsymbol{r}_q) \right\} \mathrm{d}s_q = 0 \qquad (7.9)$$

for $P \in B_-$. As the regular wavefunctions $\hat{\psi}_n^m$ are orthogonal over any sphere centred at O, it follows that each term in (7.9) must vanish, whence

$$\int_S u(q) \frac{\partial}{\partial n_q} \psi_n^m(\boldsymbol{r}_q)\, \mathrm{d}s_q = f_n^m, \qquad \begin{array}{l} n = 0, 1, 2, \dots, \\ m = -n, \dots, n, \end{array} \qquad (7.10)$$

where

$$f_n^m = \int_S f(q)\, \psi_n^m(\boldsymbol{r}_q)\, \mathrm{d}s_q \qquad (7.11)$$

are known constants. Equations (7.10) and (7.11) are known as the *null-field equations* for the exterior Neumann problem. Similar equations can be derived for the exterior Dirichlet problem.

Let X denote the infinite set of functions $\{\partial\psi_n^m/\partial n_q\}$, defined on S. Then, the null-field equations (7.10) simply give the 'moment' of $u(q)$ with respect to each function in X. Thus, $u(q)$ is to be determined from the infinite set of moment-like equations (7.10).

The null-field equations can be derived directly. Simply apply Green's second theorem (5.33) in B_e to $u(Q)$ (the solution of the exterior Neumann problem) and $\psi_n^m(\boldsymbol{r}_Q)$ (with $O \in B$). As both functions are outgoing wavefunctions, we immediately obtain (7.10) and (7.11). Some virtues of this derivation are that it is simple, it does not use analytic continuation, and it does not use the bilinear expansion of G, (7.8). A similar derivation was given by Barantsev in 1963 [62]; see also [942, 941, 1292].

The infinite set of null-field equations is equivalent to the interior integral relation (7.2) in the sense that they are both uniquely solvable for all k^2: there are no irregular frequencies. To prove uniqueness, we merely reverse Waterman's derivation of the null-field equations. Thus, suppose that $u(q)$ solves

$$\int_S u(q) \frac{\partial}{\partial n_q} \psi_n^m(\boldsymbol{r}_q)\, \mathrm{d}s_q = 0, \qquad \begin{array}{l} n = 0, 1, 2, \dots, \\ m = -n, \dots, n. \end{array}$$

Multiply each of these by $(-1)^m \hat{\psi}_n^{-m}(\boldsymbol{r}_P)$, where $P \in B_-$, and then sum over m and n, using (7.8), to give

$$U(P) \equiv \int_S u(q) \frac{\partial}{\partial n_q} G(P, q)\, \mathrm{d}s_q = 0, \qquad P \in B_-.$$

However, as U is a wavefunction that vanishes in B_-, we can use continuation arguments to assert that U vanishes everywhere in B. Theorem 6.8 then gives that $u(q) = 0$ for all $q \in S$, which proves uniqueness. We already know that the solution of the exterior Neumann problem also solves the null-field equations, so we have proved the following theorem.

Theorem 7.1 *The null-field equations for the exterior Neumann problem,* (7.10) *and* (7.11)*, are uniquely solvable for all real values of* k^2.

The same result can be proved for the exterior Dirichlet problem. Direct proofs of existence can also be given by showing that $u(q)$ solves the null-field equations if and only if $u(q)$ solves a uniquely-solvable boundary integral equation. Such equations can be constructed using modified Green's functions, as in Section 6.9.1; see [779, 781].

Having found $u(q)$ on S, somehow, the solution $u(P)$ for $P \in B_e$ can then be constructed using the integral representation (7.1). If we suppose that P is outside the circumscribed sphere S_+, we can use (7.8) in (7.1) to give the expansion

$$u(P) = \sum_{n,m} c_n^m \psi_n^m(\boldsymbol{r}_P), \quad P \in B_+, \tag{7.12}$$

where the coefficients c_n^m are given by

$$\begin{aligned} c_n^m &= -\mathrm{i}k(-1)^m \int_S \left\{ f(q)\, \hat{\psi}_n^{-m}(\boldsymbol{r}_q) - u(q) \frac{\partial}{\partial n_q} \hat{\psi}_n^{-m}(\boldsymbol{r}_q) \right\} \mathrm{d}s_q \\ &= -\mathrm{i}k \int_S \left\{ f(q)\, \overline{\hat{\psi}_n^m(\boldsymbol{r}_q)} - u(q) \frac{\partial}{\partial n_q} \overline{\hat{\psi}_n^m(\boldsymbol{r}_q)} \right\} \mathrm{d}s_q, \end{aligned}$$

using (3.91). The formula (7.12) merely states that a radiating wavefunction can be expanded as an infinite series of outgoing spherical wavefunctions, outside S_+. The assumption that (7.12) can be used everywhere in B_e, including in the region between S_+ and S, is known as the *Rayleigh hypothesis* [862].

7.6 Null-field equations for scattering problems: one obstacle

He took his Waterman in hand;
Long time the proper proof he sought;
Then rested he by the XYZ
And sat awhile in thought.

And as in inverse thought he sat
A brilliant proof, in lines of flame,
All neat and trim, it came to him.
Tangenting as it came.

(Extract from 'Plane geometry' by Emma Rounds, quoted in [333, p. 279])

Consider scattering by a sound-hard obstacle. There are two ways of obtaining null-field equations for this problem. First, we could specialise the analysis in Section 7.5, putting $f = -\partial u_{\mathrm{inc}}/\partial n$, leading to an infinite set of null-field equations for $u_{\mathrm{sc}}(q)$.

Alternatively, we can obtain null-field equations for the boundary values of the total field u, as follows.

Suppose $P \in B_-$. Then, using (7.8) in the interior integral relation (7.7), we obtain

$$u_{\text{inc}}(P) = -\mathrm{i}k \sum_{n,m} (-1)^m \hat{\psi}_n^m(\boldsymbol{r}_P) \int_S u(q) \frac{\partial}{\partial n_q} \psi_n^{-m}(\boldsymbol{r}_q)\, \mathrm{d}s_q \tag{7.13}$$

for $P \in B_-$. But, as $u_{\text{inc}}(P)$ is assumed to be a regular wavefunction in B_-, there exist coefficients d_n^m such that

$$u_{\text{inc}}(P) = \sum_{n,m} d_n^m \hat{\psi}_n^m(\boldsymbol{r}_P), \quad P \in B_-; \tag{7.14}$$

this is separation of variables inside a ball. The coefficients d_n^m can be considered as known; for example, if the incident field is a plane wave, d_n^m can be obtained from (3.69) and are given by (4.40). Equating coefficients in (7.13) and (7.14) gives

$$-\mathrm{i}k(-1)^m \int_S u(q) \frac{\partial}{\partial n_q} \psi_n^{-m}(\boldsymbol{r}_q)\, \mathrm{d}s_q = d_n^m, \quad \begin{array}{l} n = 0, 1, 2, \ldots, \\ m = -n, \ldots, n. \end{array} \tag{7.15}$$

These are the null-field equations for scattering by a sound-hard obstacle. They are uniquely solvable for all real values of k^2.

Having found $u(q)$ on S, somehow, the scattered field $u_{\text{sc}}(P)$ for $P \in B_{\text{e}}$ is given by (7.6). If P is outside S_+, we can use (7.8) to give

$$u_{\text{sc}}(P) = \sum_{n,m} c_n^m \psi_n^m(\boldsymbol{r}_P), \quad P \in B_+, \tag{7.16}$$

where the coefficients c_n^m are given by

$$c_n^m = \mathrm{i}k(-1)^m \int_S u(q) \frac{\partial}{\partial n_q} \hat{\psi}_n^{-m}(\boldsymbol{r}_q)\, \mathrm{d}s_q = \mathrm{i}k \int_S u(q) \frac{\partial}{\partial n_q} \overline{\hat{\psi}_n^m(\boldsymbol{r}_q)}\, \mathrm{d}s_q. \tag{7.17}$$

7.6.1 Literature

The three-dimensional null-field equations (7.15) were first obtained by Waterman [1288]. Bates [75] obtained the two-dimensional null-field equations. Numerical solutions for acoustic scattering by cylinders of various cross-sections can be found in [75, 113, 114, 1274, 1252, 1259]. Most applications have focussed on computing the T-matrix, which we will discuss in Section 7.9. Further remarks on solving the null-field equations themselves can be found in Section 7.8. Extension to scattering by buried obstacles (where two acoustic half-spaces are joined by a nominally flat interface and the obstacle does not intersect the interface) is made in [665].

For electromagnetic scattering, the null-field equations were first derived by Waterman [1287]. This approach has come to be called the *extended boundary condition method* (EBCM). Alternative derivations and descriptions of the EBCM can be found in [1289, 63, 1152, 881], [285, Chapter 10], [1190, Chapter 2, §8] and [870, §5.8]. Many applications can be found in the literature. Most of these are for axisymmetric scatterers, defined by

$$r = \rho(\theta), \quad 0 \le \theta \le \pi, \quad -\pi \le \phi < \pi, \tag{7.18}$$

where (r, θ, ϕ) are spherical polar coordinates and ρ is a given function of θ (only). Examples are spheroids, for which

$$\rho(\theta) = ab(a^2 \cos^2\theta + b^2 \sin^2\theta)^{-1/2},$$

and so-called 'Chebyshev particles', for which

$$\rho(\theta) = a\{1 + \varepsilon\, T_n(\cos\theta)\}.$$

Here, a, b and ε are constants, n is an integer and $T_n(x)$ is a Chebyshev polynomial of the first kind. For spheroids, see, for example, [63, 64, 65, 66, 685, 543]. Wiscombe & Mugnai [1330, 889, 1331, 890] have made extensive computations for Chebyshev particles. For osculating spheres, see [872]. The EBCM has also been used for non-axisymmetric scatterers such as triaxial ellipsoids [1076, 1075] and regular polyhedral prisms of finite length (such as cubes) [577]. For scatterers in a half-space, see [662].

The elastodynamic null-field equations were derived in [1291, 1260]. These derivations cover the scattering of elastic waves by cavities, fixed rigid obstacles, and elastic inclusions with welded interfaces. Rigid movable inclusions are treated in [936], whereas elastic inclusions with imperfect interfaces are treated in [127, 937]. Buried obstacles are considered in [132], where numerical results for scattering of a Rayleigh wave by a spheroidal cavity are given.

The null-field equations for two-dimensional water waves are derived and used in [780, 782]. The corresponding equations in three dimensions are also derived in [780].

7.7 Infinite sets of functions

Uniqueness for the null-field equations is related to the completeness of the set $X = \{\partial\psi_n^m/\partial n_q\}$; this property will be defined and discussed below. Sets of functions are also used in methods designed to solve the null-field equations. Therefore, we begin (in Section 7.7.1) by considering basic properties of infinite sets of functions. These include the following notions: complete, closed, linearly independent and minimal sets, and basis sets. Then, in Section 7.7.2, we study sets of wavefunctions, such as X, and clarify their properties; we will see that X and $W = \{\psi_n^m\}$ are both complete, linearly independent and minimal, but neither set forms a basis.

7.7.1 Basic properties

Let H be a complex separable Hilbert space. In our applications, we take $H = L^2(S)$, with inner product and norm defined by

$$(u, v) = \int_S u(q)\,\overline{v(q)}\,\mathrm{d}s_q \quad \text{and} \quad \|u\| = \sqrt{(u, u)}, \tag{7.19}$$

respectively. Let $\Phi = \{\phi_n\}$, $n = 1, 2, \ldots$, be a set of functions with each $\phi_n \in H$. We are interested in using these functions to represent (or approximate) arbitrary functions from H.

Standard textbooks discuss situations in which Φ is an orthogonal set, so that $(\phi_m, \phi_n) = 0$ when $m \neq n$. In practice, where we want to represent functions defined on a surface, we do not usually have orthogonal sets available. However, we will want Φ to have certain properties.

We begin by introducing the *linear span* of Φ, defined by

$$\operatorname{span}\Phi = \left\{ u = \sum_{n=1}^{N} \alpha_n \phi_n \ : \text{ any integer } N, \text{ any coefficients } \alpha_n \right\};$$

it is the set of functions that can be written as finite linear combinations of the functions in Φ.

Next, we want to eliminate redundancy in Φ. Thus, let

$$\Phi_N = \{\phi_1, \phi_2, \ldots, \phi_N\}$$

be a set of N functions from Φ. This set is called *linearly dependent* if there is a linear combination $\sum_{n=1}^{N} \alpha_n \phi_n = 0$ in which not all the coefficients α_n vanish; otherwise, Φ_N is *linearly independent*. If every finite set of functions from the infinite set Φ is linearly independent, then Φ is said to be **linearly independent**.

Imposing more restrictions, Φ is called **minimal** in H if no $\phi_m \in \Phi$ belongs to the closure of the linear span of all the remaining elements $\phi_n \in \Phi$. Evidently,

[Φ is minimal] implies that Φ is linearly independent.

The notion of denseness comes next, because we want to be able to represent *any* function in H. Thus, let M be a subset of H. The set M is said to be *dense* in H if, for any $u \in H$ and for any $\varepsilon > 0$, there is a $u_\varepsilon \in M$ such that $\|u - u_\varepsilon\| < \varepsilon$. Equivalently, M is dense in H if and only if for any $u \in H$, there is a sequence (u_n), with $u_n \in M$, such that $\|u - u_n\| \to 0$ as $n \to \infty$.

Suppose that M is dense in H and $u \in H$. It follows that if u is orthogonal to M, then $u = 0$. This fact involves orthogonality and density; these aspects lead to the definitions of closed and complete sets of functions, respectively.

Φ is called **closed** in H if zero is the only function in H that is orthogonal to every $\phi_n \in H$:

[$u \in H$] and [$(u, \phi_n) = 0$, $n = 1, 2, \ldots$] imply that $u = 0$.

Φ is called **complete** in H if span Φ is dense in H. Equivalently, Φ is complete in H if any $u \in H$ can be approximated by functions in Φ: for any $\varepsilon > 0$, there is an integer $N(\varepsilon)$ and a set of coefficients $a_1^N, a_2^N, \ldots, a_N^N$ such that $\|u - \sum_{n=1}^N a_n^N \phi_n\| < \varepsilon$.

It turns out that Φ is complete in H if and only if Φ is closed in H; see [243, p. 110], [285, Chapter 1, Theorem 2.1].

Given Φ, we may be able to find a *biorthonormal* set of functions $\Phi^* = \{\phi_n^*\}$ [884, p. 931], so that

$$(\phi_m, \phi_n^*) = \delta_{mn} \quad \text{for all } m \text{ and } n. \tag{7.20}$$

It is known that a biorthonormal set Φ^* exists if and only if Φ is minimal in H. Moreover, suppose that Φ is minimal in H. Then, Φ^* is *uniquely* defined if and only if Φ is also complete in H. See [251, Lemma 1.1]. Thus,

$$[\Phi \text{ is both minimal and complete] implies that exactly one } \Phi^* \text{ exists.}$$

Φ is a **Schauder basis** for H if for any $u \in H$, there is a *unique* set of coefficients $a_n(u)$ such that

$$u = \sum_{n=1}^{\infty} a_n(u)\, \phi_n; \tag{7.21}$$

here, convergence is in the norm for H, so that

$$\lim_{N\to\infty} \left\| u - \sum_{n=1}^{N} a_n(u)\, \phi_n \right\| = 0.$$

Unless stated otherwise, 'basis' means 'Schauder basis'.

The coefficient functionals $a_n(u)$ are continuous [248, Theorem 8.1]. Therefore, the Riesz representation theorem gives

$$a_n(u) = (u, \phi_n^*), \tag{7.22}$$

where $\phi_n^* \in H$ is uniquely defined. Also, it is a consequence of the uniqueness of $a_n(u)$ in (7.21) that

$$a_n(\phi_m) = \delta_{mn}. \tag{7.23}$$

Combining (7.22) and (7.23) gives (7.20). Thus,

$$[\Phi \text{ is a basis] implies that } \Phi \text{ is } \begin{cases} \text{complete, minimal} \\ \text{and linearly independent.} \end{cases}$$

Let $Y = \{\chi_n\}$, $n = 1, 2, \ldots$, be a set of functions with each $\chi_n \in H$. Suppose that Y is an *orthonormal system*, so that $(\chi_n, \chi_m) = \delta_{mn}$; it follows that Y is linearly independent. When Y is a basis, we have the (generalised) Fourier series

$$u = \sum_{n=1}^{\infty} (u, \chi_n)\, \chi_n,$$

for any $u \in H$. It is well known that a complete orthonormal system is a basis; see, for example, [248, p. 278] or [655, Theorem 1.28].

Now, let L be a bounded linear operator from H to H with a bounded inverse, and consider

$$\Phi = \{\phi_n\} \quad \text{with} \quad \phi_n = L\chi_n,$$

where $Y = \{\chi_n\}$ is an orthonormal basis. Then, Φ is a basis; in fact, it is a special kind of basis known as a **Riesz basis** [1366, p. 31]. Some characterisations of Riesz bases are given in [1366, Chapter 1, §8] and in [664, p. 2627].

7.7.2 Spherical wavefunctions

Null-field methods (usually) make use of sets of spherical wavefunctions. Four sets are often used; these are

$$W = \{\psi_n^m\}, \quad X = \{\partial\psi_n^m/\partial n_q\}, \quad \widehat{W} = \{\hat{\psi}_n^m\}, \quad \widehat{X} = \{\partial\hat{\psi}_n^m/\partial n_q\},$$

all evaluated at $q \in S$. We assume that the origin for these functions is at one point inside S.

It is well known that the sets of outgoing wavefunctions, W and X, are complete in $H = L^2(S)$, for all k^2. The sets of regular wavefunctions, $\widehat{W}$ and $\widehat{X}$, are complete in H, except when k^2 is an eigenvalue of the interior Dirichlet and Neumann problems, respectively. For proofs, see [894] and [285, Chapter III, §1.1]. These sets are also linearly independent, under the same conditions; see [631, Appendix 1] and [285, Chapter III, §2]. Moreover, these sets are also minimal, again under the same conditions on k^2 for the regular wavefunctions. These results were proved by Dallas [251]; we sketch his proof.

Let $\mathcal{A}$ be the *Neumann-to-Dirichlet map*: given f, solve the exterior Neumann problem with $\partial u/\partial n = f$ on S, and then $\mathcal{A}f = u$, the boundary values of the solution $u(P)$ on S. In particular, we have

$$\mathcal{A}_q \frac{\partial G}{\partial n_q}(P, q) = G(P, q) \quad \text{for } P \in B \tag{7.24}$$

and $\mathcal{A}(\partial\psi_n^m/\partial n_q) = \psi_n^m$.

An application of Green's theorem in B_e to any two outgoing wavefunctions gives $\mathcal{A}^* = \overline{\mathcal{A}}$ for the L^2-adjoint of $\mathcal{A}$. From (6.123), we have

$$(I + \overline{K^*} + i\eta N)\mathcal{A}f = \{S + i\eta(K - I)\}f$$

for any f and any constant η. If η is taken as real and non-zero, this formula can be taken to define $\mathcal{A}$ and $\mathcal{A}^{-1}$, for any k.

We can use $\mathcal{A}$ within the Helmholtz integral representation for outgoing wavefunctions, (5.36). For example, for the exterior Dirichlet problem with $u = g$ on S, we obtain (5.60), which we can write as

$$2u(P) = \int_S \left\{ G(P,q)\,\mathcal{A}^{-1} - \frac{\partial G}{\partial n_q}(P,q) \right\} g(q)\,\mathrm{d}s_q, \quad P \in B_\mathrm{e}.$$

In particular, we have

$$2\psi_n^m(\boldsymbol{r}_P) = \int_S \left\{ G(P,q)\,\mathcal{A}^{-1} - \frac{\partial G}{\partial n_q}(P,q) \right\} \psi_n^m(\boldsymbol{r}_q)\,\mathrm{d}s_q, \quad P \in B_\mathrm{e},$$

and, for $P \in B_\mathrm{e}$ and $P_0 \in B$,

$$2G(P,P_0) = \int_S \left\{ G(P,q)\,\mathcal{A}_q^{-1} - \frac{\partial G}{\partial n_q}(P,q) \right\} G(q,P_0)\,\mathrm{d}s_q. \tag{7.25}$$

We can obtain similar expressions for interior problems. Moreover, for $P \in B$, we can use $\mathcal{A}$ in (5.34), together with (7.24), giving

$$\begin{aligned} -2w(P) &= \int_S \left(\frac{\partial w}{\partial n} - \mathcal{A}^{-1} w \right) G(P,q)\,\mathrm{d}s_q \\ &= \int_S \left(\mathcal{A}\frac{\partial w}{\partial n} - w \right) \frac{\partial G}{\partial n_q}(P,q)\,\mathrm{d}s_q, \quad P \in B, \end{aligned}$$

which shows that regular internal wavefunctions can always be represented as single-layer or double-layer potentials. In particular, we have

$$\begin{aligned} \hat{\psi}_n^m(\boldsymbol{r}_P) &= \frac{\mathrm{i}}{2k}(-1)^m \int_S \mu_n^m(q)\,G(P,q)\,\mathrm{d}s_q, \quad P \in B, && (7.26) \\ &= \frac{\mathrm{i}}{2k}(-1)^m \int_S \nu_n^m(q)\,\frac{\partial G}{\partial n_q}(P,q)\,\mathrm{d}s_q, \quad P \in B, && (7.27) \end{aligned}$$

where the densities are given by

$$\mathcal{A}\mu_n^m = \nu_n^m(q) = \mathrm{i}k(-1)^m \left(\mathcal{A}\frac{\partial}{\partial n_q}\hat{\psi}_n^m - \hat{\psi}_n^m \right).$$

The normalisation constants have been chosen so that the following biorthonormality results hold:

$$\left(\psi_\nu^\mu, \overline{\mu_n^{-m}}\right) = \left(\partial\psi_\nu^\mu/\partial n, \overline{\nu_n^{-m}}\right) = \delta_{n\nu}\delta_{m\mu}. \tag{7.28}$$

The first equality is easily proved using properties of $\mathcal{A}$. For the second, we have

$$\begin{aligned} \int_S \frac{\partial \psi_\nu^\mu}{\partial n_q}\,\nu_n^{-m}\,\mathrm{d}s_q &= \mathrm{i}k(-1)^m \int_S \frac{\partial \psi_\nu^\mu}{\partial n_q} \left(\mathcal{A}\frac{\partial}{\partial n_q}\hat{\psi}_n^{-m} - \hat{\psi}_n^{-m} \right) \mathrm{d}s_q \\ &= \mathrm{i}k(-1)^m \int_S \left(\psi_\nu^\mu \frac{\partial}{\partial n_q}\hat{\psi}_n^{-m} - \hat{\psi}_n^{-m}\frac{\partial}{\partial n_q}\psi_\nu^\mu \right) \mathrm{d}s_q \\ &= -\mathrm{i}k(-1)^m \left[\hat{\psi}_n^{-m}, \psi_\nu^\mu \right] = \delta_{n\nu}\delta_{m\mu}, \end{aligned}$$

where we have used Green's theorem to replace S by Ω_a (a sphere of radius a, centred at the origin, where ψ_n^m is singular), and then used (6.108) and (6.109). Finally, as we have exhibited the biorthonormal sets, W^* and X^*, say, we can conclude that W and X are minimal.

In summary, the sets W and X are complete, linearly independent and minimal, for all k. However, *neither set forms a basis*, unless S is a sphere centred at the origin. This result was proved by Dallas [251]; a sketch proof is also given in [285, p. 99]. As an example, let us suppose that W is a basis. Then, we can write

$$G(q, P_0) = \sum_{n,m} G_n^m(P_0)\, \psi_n^m(\boldsymbol{r}_q), \qquad q \in S, \tag{7.29}$$

where P_0 is any point not on S, the coefficients $G_n^m(P_0)$ are given by

$$G_n^m = \left(G, \overline{\mu_n^{-m}}\right) = \int_S G(q, P_0)\, \mu_n^{-m}(q)\, \mathrm{d}s_q = -2\mathrm{i}k(-1)^m\, \hat{\psi}_n^{-m}(\boldsymbol{r}_{P_0}),$$

and we have used (7.26). The series (7.29) is supposed to converge in $L^2(S)$, and so it can be integrated term by term. Thus, if we take $P_0 \in B$, we can substitute (7.29) in (7.25), giving

$$\begin{aligned} G(P, P_0) &= -\mathrm{i}k \sum_{n,m} (-1)^m\, \hat{\psi}_n^{-m}(\boldsymbol{r}_{P_0}) \\ &\qquad \times \int_S \left\{ G(P, q)\, \mathcal{A}_q^{-1} - \frac{\partial G}{\partial n_q}(P, q) \right\} \psi_n^m(\boldsymbol{r}_q)\, \mathrm{d}s_q \\ &= -2\mathrm{i}k \sum_{n,m} (-1)^m\, \hat{\psi}_n^{-m}(\boldsymbol{r}_{P_0})\, \psi_n^m(\boldsymbol{r}_P) \end{aligned} \tag{7.30}$$

for $P \in B_\mathrm{e}$ and $P_0 \in B$. Notice that this conclusion holds for *all* $P \in B_\mathrm{e}$ and for *all* $P_0 \in B$, not just for those $P \in B_\mathrm{e}$ and $P_0 \in B$ with $|\boldsymbol{r}_P| > |\boldsymbol{r}_{P_0}|$ (for which the expansion (7.30) is assured by Theorem 6.4). So, we will obtain a contradiction by choosing $P \in B_\mathrm{e}$ and $P_0 \in B$ with $|\boldsymbol{r}_P| < |\boldsymbol{r}_{P_0}|$ (which we can always do if S is not a sphere centred at the origin) and then showing that the series on the right-hand side of (7.30) diverges for such a choice of P and P_0. But this is easily done: we have, for example,

$$\hat{\psi}_n^0(\boldsymbol{r}_{P_0})\, \psi_n^0(\boldsymbol{r}_P) \sim -\mathrm{i}(4\pi r_P)^{-1}(r_0/r_P)^n P_n(\cos\theta_P)\, P_n(\cos\theta_0) \tag{7.31}$$

as $n \to \infty$, where (r_P, θ_P) and (r_0, θ_0) are spherical polar coordinates of P and P_0, respectively; the product of Legendre polynomials decays no faster than n^{-1} as $n \to \infty$, and so the right-hand side of (7.31) is unbounded as $n \to \infty$ whenever $r_0/r_P > 1$, as assumed, which implies that the infinite series (7.30) diverges. A similar argument can be given to show that X is not a basis.

When S is a sphere centred at the origin, the properties of W and X reduce to the well-known properties of spherical harmonics $\{Y_n^m\}$: these form an orthonormal basis for $L^2(\Omega)$. For $\widehat{W}$ and $\widehat{X}$, we obtain similar results, except when ka is such that $j_n(ka)$ or $j_n'(ka)$ vanishes (where a is the radius of the sphere).

As W and X do not form bases for $L^2(S)$, they also do not form Riesz bases. This was shown numerically in [664].

7.7.3 Back to the null-field equations

Recall the null-field equations for the exterior Neumann problem, (7.10). When examining uniqueness for these equations, we consider

$$\int_S \overline{h(q)} \frac{\partial}{\partial n_q} \psi_n^m(\boldsymbol{r}_q) \, \mathrm{d}s_q = 0, \qquad \begin{array}{l} n = 0, 1, 2, \ldots, \\ m = -n, \ldots, n, \end{array}$$

where $h \in L^2(S)$. These equations state that h is orthogonal to every member of X. If they imply that $h(q) = 0$ almost everywhere on S, then X is closed and is hence complete in $L^2(S)$; see, for example, [862, 863] or [1183, §3.4]. Evidently, if we already know that X is complete, then we obtain uniqueness for the null-field equations [224]. In fact, as we saw in Section 7.7.2, the completeness of X has been proved; see [894] and [285, Chapter III, §1.1].

The completeness of vector spherical wavefunctions has been discussed in [51] and [285, Chapter VII, §1.1]. Linear independence is shown in [285, Chapter VII, §2].

7.8 Solution of the null-field equations

In this section, we shall discuss methods for solving the null-field equations. As a representative example, consider the null-field equations for scattering by a sound-hard obstacle, (7.15). The most common method for solving these equations begins by choosing a set of expansion functions, $\{\phi_n^m(q)\}$, which is suitable for representing functions defined on S; thus, suppose that we can write

$$u(q) = \sum_{\nu,\mu} u_\nu^\mu \, \phi_\nu^\mu(q), \qquad q \in S, \tag{7.32}$$

where u_ν^μ are coefficients to be found and we have used (4.39). Note that we could reorder the summation so that there is a single summation index; this is often done in the literature.

Substituting (7.32) in (7.15) gives

$$\sum_{\nu,\mu} Q_{n\nu}^{m\mu} \, u_\nu^\mu = d_n^m, \qquad \begin{array}{l} n = 0, 1, 2, \ldots, \\ m = -n, \ldots, n, \end{array} \tag{7.33}$$

where

$$Q_{n\nu}^{m\mu} = -\mathrm{i}k(-1)^m \int_S \phi_\nu^\mu \frac{\partial}{\partial n_q} \psi_n^{-m}(\boldsymbol{r}_q) \, \mathrm{d}s_q. \tag{7.34}$$

Equation (7.33) gives an infinite system of linear algebraic equations for u_ν^μ; truncating this system leads to a numerical method for solving the null-field equations.

To proceed further, we must choose a set $\{\phi_n^m\}$. In theory, we can choose any convenient basis. However, in practice, the choice may be crucial: we want the truncated system of equations to be well conditioned and to yield a good approximation to $u(q)$.

The ideal choice would be $\phi_n^m = \nu_n^m$, defined by (7.27), because of the biorthonormality relation (7.28); explicitly

$$\int_S \nu_\nu^\mu(q) \frac{\partial}{\partial n_q} \psi_n^{-m}(\boldsymbol{r}_q) \, \mathrm{d}s_q = \delta_{n\nu}\delta_{m\mu}. \tag{7.35}$$

If these conditions are satisfied, the system (7.33) will decouple, yielding $u_n^m = (\mathrm{i}/k)(-1)^m d_n^m$. However, we do not have ready access to the functions $\{\nu_n^m\}$ that satisfy the relations (7.35) for a given boundary S (the exception being when S is a sphere centred at O; see Section 7.8.3). Computationally, it is probably not worthwhile to determine ν_n^m [77, p. 57]. However, it may be possible to choose ν_n^m so that the relations (7.35) are 'almost' satisfied, giving a system of equations (7.33) that is diagonally-dominant.

Most authors have chosen wavefunctions for $\phi_n^m(q)$; there are eight obvious choices, namely,

$$\psi_n^m, \quad \partial\psi_n^m/\partial n_q, \quad \hat{\psi}_n^m, \quad \partial\hat{\psi}_n^m/\partial n_q \tag{7.36}$$

and their complex conjugates. As we saw in Section 7.7.2, those involving ψ_n^m are known to give complete sets, whereas those involving $\hat{\psi}_n^m$ do not. For example, $\{\hat{\psi}_n^m\}$ is not complete whenever k^2 is an eigenvalue of the interior Dirichlet problem for B. Thus, the use of regular wavefunctions reintroduces the difficulties at irregular frequencies. Nevertheless, regular wavefunctions do offer some advantages; see Section 7.8.1. The choice $\overline{\partial\psi_n^{-m}/\partial n_q}$ is interesting because it also arises when the underlying boundary-value problem is solved approximately by the method of least squares; see Section 7.8.2. A third choice, discussed briefly in Section 7.8.3, is to use spherical harmonics Y_n^m.

Many other choices could be made. Waterman [1293] has given a good discussion of the relative merits of the sets (7.36) (for the corresponding electromagnetic wavefunctions); see also [1294, 1295]. In practice, it is usual to use 'global bases', meaning sets of functions that are not identically zero over any portion of S. One could use 'local bases', just as element-based methods are commonly used to solve boundary integral equations, but this does not seem to be advantageous [1271].

7.8.1 Regular wavefunctions

Following Waterman [1288], take $\phi_n^m(q) = \hat{\psi}_n^m(\boldsymbol{r}_q)$, so that

$$Q_{n\nu}^{m\mu} = -\mathrm{i}k(-1)^m \int_S \hat{\psi}_\nu^\mu(\boldsymbol{r}_q) \frac{\partial}{\partial n_q} \psi_n^{-m}(\boldsymbol{r}_q) \, \mathrm{d}s_q. \tag{7.37}$$

It turns out that this Q-matrix is symmetric if S has certain properties. To show this, we first introduce

$$\widetilde{Q}_{n\nu}^{m\mu} = -\mathrm{i}k(-1)^m \int_S \psi_n^{-m}(\boldsymbol{r}_q) \frac{\partial}{\partial n_q} \hat{\psi}_\nu^\mu(\boldsymbol{r}_q)\,\mathrm{d}s_q,$$

which arises in the sound-soft problem. An application of Green's theorem in the region between S and S_- gives

$$Q_{n\nu}^{m\mu} - \widetilde{Q}_{n\nu}^{m\mu} = -\mathrm{i}k(-1)^m \left[\hat{\psi}_\nu^\mu, \psi_n^{-m}\right] = \delta_{n\nu}\,\delta_{m\mu},$$

where we have used (6.108) (with $S_- = \Omega_a$) and (6.109). We also have

$$Q_{n\nu}^{m\mu} + \widetilde{Q}_{n\nu}^{m\mu} = -\mathrm{i}k(-1)^m \int_S \frac{\partial}{\partial n_q} \left\{\psi_n^{-m}(\boldsymbol{r}_q)\,\hat{\psi}_\nu^\mu(\boldsymbol{r}_q)\right\}\,\mathrm{d}s_q,$$

whence

$$2Q_{n\nu}^{m\mu} = \delta_{n\nu}\,\delta_{m\mu} - \mathrm{i}k(-1)^m \int_S \frac{\partial}{\partial n_q} \left\{\psi_n^{-m}(\boldsymbol{r}_q)\,\hat{\psi}_\nu^\mu(\boldsymbol{r}_q)\right\}\,\mathrm{d}s_q. \tag{7.38}$$

Suppose that S is star-shaped with respect to O, so that S is given by $r = \rho(\theta, \phi)$, where (r, θ, ϕ) are spherical polar coordinates and ρ is a given function; see Section 5.5.3. Then, using (5.29) and (5.30), we obtain

$$\int_S \frac{\partial}{\partial n_q} \left\{\psi_n^{-m}\,\hat{\psi}_\nu^\mu\right\}\,\mathrm{d}s_q = \int_\Omega \Big\{\rho^2 \Phi'_{\nu n}(\rho) - \Phi_{\nu n}(\rho)\Big(\rho_\theta \frac{\partial}{\partial\theta} + \frac{\rho_\phi}{\sin^2\theta}\frac{\partial}{\partial\phi}\Big)\Big\}\, Y_n^{-m}\, Y_\nu^\mu\,\mathrm{d}\Omega,$$

where $\rho_\theta = \partial\rho/\partial\theta$, $\rho_\phi = \partial\rho/\partial\phi$ and $\Phi_{\nu n}(r) = j_\nu(kr)\,h_n(kr)$. Then, the linearisation formula for spherical harmonics, Theorem 3.20, gives

$$\int_S \frac{\partial}{\partial n} \left\{\psi_n^{-m}\,\hat{\psi}_\nu^\mu\right\}\,\mathrm{d}s = \sum_q \mathcal{G}(n, -m; \nu, \mu; q)\left(I_{\nu n q}^{\mu-m} - J_{\nu n q}^{\mu-m}\right), \tag{7.39}$$

where $\mathcal{G}$ is a Gaunt coefficient,

$$I_{\nu n q}^{m} = \int_\Omega \rho^2 \Phi'_{\nu n}(\rho)\, Y_q^m\,\mathrm{d}\Omega$$

and

$$J_{\nu n q}^{m} = \int_\Omega \Phi_{\nu n}(\rho) \left(\rho_\theta \frac{\partial}{\partial\theta} + \frac{\rho_\phi}{\sin^2\theta}\frac{\partial}{\partial\phi}\right) Y_q^m\,\mathrm{d}\Omega.$$

Next, we remove the θ and ϕ derivatives from this formula. Define

$$\Psi_{\nu n}(r) = \int_{r_0}^{r} \Phi_{\nu n}(t)\,\mathrm{d}t, \tag{7.40}$$

where r_0 is a positive constant, so that $\Psi'_{\nu n} = \Phi_{\nu n}$. Then

$$\begin{aligned} J^m_{\nu nq} &= \int_{-\pi}^{\pi}\int_0^{\pi}\Big\{\Big(\frac{\partial}{\partial\theta}\Psi_{\nu n}(\rho)\Big)\sin\theta\frac{\partial}{\partial\theta} \\ &\quad + \Big(\frac{\partial}{\partial\phi}\Psi_{\nu n}(\rho)\Big)\frac{1}{\sin\theta}\frac{\partial}{\partial\phi}\Big\}Y^m_q\,\mathrm{d}\theta\,\mathrm{d}\phi \\ &= -\int_{-\pi}^{\pi}\int_0^{\pi}\Psi_{\nu n}(\rho)\Big\{\frac{\partial}{\partial\theta}\Big(\sin\theta\frac{\partial Y^m_q}{\partial\theta}\Big)+\frac{1}{\sin\theta}\frac{\partial^2 Y^m_q}{\partial\phi^2}\Big\}\,\mathrm{d}\theta\,\mathrm{d}\phi \\ &= q(q+1)\int_\Omega \Psi_{\nu n}(\rho)\,Y^m_q\,\mathrm{d}\Omega, \end{aligned}$$

where we have integrated by parts (there are no contributions from the integrated terms) and used (3.18).

Suppose now that

$$\rho(\theta,\phi) = \rho(\theta,-\phi), \tag{7.41}$$

so that S is symmetric about the plane $y = 0$. Then, we can show that

$$I^{-m}_{\nu nq} = (-1)^m I^m_{\nu nq} \quad \text{and} \quad J^{-m}_{\nu nq} = (-1)^m J^m_{\nu nq}.$$

Furthermore, using (7.38), (7.39) and the symmetry relation (3.75), we obtain

$$Q^{m\mu}_{n\nu} - Q^{\mu m}_{\nu n} = \frac{k(-1)^m}{2\mathrm{i}\sqrt{\pi}}\sum_q \mathcal{G}(n,-m;\nu,\mu;q)\,A^{\mu-m}_q\,L^{\mu-m}_{n\nu q}, \tag{7.42}$$

where A^m_n is the normalisation constant in the definition of Y^m_n (see just above (3.12)),

$$L^m_{n\nu q} = \int_0^{\pi}\int_0^{\pi} K^q_{n\nu}(\rho)\,P^m_q(\cos\theta)\,\cos m\phi\,\sin\theta\,\mathrm{d}\theta\,\mathrm{d}\phi$$

and

$$K^q_{n\nu}(\rho) = \rho^2\,(\Phi'_{\nu n} - \Phi'_{n\nu}) - q(q+1)\,(\Psi_{\nu n} - \Psi_{n\nu}). \tag{7.43}$$

We are going to show that the right-hand side of (7.42) vanishes for a certain class of surfaces. Thus, let us assume that S has *three* orthogonal symmetry planes, augmenting (7.41) to

$$\rho(\theta,\phi) = \rho(\theta,-\phi) = \rho(\theta,\pi-\phi) = \rho(\pi-\theta,\phi). \tag{7.44}$$

Also, for any function $F(\theta,\phi)$, we have

$$\begin{aligned} \int_0^{\pi}\int_0^{\pi} F(\theta,\phi)\,\mathrm{d}\theta\,\mathrm{d}\phi &= \int_0^{\pi/2}\int_0^{\pi/2}\{F(\theta,\phi)+F(\pi-\theta,\phi) \\ &\quad + F(\theta,\pi-\phi)+F(\pi-\theta,\pi-\phi)\}\,\mathrm{d}\theta\,\mathrm{d}\phi. \end{aligned}$$

So, as $\cos m(\pi-\phi)=(-1)^m\cos m\phi$ and $P_n^m(-x)=(-1)^{n+m}P_n^m(x)$, we find that

$$L_{n\nu q}^{\mu-m}=[1+(-1)^q]\left[1+(-1)^{m+\mu}\right]$$
$$\times\int_{\Omega_8}K_{n\nu}^q(\rho)\,P_q^{\mu-m}(\cos\theta)\,\cos\left[(\mu-m)\phi\right]\mathrm{d}\Omega,$$

where Ω_8 is the first octant of Ω, defined by $0\le\theta\le\frac{1}{2}\pi$ and $0\le\phi\le\frac{1}{2}\pi$. Thus, $L_{n\nu q}^{m-\mu}$ vanishes unless both q and $(m+\mu)$ are even. As q runs from $|n-\nu|$ to $(n+\nu)$ in steps of 2 (see Theorem 3.20), we can assume that

$$\nu=n+2\ell \text{ and } q=2(s+\ell) \text{ with } \ell\ge 1 \text{ and } s=0,1,2,\ldots n.$$

With these values,

$$Q_{n\nu}^{m\mu}-Q_{\nu n}^{\mu m}=\frac{k(-1)^m}{2\mathrm{i}}\sum_q \mathcal{G}(n,-m;\nu,\mu;q)\int_\Omega K_{n\nu}^q(\rho)\,Y_q^{\mu-m}\,\mathrm{d}\Omega, \tag{7.45}$$

and then we are interested in evaluating $K_{n,n+2\ell}^{2s+2\ell}$. We have

$$\Phi_{n+2\ell,n}(\rho)-\Phi_{n,n+2\ell}(\rho)=\mathrm{i}W_{2\ell}^n(k\rho), \tag{7.46}$$

where

$$W_\ell^n(w)=j_{n+\ell}(w)\,y_n(w)-j_n(w)\,y_{n+\ell}(w). \tag{7.47}$$

Some properties of this function are collected next.

Lemma 7.2 *The function $W_\ell^n(w)$, defined by* (7.47), *satisfies the recurrence relation*

$$W_{\ell+1}^n(w)=(2n+2\ell+1)w^{-1}W_\ell^n(w)-W_{\ell-1}^n(w) \tag{7.48}$$

for $\ell\ge 1$, with $W_0^n=0$ and $W_1^n(w)=w^{-2}$. The function $W_{2\ell}^n(w)$ is a polynomial in w^{-1}, given by

$$W_{2\ell}^n(w)=\sum_{j=1}^{\ell}\alpha_j(\ell,n)\,w^{-2j-1}, \tag{7.49}$$

where the coefficients $\alpha_j(\ell,n)$ can be found using (7.48).

Proof Equation (7.48) follows from $z_{n+1}(w)=(2n+1)w^{-1}z_n(w)-z_{n-1}(w)$, where z_n is any spherical Bessel function. The form of (7.49) follows immediately. In fact, we have

$$W_{2\ell}^n(w)=(-1)^{\ell+1}w^{-1}\Big\{P(n+\frac{1}{2},w)\,Q(n+2\ell+\frac{1}{2},w)$$
$$-P(n+2\ell+\frac{1}{2},w)\,Q(n+\frac{1}{2},w)\Big\},$$

where P and Q are the polynomials in w^{-1} defined on p. 437 of [1], whereas the small-w asymptotic approximations, $j_n(w)=O(w^n)$ and $y_n(w)=O(w^{-n-1})$ as $w\to 0$, show that the summation in (7.49) stops at $j=\ell$. □

Substituting (7.49) in (7.43), using (7.40) and (7.46), we obtain

$$K^q_{n,n+2\ell}(\rho) = \frac{1}{\mathrm{i}k}\sum_{j=1}^{\ell}\frac{\alpha_j(\ell,n)}{2j}\{2j(2j+1)-q(q+1)\}(k\rho)^{-j}, \tag{7.50}$$

where $q = 2s + 2\ell$ and $s \geq 0$. Note that the expression inside the braces vanishes when $j = \ell$ and $q = 2\ell$. Note also that the contribution from r_0 in (7.40) can be discarded because additive constants in $K^q_{n\nu}$ integrate to zero when substituted in (7.45).

Now, suppose that S is an *ellipsoid*, satisfying (7.44), given by

$$(x/a)^2 + (y/b)^2 + (z/c)^2 = 1. \tag{7.51}$$

It follows that the corresponding $\rho(\theta, \phi)$ is given by

$$\rho^{-2} = \sum_{N=0}^{1}\sum_{M=-N}^{N} e^{(1)}_{NM}\, Y^{2M}_{2N}(\theta, \phi),$$

whence

$$\rho^{-2j} = \sum_{N=0}^{j}\sum_{M=-N}^{N} e^{(j)}_{NM}\, Y^{2M}_{2N}(\theta, \phi)$$

for some coefficients $e^{(j)}_{NM}$. Thus, $K^q_{n,n+2\ell}$ has a spherical-harmonic expansion involving Y^M_N with $N \leq 2\ell$, whereas the integrand in (7.45) involves Y^m_q with $q \geq 2\ell$. Hence, the orthogonality of spherical harmonics shows that the integral in (7.45) vanishes when $q > 2\ell$, whereas the integrand vanishes when $j = \ell$ and $q = 2\ell$ (see the remark below (7.50)). Thus, we have proved the following result.

Theorem 7.3 *Let S be an ellipsoid, defined by* (7.51). *Then, the Q-matrix with entries $Q^{m\mu}_{n\nu}$, defined by* (7.37), *is symmetric, so that*

$$Q^{m\mu}_{n\nu} = Q^{\mu m}_{\nu n}.$$

This result and its proof are due to Waterman [1288]. He also proved the corresponding result in two dimensions for ellipses. Kvyatkovskii [677] has examined the loss of symmetry in the Q-matrix when S is approximately spheroidal.

7.8.2 Outgoing wavefunctions

Outgoing wavefunctions may be used for ϕ^m_n. One choice is

$$\phi^m_n(q) = \mathrm{i}(-1)^m \overline{\partial\psi^{-m}_n/\partial n_q},$$

whence (7.34) gives

$$Q^{m\mu}_{n\nu} = k\int_S \phi^\mu_\nu(q)\,\overline{\phi^m_n(q)}\,\mathrm{d}s_q = \overline{Q^{\mu m}_{\nu n}}, \tag{7.52}$$

so that this Q-matrix is Hermitian.

The matrix (7.52) also arises when the boundary-value problem is solved approximately by the *method of least squares*. To see this, suppose that

$$u(P) \simeq u_N(P) = u_{\text{inc}}(P) + \sum_{n=0}^{N} \sum_{m=-n}^{n} c_n^m(N)\, \psi_n^m(\boldsymbol{r}_P), \qquad P \in B_{\text{e}}, \tag{7.53}$$

so that we have truncated the expansion (7.16). Note that (7.16) is an infinite series, convergent in B_+; usually, (7.16) does not converge for all $P \in B_{\text{e}}$ (Rayleigh hypothesis). On the other hand, (7.53) is a *finite* series, and we do not consider the limit $N \to \infty$. Instead, we determine the coefficients $c_n^m(N)$ by minimising

$$\int_S \left| \frac{\partial u_N(q)}{\partial n_q} \right|^2 \mathrm{d}s_q,$$

which means that the boundary condition on the sound-hard surface S is satisfied in a least-squares sense. This is a standard procedure [851, p. 492]. Thus, we find that $c_n^m(N)$ must satisfy

$$\sum_{\nu=0}^{N} \sum_{\mu=-\nu}^{\nu} (-1)^\mu\, c_\nu^{-\mu}(N)\, Q_{\nu n}^{\mu m} = \mathrm{i}k \int_S \phi_n^m \frac{\partial u_{\text{inc}}}{\partial n_q} \, \mathrm{d}s_q, \qquad \begin{array}{l} n = 0, 1, 2, \dots, N, \\ m = -n, \dots, n, \end{array}$$

where $Q_{n\nu}^{m\mu}$ is defined by (7.52). This method has been used extensively for a variety of scattering problems; see, for example, [839, 528, 862, 315, 1266, 1237], [851, Chapter X] and [353, Chapter 2]

7.8.3 Spherical harmonics

The expansion functions $\phi_n^m(q)$ do not have to be wavefunctions. For example, if S is star-shaped, we can write (7.34) as an integral over the unit sphere Ω, and then use spherical harmonics. Thus, if S is given by $r = \rho(\theta, \phi) \equiv \rho(\hat{\boldsymbol{r}})$, with $\hat{\boldsymbol{r}} \in \Omega$, we can use (5.29) and (5.30) to obtain

$$\begin{aligned} Q_{n\nu}^{m\mu} = -\mathrm{i}k \int_\Omega \tilde{\phi}_\nu^\mu(\hat{\boldsymbol{r}}) \Big\{ & \rho^2 k\, h_n'(k\rho)\, \overline{Y_n^m} \\ & - h_n(k\rho) \Big(\rho_\theta \frac{\partial}{\partial \theta} + \frac{\rho_\phi}{\sin^2\theta} \frac{\partial}{\partial \phi} \Big) \overline{Y_n^m} \Big\} \, \mathrm{d}\Omega, \end{aligned}$$

where $\tilde{\phi}_n^m(\hat{\boldsymbol{r}}) = \phi_n^m(\rho(\hat{\boldsymbol{r}})\hat{\boldsymbol{r}}) = \phi_n^m(q)$. Then, it is natural to choose $\tilde{\phi}_n^m = Y_n^m$, given that much is known about approximating functions defined on Ω; for references, see [1117, 430] and [49, §5.5].

Numerical results obtained using Y_n^m have been published. Thus, acoustic scattering by prolate spheroids has been considered in [1177]. Boström [129] has computed far-field results for acoustic scattering by elastic 'superspheroids', defined by (7.18) with

$$\rho(\theta) = ab\,(a^p \cos^p\theta + b^p \sin^p\theta)^{-1/p},$$

where p is a parameter; $p = 2$ gives a spheroid. He compared his results with those obtained using a similar method [128] in which regular wavefunctions were used. He found that regular wavefunctions were better for spheroids ($p = 2$) but spherical harmonics were better for $p = 3$. Here, 'better' means more stable results with respect to truncation. The use of vector spherical harmonics for electromagnetic problems has been explored in [1154].

Note that if S is a sphere of radius a, centred at O, we have $\rho = a$ and

$$Q_{n\nu}^{m\mu} = -\mathrm{i}(ka)^2 h_n'(ka)\,\delta_{n\nu}\,\delta_{m\mu},$$

so that the Q-matrix is diagonal, whence (7.32) and (7.33) give

$$u(q) \equiv u(a, \theta, \phi) = \frac{\mathrm{i}}{(ka)^2} \sum_{n,m} \frac{d_n^m}{h_n'(ka)}\, Y_n^m(\theta, \phi),$$

which is (4.43); this shows that the null-field method is a generalisation of the method of separation of variables (in spherical polar coordinates), as described in Chapter 4.

Ramm [1001] has shown that the set of functions $\phi_n^m(q) = Y_n^m(\hat{\boldsymbol{r}})$ forms a Riesz basis for $L^2(S)$ when S is star-shaped; see also [1005, p. 177].

7.8.4 Numerical experience and convergence

> *One present* [i.e. in 1965] *drawback of the method is the poor numerical convergence of the truncation procedure which has been found when dealing with more elongated shapes. This is of course not surprising, since one is departing from the nearly spherical shapes most ideally suited to the* [spherical] *vector wave functions employed.*
>
> (Waterman [1287, p. 811])

In his 1965 paper on electromagnetic scattering, Waterman [1287] used outgoing spherical vector wavefunctions for ϕ_n^m; the scalar equivalent would be to use ψ_n^m. In his 1969 paper on acoustic scattering [1288], he showed that it is better to use regular spherical wavefunctions $\hat{\psi}_n^m$, because this choice leads to a symmetric Q-matrix if S is an ellipse in two dimensions or an ellipsoid in three dimensions (Theorem 7.3). As he emphasised in his 1999 paper, this offers

> *a tremendous advantage: in the* [low-frequency] *Rayleigh limit, for example, what are apparently large numerical values below the diagonal are replaced by their small counterparts above the diagonal. Note that this replacement* must *be carried out to avoid serious precision problems.*
>
> (Waterman [1295, p. 2970])

To understand this difficulty, observe that, for fixed w,

$$h_n(w) \sim \left(\frac{2}{w}\right)^n \frac{\Gamma(n+\frac{1}{2})}{\mathrm{i}w\sqrt{\pi}} \quad \text{and} \quad j_n(w) \sim \left(\frac{w}{2}\right)^n \frac{\sqrt{\pi}}{2\,\Gamma(n+\frac{3}{2})}$$

as $n \to \infty$, so that $h_n(w)$ grows rapidly with n whereas $j_n(w)$ decays rapidly. Thus, the matrix $Q_{n\nu}^{m\mu}$, defined by (7.34), seems to grow as $n \to \infty$. Also, Waterman's $\mathcal{Q}_{n\nu}^{m\mu}$, defined by (7.37), seems to grow as $n \to \infty$ but seems to decay as $\nu \to \infty$; however, the growth with n is illusory when S is an ellipsoid because we know then that $\mathcal{Q}_{n\nu}^{m\mu} = \mathcal{Q}_{\nu n}^{\mu m}$, and this fact can be used to construct a diagonally-dominant $\mathcal{Q}$-matrix. This explains why regular wavefunctions should be used for ϕ_n^m for ellipsoids of *any* eccentricity.

If S is not an ellipsoid, other methods may help to alleviate the rapid growth of $h_n(kr)$. One way is to extract a factor that grows with n, writing, say,

$$h_n(w) = \left(\frac{2}{w}\right)^n \frac{\Gamma(n+\frac{1}{2})}{\mathrm{i}w\sqrt{\pi}} \mathcal{H}_n(w),$$

and then rewriting the null-field equations in terms of $\mathcal{H}_n$; see [1271].

Another way is to replace the spherical wavefunction ψ_n^m in (7.34) by a *spheroidal* wavefunction. This method was introduced by Bates & Wall [77]. They implemented it for two-dimensional problems, using elliptical wavefunctions (see Section 2.9). Subsequently, spheroidal wavefunctions were used by Hackman [461]; extensions to elastodynamics and electromagnetism are described in [463] and [578], respectively. The use of elliptical or spheroidal wavefunctions seems to give a Q-matrix that is better conditioned when S is elongated, and this leads to a better numerical scheme. The precise reason for this has not been investigated, but, intuitively, it is because the region in which the interior integral relation (7.7) is imposed 'explicitly' (rather than by analytic continuation) has been enlarged from the inscribed ball B_- to an inscribed spheroid. This idea has motivated several variants of the null-field method, such as an iterative form in which the interior B is partitioned into several overlapping subdomains [544, 680]. For additional information and comparisons, see [1254, 1336], [285, p. 100 & p. 267], [866, p. 159] and [870, §5.8.4].

Yet another way to improve the conditioning of the Q-matrix is simply to compute with higher precision, thus reducing the rounding errors by carrying more digits in the floating-point arithmetic. This 'brute-force' method was shown to be very effective by Mishchenko & Travis [869], who used complex arithmetic with approximately 31 decimal digits; see also [871, p. 544], [866, p. 160] and [870, p. 150].

7.8.4.1 Analysis of convergence

In a numerical scheme, we truncate (7.32), and define

$$u_N(q) = \sum_{\nu=0}^{N} \sum_{\mu=-\nu}^{\nu} u_\nu^\mu(N)\, \phi_\nu^\mu(q). \tag{7.54}$$

The coefficients $u_\nu^\mu(N)$ (which depend on N, in general) are to be found by solving

$$\sum_{\nu=0}^{N} \sum_{\mu=-\nu}^{\nu} Q_{n\nu}^{m\mu} u_\nu^\mu(N) = d_n^m, \qquad \begin{array}{l} n = 0, 1, 2, \ldots, N, \\ m = -n, \ldots, n, \end{array} \tag{7.55}$$

where d_n^m are known constants and $Q_{n\nu}^{m\mu}$ is defined by (7.34).

According to Kristensson *et al.* [664] and Ramm [1005, p. 146], justification of the above numerical scheme requires positive answers to the following questions.

Q1: Is (7.55) solvable for sufficiently large N?

Q2: Does $u_N(q) \to u(q)$ as $N \to \infty$, where convergence is with respect to the norm in $L^2(S)$ and $u(q)$ is the exact solution?

Q3: Does $u_\nu^\mu(N) \to u_\nu^\mu$ as $N \to \infty$, and is the convergence uniform with respect to ν and μ?

Q4: Does the equality (7.32) hold, where u_ν^μ are defined by Q3 (assuming that the limits exist)?

Q5: Is the numerical scheme based on (7.55) stable with respect to small perturbations of d_n^m and of the matrix entries $Q_{n\nu}^{m\mu}$?

In [664] and [1005, Chapter IV, §§2.4–2.6], sufficient conditions are given which ensure that all five questions have positive answers. These conditions include the requirement that the set $X = \{\partial\psi_n^m/\partial n_q\}$ be a Riesz basis. However, we know that X is not a Riesz basis, and so the analysis in [664, 1005] does not shed much light on the numerical behaviour of the null-field method [251]: 'One can prove that the convergence criteria of Kristensson, Ramm, and Ström [664] are never fulfilled if the obstacle is nonspherical and the spherical wavefunctions have their usual normalization' [1057].

The extension of the analysis in [664, 1005] to electromagnetic scattering is given in [284] and [285, Chapter X].

7.8.5 A connection with boundary integral equations

Suppose that k^2 is not an eigenvalue of the interior Dirichlet problem, so that the Helmholtz integral equation (5.68) is uniquely solvable. A general method for solving such equations is the *Petrov–Galerkin method* ([49, §3.1.2], [616, §3.2]); this is also called the *method of moments*. It begins by substituting the approximation (7.54) in (5.68). Then, determine the coefficients u_ν^μ by solving

$$\left((I+\overline{K^*})u_N - 2u_{\text{inc}},\, \chi_n^m\right) = 0, \quad n = 0, 1, 2, \ldots, N, \quad m = -n, \ldots, n,$$

where the test functions $\{\chi_n^m(q)\}$ are chosen and we have used the inner product in $L^2(S)$, (7.19). (The choice $\chi_n^m = \psi_n^m$ gives Galerkin's method.) Explicitly, we have the system

$$\sum_{\nu=0}^{N}\sum_{\mu=-\nu}^{\nu} A_{n\nu}^{m\mu}\, u_\nu^\mu = b_n^m, \qquad \begin{array}{l} n = 0, 1, 2, \ldots, N \\ m = -n, \ldots, n, \end{array}$$

where

$$A_{n\nu}^{m\mu} = \left((I+\overline{K^*})\phi_\nu^\mu,\, \chi_n^m\right) \quad \text{and} \quad b_n^m = 2(u_{\text{inc}},\, \chi_n^m). \tag{7.56}$$

The Petrov–Galerkin method is a well-studied projection method. For example, necessary and sufficient conditions on ϕ_n^m and χ_n^m can be found which guarantee that $u_N \to u$ as $N \to \infty$; see [651, §16.2].

Now, we are going to make a specific choice for χ_n^m [783, 795]. Write

$$\psi_n^{-m}(\boldsymbol{r}_P) = \frac{\mathrm{i}}{2k}(-1)^m \int_S \tau_n^m(q)\, G(P,q)\, \mathrm{d}s_q, \quad P \in B_\mathrm{e}, \tag{7.57}$$

where τ_n^m is a source density; cf. (7.26). As k^2 is not an eigenvalue of the interior Dirichlet problem, we could determine τ_n^m by solving the integral equation

$$-2\mathrm{i}k(-1)^m \frac{\partial}{\partial n_p} \psi_n^{-m}(\boldsymbol{r}_p) = \tau_n^m(p) + \int_S \tau_n^m(q) \frac{\partial}{\partial n_p} G(p,q)\, \mathrm{d}s_q \tag{7.58}$$

for $p \in S$, which is obtained by calculating the normal derivative of (7.57) at $p \in S$; it is a special case of the indirect integral equation (5.51). Also, use of (7.8) in (7.57) for $P \in B_+$ shows that we have biorthonormality:

$$\left(\hat{\psi}_\nu^\mu, \overline{\tau_n^m}\right) = \delta_{n\nu}\, \delta_{m\mu}. \tag{7.59}$$

Next, we make the choice $\chi_n^m = \overline{\tau_n^m}$ in (7.56). Changing the order of integration, using (5.10) and (7.58), and the symmetry of G gives $A_{n\nu}^{m\mu} = 2Q_{n\nu}^{m\mu}$, where $Q_{n\nu}^{m\mu}$ is defined by (7.34). Combining (7.14) and (7.59) gives $b_n^m = 2d_n^m$. Hence, solving the Helmholtz integral equation using a particular Petrov–Galerkin method leads to precisely the equations, (7.33), obtained when we solved the null-field equations.

This connection does not seem to have any practical value, but it may lead to a numerical analysis of the null-field method. It can be modified to accommodate irregular frequencies by using modified integral equations; see Section 6.10.

7.8.6 The scattered field

In the region outside S_+, the scattered field is given by (7.16) in terms of the coefficients c_n^m, defined by (7.17). Substituting (7.32) gives

$$c_n^m = -\sum_{\nu,\mu} \widehat{Q}_{n\nu}^{m\mu} u_\nu^\mu, \tag{7.60}$$

where

$$\widehat{Q}_{n\nu}^{m\mu} = -\mathrm{i}k(-1)^m \int_S \phi_\nu^\mu \frac{\partial}{\partial n_q} \hat{\psi}_n^{-m}(\boldsymbol{r}_q)\, \mathrm{d}s_q; \tag{7.61}$$

this formula should be compared with (7.34).

7.9 The T-matrix for one obstacle

Recall the two expansions (7.14) and (7.16), which we rewrite here:

$$u_{\text{inc}}(P) = \sum_{n,m} d_n^m \, \hat{\psi}_n^m(\boldsymbol{r}_P), \quad P \in B_-, \tag{7.62}$$

$$u_{\text{sc}}(P) = \sum_{n,m} c_n^m \, \psi_n^m(\boldsymbol{r}_P), \quad P \in B_+. \tag{7.63}$$

As the underlying problem is linear, we must be able to write

$$c_n^m = \sum_{\nu,\mu} T_{n\nu}^{m\mu} \, d_\nu^\mu \tag{7.64}$$

for some (infinite) matrix $\mathbf{T}$ with entries $T_{n\nu}^{m\mu}$. This is usually called the *T-matrix.*

We may use the null-field method to calculate $\mathbf{T}$ (although other methods could be used). From (7.33), we have

$$\sum_{\nu,\mu} Q_{n\nu}^{m\mu} \, u_\nu^\mu = d_n^m, \tag{7.65}$$

where $Q_{n\nu}^{m\mu}$ is defined by (7.34),

$$u(q) = \sum_{\nu,\mu} u_\nu^\mu \, \phi_\nu^\mu(q), \quad q \in S,$$

and $\{\phi_n^m(q)\}$ is a set of basis functions. Similarly, the coefficients c_n^m are related to u_ν^μ by (7.60).

Let us use an obvious matrix notation and write (7.65) and (7.60) as

$$\mathbf{Qu} = \mathbf{d} \quad \text{and} \quad \mathbf{c} = -\widehat{\mathbf{Q}}\mathbf{u},$$

respectively. Eliminating $\mathbf{u}$ gives (7.64), which we write as

$$\mathbf{c} = \mathbf{Td} \quad \text{with} \quad \mathbf{T} = -\widehat{\mathbf{Q}}\mathbf{Q}^{-1}. \tag{7.66}$$

The first of these equations is the *definition* of the T-matrix, whereas the second gives a prescription for calculating the T-matrix.

The T-matrix characterises the scattering properties of the obstacle: it depends on the scatterer's shape and other properties (boundary condition or internal properties) and on the frequency, but not on the incident field (which is represented by the coefficients d_n^m in $\mathbf{d}$). This makes the T-matrix useful as a 'building block' for multiple-scattering problems, as we shall see.

7.9.1 Properties of the T-matrix

It is clear that the unique-solvability of the underlying boundary-value problem implies that the T-matrix exists and is unique. This in turn implies that the T-matrix must be independent of the choice of $\{\phi_n^m(q)\}$. However, this choice may be important in numerical calculations, when $\mathbf{T}$ must necessarily be truncated.

The T-matrix must also satisfy additional constraints. These follow from considerations of reciprocity and energy, and they are contained in the next theorem.

Theorem 7.4 *For a bounded three-dimensional scatterer, we have*

$$T_{n\nu}^{m\mu} = (-1)^{m+\mu} T_{\nu n}^{-\mu,-m} \tag{7.67}$$

and

$$T_{n\nu}^{m\mu} + \overline{T_{\nu n}^{\mu m}} + 2 \sum_{L,M} \overline{T_{Ln}^{Mm}}\, T_{L\nu}^{M\mu} = 0. \tag{7.68}$$

Proof Consider scattering by a single three-dimensional obstacle with boundary S. We consider two different incident fields, u_{inc} and $\tilde{u}_{\text{inc}}$, with corresponding scattered and total fields. Suppose that we can write

$$u(\boldsymbol{r}_P) = \sum_{n,m} \left\{ d_n^m \hat{\psi}_n^m(\boldsymbol{r}_P) + c_n^m \psi_n^m(\boldsymbol{r}_P) \right\},$$

$$\tilde{u}(\boldsymbol{r}_P) = \sum_{n,m} \left\{ \tilde{d}_n^m \hat{\psi}_n^m(\boldsymbol{r}_P) + \tilde{c}_n^m \psi_n^m(\boldsymbol{r}_P) \right\},$$

for $P \in \Omega_a$, a sphere of radius a, centred at the origin; we can take $\Omega_a = S_+$, the escribed sphere.

An application of Green's theorem to the total fields u and $\tilde{u}$ in the region between S and Ω_a gives $[u, \tilde{u}] = 0$, after using the boundary condition on S, where we have used the notation (6.108). Then, (6.106) gives

$$0 = \sum_{n,m} (-1)^m \left(d_n^m \tilde{c}_n^{-m} - \tilde{d}_n^m c_n^{-m} \right).$$

Substituting from (7.64), we obtain

$$0 = \sum_{n,m} \sum_{\nu,\mu} \{ (-1)^m T_{n\nu}^{-m,\mu} - (-1)^\mu T_{\nu n}^{-\mu,m} \} \, d_n^m \tilde{d}_\nu^\mu.$$

As this equation must hold for all possible pairs of incident fields, we deduce (7.67).

A similar argument, beginning with $[u, \overline{\tilde{u}}] = 0$ and making use of (6.107), leads to (7.68). □

The relations (7.67) and (7.68) can be used as a check on computations of $\mathbf{T}$, or they can be incorporated into schemes for computing $\mathbf{T}$ [1290, 1310, 681]; see [1254] and [866, p. 160] for further references.

7.9.2 A translation of origin

The T-matrix relates two expansions about an origin O. We can use $\mathbf{T}$ to obtain a similar relation about another origin, $\widetilde{O}$, say. Thus, suppose that $\tilde{\boldsymbol{r}}_P$ and $\tilde{\boldsymbol{b}}$ are the position vectors of P and O, respectively, with respect to $\widetilde{O}$, so that $\tilde{\boldsymbol{r}}_P = \boldsymbol{r}_P + \tilde{\boldsymbol{b}}$. We want to construct the T-matrix, $\widetilde{\mathbf{T}}$, defined by

$$\tilde{\mathbf{c}} = \widetilde{\mathbf{T}}\tilde{\mathbf{d}}, \tag{7.69}$$

where

$$u_{\text{inc}}(P) = \sum_{n,m} \tilde{d}_n^m \, \hat{\psi}_n^m(\tilde{\boldsymbol{r}}_P) \tag{7.70}$$

and

$$u_{\text{sc}}(P) = \sum_{n,m} \tilde{c}_n^m \, \psi_n^m(\tilde{\boldsymbol{r}}_P) \tag{7.71}$$

are expansions centred at $\widetilde{O}$. For $|\tilde{\boldsymbol{r}}_P| > |\tilde{\boldsymbol{b}}|$, the addition theorem (3.85) gives

$$\psi_n^m(\boldsymbol{r}_P) = \sum_{\nu,\mu} \widehat{S}_{n\nu}^{m\mu}(-\tilde{\boldsymbol{b}}) \, \psi_\nu^\mu(\tilde{\boldsymbol{r}}_P).$$

Substituting this expression in (7.63) and comparison with (7.71) gives

$$\tilde{c}_n^m = \sum_{\nu,\mu} \widehat{S}_{\nu n}^{\mu m}(-\tilde{\boldsymbol{b}}) \, c_\nu^\mu.$$

Similarly, use of (3.78), (7.62) and (7.70) gives

$$d_n^m = \sum_{\nu,\mu} \widehat{S}_{\nu n}^{\mu m}(\tilde{\boldsymbol{b}}) \, \tilde{d}_\nu^\mu.$$

Using matrix notation, we obtain

$$\tilde{\mathbf{c}} = \overline{\widehat{\mathbf{S}}(\tilde{\boldsymbol{b}})}\,\mathbf{c} \quad \text{and} \quad \mathbf{d} = \overline{\widehat{\mathbf{S}}(-\tilde{\boldsymbol{b}})}\,\tilde{\mathbf{d}},$$

where we have used (3.95), namely

$$\left[\widehat{\mathbf{S}}(-\boldsymbol{b})\right]^T = \overline{\widehat{\mathbf{S}}(\boldsymbol{b})}. \tag{7.72}$$

Hence, using $(7.66)_1$ and (7.69), we obtain the desired formula:

$$\widetilde{\mathbf{T}} = \overline{\widehat{\mathbf{S}}(\tilde{\boldsymbol{b}})}\,\mathbf{T}\,\overline{\widehat{\mathbf{S}}(-\tilde{\boldsymbol{b}})}. \tag{7.73}$$

7.9.3 A connection with modified fundamental solutions

In Section 6.9, we saw that boundary integral equations can be derived using a different fundamental solution instead of G. These fundamental solutions usually have the form (6.92). In [631], the authors generalise (6.92), and consider

$$G_1(P,Q) = G(P,Q) - 2ik \sum_{n,m} \sum_{\nu,\mu} (-1)^m \alpha_{n\nu}^{m\mu} \psi_n^m(\boldsymbol{r}_P) \, \psi_\nu^{-\mu}(\boldsymbol{r}_Q),$$

where $\alpha_{n\nu}^{m\mu}$ are constants. They show that

$$\int_{S_-}\int_S \left|\frac{\partial}{\partial n_q} G_1(p,q)\right|^2 \mathrm{d}s_q\,\mathrm{d}s_p$$

will be minimised if $\alpha_{n\nu}^{m\mu} = T_{n\nu}^{m\mu}$, which means that G_1 will then be the best L^2-approximation to the exact Green's function for the exterior Neumann problem, G^{E}; see Section 6.12.1. Ahner [15] has discussed further connections with exact Green's functions.

7.9.4 Using a boundary integral equation to calculate the T-matrix

It is obvious that we can compute the field scattered by an object in (at least) two ways. We could solve a boundary integral equation, such as (5.68) or (6.124), for $u(q)$, and then use (7.6) to calculate $u_{\mathrm{sc}}(P)$. Alternatively, we could use the T-matrix, combining (7.14), (7.16) and the first of (7.66). As the underlying boundary-value problem has exactly one solution, both ways must give the same result.

The idea of using an integral equation to obtain the T-matrix is explicit in a short note of Gurel & Chew [451]; see also [452]. In the context of electromagnetic scattering by thin strips, they show

> *how to obtain a TMM* [T-matrix method] *solution from an* IE [integral-equation] *solution, i.e.,*
>
> IE $\Longrightarrow$ [Surface] Current Distribution
> $\Longrightarrow$ T Matrix ($\Longrightarrow$ Scattered Field). (∗)
>
> *...Although both of the methods have long histories, the idea of bridging the two has not attracted much attention ... The use of (∗) is best appreciated when the individual T matrix of a particular ... scatterer is difficult, complicated, or simply impossible to compute using a conventional TMM* [based on a null-field approach].
>
> (Gurel & Chew [451, p. 1624])

Let us explain how to realise this idea [795]. First, suppose that

$$d_n^m = \delta_{nN}\,\delta_{mM}$$

in (7.14), so that

$$u_{\mathrm{inc}}(P) = \hat{\psi}_N^M(\boldsymbol{r}_P) \quad \text{and} \quad c_n^m = T_{nN}^{mM}.$$

Denote the corresponding solution of (5.68) by $U_N^M(q)$, so that

$$(I + \overline{K^*})U_N^M = 2\hat{\psi}_N^M. \tag{7.74}$$

(More generally, we should solve a modified integral equation, such as (6.124), if we want to avoid irregular frequencies.) Then, having found U_N^M, the scattered field is given by (7.16) and (7.17). In particular, we can calculate c_n^m, whence

$$T_{nN}^{mM} = \mathrm{i}k(-1)^m \int_S U_N^M(q) \frac{\partial}{\partial n_q} \hat{\psi}_n^{-m}(\boldsymbol{r}_q)\, \mathrm{d}s_q. \qquad (7.75)$$

This is a formula for the entries in the T-matrix.

Notice that the integral equations to be solved, (7.74), have the same operator on the left-hand side but many different right-hand sides; their solutions, U_N^M, can be obtained efficiently.

The formula (7.75) is simple and explicit. It bypasses all the problems associated with the ill-conditioning of the Q-matrix, and it permits the use of sophisticated boundary-element methods in the computation of $U_N^M(q)$ for $q \in S$. As far as we know, this connection between boundary integral equations and T-matrix methods has not been exploited.

Another formula for the entries in the T-matrix is given in the next section.

7.9.5 The far field

The far-field pattern f is specified by Definition 4.2 and it is given by the formula (4.46). In particular, for plane-wave incidence, the coefficients d_n^m are given by (4.40), whence

$$f(\hat{\boldsymbol{r}}; \hat{\boldsymbol{\alpha}}) = 4\pi \sum_{n,m} \sum_{\nu,\mu} (-\mathrm{i})^n \, \mathrm{i}^\nu \, T_{n\nu}^{m\mu} \, Y_n^m(\hat{\boldsymbol{r}}) \, \overline{Y_\nu^\mu(\hat{\boldsymbol{\alpha}})}.$$

We can invert this formula, using the orthogonality of the spherical harmonics, (3.9); the result is

$$T_{n\nu}^{m\mu} = \frac{\mathrm{i}^n(-\mathrm{i})^\nu}{4\pi} \int_\Omega \int_\Omega f(\hat{\boldsymbol{r}}; \hat{\boldsymbol{\alpha}}) \, \overline{Y_n^m(\hat{\boldsymbol{r}})} \, Y_\nu^\mu(\hat{\boldsymbol{\alpha}}) \, \mathrm{d}\Omega(\hat{\boldsymbol{r}}) \, \mathrm{d}\Omega(\hat{\boldsymbol{\alpha}}). \qquad (7.76)$$

This formula can be found in [249, eqn (2.5)].

The formula (7.76) shows that the T-matrix can be constructed explicitly from the far-field pattern. Thus, we can reconstruct the scattered field everywhere outside the circumscribed sphere S_+ from the scattered field at infinity. In fact, there is an alternative reconstruction of u_{sc}, known as the *Atkinson–Wilcox theorem*; see [262, §2.A.3]. Thus,

$$u_{\mathrm{sc}}(P) = \frac{\mathrm{e}^{\mathrm{i}kr}}{\mathrm{i}kr} \sum_{n=0}^{\infty} \frac{f_n(\hat{\boldsymbol{r}})}{r^n}, \qquad P \in B_+, \qquad (7.77)$$

where $f_0 \equiv f$. For $n = 1, 2, \ldots, f_n$ is obtained by applying a second-order differential operator (essentially, the angular part of the Laplacian (3.2)) to f_{n-1}. Note that $f_0(\hat{\boldsymbol{r}})$ is the far-field pattern corresponding to the incident field u_{inc} that produced the scattered field u_{sc}, whereas the T-matrix does not depend on the incident field.

On the other hand, the formula (7.76) requires the far-field pattern for all incidence directions $\hat{\boldsymbol{\alpha}}$ and for all observation directions $\hat{\boldsymbol{r}}$.

It is well known that

$$f(\hat{\boldsymbol{r}}; \hat{\boldsymbol{\alpha}}) = f(-\hat{\boldsymbol{\alpha}}; -\hat{\boldsymbol{r}}), \tag{7.78}$$

which is a *reciprocity theorem*; see, for example, [225, Theorem 3.13] or [262, eqn (2.46)]. When (7.78) is used in (7.76), one can easily rederive (7.67).

In two dimensions, we combine (4.12) and (4.18) to obtain

$$f(\theta; \alpha) = \sum_n \sum_m (-\mathrm{i})^n \mathrm{i}^m T_{nm} \, \mathrm{e}^{\mathrm{i}n\theta} \, \mathrm{e}^{-\mathrm{i}m\alpha},$$

whence

$$T_{nm} = \frac{\mathrm{i}^n (-\mathrm{i})^m}{(2\pi)^2} \int_{-\pi}^{\pi} \int_{-\pi}^{\pi} f(\theta; \alpha) \, \mathrm{e}^{-\mathrm{i}n\theta} \, \mathrm{e}^{\mathrm{i}m\alpha} \, \mathrm{d}\theta \, \mathrm{d}\alpha. \tag{7.79}$$

The two-dimensional analogue of (7.77) is more complicated, and was obtained by Karp [588]:

$$u_{\mathrm{sc}}(P) = H_0(kr) \sum_{n=0}^{\infty} \frac{p_n(\theta)}{r^n} + H_1(kr) \sum_{n=0}^{\infty} \frac{q_n(\theta)}{r^n}, \qquad P \in B_+. \tag{7.80}$$

Here, the coefficients p_n and q_n can be determined recursively in terms of derivatives of the far-field pattern $f(\theta)$ with respect to θ. For explicit results, see [161].

7.9.6 Literature

The T-matrix for acoustic scattering problems was introduced by Waterman [1288]. Computations for rigid spheroids are given in [1249, 1177]. Results for a penetrable sphere containing one or two smaller spherical scatterers are given in [965]. Low-frequency approximations for the T-matrix are given in Section 8.2.

The T-matrix for electromagnetic scattering problems is discussed in detail in Waterman's 1971 paper [1290]. This paper also gives connections with the so-called *scattering matrix*, $\mathcal{S}$, defined by $\mathcal{S} = I + 2T$, and contains numerical results for various axisymmetric geometries. Since then, the T-matrix has been used in many applications; see, for example, [964, 1154, 1155, 869, 871], [866, Chapter 6], [870, Chapter 5] and references therein. For obstacles in half-spaces or near plane interfaces, see, for example, [662, 666, 1263, 283]. Hunka & Mei [524] have given results for scattering by two axisymmetric objects, using the T-matrix for each object and an orders-of-scattering approach. The 2004 paper [873] contains over 700 references on T-matrix methods for electromagnetic scattering problems.

The T-matrix as defined is very useful if one is interested in the effects due to rotations of the scatterer. Indeed, suppose that we have two systems of spherical polar coordinates, (r, θ, ϕ) and (r', θ', ϕ'), with a common origin. Suppose that (r, θ, ϕ) is fixed in space and (r', θ', ϕ') is fixed in the scatterer. Then, one can compute the

T-matrix in one system from the T-matrix in the other system [870, §5.2.1]. This permits the analytical calculation of averages over orientation [870, §§5.3–5.6].

The T-matrix for elastic waves is discussed in [1291, 1260]. Further applications are given in [1253, 1256]. Fluid inclusions are considered in [966].

For an elastic half-space $y > 0$, one can construct outgoing cylindrical wavefunctions that are singular along the y-axis and that satisfy the boundary conditions on the surface $y = 0$. These can then be used to derive null-field equations for scatterers that enclose the y-axis, such as vertical cylinders. Bostock [124] has studied the associated T-matrix and given numerical results.

The fluid–solid problem of scattering of acoustic waves by an elastic obstacle has received considerable attention in the literature. Examples are [128, 968, 970, 1159, 352, 969]. The T-matrix for two-dimensional water waves is discussed in [784, 787].

The T-matrix has been obtained for straight thin strips [1300, 1235] and cracks [1236] in two dimensions, and for circular screens [667], elliptical screens [663, 104] and penny-shaped cracks [131] in three dimensions.

We have already mentioned that one can obtain the T-matrix by using elliptical or spheroidal wavefunctions [77, 461, 463]. For additional applications to fluid–solid problems, see [465, 464]. In [1080], the authors compute the T-matrix for electromagnetic scattering by a spheroid, using spheroidal coordinates, and then convert the result into spherical polar coordinates, thus facilitating orientational averaging.

See also the references listed in Section 7.6.1.

7.10 The T-matrix for two obstacles

Consider acoustic scattering by two sound-hard obstacles in three dimensions. Thus, from (7.7), we have the interior relation

$$\int_S u(q)\,\frac{\partial}{\partial n_q}G(P,q)\,\mathrm{d}s_q = 2u_{\mathrm{inc}}(P), \qquad P \in B, \tag{7.81}$$

where $S = S_1 \cup S_2$ and $B = B_1 \cup B_2$. From (7.6), we also have the integral representation

$$2u_{\mathrm{sc}}(P) = -\int_S u(q)\,\frac{\partial}{\partial n_q}G(P,q)\,\mathrm{d}s_q, \qquad P \in B_{\mathrm{e}}. \tag{7.82}$$

Our goal is to construct the T-matrix for the two-obstacle scatterer, from (7.81) and (7.82), in terms of the T-matrices for each obstacle in isolation.

Choose an origin $O_j \in B_j$, and then let S_j^- and S_j^+ be the inscribed sphere and the escribed sphere, respectively, to S_j, centred at O_j, $j = 1, 2$. The interior of S_j^- is the ball B_j^-, $j = 1, 2$. The region exterior to $S_1^+ \cup S_2^+$ is $B_{\mathrm{e}}^+ \subset B_{\mathrm{e}}$. See Fig. 7.3.

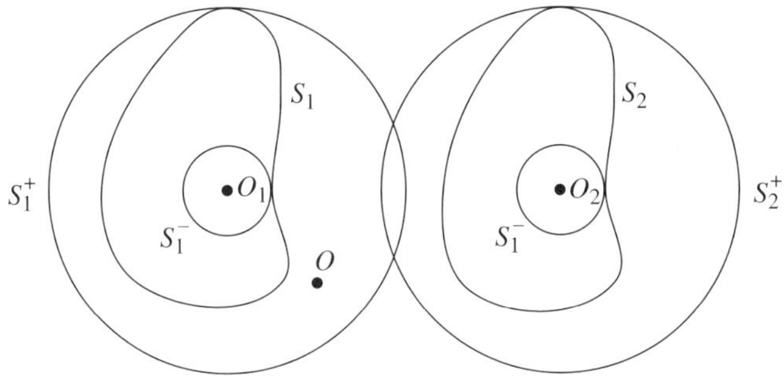

Fig. 7.3. Notation for two scatterers and the null-field method.

Suppose that $P \in B_e^+$. Then, writing $S = S_1 \cup S_2$ in (7.82), we obtain

$$u_{sc}(P) = \sum_{j=1}^{2} \sum_{n,m} c_{nj}^m \psi_n^m(\boldsymbol{r}_j^P), \quad P \in B_e^+, \tag{7.83}$$

where

$$c_{nj}^m = \mathrm{i}k(-1)^m \int_{S_j} u(q) \frac{\partial}{\partial n_q} \hat{\psi}_n^{-m}(\boldsymbol{r}_j^q) \, \mathrm{d}s_q \tag{7.84}$$

and we have used (7.8) twice. The notation $\boldsymbol{r}_j^Q$ indicates the position vector of the point Q with respect to the origin O_j.

The formula (7.83) gives an expansion of the scattered field in B_e^+ in terms of two sets of outgoing spherical wavefunctions, one centred at O_1 and one centred at O_2. Evidently, if S_1 and S_2 are spheres centred at O_1 and O_2, respectively, then $B_e^+ \equiv B_e$ and (7.83) reduces to the expansion (4.51) used in Section 4.7.

In order to construct the T-matrix for the two-obstacle scatterer, we have to expand the scattered and incident fields about a single origin, O, which we choose in the vicinity of the two obstacles. (We could choose $O = O_1$ or $O = O_2$, but this would lead to an asymmetric formulation.) Let S_+ be the smallest escribed sphere to S, centred at O, so that S_+ encloses both scatterers. The region exterior to S_+ is $B_+ \subset B_e$. Then, we have

$$u_{sc}(P) = \sum_{n,m} c_n^m \psi_n^m(\boldsymbol{r}_P), \quad P \in B_+, \tag{7.85}$$

exactly as for one scatterer; see (7.16) and (7.17).

The expansions (7.83) and (7.85) are related. As usual, let $\boldsymbol{b}_j$ be the position vector of O_j with respect to O, so that $\boldsymbol{r}_j^P = \boldsymbol{r}_P - \boldsymbol{b}_j$, $j = 1, 2$. As P is outside S_+, we have $|\boldsymbol{r}_P| > |\boldsymbol{b}_j|$. Then, from Theorem 3.27, we have

$$\psi_n^m(\boldsymbol{r}_j^P) = \sum_{\nu,\mu} \widehat{S}_{n\nu}^{m\mu}(-\boldsymbol{b}_j) \, \hat{\psi}_\nu^\mu(\boldsymbol{r}_P), \quad j = 1, 2,$$

where $\widehat{\mathbf{S}}$ is a separation matrix. Substituting in (7.83) and rearranging gives (7.85) with

$$c_n^m = \sum_{j=1}^{2} \sum_{\nu,\mu} c_{\nu j}^{\mu} \, \widehat{S}_{\nu n}^{\mu m}(-\boldsymbol{b}_j). \tag{7.86}$$

Next, suppose that u_{inc} is a regular wavefunction inside S_+, so that

$$u_{\text{inc}}(P) = \sum_{n,m} d_n^m \, \hat{\psi}_n^m(\boldsymbol{r}_P). \tag{7.87}$$

(S_+ can be replaced by any smaller sphere that contains both O_1 and O_2.) We want to express the coefficients c_n^m in (7.85) in terms of the (known) coefficients d_n^m in (7.87), as this will give us the T-matrix for the two-obstacle scatterer.

We begin by re-expanding u_{inc} about O_1 and O_2. As $\boldsymbol{r}_P = \boldsymbol{b}_j + \boldsymbol{r}_j^P$, $j = 1, 2$, we have

$$\hat{\psi}_n^m(\boldsymbol{r}_P) = \sum_{\nu,\mu} \widehat{S}_{n\nu}^{m\mu}(\boldsymbol{b}_j) \, \hat{\psi}_\nu^\mu(\boldsymbol{r}_j^P), \quad j = 1, 2,$$

whence (7.87) gives

$$u_{\text{inc}}(P) = \sum_{n,m} d_{nj}^m \, \hat{\psi}_n^m(\boldsymbol{r}_j^P), \quad j = 1, 2, \tag{7.88}$$

where

$$d_{nj}^m = \sum_{\nu,\mu} d_\nu^\mu \, \widehat{S}_{\nu n}^{\mu m}(\boldsymbol{b}_j). \tag{7.89}$$

Now, let us relate c_{nj}^m and d_{nj}^m in (7.83) and (7.88), respectively. We shall do this in two alternative ways. The first way proceeds via the null-field equations, and is the derivation given by Peterson & Ström [963]. The second way is perhaps more direct, and has its roots in Rayleigh's 1892 paper; see Section 1.1.4.

7.10.1 The derivation of Peterson and Ström

We shall derive the null-field equations for a pair of sound-hard obstacles. Thus, suppose that $P \in B_1^-$. Then, as $|\boldsymbol{r}_1^q| > |\boldsymbol{r}_1^P|$ for all $q \in S = S_1 \cup S_2$, Theorem 6.4 gives

$$G(P, q) = -2ik \sum_{n\,m} (-1)^m \, \hat{\psi}_n^m(\boldsymbol{r}_1^P) \, \psi_n^{-m}(\boldsymbol{r}_1^q).$$

This bilinear expansion can be inserted in (7.81) for $P \in B_1^-$, together with (7.88) for $j = 1$; equating coefficients of $\hat{\psi}_n^m(\boldsymbol{r}_1^P)$ then gives

$$-ik(-1)^m \int_S u(q) \, \frac{\partial}{\partial n_q} \psi_n^{-m}(\boldsymbol{r}_1^q) \, ds_q = d_{n1}^m, \qquad \begin{aligned} &n = 0, 1, 2, \ldots, \\ &m = -n, \ldots, n. \end{aligned} \tag{7.90}$$

The left-hand side of this equation is the sum of an integral over S_1 and an integral over S_2. The first of these is in the same form as when S_1 is an isolated scatterer; see

(7.15). However, the integral over S_2 is not in this form. To decompose this integral, we use $\boldsymbol{r}_1^q = \boldsymbol{r}_2^q - \boldsymbol{b}$, where $\boldsymbol{b} = \boldsymbol{b}_1 - \boldsymbol{b}_2$ is the position vector of O_1 with respect to O_2. Then, *assuming that O_1 is outside S_2^+*, we have $|\boldsymbol{r}_2^q| < |\boldsymbol{b}|$. Hence, (3.84) gives

$$\psi_n^{-m}(\boldsymbol{r}_1^q) = \sum_{\nu,\mu} S_{n\nu}^{-m,\mu}(-\boldsymbol{b})\,\hat{\psi}_\nu^{\mu}(\boldsymbol{r}_2^q) = \sum_{\nu,\mu}(-1)^{m+\mu} S_{\nu n}^{\mu m}(\boldsymbol{b})\,\hat{\psi}_\nu^{-\mu}(\boldsymbol{r}_2^q),$$

where the second equality follows from (3.99). Substituting this expansion in (7.90) gives

$$\begin{aligned} -\mathrm{i}k(-1)^m &\int_{S_1} u(q)\,\frac{\partial}{\partial n_q}\psi_n^{-m}(\boldsymbol{r}_1^q)\,\mathrm{d}s_q \\ &- \mathrm{i}k\sum_{\nu,\mu}(-1)^{\mu} S_{\nu n}^{\mu m}(\boldsymbol{b}) \int_{S_2} u(q)\,\frac{\partial}{\partial n_q}\hat{\psi}_n^{-m}(\boldsymbol{r}_2^q)\,\mathrm{d}s_q = d_{n1}^m, \end{aligned} \tag{7.91}$$

for $n = 0, 1, 2, \ldots$ and $m = -n, \ldots, n$. Similarly, if we also assume that O_2 is outside S_1^+, we obtain

$$\begin{aligned} -\mathrm{i}k(-1)^m &\int_{S_2} u(q)\,\frac{\partial}{\partial n_q}\psi_n^{-m}(\boldsymbol{r}_2^q)\,\mathrm{d}s_q \\ &- \mathrm{i}k\sum_{\nu,\mu}(-1)^{\mu} S_{\nu n}^{\mu m}(-\boldsymbol{b}) \int_{S_1} u(q)\,\frac{\partial}{\partial n_q}\hat{\psi}_n^{-m}(\boldsymbol{r}_1^q)\,\mathrm{d}s_q = d_{n2}^m, \end{aligned} \tag{7.92}$$

for $n = 0, 1, 2, \ldots$ and $m = -n, \ldots, n$. Equations (7.91) and (7.92) are the null-field equations for scattering by two sound-hard obstacles.

To solve the null-field equations, we proceed as in Section 7.8 and write

$$u(q) = \sum_{\nu,\mu} u_{\nu j}^{\mu}\,\phi_{\nu j}^{\mu}(q), \qquad q \in S_j, \qquad j = 1, 2,$$

where $\phi_{\nu j}^{\mu}$ are expansion functions and $u_{\nu j}^{\mu}$ are unknown coefficients. Substitution in (7.91) gives

$$d_{n1}^m = \sum_{\nu,\mu} Q_{n\nu}^{m\mu}(1)\,u_{\nu 1}^{\mu} + \sum_{\nu,\mu}\sum_{N,M} S_{\nu n}^{\mu m}(\boldsymbol{b})\,\widehat{Q}_{\nu N}^{\mu M}(2)\,u_{N2}^{M}, \tag{7.93}$$

where (cf. (7.61) and (7.34))

$$Q_{n\nu}^{m\mu}(j) = -\mathrm{i}k(-1)^m \int_{S_j} \phi_{\nu j}^{\mu}\,\frac{\partial}{\partial n_q}\psi_n^{-m}(\boldsymbol{r}_j^q)\,\mathrm{d}s_q,$$

$$\widehat{Q}_{n\nu}^{m\mu}(j) = -\mathrm{i}k(-1)^m \int_{S_j} \phi_{\nu j}^{\mu}\,\frac{\partial}{\partial n_q}\hat{\psi}_n^{-m}(\boldsymbol{r}_j^q)\,\mathrm{d}s_q.$$

Using an obvious matrix notation, we can write (7.93) as

$$\mathbf{Q}_1\mathbf{u}_1 + [\mathbf{S}(\boldsymbol{b})]^T\widehat{\mathbf{Q}}_2\mathbf{u}_2 = \mathbf{d}_1. \tag{7.94}$$

Similarly, (7.92) gives

$$\mathbf{Q}_2\mathbf{u}_2 + [\mathbf{S}(-\boldsymbol{b})]^T\widehat{\mathbf{Q}}_1\mathbf{u}_1 = \mathbf{d}_2. \tag{7.95}$$

Before solving these, it is convenient to define

$$\mathbf{v}_j = \mathbf{Q}_j \mathbf{u}_j, \quad j = 1, 2.$$

Then, we can write (7.94) and (7.95) as

$$\mathbf{v}_1 - [\mathbf{S}(\boldsymbol{b})]^T \mathbf{T}_2 \mathbf{v}_2 = \mathbf{d}_1,$$

$$\mathbf{v}_2 - [\mathbf{S}(-\boldsymbol{b})]^T \mathbf{T}_1 \mathbf{v}_1 = \mathbf{d}_2,$$

where

$$\mathbf{T}_j = -\widehat{\mathbf{Q}}_j \mathbf{Q}_j^{-1}$$

is the T-matrix for the jth scatterer in isolation.

Solving formally for $\mathbf{v}_j$, we obtain

$$\mathbf{v}_1 = \mathbf{A}_1 \{\mathbf{d}_1 + [\mathbf{S}(\boldsymbol{b})]^T \mathbf{T}_2 \mathbf{d}_2\},$$

$$\mathbf{v}_2 = \mathbf{A}_2 \{\mathbf{d}_2 + [\mathbf{S}(-\boldsymbol{b})]^T \mathbf{T}_1 \mathbf{d}_1\},$$

where

$$\mathbf{A}_1 = \left(\mathbf{I} - [\mathbf{S}(\boldsymbol{b})]^T \mathbf{T}_2 [\mathbf{S}(-\boldsymbol{b})]^T \mathbf{T}_1\right)^{-1}, \tag{7.96}$$

$$\mathbf{A}_2 = \left(\mathbf{I} - [\mathbf{S}(-\boldsymbol{b})]^T \mathbf{T}_1 [\mathbf{S}(\boldsymbol{b})]^T \mathbf{T}_2\right)^{-1} \tag{7.97}$$

and $\mathbf{I}$ is the identity matrix. But, from (7.84), (7.86) and (7.89), we have

$$\mathbf{c} = \overline{\widehat{\mathbf{S}}(\boldsymbol{b}_1)}\, \mathbf{c}_1 + \overline{\widehat{\mathbf{S}}(\boldsymbol{b}_2)}\, \mathbf{c}_2, \tag{7.98}$$

$$\mathbf{c}_j = -\widehat{\mathbf{Q}}_j \mathbf{u}_j = \mathbf{T}_j \mathbf{v}_j$$

and

$$\mathbf{d}_j = \overline{\widehat{\mathbf{S}}(-\boldsymbol{b}_j)}\, \mathbf{d}, \tag{7.99}$$

where we have used (7.72). Hence,

$$\mathbf{c} = \mathbf{T}\mathbf{d}, \tag{7.100}$$

where

$$\begin{aligned} \mathbf{T} = {} & \overline{\widehat{\mathbf{S}}(\boldsymbol{b}_1)}\, \mathbf{T}_1 \mathbf{A}_1 \left\{ \overline{\widehat{\mathbf{S}}(-\boldsymbol{b}_1)} + [\mathbf{S}(\boldsymbol{b})]^T \mathbf{T}_2 \overline{\widehat{\mathbf{S}}(-\boldsymbol{b}_2)} \right\} \\ & + \overline{\widehat{\mathbf{S}}(\boldsymbol{b}_2)}\, \mathbf{T}_2 \mathbf{A}_2 \left\{ \overline{\widehat{\mathbf{S}}(-\boldsymbol{b}_2)} + [\mathbf{S}(-\boldsymbol{b})]^T \mathbf{T}_1 \overline{\widehat{\mathbf{S}}(-\boldsymbol{b}_1)} \right\}. \end{aligned} \tag{7.101}$$

This expression for the total two-body T-matrix was first given by Peterson & Ström [963, eqn (25)]. It 'has the advantages that it treats the two scatterers in a symmetric way and that the $Q(j)$ matrices do not appear separately but only in combinations which give the $T(j)$ matrices' [963, p. 774]. Consequently, (7.101) does not require the use of the null-field method: the individual T-matrices can be obtained by any convenient method. The fact that (7.101) does not involve any Q-matrices suggests that a more direct derivation exists: this is given next.

7.10.2 An alternative derivation

Consider the scatterer S_1. From (7.83), the field scattered by S_1 is

$$\sum_{n,m} c_{n1}^m \, \psi_n^m(\boldsymbol{r}_1^P),$$

whereas the field 'incident' on S_1 is

$$u_{\rm inc} + \sum_{n,m} c_{n2}^m \, \psi_n^m(\boldsymbol{r}_2^P) = \sum_{n,m} \hat{\psi}_n^m(\boldsymbol{r}_1^P) \left\{ d_{n1}^m + \sum_{\nu,\mu} S_{\nu n}^{\mu m}(\boldsymbol{b}) \, c_{\nu 2}^{\mu} \right\};$$

here, we have used (7.88) with $j = 1$, and

$$\psi_n^m(\boldsymbol{r}_2^P) = \sum_{\nu,\mu} S_{n\nu}^{m\mu}(\boldsymbol{b}) \, \hat{\psi}_\nu^\mu(\boldsymbol{r}_1^P),$$

which is valid for $|\boldsymbol{r}_1^P| < |\boldsymbol{b}|$. Hence, as $\mathbf{T}_1$ characterises scattering by S_1, we must have

$$\mathbf{c}_1 = \mathbf{T}_1 \left\{ \mathbf{d}_1 + [\mathbf{S}(\boldsymbol{b})]^T \mathbf{c}_2 \right\}. \tag{7.102}$$

Similarly, consideration of S_2 leads to

$$\mathbf{c}_2 = \mathbf{T}_2 \left\{ \mathbf{d}_2 + [\mathbf{S}(-\boldsymbol{b})]^T \mathbf{c}_1 \right\}. \tag{7.103}$$

Solving (7.102) and (7.103), we obtain

$$\mathbf{c}_1 = \mathbf{B}_1 \, \mathbf{T}_1 \{ \mathbf{d}_1 + [\mathbf{S}(\boldsymbol{b})]^T \mathbf{T}_2 \mathbf{d}_2 \},$$

$$\mathbf{c}_2 = \mathbf{B}_2 \, \mathbf{T}_2 \{ \mathbf{d}_2 + [\mathbf{S}(-\boldsymbol{b})]^T \mathbf{T}_1 \mathbf{d}_1 \},$$

where

$$\mathbf{B}_1 = \left(\mathbf{I} - \mathbf{T}_1 [\mathbf{S}(\boldsymbol{b})]^T \mathbf{T}_2 [\mathbf{S}(-\boldsymbol{b})]^T \right)^{-1}, \tag{7.104}$$

$$\mathbf{B}_2 = \left(\mathbf{I} - \mathbf{T}_2 [\mathbf{S}(-\boldsymbol{b})]^T \mathbf{T}_1 [\mathbf{S}(\boldsymbol{b})]^T \right)^{-1}. \tag{7.105}$$

Notice that

$$\mathbf{B}_1 \, \mathbf{T}_1 = \mathbf{T}_1 \, \mathbf{A}_1 \quad \text{and} \quad \mathbf{B}_2 \, \mathbf{T}_2 = \mathbf{T}_2 \, \mathbf{A}_2,$$

where $\mathbf{A}_1$ and $\mathbf{A}_2$ are defined by (7.96) and (7.97), respectively.

Finally, construction of $\mathbf{T}$ follows as before, using (7.98), (7.99) and (7.100); the result is (7.101).

One virtue of this derivation is that it does not make explicit use of the boundary conditions on each scatterer. This information is all contained in the individual T-matrices, $\mathbf{T}_1$ and $\mathbf{T}_2$: these matrices are used as building blocks in a natural way.

7.10.3 Discussion

The expression (7.101) has several attractive properties. First, notice that we can remove the second scatterer, say, by setting $\mathbf{T}_2 = \mathbf{0}$; the result is

$$\mathbf{T} = \overline{\widehat{\mathbf{S}}(\boldsymbol{b}_1)}\, \mathbf{T}_1\, \overline{\widehat{\mathbf{S}}(-\boldsymbol{b}_1)},$$

which is the known result for translating the origin, given previously in Section 7.9.2; see (7.73).

Second, if all of the scatterers are spheres, then the method reduces to the method of Section 4.7. However, even for spheres, the construction of the total T-matrix still has value.

Third, 'if one makes a formal expansion of the inverses which appear in (7.96) and (7.97), the terms which appear in these expansions can be interpreted as multiple-scattering contributions as follows: each factor $\mathbf{T}_1$ [$\mathbf{T}_2$] represents a scattering by scatterer 1 [2] and each factor $[\mathbf{S}(-\boldsymbol{b})]^T$ ($[\mathbf{S}(\boldsymbol{b})]^T$) represents a propagation from 1 to 2 (2 to 1)' [963, p. 774]. Thus, (7.101) yields the orders-of-scattering view of multiple scattering; but, we emphasise that (7.101) is exact.

Fourth, (7.101) can be used to derive approximations in a systematic way. Thus, wide-spacing approximations can be deduced by inserting approximations for the separation matrices, valid for large kb; see Section 3.13.2. Alternatively, 'small scatterers' can be treated by inserting low-frequency (small ka) approximations for $\mathbf{T}_1$ and/or $\mathbf{T}_2$; such approximations can be found in [1255] and in Section 8.2.

The T-matrix (7.101) is only defined if the incident field is a regular wavefunction inside the circumscribed sphere S_+. If u_{inc} is not regular within this region (for example, u_{inc} could be generated by a point source located between S and S_+), the expansion (7.87) is not valid. Nevertheless, provided u_{inc} is regular inside S_1^+ and S_2^+, then the expansions (7.88) will be valid, and so we can still solve for $\mathbf{c}_1$ and $\mathbf{c}_2$. The scattered field can then be calculated using (7.83), or by using (7.85) and (7.86).

7.10.4 Literature

The paper of Peterson & Ström [963] contains numerical results for plane-wave scattering by a pair of identical (soft or hard) spheres; computational details were given in an unpublished report [960]. Bates & Wall [77] have developed the corresponding theory in two dimensions and given numerical results for scattering by elliptic and square cylinders. For scattering by two coaxial spheroids, see [1309].

For an extension to systems governed by the Lippmann–Schwinger equation (see Section 6.3.3), see [1153]. Periodic arrays of scatterers are considered in [961].

Recall (7.90), obtained by considering S_1. One can write down similar equations for each scatterer. Two-dimensional versions of these equations were derived by Okamoto [930, 931] and solved numerically for various configurations of circular and elliptic cylinders.

The T-matrix for electromagnetic scattering by two obstacles was also obtained by Peterson & Ström [962]. Numerical results for plane-wave scattering by a pair of identical, perfectly conducting spheres were given. The T-matrix for two dielectric spheres has been computed by several authors; see, for example, [246, 868, 867, 991, 992] and [209, §4.5.2]. For scattering by two coaxial spheroids, see [246].

For an extension to inhomogeneous obstacles, see [1151]. For periodic arrays, see [1296].

The T-matrix for the scattering of elastic waves by two obstacles was obtained by Boström [126] and by Varadan & Varadan [1258]. Numerical results for scattering by a pair of coaxial spheroidal cavities were given. Multiple scattering of acoustic waves by elastic obstacles is considered in [967]. In [125], the authors have given results for the scattering of waves in a layered half-space $y > 0$ containing two circular cylinders (with their axes parallel to the y-axis), using the T-matrix method of Bostock [124] (see Section 7.9.6).

The T-matrix for two-dimensional waves on deep water with two surface-piercing horizontal cylinders is given in [784].

7.11 The T-matrix for N obstacles

Formally, it is straightforward to construct the T-matrix for N obstacles, B_j, $j = 1, 2, \ldots, N$. Choose an origin $O_j \in B_j$ and, for simplicity, suppose that all the escribed spheres S_j^+, $j = 1, 2, \ldots, N$, do not intersect. (This restriction can be relaxed.) Then, the scattered field can be written as (7.85) or as

$$u_{\mathrm{sc}}(P) = \sum_{j=1}^{N} \sum_{n,m} c_{nj}^m \, \psi_n^m(\boldsymbol{r}_j^P), \quad P \text{ outside } \bigcup_{j=1}^{N} S_j^+.$$

The incident field can be written as (7.87) or as (7.88) for each j. The generalisation of (7.102) and (7.103) is

$$\mathbf{c}_l = \mathbf{T}_l \left\{ \mathbf{d}_l + \sum_{\substack{j=1 \\ j \neq l}}^{N} \left[\mathbf{S}(\boldsymbol{b}_{lj}) \right]^T \mathbf{c}_j \right\}, \quad l = 1, 2, \ldots, N, \tag{7.106}$$

where $\boldsymbol{b}_{lj}$ is the position vector of O_l with respect to O_j. Solve this system for the vectors $\mathbf{c}_1, \mathbf{c}_2, \ldots, \mathbf{c}_N$; the solution can be written in the form

$$\mathbf{c}_l = \sum_{j=1}^{N} \mathbf{T}_{lj} \, \mathbf{d}_j, \quad l = 1, 2, \ldots, N.$$

Then, to obtain the N-body T-matrix, we use (7.99), (7.100) and

$$\mathbf{c} = \sum_{j=1}^{N} \overline{\widehat{\mathbf{S}}(\boldsymbol{b}_j)} \, \mathbf{c}_j;$$

the result is

$$\mathbf{T} = \sum_{l=1}^{N} \overline{\widehat{\mathbf{S}}(\boldsymbol{b}_l)} \sum_{j=1}^{N} \mathbf{T}_{lj} \overline{\widehat{\mathbf{S}}(-\boldsymbol{b}_j)}. \tag{7.107}$$

The matrix $\mathbf{T}$ is sometimes called the *cluster T-matrix* or the *T-supermatrix*.

7.11.1 Discussion and literature

The derivation given above can be found in a paper of Varadan & Varadan [1258]; see also the earlier papers [1244, 1257] on electromagnetic scattering. Subsequently, a similar derivation was given by Lim & Hackman [717]; in fact, these authors do not solve (7.106) for $\mathbf{c}_l$ but an equivalent system for $\mathbf{T}_l^{-1}\mathbf{c}_l$. Numerical results are given in [717] for acoustic scattering by a row of ten steel spherical shells in water, and for the scattering of a P-wave by a row of five elastic spherical inclusions.

The T-matrix for $N = 3$ is given in [963, eqn (32)].

Note that the structure of (7.106) is exactly the same as (4.60). Indeed, (7.106) reduces to (4.60) when all the scatterers are hard spheres.

The two-dimensional version of (7.106) is in [340, eqn (27)]; this paper contains applications to scattering by clusters of ellipses, squares and circles. Equation (7.106) is also derived in [640]. For scatterers in layered acoustic media, see [716].

The electromagnetic version of (7.107) is derived in [871], in [866, pp. 161–165] and in [870, §5.9]. The electromagnetic version of (7.106) can also be found in [380, eqn (18)]. These references and [557] contain examples and further references for scattering by clusters of spheres. For other geometries, see, for example, [1376, 286].

The direct computation of the cluster T-matrix for N scatterers is expensive. Consequently, various iterative and recursive methods have been developed for computing $\mathbf{T}$. For example, one can partition the N scatterers into two clusters, one of $N - N_1$ scatterers and one of N_1 scatterers, with $N_1 \geq 1$. Each cluster can be regarded as a single scatterer. (Partitioning of this kind was used in 1963 by Millar: 'What may be novel in the present approach is to regard certain problems as two- or three-body problems, although at least one of these bodies may contain more than one scatterer' [857, p. 2107].) Each cluster has its own T-matrix; call them $\mathbf{T}(N - N_1)$ and $\mathbf{T}(N_1)$. The formula for the two-body T matrix (given in Section 7.10) can then be used to combine $\mathbf{T}(N - N_1)$ and $\mathbf{T}(N_1)$ so as to obtain the N-scatterer T-matrix. Chew and his students have exploited this idea so as to build T-matrices for large N. For example, Wang & Chew [1283] have given results for electromagnetic scattering by 6859 dielectric spheres. For reviews and other results, see [196, §8.6.2], [209, §4.5.3] and [195, 199, 200, 204, 203, 640, 450]. For another application, see [1218]. For two-dimensional problems, see [169]; extensive results for scattering by regular arrays of up to 2880 identical circular cylinders were given [170, 171].

It turns out that the recursive algorithm used in [1283] does not always converge, so that care is needed; recall the geometrical constraints on the derivations in Section 7.10. For more information, see [1048, 1109]. For an improved algorithm, see [1148, 50].

Kagemoto & Yue [576] considered water waves in a three-dimensional ocean of constant finite depth, with N immersed rigid bodies. Assume that the jth body can be enclosed by a vertical circular cylinder, $\mathcal{C}_j$. Outside $\mathcal{C}_j$, radiating potentials can be expanded using (4.67), including evanescent contributions. Inside $\mathcal{C}_j$, regular potentials can be expanded similarly, with H_n and K_n replaced by J_n and I_n, respectively. Evidently, the coefficients in these two expansions are related by a T-matrix, denoted by $\mathbf{B}_j$ and called the 'diffraction transfer matrix' in [576]. Then, the multiple-scattering problem can be solved as above, using two-dimensional addition theorems. This was done in [576], with several applications. Since then, many applications and refinements have been made; see, for example, [607, 1359, 1361, 895, 182, 590, 575]. For a description of the method, see also [731, §6.2].

8
Approximations

I owe a lot to my engineering training because it [taught] *me to tolerate approximations. Previously to that I thought... one should just concentrate on exact equations all the time. Then I got the idea that in the actual world all our equations are only approximate. We must just tend to greater and greater accuracy. In spite of the equations being approximate, they can be beautiful.*

(P.A.M. Dirac, quoted in [89, p. 36])

8.1 Introduction

In all previous chapters, we have concentrated on exact methods: for a given configuration of known scattering obstacles, solve the boundary-value problem for the scattered field. However, it may be possible to simplify the problem by exploiting, for example, the small size of (some of) the scatterers or the large spacing between the scatterers. On the other hand, it may not be feasible to have complete knowledge of the scatterers; for example, their locations could be random, and this leads to other approximations. Several of these approximations are discussed below.

8.2 Small scatterers

In many applications, individual scatterers are small compared with the length of the incident waves, meaning that

$$0 < ka \ll 1, \tag{8.1}$$

where $2a$ is a typical diameter of the scatterers. The assumption (8.1) leads to approximations known as *long-wave* or *low-frequency* approximations. Roughly speaking, we may expect that the acoustic field near a small scatterer is given approximately by the solution of a related problem for Laplace's equation. In many situations, this expectation can be realised and justified rigorously, yielding asymptotic approximations that hold as $ka \to 0$.

It turns out that three-dimensional problems are the most straightforward: typical wavefunctions are analytic functions of k, so that power series in k can be found, and these series converge in a neighbourhood of $k = 0$. Two-dimensional problems are often more difficult, because terms in $\log k$ usually arise.

For many problems, complete low-frequency asymptotic expansions can be found. However, we are mainly interested in *Rayleigh approximations*, defined as 'the first non-trivial terms in the low-frequency expansions' [262, p. 161]. These approximations can then be incorporated into schemes for solving multiple-scattering problems.

8.2.1 Three dimensions: sound-soft obstacles

For scattering by a sound-soft obstacle, the scattered field is given exactly by (5.84). The far field is given by (4.44) as

$$u_{\rm sc}(\boldsymbol{r}) \sim (\mathrm{i}kr)^{-1}\,\mathrm{e}^{\mathrm{i}kr}\,f(\hat{\boldsymbol{r}}) \quad \text{as } r \to \infty, \tag{8.2}$$

where, from (5.85),

$$f(\hat{\boldsymbol{r}}) = -\frac{\mathrm{i}k}{4\pi}\int_S \frac{\partial u}{\partial n_q}\exp(-\mathrm{i}k\hat{\boldsymbol{r}}\cdot\boldsymbol{r}_q)\,\mathrm{d}s_q. \tag{8.3}$$

The incident field can be expanded as

$$u_{\rm inc}(\boldsymbol{r}) = \sum_{n=0}^{\infty}(\mathrm{i}k)^n\,u_{\rm inc}^n(\boldsymbol{r}) \tag{8.4}$$

in a region that includes the scatterer. Then, it can be shown [262, §5.A.1] that, near S,

$$u(\boldsymbol{r}) = u_{\rm inc}^0(\boldsymbol{r}) + v_{\rm s}(\boldsymbol{r}) + O(k) \quad \text{as } k \to 0,$$

where $v_{\rm s}$ is the unique solution of the following Dirichlet problem:

$$\nabla^2 v_{\rm s} = 0 \quad \text{in } B_{\rm e},$$
$$v_{\rm s} = -u_{\rm inc}^0 \quad \text{on } S \text{ and}$$
$$v_{\rm s} = O(r^{-1}) \quad \text{as } r \to \infty.$$

It follows from (8.3) (after expanding the exponential therein) that

$$f(\hat{\boldsymbol{r}}) = -\frac{\mathrm{i}k}{4\pi}\int_S \frac{\partial}{\partial n_q}\left\{u_{\rm inc}^0 + v_{\rm s}\right\}\mathrm{d}s_q + O(k^2) \quad \text{as } k \to 0. \tag{8.5}$$

Thus, the far field of a small soft scatterer does not depend on the direction of observation, $\hat{\boldsymbol{r}}$: the scattering is *isotropic*.

For the special case of an incident plane wave, given by

$$u_{\rm inc}(\boldsymbol{r}) = \exp(\mathrm{i}k\boldsymbol{r}\cdot\hat{\boldsymbol{\alpha}}), \tag{8.6}$$

we have $u^0_{\text{inc}} \equiv 1$, and so the formulae above simplify. Thus, $v_s = -\phi_c$, where

$$\begin{aligned} \nabla^2 \phi_c &= 0 \quad \text{in } B_e, \\ \phi_c &= 1 \quad \text{on } S \text{ and} \\ \phi_c &= O(r^{-1}) \quad \text{as } r \to \infty. \end{aligned}$$

The harmonic function ϕ_c is called the *conductor potential*; see [262, eqn (3.42)]. It can be found by solving a boundary integral equation, using $|\boldsymbol{r}_P - \boldsymbol{r}_Q|^{-1}$ as a fundamental solution. For example, we can write

$$\phi_c(\boldsymbol{r}) = \int_S \mu_0(q) \frac{\mathrm{d}s_q}{|\boldsymbol{r} - \boldsymbol{r}_q|}, \tag{8.7}$$

whence μ_0 solves

$$\int_S \mu_0(q) \frac{\mathrm{d}s_q}{|\boldsymbol{r}_p - \boldsymbol{r}_q|} = 1 \quad \text{for } p \in S. \tag{8.8}$$

Making use of ϕ_c, (8.5) gives the far-field pattern as

$$f(\hat{\boldsymbol{r}}; \hat{\boldsymbol{\alpha}}) = -\mathrm{i}k\mathcal{C} + O(k^2) \quad \text{as } k \to 0, \tag{8.9}$$

where the constant

$$\mathcal{C} = -\frac{1}{4\pi} \int_S \frac{\partial \phi_c}{\partial n} \, \mathrm{d}s \tag{8.10}$$

is the *capacity* of S [597, p. 330], [405, §2.9]; if ϕ_c is sought in the form (8.7), (8.10) gives

$$\mathcal{C} = \int_S \mu_0(q) \, \mathrm{d}s_q. \tag{8.11}$$

(For a sphere of radius a, $\phi_c = a/r$ and $\mathcal{C} = a$.) Thus, the far field of a small soft scatterer is independent of both the observation direction and the direction of incidence.

We can use (8.9) to calculate the T-matrix. Substituting in (7.76), we have

$$T^{m\mu}_{n\nu} = -\mathrm{i}k\mathcal{C}\, \overline{y^m_n}\, y^\mu_\nu + O(k^2) \quad \text{as } k \to 0,$$

where

$$y^m_n = \frac{(-\mathrm{i})^n}{\sqrt{4\pi}} \int_\Omega Y^m_n(\hat{\boldsymbol{r}}) \, \mathrm{d}\Omega(\hat{\boldsymbol{r}}) = \delta_{n0}\delta_{m0} \tag{8.12}$$

(using $Y^0_0 = (4\pi)^{-1/2}$). Thus, we find that every entry of the T-matrix is $O(k^2)$ except that

$$T^{00}_{00} = -\mathrm{i}k\mathcal{C} + O(k^2) \quad \text{as } k \to 0.$$

Consequently, for any incident field $u_{\text{inc}}(\boldsymbol{r})$, we have

$$u_{\text{sc}}(\boldsymbol{r}) \simeq T^{00}_{00}\, d^0_0\, \psi^0_0(\boldsymbol{r}),$$

where

$$u_{\text{inc}}(\boldsymbol{r}) = \sum_{n,m} d_n^m \, \hat{\psi}_n^m(\boldsymbol{r}), \tag{8.13}$$

whence $d_0^0 = \sqrt{4\pi}\, u_{\text{inc}}(\boldsymbol{0})$. Thus,

$$u_{\text{sc}}(\boldsymbol{r}) \simeq -\mathrm{i}k\mathcal{C}\, u_{\text{inc}}(\boldsymbol{0})\, h_0(kr), \tag{8.14}$$

where $h_0(w) = \mathrm{e}^{\mathrm{i}w}/(\mathrm{i}w)$. In fact, to be consistent, we should replace $u_{\text{inc}}(\boldsymbol{0})$ in (8.14) by $u_{\text{inc}}^0(\boldsymbol{0})$.

Notice that the approximation (8.14) holds everywhere outside the escribed sphere to S, not just in the far field.

8.2.2 Three dimensions: sound-hard obstacles

For sound-hard obstacles, the scattered field is given exactly by (5.80) and the far-field pattern (defined by (8.2)) is given by (5.81) as

$$f(\hat{\boldsymbol{r}}) = \frac{k^2}{4\pi} \int_S u(q)\, [\hat{\boldsymbol{r}} \cdot \boldsymbol{n}(q)] \exp(-\mathrm{i}k\hat{\boldsymbol{r}} \cdot \boldsymbol{r}_q)\, \mathrm{d}s_q. \tag{8.15}$$

For low frequencies, it can be shown [262, §5.A.2] that, near S,

$$u(\boldsymbol{r}) = \left(u_{\text{inc}}^0 + v_{\text{h}}^0\right) + \mathrm{i}k\left(u_{\text{inc}}^1 + v_{\text{h}}^1\right) + O(k^2) \quad \text{as } k \to 0, \tag{8.16}$$

where u_{inc}^0 and u_{inc}^1 appear in the expansion (8.4), and v_{h}^0 and v_{h}^1 solve Neumann problems for Laplace's equation:

$$\nabla^2 v_{\text{h}}^m = 0 \quad \text{in } B_{\text{e}},$$
$$\frac{\partial v_{\text{h}}^m}{\partial n} = -\frac{\partial u_{\text{inc}}^m}{\partial n} \quad \text{on } S \text{ and}$$
$$v_{\text{h}}^m = O(r^{-2}) \quad \text{as } r \to \infty,$$

with $m = 0, 1$. These problems are uniquely solvable. Hence, (8.15) gives

$$f(\hat{\boldsymbol{r}}) = \frac{k^2}{4\pi} \int_S [\hat{\boldsymbol{r}} \cdot \boldsymbol{n}(q)] \{f_0 + \mathrm{i}kf_1\}\, \mathrm{d}s_q + O(k^4) \quad \text{as } k \to 0, \tag{8.17}$$

where

$$f_0(q) = u_{\text{inc}}^0 + v_{\text{h}}^0, \quad f_1(q) = u_{\text{inc}}^1 + v_{\text{h}}^1 - (\hat{\boldsymbol{r}} \cdot \boldsymbol{r}_q)(u_{\text{inc}}^0 + v_{\text{h}}^0)$$

and we have expanded the exponential in (8.15).

For incident plane waves, we have $u_{\text{inc}}^0 \equiv 1$, whence $v_{\text{h}}^0 \equiv 0$, and so the leading term in (8.17) reduces to

$$\frac{k^2}{4\pi} \int_S [\hat{\boldsymbol{r}} \cdot \boldsymbol{n}(q)]\, \mathrm{d}s_q = 0, \tag{8.18}$$

by the divergence theorem. (This explains why we retained two terms in the expansion (8.16).) For the next term, we have $u_{\text{inc}}^1(\boldsymbol{r}) = \boldsymbol{r} \cdot \hat{\boldsymbol{\alpha}}$ and so $\partial v_{\text{h}}^1/\partial n = -\boldsymbol{n} \cdot \hat{\boldsymbol{\alpha}}$ on S. Therefore, it is convenient to introduce the vector field $\boldsymbol{\Psi}$, defined by

$$\nabla^2 \boldsymbol{\Psi} = \boldsymbol{0} \quad \text{in } B_{\text{e}}, \tag{8.19}$$

$$\frac{\partial \boldsymbol{\Psi}}{\partial n} = \boldsymbol{n} \quad \text{on } S \text{ and} \tag{8.20}$$

$$\boldsymbol{\Psi} = O(r^{-2}) \quad \text{as } r \to \infty; \tag{8.21}$$

see [262, eqn (5.20)]. Hence

$$f_1(q) = \boldsymbol{r}_q \cdot (\hat{\boldsymbol{\alpha}} - \hat{\boldsymbol{r}}) - \hat{\boldsymbol{\alpha}} \cdot \boldsymbol{\Psi}(q). \tag{8.22}$$

As $\operatorname{div}_Q \{\hat{\boldsymbol{r}}[\boldsymbol{r}_Q \cdot (\hat{\boldsymbol{\alpha}} - \hat{\boldsymbol{r}})]\} = \hat{\boldsymbol{r}} \cdot \hat{\boldsymbol{\alpha}} - 1$,

$$f(\hat{\boldsymbol{r}}; \hat{\boldsymbol{\alpha}}) = \frac{\mathrm{i}k^3}{4\pi} \left\{ V_B (\hat{\boldsymbol{r}} \cdot \hat{\boldsymbol{\alpha}} - 1) - \int_S [\hat{\boldsymbol{r}} \cdot \boldsymbol{n}(q)] [\hat{\boldsymbol{\alpha}} \cdot \boldsymbol{\Psi}(q)] \, \mathrm{d}s_q \right\} + O(k^4)$$

as $k \to 0$, where V_B is the volume of the scatterer B.

Now, following Dassios & Kleinman [262, p. 166], we define the *virtual mass tensor* $\mathbf{W}$ by

$$W_{ij} = -\int_S n_i \Psi_j \, \mathrm{d}s = W_{ji}, \tag{8.23}$$

and the *magnetic polarizability tensor* $\mathbf{M}$ by

$$M_{ij} = W_{ij} + V_B \delta_{ij} = M_{ji}. \tag{8.24}$$

(For information on properties of $\mathbf{M}$, see [1002, Chapter 5], [632] and [262, §5.0]. In particular, for the special case of a sphere, $M_{ij} = \frac{3}{2} V_B \delta_{ij}$.) Then, we can express the far-field pattern concisely by

$$f(\hat{\boldsymbol{r}}; \hat{\boldsymbol{\alpha}}) = \frac{\mathrm{i}k^3}{4\pi} \{\hat{\boldsymbol{r}} \cdot \mathbf{M} \cdot \hat{\boldsymbol{\alpha}} - V_B\} + O(k^4) \quad \text{as } k \to 0. \tag{8.25}$$

Thus, the far field of a small hard scatterer depends linearly on both the observation direction and the incident direction, and it is much smaller than the far field of a small soft scatterer. This result was known to Lord Rayleigh; see [1010, §334, eqn (21)].

It turns out that the far field of a small, penetrable, homogeneous obstacle (transmission problem) has a form very similar to (8.25) [262, §5.A.4]:

$$f(\hat{\boldsymbol{r}}; \hat{\boldsymbol{\alpha}}) = -\frac{\mathrm{i}k^3}{4\pi} \{\hat{\boldsymbol{r}} \cdot \mathbf{X}(\gamma) \cdot \hat{\boldsymbol{\alpha}} + [1 - \gamma (k_0/k)^2] V_B\} + O(k^4) \tag{8.26}$$

as $k \to 0$. Here, k_0 is the wavenumber inside B, γ is the parameter occurring in the transmission conditions across S and $\mathbf{X}(\gamma)$ is known as the *general polarisability tensor* [262, p. 167]; it can be shown that $\mathbf{X}(0) = -\mathbf{M}$, so that (8.26) reduces to (8.25) as $\gamma \to 0$.

We can use (8.25) to calculate the T-matrix for a small sound-hard scatterer. Substituting in (7.76), we obtain

$$T_{n\nu}^{m\mu} = \frac{\mathrm{i}k^3}{4\pi}\left\{\overline{\mathbf{z}_n^m}\cdot\mathbf{M}\cdot\mathbf{z}_\nu^\mu - V_B\,\overline{y_n^m}\,y_\nu^\mu\right\} + O(k^4) \quad \text{as } k\to 0, \tag{8.27}$$

where $y_n^m = \delta_{n0}\delta_{m0}$ is defined by (8.12) and

$$\mathbf{z}_n^m = \frac{(-\mathrm{i})^n}{\sqrt{4\pi}}\int_\Omega \hat{\boldsymbol{r}}\,Y_n^m(\hat{\boldsymbol{r}})\,\mathrm{d}\Omega(\hat{\boldsymbol{r}}).$$

Now, from (3.24), we have

$$Y_1^0 = \sqrt{\frac{3}{4\pi}}\cos\theta, \quad Y_1^1 = -\sqrt{\frac{3}{8\pi}}\,\mathrm{e}^{\mathrm{i}\phi}\sin\theta, \quad Y_1^{-1} = \sqrt{\frac{3}{8\pi}}\,\mathrm{e}^{-\mathrm{i}\phi}\sin\theta. \tag{8.28}$$

Hence, using $\hat{\boldsymbol{r}} = (\sin\theta\cos\phi,\ \sin\theta\sin\phi,\ \cos\theta)$, we find that

$$\mathbf{z}_1^0 = (0,0,-\mathrm{i})/\sqrt{3}, \quad \mathbf{z}_1^1 = (\mathrm{i},-1,0)/\sqrt{6} \quad \text{and} \quad \mathbf{z}_1^{-1} = \overline{\mathbf{z}_1^1},$$

with $\mathbf{z}_n^m = \mathbf{0}$ otherwise.

Substitution in (8.27) shows that the T-matrix has ten entries that are $O(k^3)$ as $k\to 0$:

$$\begin{aligned}
T_{00}^{00} &= -\mathrm{i}k^3 V_B/(4\pi),\\
T_{11}^{00} &= \mathrm{i}k^3 M_{33}/(12\pi),\\
T_{11}^{01} &= -\mathrm{i}k^3(M_{31}+\mathrm{i}M_{32})/(12\pi\sqrt{2}) = -T_{11}^{-1,0},\\
T_{11}^{10} &= -\mathrm{i}k^3(M_{31}-\mathrm{i}M_{32})/(12\pi\sqrt{2}) = -T_{11}^{0,-1},\\
T_{11}^{11} &= \mathrm{i}k^3(M_{11}+M_{22})/(24\pi) = T_{11}^{-1,-1},\\
T_{11}^{1,-1} &= \mathrm{i}k^3(M_{22}-M_{11}+2\mathrm{i}M_{12})/(24\pi),\\
T_{11}^{-1,1} &= \mathrm{i}k^3(M_{22}-M_{11}-2\mathrm{i}M_{12})/(24\pi).
\end{aligned}$$

Notice that these results are in accord with Theorem 7.4. In particular, (7.68) shows that

$$T_{11}^{m\mu} + \overline{T_{11}^{\mu m}} = O(k^6) \quad \text{as } k\to 0.$$

Let us calculate the scattered field for any incident field, $u_{\mathrm{inc}}(\boldsymbol{r})$. We introduce a vector $\boldsymbol{U}$ with components

$$U_j = \frac{1}{k}\frac{\partial u_{\mathrm{inc}}}{\partial x_j} \quad \text{evaluated at } \boldsymbol{r}=\mathbf{0}. \tag{8.29}$$

Then, applying $\partial/\partial x_j$ to (8.13), making use of Theorems 3.13 and 3.14, we find

$$d_1^0 = \sqrt{12\pi}\,U_3, \quad d_1^1 = -\sqrt{6\pi}(U_1-\mathrm{i}U_2) \quad \text{and} \quad d_1^{-1} = \sqrt{6\pi}(U_1+\mathrm{i}U_2).$$

Also, as before, $d_0^0 = \sqrt{4\pi}\,u_{\mathrm{inc}}(\mathbf{0})$.

Next, we calculate

$$c_n^m = \sum_{\nu,\mu} T_{n\nu}^{m\mu} \, d_\nu^\mu.$$

For example, we have

$$\begin{aligned} c_1^1 &= T_{11}^{11} \, d_1^1 + T_{11}^{10} \, d_1^0 + T_{11}^{1,-1} \, d_1^{-1} \\ &= \frac{1}{2}(T_{11}^{11} + T_{11}^{1,-1})(d_1^1 + d_1^{-1}) + \frac{1}{2}(T_{11}^{11} - T_{11}^{1,-1})(d_1^1 - d_1^{-1}) + T_{11}^{10} d_1^0 \\ &= \frac{\mathrm{i}k^3}{\sqrt{24\pi}}(\mathrm{i}M_{2j} - M_{1j})U_j, \end{aligned}$$

where summation over j from 1 to 3 is implied. Similarly,

$$c_1^{-1} = \frac{\mathrm{i}k^3}{\sqrt{24\pi}}(\mathrm{i}M_{2j} + M_{1j})U_j,$$

$$c_1^0 = \frac{\mathrm{i}k^3}{\sqrt{12\pi}} M_{3j} U_j \quad \text{and} \quad c_0^0 = -\frac{\mathrm{i}k^3}{\sqrt{4\pi}} V_B u_{\mathrm{inc}}(\mathbf{0}).$$

All the other coefficients c_n^m are asymptotically smaller. Hence, the scattered field is given by

$$\begin{aligned} u_{\mathrm{sc}}(\boldsymbol{r}) &\simeq c_0^0 \psi_0^0 + c_1^1 \psi_1^1 + c_1^0 \psi_1^0 + c_1^{-1} \psi_1^{-1} \\ &\simeq -\frac{\mathrm{i}k^3}{4\pi} V_B u_{\mathrm{inc}}(\mathbf{0}) \, h_0(kr) + \frac{\mathrm{i}k^3}{\sqrt{24\pi}} X_j U_j \, h_1(kr), \end{aligned}$$

where

$$X_j = (Y_1^{-1} - Y_1^1)M_{1j} + \mathrm{i}(Y_1^{-1} + Y_1^1)M_{2j} + \sqrt{2} Y_1^0 M_{3j}.$$

Making use of (8.28), we finally obtain

$$u_{\mathrm{sc}}(\boldsymbol{r}) \simeq \frac{\mathrm{i}k^3}{4\pi} \{\hat{\boldsymbol{r}} \cdot \mathbf{M} \cdot \boldsymbol{U} \, h_1(kr) - V_B u_{\mathrm{inc}}(\mathbf{0}) \, h_0(kr)\}. \tag{8.30}$$

The far-field form of this expression is essentially [1002, p. 90, eqn (39)].

8.2.3 Two dimensions: introduction

The analysis for two-dimensional problems is more involved. To fix ideas, let us consider scattering by a sound-soft cylindrical obstacle. The scattered field is then given exactly by (5.78) and the far field is given by (4.9). From (5.79), the far-field pattern is

$$f(\hat{\boldsymbol{r}}) = -\frac{\mathrm{i}}{4} \int_S \frac{\partial u}{\partial n_q} \exp(-\mathrm{i}k\hat{\boldsymbol{r}} \cdot \boldsymbol{r}_q) \, \mathrm{d}s_q. \tag{8.31}$$

As $\hat{\boldsymbol{r}} = (\cos\theta, \sin\theta)$, we can write $f(\hat{\boldsymbol{r}}) = f(\theta)$.

For low frequencies, we expand u_{inc} as (8.4). Then, near S, we have

$$u(\boldsymbol{r}) = u_{\text{inc}}^0(\boldsymbol{r}) + w_s(\boldsymbol{r}) + O(|\log k|^{-1}) \quad \text{as } k \to 0, \tag{8.32}$$

where w_s solves the following Dirichlet problem:

$$\nabla^2 w_s = 0 \quad \text{in } B_e,$$
$$w_s = -u_{\text{inc}}^0 \quad \text{on } S \text{ and}$$
$$w_s = O(1) \quad \text{as } r \to \infty.$$

The unique solution of this problem can be found by solving a boundary integral equation over S [1055, Theorem 3.2.7].

The result (8.32) is given in [761] and [1006, p. 330], for example. However, for an incident plane wave,

$$u_{\text{inc}} = e^{ikr\cos(\theta-\alpha)}, \tag{8.33}$$

we have $u_{\text{inc}}^0 = 1$ and $w_s = -1$, so that the leading term in (8.32) gives a zero contribution to the integral in (8.31): therefore, we have to determine the term of order $|\log k|^{-1}$. We do this next, using the results of Kleinman & Vainberg [626].

8.2.4 Two dimensions: Kleinman–Vainberg theory

Kleinman & Vainberg [626] have given a systematic method for obtaining low-frequency asymptotic expansions in two dimensions for the following problem:

$$(\nabla^2 + k^2)v = F \text{ in } B_e, \ \mathcal{B}v = 0 \text{ on } S \text{ and}$$
$$v \text{ satisfies the Sommerfeld radiation condition as } r \to \infty.$$

Here, $\mathcal{B}v$ denotes v or $\partial v/\partial n$, and F has compact support. We shall refer to this problem as the *K–V problem*.

Our scattering problems can be formulated as

$$(\nabla^2 + k^2)u_{\text{sc}} = 0 \text{ in } B_e, \ \mathcal{B}u_{\text{sc}} = -\mathcal{B}u_{\text{inc}} \text{ on } S \text{ and}$$
$$u_{\text{sc}} \text{ satisfies the Sommerfeld radiation condition as } r \to \infty.$$

Let us convert this problem into the K–V problem. Choose c so that B is contained inside the disc $r < c$. Let χ be a C^∞ cut-off function, so that $\chi = 0$ for $r > 2c$ and $\chi = 1$ for $0 \le r < c$. Put

$$u_{\text{sc}} = -u_{\text{inc}}\chi + v.$$

Then, v solves the K–V problem with

$$F = (\nabla^2 + k^2)(u_{\text{inc}}\chi) = u_{\text{inc}}\nabla^2\chi + 2\,\text{grad}\,\chi \cdot \text{grad}\,u_{\text{inc}}.$$

Notice that the total field

$$u = u_{sc} + u_{inc} = (1-\chi)u_{inc} + v,$$

so that v is the total field near S ($r < c$) but v is the scattered field far from S ($r > 2c$).

The results of [626] assume that F does not depend on k, so we write

$$F(\boldsymbol{r}) = \sum_{n=0}^{\infty} (\mathrm{i}k)^n F_n(\boldsymbol{r}),$$

where

$$F_n = u_{inc}^n \nabla^2 \chi + 2 \operatorname{grad} \chi \cdot \operatorname{grad} u_{inc}^n. \tag{8.34}$$

We write v_n for the solution of the K–V problem with $F = F_n$, so that

$$v(\boldsymbol{r}) = \sum_{n=0}^{\infty} (\mathrm{i}k)^n v_n(\boldsymbol{r}). \tag{8.35}$$

We are going to approximate v_n near S using solutions of related static problems. The results are different depending on the boundary condition on S. Therefore, we now consider soft and hard cylinders separately.

8.2.5 Two dimensions: sound-soft cylinders

Consider the following uniquely solvable problems,

$$\begin{aligned} \nabla^2 u_0^n &= F_n \quad \text{in } B_e, \\ u_0^n &= 0 \quad \text{on } S \text{ and} \\ u_0^n &= O(1) \quad \text{as } r \to \infty \end{aligned}$$

and

$$\begin{aligned} \nabla^2 u_1 &= 0 \quad \text{in } B_e, \\ u_1 &= 0 \quad \text{on } S \text{ and} \\ u_1 &= \log(r/\ell) + o(1) \quad \text{as } r \to \infty. \end{aligned} \tag{8.36}$$

Here, the constant ℓ is unknown, but it is to be determined by solving the problem for u_1 with the specified logarithmic growth at infinity. (In [626], there is a parameter $\lambda_0 = -\log \ell$.) Also, let β be the complex constant occurring in the asymptotic approximation

$$H_0^{(1)}(w) = (2\mathrm{i}/\pi)(\log w - \beta) + O(w^2 \log w) \quad \text{as } w \to 0;$$

thus, $\beta = \log 2 - \gamma + \mathrm{i}\pi/2$, where $\gamma = 0.5772\ldots$ is Euler's constant. Then [626, Theorem 1], near S,

$$v_n(\boldsymbol{r}) = u_0^n(\boldsymbol{r}) + \frac{C_0^n\, u_1(\boldsymbol{r})}{\log k\ell - \beta} + O(k^2 \log k) \quad \text{as } k \to 0,$$

where $C_0^n = \lim_{r\to\infty} u_0^n(\boldsymbol{r})$. It follows from (8.35) that we only need v_0 here. Then, the far-field pattern is given by (8.31) as

$$f(\theta) = -\frac{\mathrm{i}}{4}\int_S \frac{\partial v_0}{\partial n_q}\,\mathrm{d}s_q + O(k) \quad \text{as } k \to 0. \tag{8.37}$$

For a plane wave, given by (8.33), $u_{\mathrm{inc}}^0 \equiv 1$ and $F_0 = \nabla^2\chi$. Thus $u_0^0(\boldsymbol{r}) = \chi(\boldsymbol{r}) - 1$, which is constant (in fact, zero) near S, and so (as expected) the leading term does not contribute to (8.37). The next term then gives

$$f(\theta;\alpha) = -\frac{\mathrm{i}}{4}\frac{C_0^0}{\log k\ell - \beta}\int_S \frac{\partial u_1}{\partial n_q}\,\mathrm{d}s_q + O(k) \quad \text{as } k \to 0. \tag{8.38}$$

However, we have $C_0^0 = \lim_{r\to\infty} u_0^0 = -1$ and Green's theorem gives

$$\int_S \frac{\partial u_1}{\partial n_q}\,\mathrm{d}s_q = \lim_{r\to\infty}\int_0^{2\pi} \frac{\partial u_1}{\partial r}\, r\,\mathrm{d}\theta = 2\pi.$$

Hence, (8.38) reduces to

$$f(\theta;\alpha) = \frac{1}{2}\pi\mathrm{i}\,(\log k\ell - \beta)^{-1} + O(k) \quad \text{as } k \to 0. \tag{8.39}$$

The parameter ℓ can be obtained explicitly for simple shapes. For a circle of radius a, $u_1 = \log(r/a)$ and so $\ell = a$. For an ellipse, we can use elliptic cylindrical coordinates, (ξ, η), defined by (2.49), giving $u_1 = \xi - \xi_0$ (where $\xi = \xi_0$ defines S); then (2.50) gives (8.36) with

$$\ell = \frac{1}{2}c\,\mathrm{e}^{\xi_0} = \frac{1}{2}(a+b).$$

The corresponding low-frequency approximation for scattering by a soft elliptic cylinder can be found in [916, eqn (48a)] and [135, §3.2.1.2].

We could simplify the formula (8.39) further, obtaining

$$\begin{aligned} f(\theta;\alpha) &= \frac{1}{2}\pi\mathrm{i}\,(\log k\ell)^{-1} + O(|\log k|^{-2}) \\ &= \frac{1}{2}\pi\mathrm{i}\,(\log kL)^{-1} + O(|\log k|^{-2}) \end{aligned}$$

as $k \to 0$, where L is *any* constant length; for example, we could choose $L = a$, where $2a$ is the diameter of B, or we could choose L to be the length of the boundary S. With this latter choice, we obtain

$$|f(\theta;\alpha)| \simeq -\frac{1}{2}\pi\,(\log kL)^{-1},$$

a formula that was obtained by Van Bladel [1230, p. 200] in 1963; see also [475].

We can use (8.39) to calculate the T-matrix. Substituting in (7.79), we find that

$$T_{00} = \frac{1}{2}\mathrm{i}\pi(\log k\ell - \beta)^{-1} + O(k) \quad \text{as } k \to 0;$$

all other entries T_{nm} are asymptotically smaller. Consequently, for any incident field $u_{\text{inc}}(\boldsymbol{r})$, we obtain

$$u_{\text{sc}}(\boldsymbol{r}) \simeq \frac{1}{2}\mathrm{i}\pi(\log k\ell - \beta)^{-1} u_{\text{inc}}(\boldsymbol{0})\, H_0(kr). \tag{8.40}$$

8.2.6 Two dimensions: sound-hard cylinders

For a sound-hard cylinder, the scattered field is given by (5.76) and the far-field pattern (see Definition 4.1) is given by (5.77) as

$$f(\hat{\boldsymbol{r}}) = \frac{k}{4}\int_S u(q)\ [\hat{\boldsymbol{r}}\cdot\boldsymbol{n}(q)]\exp(-\mathrm{i}k\hat{\boldsymbol{r}}\cdot\boldsymbol{r}_q)\,\mathrm{d}s_q. \tag{8.41}$$

For low-frequency approximations, we are led to consider the following uniquely solvable problems,

$$\nabla^2 w_m = F_m \quad \text{in } B_{\text{e}},$$
$$\frac{\partial w_m}{\partial n} = 0 \quad \text{on } S \text{ and}$$
$$w_m = \alpha_m(\log r - \beta) + o(1) \quad \text{as } r \to \infty,$$

where $m = 0, 1, 2, \ldots$ and

$$\alpha_m = \frac{1}{2\pi}\int_{B_{\text{e}}} F_m\,\mathrm{d}V.$$

Then [626, Theorem 2], near S,

$$v_n(\boldsymbol{r}) = \alpha_n \log k + w_n(\boldsymbol{r}) + O(k^2\log^3 k) \quad \text{as } k \to 0.$$

For a plane wave, $u_{\text{inc}}^0 \equiv 1$ and $u_{\text{inc}}^1(\boldsymbol{r}) = \boldsymbol{r}\cdot\hat{\boldsymbol{\alpha}}$. Then, (8.34) gives $F_0 = \nabla^2\chi$ and $F_1 = \nabla^2(\chi\boldsymbol{r}\cdot\hat{\boldsymbol{\alpha}})$. Hence,

$$2\pi\alpha_0 = \int_{B_{\text{e}}} \nabla^2\chi\,\mathrm{d}V = -\int_S \frac{\partial\chi}{\partial n}\,\mathrm{d}s = 0$$

and

$$2\pi\alpha_1 = -\int_S \frac{\partial}{\partial n}\{\chi\boldsymbol{r}\cdot\hat{\boldsymbol{\alpha}}\}\,\mathrm{d}s = -\int_S \boldsymbol{n}\cdot\hat{\boldsymbol{\alpha}}\,\mathrm{d}s = 0,$$

where the last equality follows from an application of the divergence theorem in B, and we have used $\chi = 1$ near S. Thus,

$$v = \sum_{n=0}^{\infty}(\mathrm{i}k)^n v_n = w_0 + \mathrm{i}kw_1 + O(k^2\log^3 k) \quad \text{as } k \to 0.$$

Then, as $u \equiv v$ near S, (8.41) gives

$$f(\hat{\boldsymbol{r}};\hat{\boldsymbol{\alpha}}) = \frac{k}{4}\int_S [\hat{\boldsymbol{r}}\cdot\boldsymbol{n}(q)]\ (w_0 + \mathrm{i}kf_1)\,\mathrm{d}s_q + O(k^3\log^3 k) \tag{8.42}$$

as $k \to 0$, where $f_1(q) = w_1 - w_0\hat{\boldsymbol{r}}\cdot\boldsymbol{r}_q$.

As $\alpha_0 = 0$ and $F_0 = \nabla^2\chi$, we obtain $w_0 = \chi$. Then, (8.18) shows that the leading contribution to (8.42) vanishes. As $\alpha_1 = 0$ and $F_1 = \nabla^2(\chi \boldsymbol{r} \cdot \hat{\boldsymbol{\alpha}})$, we obtain

$$w_1(\boldsymbol{r}) = \chi \boldsymbol{r} \cdot \hat{\boldsymbol{\alpha}} - \hat{\boldsymbol{\alpha}} \cdot \boldsymbol{\Psi},$$

where the two-dimensional vector field $\boldsymbol{\Psi}$ is defined by (8.19), (8.20) and $\boldsymbol{\Psi} = O(r^{-1})$ as $r \to \infty$. It follows that $f_1(q)$ is given precisely by (8.22) and then

$$f(\hat{\boldsymbol{r}}; \hat{\boldsymbol{\alpha}}) = \frac{1}{4}\mathrm{i}k^2 \{\hat{\boldsymbol{r}} \cdot \mathbf{M} \cdot \hat{\boldsymbol{\alpha}} - A_B\} + O(k^3 \log^3 k) \quad \text{as } k \to 0, \tag{8.43}$$

where A_B is the area of B,

$$M_{ij} = W_{ij} + A_B \delta_{ij} = M_{ji}$$

and W_{ij} is defined by (8.23). Thus, the far-field pattern is given by a formula that is similar to the analogous formula in three dimensions, namely (8.25). A formula similar to (8.43) can be found in [475].

For a circle of radius a, $M_{ij} = 2A_B\delta_{ij}$ and then (8.43) reduces to

$$f(\theta; \alpha) \simeq \frac{1}{4}\mathrm{i}\pi(ka)^2\{2\cos(\theta - \alpha) - 1\},$$

in agreement with the known result [135, eqn (2.43)].

Substitution of (8.43) in (7.79) gives the T-matrix for a small sound-hard cylinder. We obtain

$$T_{nm} = \frac{1}{4}\mathrm{i}k^2 \{\overline{\mathbf{z}_n} \cdot \mathbf{M} \cdot \mathbf{z}_m - A_B\delta_{n0}\delta_{m0}\} + O(k^3 \log^3 k) \quad \text{as } k \to 0,$$

where

$$\mathbf{z}_n = \frac{(-\mathrm{i})^n}{2\pi}\int_{-\pi}^{\pi} \hat{\boldsymbol{r}}\, \mathrm{e}^{\mathrm{i}n\theta}\, \mathrm{d}\theta.$$

Thus,

$$\mathbf{z}_1 = \frac{1}{2}(-\mathrm{i}, 1) = \overline{\mathbf{z}_{-1}} \quad \text{and} \quad \mathbf{z}_n = \mathbf{0} \quad \text{for } n \neq \pm 1.$$

Hence, the T-matrix has five entries that are $O(k^2)$ as $k \to 0$:

$$T_{00} = -\frac{1}{4}\mathrm{i}k^2 A_B,$$

$$T_{11} = \frac{1}{16}\mathrm{i}k^2(M_{11} + M_{22}) = T_{-1,-1},$$

$$T_{1,-1} = \frac{1}{16}\mathrm{i}k^2(M_{22} - M_{11} + 2\mathrm{i}M_{12}),$$

$$T_{-1,1} = \frac{1}{16}\mathrm{i}k^2(M_{22} - M_{11} - 2\mathrm{i}M_{12}).$$

Next, we calculate the scattered field for any incident field, $u_{\text{inc}}(\boldsymbol{r})$. We have

$$d_0 = u_{\text{inc}}(\mathbf{0}), \quad d_1 = U_1 - \mathrm{i}U_2 \quad \text{and} \quad d_{-1} = -U_1 - \mathrm{i}U_2,$$

where $\boldsymbol{U} = (U_1, U_2)$ is defined by (8.29) with $x_1 = x$ and $x_2 = y$,

$$u_{\text{inc}}(\boldsymbol{r}) = \sum_n d_n J_n(kr)\,\mathrm{e}^{\mathrm{i}n\theta}$$

and we have used Theorem 2.9 with $m = 1$. For the scattered field, we calculate $c_n = \sum_m T_{nm} d_m$ and find

$$c_0 = -\frac{1}{4}\mathrm{i}k^2 A_B u_{\text{inc}}(\mathbf{0}),$$

$$c_1 = \frac{1}{8}\mathrm{i}k^2 (M_{1j} - \mathrm{i}M_{2j})U_j,$$

$$c_{-1} = -\frac{1}{8}\mathrm{i}k^2 (M_{1j} + \mathrm{i}M_{2j})U_j,$$

where summation over $j = 1$ and $j = 2$ is implied. All the other coefficients c_n are asymptotically smaller. Hence, the scattered field is

$$\begin{aligned} u_{\text{sc}} &\simeq c_0\psi_0 + c_1\psi_1 + c_{-1}\psi_{-1} \\ &\simeq \frac{1}{4}\mathrm{i}k^2 \left\{\hat{\boldsymbol{r}}\cdot\mathbf{M}\cdot\boldsymbol{U}\,H_1(kr) - A_B u_{\text{inc}}(\mathbf{0})\,H_0(kr)\right\}. \end{aligned} \tag{8.44}$$

8.2.7 Literature

There is an extensive literature on low-frequency scattering. For the most comprehensive bibliography, see the book by Dassios & Kleinman [262]. This book also develops the three-dimensional theory in detail, in the contexts of acoustic, electromagnetic and elastodynamic scattering. For Rayleigh scattering, in particular, see [1205, 632].

The literature on two-dimensional problems is more fragmented. The definitive theoretical analysis is probably [626]. Earlier analyses are [916, 1230, 761, 475]. See also [1006, Appendix 1].

For scattering by specific simple shapes, see [135, 262].

Low-frequency approximations for the T-matrix have been given in [1255], but only for ellipses and spheroids.

For water waves, the definitive review is by McIver [831].

For the scattering of long waves by a cluster of obstacles, one can regard the cluster as a single obstacle, and then the general theory applies. However, this theory requires the solution of an associated harmonic problem for the actual geometry of the cluster, and this is difficult, in general. For acoustic scattering by two soft spheres, see [1013, 183], where the associated harmonic problem is solved using bispherical coordinates [884, p. 1298]. Nevertheless, further approximations can be made in some situations.

To fix ideas, consider scattering by two spheres, each of radius a. Three (related) dimensionless parameters arise,

$$ka, \quad kb \quad \text{and} \quad a/b,$$

where b is the distance between the centres of the spheres. Standard low-frequency theory gives an approximation for a fixed geometry (a/b fixed) and long waves ($kb \ll 1$). However, in many applications, the individual scatterers are small compared with both the spacing and the wavelength, so that

$$ka \ll 1 \quad \text{and} \quad a/b \ll 1.$$

Under these conditions, we can obtain two different approximations, depending on the size of kb. If

$$kb = O(1) \quad \text{and} \quad a/b = O(ka) \quad \text{as } ka \to 0,$$

we obtain a 'small-body' solution [830]. This corresponds to $a \to 0$ with k and b both fixed. With this approximation, there is really only one small parameter, ka. Approximations of this kind are often employed; for example, wide-spacing approximations fall into this class.

Alternatively, suppose that

$$ka \ll a/b \ll 1,$$

so that the waves are long compared with the spacing. Then, there are two independent small parameters, kb and a/b. Exploiting this fact requires the *method of matched asymptotic expansions*; see [56, 829, 830] and [831, §11]. In particular, acoustic scattering by N identical hard circular cylinders is considered in [829]. Analogous water-wave problems for floating hemispheres and horizontal submerged circular cylinders are considered in [829] and [830], respectively.

8.3 Foldy's method

In 1945, Foldy [354] published a well-known paper, entitled 'The multiple scattering of waves I. General theory of isotropic scattering by randomly distributed scatterers'. Thus: 'the problem of the multiple scattering of scalar waves by a random distribution of isotropic scatterers is considered in detail on the basis of a consistent wave treatment' [354, Abstract]. Most of the paper is concerned with aspects of randomness, but, in this section, we are interested in the 'consistent wave treatment': it fits on p. 110 of the paper.

Foldy considers 'isotropic point scatterers', meaning that 'in the neighborhood of the j-th scatterer', the scattered field 'will behave like'

$$A_j G(\boldsymbol{r} - \boldsymbol{r}_j),$$

where the jth scatterer is centred at $\boldsymbol{r}_j$, A_j is an unknown amplitude and G is the free-space Green's function; in two dimensions, we can take

$$G(\boldsymbol{r}) = H_0(k|\boldsymbol{r}|),$$

whereas in three dimensions, we can take

$$G(\boldsymbol{r}) = h_0(k|\boldsymbol{r}|) = \exp(\mathrm{i}k|\boldsymbol{r}|)/(\mathrm{i}k|\boldsymbol{r}|). \tag{8.45}$$

(The chosen normalisation factors are unimportant, but we shall use (8.45) later in this chapter.)

Foldy represents the total field as

$$u(\boldsymbol{r}) = u_{\mathrm{inc}}(\boldsymbol{r}) + \sum_j A_j G(\boldsymbol{r} - \boldsymbol{r}_j), \tag{8.46}$$

where the sum is over all the scatterers. The so-called 'external field' is

$$\begin{aligned} u_n(\boldsymbol{r}) &\equiv u(\boldsymbol{r}) - A_n G(\boldsymbol{r} - \boldsymbol{r}_n) \\ &= u_{\mathrm{inc}}(\boldsymbol{r}) + \sum_{\substack{j \\ j \neq n}} A_j G(\boldsymbol{r} - \boldsymbol{r}_j). \end{aligned} \tag{8.47}$$

It can be regarded as the field incident on the nth scatterer, in the presence of all the other scatterers.

Now, 'characterize the scattering properties of the scatterers by'

$$A_n = g_n u_n(\boldsymbol{r}_n), \quad \text{(no sum!)} \tag{8.48}$$

'making the strength of the scattered wave from a scatterer [namely, A_n] proportional to the external field acting on it' (namely, $u_n(\boldsymbol{r}_n)$). Foldy refers to g_n 'as the *scattering coefficient* for the n-th scatterer'. Thus, the scattered field is determined by the value of the 'external field' at the centre of the scatterer, $\boldsymbol{r}_n$, together with the quantity g_n (which we will come back to later).

So, evaluating (8.47) at $\boldsymbol{r}_n$ gives, after using (8.48),

$$u_n(\boldsymbol{r}_n) = u_{\mathrm{inc}}(\boldsymbol{r}_n) + \sum_{\substack{j \\ j \neq n}} g_j\, u_j(\boldsymbol{r}_j)\, G(\boldsymbol{r}_n - \boldsymbol{r}_j), \tag{8.49}$$

which is a linear system of algebraic equations for $u_j(\boldsymbol{r}_j)$. Then, the total field is given by (8.46) and (8.48) as

$$u(\boldsymbol{r}) = u_{\mathrm{inc}}(\boldsymbol{r}) + \sum_j g_j\, u_j(\boldsymbol{r}_j)\, G(\boldsymbol{r} - \boldsymbol{r}_j). \tag{8.50}$$

Thus: 'These represent the fundamental equations of multiple scattering' [354, p. 110].

Equivalently, as we want the amplitudes A_j in (8.46), we can rewrite (8.49) as

$$g_n^{-1} A_n = u_{\mathrm{inc}}(\boldsymbol{r}_n) + \sum_{\substack{j \\ j \neq n}} A_j\, G(\boldsymbol{r}_n - \boldsymbol{r}_j). \tag{8.51}$$

The direct numerical solution of Foldy's linear system of equations was attempted by Fikioris in 1966 [345]; he envisaged a Monte Carlo approach for calculating average properties of random arrangements of spheres, but was severely limited by the computing power available. Later, Foldy's system was solved numerically (in two dimensions) for many scatterers; see [442]. Electromagnetic versions of Foldy's method have been used in [1299, 562, 563].

For an incident plane wave, given by (8.6), and N scatterers, we can write (8.51) as

$$g_n^{-1} D_n = 1 + \sum_{\substack{j=1 \\ j\neq n}}^{N} D_j \, h_0(k|\boldsymbol{r}_n - \boldsymbol{r}_j|) \, \mathrm{e}^{\mathrm{i}(\delta_j - \delta_n)}, \tag{8.52}$$

where $D_n = A_n \mathrm{e}^{-\mathrm{i}\delta_n}$ and $\delta_n = k\boldsymbol{r}_n \cdot \hat{\boldsymbol{\alpha}}$. The far field of the N-scatterer cluster is given by

$$u_{\mathrm{sc}}(\boldsymbol{r}) \sim (\mathrm{i}kr)^{-1} \mathrm{e}^{\mathrm{i}kr} F(\hat{\boldsymbol{r}}; \hat{\boldsymbol{\alpha}}) \quad \text{as } r \to \infty, \tag{8.53}$$

where

$$F(\hat{\boldsymbol{r}}; \hat{\boldsymbol{\alpha}}) = \sum_{j=1}^{N} D_j \, \mathrm{e}^{\mathrm{i}(\delta_j - \Delta_j)}$$

is the far-field pattern and $\Delta_n = k\boldsymbol{r}_n \cdot \hat{\boldsymbol{r}}$. Twersky [1216] has examined solutions of (8.52) when the scatterers are identical and located at the vertices of regular polygons or polyhedra, with $2 \le N \le 6$; see also [1181, 1215].

Foldy's method can be seen as perhaps the simplest realisation of the self-consistent method outlined in Section 1.1.3. For example, we can identify $u_{\mathrm{sc}}^n(\boldsymbol{r}) = A_n G(\boldsymbol{r} - \boldsymbol{r}_n)$ in (1.2) and then the operator $\mathcal{T}_j$ in (1.4) is given by

$$\left(\mathcal{T}_j u_j\right)(\boldsymbol{r}) = g_j \, G(\boldsymbol{r} - \boldsymbol{r}_j) \, u_j(\boldsymbol{r}_j).$$

Also, the systems (8.49) and (8.51) correspond to (1.5) and (1.6), respectively.

In order to use Foldy's method, we have to specify the scattering coefficient g_j for the jth scatterer, for each scatterer. We discuss this specification next.

8.3.1 The scattering coefficient

Consider the jth scatterer and assume, without loss of generality, that it is located at the origin ($\boldsymbol{r}_j = \boldsymbol{0}$). The coefficient g_j characterises how this scatterer scatters waves, given that we have already assumed that the scattering is isotropic: for any incident field $u_{\mathrm{inc}}(\boldsymbol{r})$, we have assumed that the total field near the scatterer is given by

$$u(\boldsymbol{r}) \simeq u_{\mathrm{inc}}(\boldsymbol{r}) + g_j u_{\mathrm{inc}}(\boldsymbol{0}) \, G(\boldsymbol{r}). \tag{8.54}$$

There are several ways of thinking about g_j. For example, we could simply choose numbers for g_j, without connecting them to the geometry or composition of the scatterer.

Another way comes from enforcing energy conservation. Thus, in three dimensions, let us use (8.13) and write (8.54) as

$$u(\boldsymbol{r}) = \sum_{n,m} d_n^m \hat{\psi}_n^m(\boldsymbol{r}) + g_j d_0^0 \psi_0^0(\boldsymbol{r}). \tag{8.55}$$

We can assume that $d_0^0 = \sqrt{4\pi}\, u_{\text{inc}}(\boldsymbol{0}) \neq 0$. An application of Green's theorem to u and its complex conjugate, $\overline{u}$, gives

$$\int_{S_j} \left(u \frac{\partial \overline{u}}{\partial n} - \overline{u} \frac{\partial u}{\partial n} \right) \mathrm{d}s = \int_{\Omega_c} \left(u \frac{\partial \overline{u}}{\partial r} - \overline{u} \frac{\partial u}{\partial r} \right) \mathrm{d}s, \tag{8.56}$$

where Ω_c denotes the surface of a sphere, $r = c$, containing the jth scatterer, S_j. Corollary 6.6 enables the evaluation of the right-hand side of (8.56), whence

$$|g_j|^2 + \mathrm{Re}(g_j) = \frac{\mathrm{i}k}{2|d_0^0|^2} \int_{S_j} \left(u \frac{\partial \overline{u}}{\partial n} - \overline{u} \frac{\partial u}{\partial n} \right) \mathrm{d}s, \tag{8.57}$$

exactly, for any c.

Now, suppose that u is the exact solution of a scattering problem so that, for example, $u = 0$ on S_j. Then, the right-hand side of (8.57) vanishes, giving

$$|g_j|^2 + \mathrm{Re}(g_j) = 0. \tag{8.58}$$

The solutions of this equation are given by

$$g_j = -\frac{1}{2} + \frac{1}{2} \mathrm{e}^{\mathrm{i}\varphi}, \tag{8.59}$$

where φ is real but unknown. Arguments of this kind have been made in [1038]; for two dimensions, see [1037, 1039].

If S_j is a sound-soft sphere of radius a, Ω_a, we can impose the boundary condition $u = 0$ on average, $\int_{\Omega_a} u \, \mathrm{d}\Omega = 0$, giving

$$g_j = -[j_0(ka)]/[h_0(ka)] = -\frac{1}{2} + \frac{1}{2} \mathrm{e}^{-2\mathrm{i}ka},$$

in agreement with (8.59) when $\varphi = -2ka$.

Let us now assume that S_j is a small three-dimensional sound-soft obstacle. In Section 8.2.1, we obtained the low-frequency Rayleigh approximation (8.14). When this is compared with (8.54), we deduce that

$$g_j = -\mathrm{i}k\mathcal{C}_j, \tag{8.60}$$

where $\mathcal{C}_j$ is the capacity of S_j, defined by (8.10). Note that the approximation (8.60) does not satisfy (8.58). However, if we replace (8.60) by

$$g_j = \frac{-\mathrm{i}k\mathcal{C}_j}{1 + \mathrm{i}k\mathcal{C}_j}, \tag{8.61}$$

we obtain an approximation that satisfies the energy-conservation constraint (8.58) identically, and agrees with the known leading-order low-frequency approximation.

In two dimensions, a similar argument leads to (8.58), whereas comparison of (8.54) with (8.40) suggests

$$g_j = \frac{1}{2}\mathrm{i}\pi(\log k\ell_j - \beta)^{-1} = -\mathrm{i}\lambda(1+\mathrm{i}\lambda)^{-1}, \tag{8.62}$$

exactly, with $\lambda = -\frac{1}{2}\pi\{\log(k\ell_j/2)+\gamma\}^{-1}$. (The length ℓ_j is defined by (8.36).) Thus, the approximation (8.62) satisfies (8.58) identically.

8.3.2 Ramm's method

Ramm [1004, 1007] gave a method for acoustic scattering by N small sound-soft obstacles, which is based on solving the first-kind integral equation (5.55). Thus, look for a solution in the form of a single-layer potential,

$$u_{\mathrm{sc}}(\boldsymbol{r}) = \sum_{j=1}^{N} \int_{S_j} \mu_j(q)\, G(\boldsymbol{r}-\boldsymbol{q}_j)\,\mathrm{d}s_q,$$

where $\boldsymbol{q}_j \in S_j$, $G(\boldsymbol{r}) = h_0(k|\boldsymbol{r}|)$ and the source densities μ_j are to be found. The boundary condition, $u_{\mathrm{sc}} + u_{\mathrm{inc}} = 0$ on S, gives

$$0 = u_{\mathrm{inc}}(\boldsymbol{p}_n) + \sum_{j=1}^{N} \int_{S_j} \mu_j(q)\, G(\boldsymbol{p}_n-\boldsymbol{q}_j)\,\mathrm{d}s_q, \quad \boldsymbol{p}_n \in S_n, \tag{8.63}$$

which is a boundary integral equation for μ_j. The far-field pattern is given by (8.53) as

$$F(\hat{\boldsymbol{r}}) = \sum_{j=1}^{N} A_j\, \mathrm{e}^{-\mathrm{i}\Delta_j},$$

where

$$A_j = \int_{S_j} \mu_j(q)\,\mathrm{d}s_q \tag{8.64}$$

and we have made the approximation

$$\exp(-\mathrm{i}k\hat{\boldsymbol{r}}\cdot\boldsymbol{q}_j) \simeq \exp(-\mathrm{i}k\hat{\boldsymbol{r}}\cdot\boldsymbol{r}_j) \equiv \mathrm{e}^{-\mathrm{i}k\Delta_j};$$

as usual, $\boldsymbol{r}_j$ locates the centre of the jth scatterer. Let us make similar approximations in (8.63): we have

$$\begin{aligned} u_{\mathrm{inc}}(\boldsymbol{p}_n) &\simeq u_{\mathrm{inc}}(\boldsymbol{r}_n), \\ G(\boldsymbol{p}_n-\boldsymbol{q}_j) &\simeq G(\boldsymbol{r}_n-\boldsymbol{r}_j) \quad \text{for } j \neq n \text{ and} \\ G(\boldsymbol{p}_n-\boldsymbol{q}_n) &\simeq (\mathrm{i}k|\boldsymbol{p}_n-\boldsymbol{q}_n|)^{-1}, \end{aligned}$$

whence (8.63) is approximated by

$$\lambda_n(\boldsymbol{p}_n) = u_{\mathrm{inc}}(\boldsymbol{r}_n) + \sum_{\substack{j=1 \\ j\neq n}}^{N} A_j\, G(\boldsymbol{r}_n-\boldsymbol{r}_j), \tag{8.65}$$

where

$$\lambda_n(\boldsymbol{p}_n) = \frac{\mathrm{i}}{k} \int_{S_n} \mu_n(q) \frac{\mathrm{d}s_q}{|\boldsymbol{p}_n - \boldsymbol{q}_n|}, \quad \boldsymbol{p}_n \in S_n.$$

Now, (8.65) shows that λ_n is a constant, so that $(\mathrm{i}/k)\mu_n = \lambda_n \mu_0$, where μ_0 solves (8.8). Then, integrating over S_n gives

$$(\mathrm{i}/k)A_n = \lambda_n \mathcal{C}_n,$$

where $\mathcal{C}_n$ is the capacity of S_n, and we have used (8.11) and (8.64). Hence, (8.65) reduces to (8.51) in which

$$g_n = -\mathrm{i}k\mathcal{C}_n, \tag{8.66}$$

which is (8.60).

We conclude that Ramm's method [1004, 1007] gives exactly the same linear system of algebraic equations as Foldy's method, with the choice (8.66) for the scattering coefficient.

8.3.3 Hard scatterers

Foldy's method (described above) is concerned with the multiple scattering of scalar waves by a distribution of (small) isotropic scatterers. The key word is *isotropic*: the waves scattered by any one scatterer are represented by a point source located at some point within the scatterer; see (8.46). This assumption is justified for soft scatterers but it is not justified for hard scatterers or for penetrable scatterers, as such scatterers always induce a dipole field; see (8.30).

Let us generalise Foldy's method. We suppose that, in a neighbourhood of the jth scatterer, the scattered field is given by

$$A_j G(\boldsymbol{r} - \boldsymbol{r}_j) + \boldsymbol{a}_j \cdot \boldsymbol{g}(\boldsymbol{r} - \boldsymbol{r}_j), \tag{8.67}$$

where the jth scatterer is centred at $\boldsymbol{r}_j$, A_j is an unknown amplitude and $\boldsymbol{a}_j$ is an unknown vector. For definiteness, we consider three-dimensional problems (the two-dimensional analysis is very similar); thus, the free-space Green's function, G, is defined by (8.45). Also,

$$\boldsymbol{g}(\boldsymbol{r}) = -k^{-1}\operatorname{grad} G(\boldsymbol{r}) = -\frac{\hat{\boldsymbol{r}}}{k} \frac{\mathrm{d}}{\mathrm{d}r} G(\boldsymbol{r}),$$

where $\hat{\boldsymbol{r}} = \boldsymbol{r}/r$, so that $\boldsymbol{g} = \hat{\boldsymbol{r}}\, h_1(kr)$. Notice that (each component of) $\boldsymbol{g}$ is an outgoing solution of the Helmholtz equation.

The first term in (8.67) is a source at $\boldsymbol{r}_j$; the strength of the source (given by A_j) is unknown. The second term is a dipole at $\boldsymbol{r}_j$; the direction and strength of the dipole (given by $\boldsymbol{a}_j$) are unknown. Foldy assumes that $\boldsymbol{a}_j \equiv \boldsymbol{0}$.

Next, we represent the total field as

$$u(\boldsymbol{r}) = u_{\text{inc}}(\boldsymbol{r}) + \sum_j \{A_j G(\boldsymbol{r}-\boldsymbol{r}_j) + \boldsymbol{a}_j \cdot \boldsymbol{g}(\boldsymbol{r}-\boldsymbol{r}_j)\}, \tag{8.68}$$

where the sum is over all the scatterers. The 'exciting field' is

$$\begin{aligned} u_n(\boldsymbol{r}) &\equiv u(\boldsymbol{r}) - A_n G(\boldsymbol{r}-\boldsymbol{r}_n) - \boldsymbol{a}_n \cdot \boldsymbol{g}(\boldsymbol{r}-\boldsymbol{r}_n) \\ &= u_{\text{inc}}(\boldsymbol{r}) + \sum_{\substack{j \\ j\neq n}} \{A_j G(\boldsymbol{r}-\boldsymbol{r}_j) + \boldsymbol{a}_j \cdot \boldsymbol{g}(\boldsymbol{r}-\boldsymbol{r}_j)\}. \end{aligned} \tag{8.69}$$

This field can be regarded as the field incident on the nth scatterer, in the presence of all the other scatterers.

Now, let us characterise the scattering properties of the scatterers by writing

$$A_n = g_n u_n(\boldsymbol{r}_n) \quad \text{and} \quad \boldsymbol{a}_n = \mathbf{A}_n \cdot \boldsymbol{v}_n(\boldsymbol{r}_n),$$

where

$$\boldsymbol{v}_n(\boldsymbol{r}) = k^{-1} \operatorname{grad} u_n. \tag{8.70}$$

As in Section 8.3, g_n is a scalar. The quantity $\mathbf{A}_n$ is a 3×3 matrix. Thus, A_n is proportional to the value of the exciting field at $\boldsymbol{r}_n$, whereas $\boldsymbol{v}_n$ is related to the gradient of the exciting field at $\boldsymbol{r}_n$. We will discuss the choice of g_n and $\mathbf{A}_n$ later.

Evaluating (8.69) at $\boldsymbol{r}_n$ gives

$$g_n^{-1} A_n = u_{\text{inc}}(\boldsymbol{r}_n) + \sum_{\substack{j \\ j\neq n}} \{\boldsymbol{a}_j \cdot \boldsymbol{g}(\boldsymbol{R}_{nj}) + A_j G(\boldsymbol{R}_{nj})\}, \tag{8.71}$$

where $\boldsymbol{R}_{nj} = \boldsymbol{r}_n - \boldsymbol{r}_j$. Also, from (8.69) and (8.70), we have

$$\boldsymbol{v}_n(\boldsymbol{r}) = \boldsymbol{v}_{\text{inc}}(\boldsymbol{r}) + \sum_{\substack{j \\ j\neq n}} \{-A_j \boldsymbol{g}(\boldsymbol{r}-\boldsymbol{r}_j) + k^{-1} \operatorname{grad} [\boldsymbol{a}_j \cdot \boldsymbol{g}(\boldsymbol{r}-\boldsymbol{r}_j)]\}, \tag{8.72}$$

where $\boldsymbol{v}_{\text{inc}}(\boldsymbol{r}) = k^{-1} \operatorname{grad} u_{\text{inc}}$. Explicit calculation gives

$$(kR_{nj})^{-1} h_1(kR_{nj}) \boldsymbol{a}_j - \widehat{\boldsymbol{R}}_{nj} (\boldsymbol{a}_j \cdot \widehat{\boldsymbol{R}}_{nj}) h_2(kR_{rj})$$

for the value of $k^{-1} \operatorname{grad} [\boldsymbol{a}_j \cdot \boldsymbol{g}(\boldsymbol{r}-\boldsymbol{r}_j)]$ at $\boldsymbol{r} = \boldsymbol{r}_n$, where $R_{nj} = |\boldsymbol{R}_{nj}|$ and $\widehat{\boldsymbol{R}}_{nj} = \boldsymbol{R}_{nj}/R_{nj}$. Hence, evaluating (8.72) at $\boldsymbol{r}_n$, we obtain

$$\begin{aligned} \mathbf{A}_n^{-1} \cdot \boldsymbol{a}_n = \boldsymbol{v}_{\text{inc}}(\boldsymbol{r}_n) + \sum_{\substack{j \\ j\neq n}} \bigg\{ &\frac{h_1(kR_{nj})}{kR_{nj}} \boldsymbol{a}_j \\ &- \widehat{\boldsymbol{R}}_{nj} (\boldsymbol{a}_j \cdot \widehat{\boldsymbol{R}}_{nj}) h_2(kR_{nj}) - A_j \boldsymbol{g}(\boldsymbol{R}_{nj}) \bigg\}. \end{aligned} \tag{8.73}$$

Equations (8.71) and (8.73) give a system of linear algebraic equations for A_n and the components of $\boldsymbol{a}_n$. For N scatterers (in three dimensions), there are $4N$ equations for the $4N$ scalar unknowns; in two dimensions, there are $3N$ equations in $3N$ unknowns.

We are not aware of any computations based on solving the system comprising (8.71) and (8.73).

In order to use (8.71) and (8.73), we have to specify the coefficient g_j and the matrix $\mathbf{A}_j$ for the jth scatterer, for each scatterer. Consider the jth scatterer and assume, without loss of generality, that it is located at the origin ($\boldsymbol{r}_j = \mathbf{0}$). For any incident field $u_{\text{inc}}(\boldsymbol{r})$, we have assumed that the total field near the scatterer is given by

$$u(\boldsymbol{r}) \simeq u_{\text{inc}}(\boldsymbol{r}) + g_j u_{\text{inc}}(\mathbf{0})\, G(\boldsymbol{r}) + [\boldsymbol{g}(\boldsymbol{r})] \cdot \mathbf{A}_j \cdot [\boldsymbol{v}_{\text{inc}}(\mathbf{0})]. \tag{8.74}$$

Thus, the scattered field is assumed to take the form

$$u_{\text{sc}}(\boldsymbol{r}) \simeq g_j u_{\text{inc}}(\mathbf{0})\, h_0(kr) + \hat{\boldsymbol{r}} \cdot \mathbf{A}_j \cdot \boldsymbol{U}\, h_1(kr), \tag{8.75}$$

where $\boldsymbol{U} \equiv \boldsymbol{v}_{\text{inc}}(\mathbf{0})$ also appeared in (8.29). Now, let us assume further that the scatterer is small and sound-hard. Then, the formula (8.75) can be compared with the known low-frequency Rayleigh approximation, which is (8.30). This comparison suggests the following choices:

$$g_j = -\frac{\mathrm{i}k^3}{4\pi}\,\mathrm{vol}_j \quad \text{and} \quad \mathbf{A}_j = \frac{\mathrm{i}k^3}{4\pi}\mathbf{M}_j.$$

Here, vol_j is the volume of the jth scatterer and $\mathbf{M}_j$ is the magnetic polarisability tensor, defined by (8.24). In particular, for small hard spheres, we have $\mathbf{M}_j = \frac{3}{2}\,\mathrm{vol}_j\,\mathbf{I}$, and then the left-hand side of (8.73) reduces to $-(3g_n/2)^{-1}\boldsymbol{a}_n$.

In [19, 1354], the authors consider widely-spaced scatterers. Their model includes directional (anisotropic) effects and gives a system of $N(N-1)$ linear algebraic equations. However, their model does not involve gradient quantities such as $\boldsymbol{v}_{\text{inc}}(\boldsymbol{r}_n)$, and so it is not consistent with the known low-frequency results for small hard scatterers, for example.

8.4 Point scatterers

There is an extensive literature on 'point scatterers', in which scatterers are represented by concentrated forcings to a partial differential equation. We will outline some of this work, starting from the equations derived in Section 6.3 for acoustic scattering by inhomogeneities.

The governing equation is (6.7), which we write as

$$(\nabla^2 + k^2)U = \theta_B V U + f_{\text{inc}}, \tag{8.76}$$

where $f_{\text{inc}} = \theta_B V p_{\text{inc}}$, the operator V is defined by (6.4) and θ_B is the characteristic function for B: $\theta_B(P) = 1$ for $P \in B$ and $\theta_B(P) = 0$ for $P \notin B$. The function $U = p_{\text{sc}}$ in B_{e}, so that U is an outgoing wavefunction.

We are going to simplify the first term on the right-hand side of (8.76). In that term, assume that the density ρ_0 is constant, so that $\theta_B VU = k^2\theta_B mU$, where

$$m(P) = 1 - ([k_0(P)]/k)^2,$$

whence (8.76) becomes

$$(\nabla^2 + k^2)U = k^2\theta_B mU + f_{\text{inc}}. \tag{8.77}$$

Suppose that U_{inc} is the outgoing solution of (8.76) when the term $\theta_B VU$ is deleted. Then, as in Section 6.3, we obtain

$$U(P) = U_{\text{inc}}(P) + \frac{k^2}{2}\int_B m(Q)\, U(Q)\, G(P,Q)\, \mathrm{d}V_Q. \tag{8.78}$$

Now, suppose that we have N 'point scatterers', located ar $\boldsymbol{r} = \boldsymbol{r}_j$, $j = 1, 2, \ldots, N$. In their substantial review paper, de Vries *et al.* 'define a point scatterer located at $r = 0$ according to' [282, p. 451]

$$m(\boldsymbol{r}) = \alpha_B\, \delta(\boldsymbol{r}), \tag{8.79}$$

where prescriptions for the constant α_B are given. Hence

$$m(\boldsymbol{r})\, U(\boldsymbol{r}) \simeq \alpha_B\, \delta(\boldsymbol{r})\, U(\boldsymbol{0}). \tag{8.80}$$

Thus, for N scatterers, we have

$$m(\boldsymbol{r})\, U(\boldsymbol{r}) = \sum_{j=1}^{N} \alpha_j\, \delta(\boldsymbol{r} - \boldsymbol{r}_j)\, U(\boldsymbol{r}_j), \tag{8.81}$$

where $\alpha_1, \alpha_2, \ldots, \alpha_N$ are given constants. Substitution in (8.78) gives

$$U(\boldsymbol{r}) = U_{\text{inc}}(\boldsymbol{r}) + \frac{k^2}{2}\sum_{j=1}^{N} \alpha_j\, U(\boldsymbol{r}_j)\, G(\boldsymbol{r} - \boldsymbol{r}_j), \tag{8.82}$$

which is reminiscent of the Foldy representation, (8.50). However, there is no easy way to calculate the numbers $U(\boldsymbol{r}_j)$, $j = 1, 2, \ldots, N$: we cannot evaluate (8.82) at $\boldsymbol{r} = \boldsymbol{r}_n$ because $G(\boldsymbol{r})$ is singular at $\boldsymbol{r} = \boldsymbol{0}$. Therefore, the authors of [282] are obliged to 'regularize' G in order to make progress; see also [1241]. Clearly, the mathematical difficulties caused by the singularity in G are a consequence of the approximation (8.79).

An alternative approximation, which Lax [689] calls the 'localized perturbation method', is

$$m(\boldsymbol{r})\, U(\boldsymbol{r}) \simeq m(\boldsymbol{r})\, U(\boldsymbol{0}); \tag{8.83}$$

this has been used by Bass & Fix [71] for acoustic scattering, and in [72] for electromagnetic scattering. The approximation (8.83) is expected to be reasonable when the function $m(\boldsymbol{r})$ 'has a sharp maximum at [the point] $\boldsymbol{r}=\boldsymbol{0}$ and dies to zero in a finite... vicinity of this point' [72, p. 271]. Thus, for N scatterers, we have

$$m(\boldsymbol{r})\,U(\boldsymbol{r}) = m(\boldsymbol{r})\,U(\boldsymbol{r}_j) \quad \text{for } \boldsymbol{r}\in B_j,\ j=1,2,\ldots,N,$$

and then (8.78) becomes

$$U(\boldsymbol{r}) = U_{\text{inc}}(\boldsymbol{r}) + \sum_{j=1}^{N} M_j(\boldsymbol{r})\,U(\boldsymbol{r}_j), \tag{8.84}$$

where

$$M_j(\boldsymbol{r}) = \frac{k^2}{2}\int_{B_j} m(\boldsymbol{r}')\,G(\boldsymbol{r}-\boldsymbol{r}')\,\mathrm{d}V(\boldsymbol{r}').$$

Evaluating (8.84) at $\boldsymbol{r}=\boldsymbol{r}_n$ then gives a linear system for $U(\boldsymbol{r}_j)$, $j=1,2,\ldots,N$:

$$\sum_{j=1}^{N} A_{nj}\,U(\boldsymbol{r}_j) = U_{\text{inc}}(\boldsymbol{r}_n), \quad n=1,2,\ldots,N,$$

with $A_{nj}=\delta_{nj}-M_j(\boldsymbol{r}_n)$. In some applications, the determinant of the frequency-dependent matrix $A=(A_{nj})$ can have (complex) zeros, corresponding to resonances; for further information, see [71, 72, 282] and references therein.

8.5 Wide-spacing approximations

At the beginning of Chapter 1, we discussed Heaviside's notion of 'orders of scattering', which is a recursive way of calculating multiple scattering. Indeed, this approach has been used by many authors. Its main virtue is that it requires a method for scattering by one obstacle in isolation, and nothing more. It leads to a sequence of approximations, comprising increasing orders of scattering. Typically, the sequence is truncated at a low order: when and why can this be expected to give a good approximation? The answer depends on the number of spatial dimensions. Thus, in three dimensions, scattered (spherical) waves decay as r^{-1}, where r is the distance from a scatterer. In two dimensions, scattered (cylindrical) waves decay as $r^{-1/2}$, whereas waves do not decay at all in one dimension. Consequently, we should expect that calculating orders of scattering will work best in three dimensions and not at all for one-dimensional problems.

For the spatial decay to be useful, we assume that the scatterers are widely spaced. Strictly, this means that

$$a/b \ll 1 \quad \text{and} \quad kb \gg 1, \tag{8.85}$$

where, as before, a is a length-scale characterising the size of the scatterers and b characterises the spacing between the scatterers.

In the sequel, we discuss a number of wide-spacing approximations. We start with an exact analysis for the one-dimensional Helmholtz equation. This is a model for waves on a string with a number of point constraints (such as masses or springs), which act as 'scatterers' [508]. Similar problems arise in the study of waves in layered media [138]. However, we are mainly interested in applications to the scattering of (one-dimensional) surface water waves by horizontal cylinders; see Section 8.5.2.

Then, in Section 8.5.3, we describe a paper by Karp [587], in which the integral equations for acoustic scattering by two obstacles are solved approximately. The method used applies in two and three dimensions. Three-dimensional water waves are considered in Section 8.5.4; the waves themselves can be viewed as two-dimensional, although there are three-dimensional local potential fields near each scatterer.

8.5.1 One-dimensional problems: exact theory

The governing equation is $u''(x)+k^2u(x)=0$. We consider two 'scatterers', one located at $x=0$ and one at $x=b$.

First, consider an isolated scatterer at $x=0$, with a wave incident from the left. Thus, to the left, we have $u=\mathrm{e}^{\mathrm{i}kx}+r\,\mathrm{e}^{-\mathrm{i}kx}$, and to the right, we have $u=t\,\mathrm{e}^{\mathrm{i}kx}$; r is the (complex) reflection coefficient and t is the (complex) transmission coefficient. For simplicity, assume that the scatterer is 'symmetric' so that we have the same r and t when the incident wave comes from the right as when it comes from the left.

It is easy to see that if the scatterer is moved to $x=b$, then the new reflection coefficient is $r\,\mathrm{e}^{2\mathrm{i}kb}$ but the transmission coefficient is unchanged.

Now, consider two scatterers; again, for simplicity, assume that the scatterers are identical. To the left, we have $u=\mathrm{e}^{\mathrm{i}kx}+R\,\mathrm{e}^{-\mathrm{i}kx}$, and to the right, we have $u=T\,\mathrm{e}^{\mathrm{i}kx}$; the goal is to calculate the two-body (complex) reflection and transmission coefficients, R and T, respectively.

Between the scatterers, we have $u=A\,\mathrm{e}^{\mathrm{i}kx}+B\,\mathrm{e}^{-\mathrm{i}kx}$, where A and B are unknown; these give the amplitudes of the waves bouncing back and forth between the two scatterers.

When the wave $A\,\mathrm{e}^{\mathrm{i}kx}$ meets the scatterer at $x=b$, it is transmitted as $tA\,\mathrm{e}^{\mathrm{i}kx}$ and reflected as $r\,\mathrm{e}^{2\mathrm{i}kb}\,A\,e^{-\mathrm{i}kx}$. Thus

$$T=tA \quad \text{and} \quad B=r\,\mathrm{e}^{2\mathrm{i}kb}\,A.$$

Next, consider the scatterer at $x=0$; it has two waves incident upon it, namely, $\mathrm{e}^{\mathrm{i}kx}$ from the left and $B\,\mathrm{e}^{-\mathrm{i}kx}$ from the right. So, to the left, the total field is

$$\mathrm{e}^{\mathrm{i}kx}+r\,\mathrm{e}^{-\mathrm{i}kx}+tB\,\mathrm{e}^{-\mathrm{i}kx} \quad \text{giving} \quad R=r+tB,$$

whereas to the right, the total field is

$$B\,\mathrm{e}^{-\mathrm{i}kx}+t\,\mathrm{e}^{\mathrm{i}kx}+rB\,\mathrm{e}^{\mathrm{i}kx} \quad \text{giving} \quad A=t+rB.$$

At this stage, we have four equations for the four unknowns R, T, A and B. Solving these equations gives

$$R = r + rt^2\Delta\,\mathrm{e}^{2ikb}, \quad T = t^2\Delta, \quad A = t\Delta \quad \text{and} \quad B = rt\Delta\,\mathrm{e}^{2ikb},$$

where $\Delta = (1 - r^2\mathrm{e}^{2ikb})^{-1}$.

These calculations can be extended to N scatterers. Instead of doing this, we consider two-dimensional scattering problems.

8.5.2 Two-dimensional water waves

Consider two immersed bodies; a prototype is a pair of half-immersed, fixed, horizontal, parallel circular cylinders. A regular wavetrain is incident from the left, say, inducing a reflected wave and a transmitted wave from the left-hand cylinder; these waves are not attenuated. In addition, there is a 'local field' near the cylinder. This field decays with distance from the cylinder; in fact, it decays like r^{-2} along the free surface. So, if the right-hand cylinder is sufficiently far away, the local field can be ignored, leaving only the wave transmitted from the left-hand cylinder. This wave will be partly reflected and partly transmitted by the right-hand cylinder, again with an associated local field. Thus, we can envisage waves bouncing back and forth along the free surface between the two cylinders. If we assume that we know the reflection and transmission coefficients for each cylinder in isolation, and for waves incident from the left and from the right, we can match the propagating waves so as to derive approximations to the two-body reflection and transmission coefficients, exactly as in Section 8.5.1. For detailed derivations, see [905, 1138, 784, 329] and [731, §6.3].

The description above gives the *two-dimensional wide-spacing approximation*. It is essentially one-dimensional; all that matters are the waves propagating along the free surface. Note that the notion of 'orders of scattering' is not appropriate; in effect, we have taken account of *all* orders, with respect to the propagating waves (this is necessary, because these waves do not attenuate). As a consequence, the wide-spacing approximation is known to be extremely effective and accurate (even when the conditions (8.85) are violated). Indeed, it can be shown that the wide-spacing approximation can be derived rationally by making appropriate asymptotic approximations in a rigorous exact formulation, namely the null-field/T-matrix formulation [784].

8.5.3 Karp's 1953 paper

In the proceedings of a 1953 conference, Karp [587] gave an approximate method for scattering by two obstacles: 'Let them be spaced far apart compared with wavelength, and let each subtend a small angle when viewed from a point on the other' [587, p. 199]; thus, in our notation, we have (8.85). Karp considered scattering by two identical sound-soft cylinders, and he used the first-kind integral equation, (5.74).

We shall use the Helmholtz integral equation in three dimensions for two sound-hard obstacles (which need not be identical in shape). Partitioning as in Section 6.12.2.1, we obtain

$$A_{11}u_1 + A_{12}u_2 = 2u_{\text{inc}}(p_1), \tag{8.86}$$

$$A_{21}u_1 + A_{22}u_2 = 2u_{\text{inc}}(p_2), \tag{8.87}$$

where $p_j \in S_j$, u_j is the total field on S_j, and the operator A_{ij} is defined by (6.139). The terms $A_{12}u_2$ and $A_{21}u_1$ give the interaction between the scatterers. Their far-field approximations are given by (6.163) and (6.164), respectively:

$$(A_{12}u_2)(p_1) \simeq \mathcal{A}_1 \exp(\mathrm{i}k\hat{\boldsymbol{b}}\cdot\boldsymbol{p}_1), \quad (A_{21}u_1)(p_2) \simeq \mathcal{A}_2 \exp(-\mathrm{i}k\hat{\boldsymbol{b}}\cdot\boldsymbol{p}_2). \tag{8.88}$$

Here, the notation is as in Section 6.12.4. Thus, we have chosen two origins, O_1 inside S_1 and O_2 inside S_2; $\boldsymbol{b}$ is the position vector of O_1 with respect to O_2; $b = |\boldsymbol{b}|$ and $\hat{\boldsymbol{b}} = \boldsymbol{b}/b$; and p_j and q_j have position vectors $\boldsymbol{p}_j$ and $\boldsymbol{q}_j$, respectively, with respect to O_j, $j = 1, 2$. The constants $\mathcal{A}_1$ and $\mathcal{A}_2$ are given by

$$\mathcal{A}_1 = -\frac{\mathrm{e}^{\mathrm{i}kb}}{2\pi b}\int_{S_2} u_2(q)\,\frac{\partial}{\partial n_q}\exp(-\mathrm{i}k\hat{\boldsymbol{b}}\cdot\boldsymbol{q}_2)\,\mathrm{d}s_q, \tag{8.89}$$

$$\mathcal{A}_2 = -\frac{\mathrm{e}^{\mathrm{i}kb}}{2\pi b}\int_{S_1} u_1(q)\,\frac{\partial}{\partial n_q}\exp(\mathrm{i}k\hat{\boldsymbol{b}}\cdot\boldsymbol{q}_1)\,\mathrm{d}s_q. \tag{8.90}$$

Substituting the approximations (8.88) in (8.86) and (8.87) gives

$$A_{11}u_1 = 2u_{\text{inc}}(p_1) - \mathcal{A}_1 \exp(\mathrm{i}k\hat{\boldsymbol{b}}\cdot\boldsymbol{p}_1), \tag{8.91}$$

$$A_{22}u_2 = 2u_{\text{inc}}(p_2) - \mathcal{A}_2 \exp(-\mathrm{i}k\hat{\boldsymbol{b}}\cdot\boldsymbol{p}_2). \tag{8.92}$$

Now, suppose we can solve the following four equations,

$$A_{11}U_1 = 2u_{\text{inc}}(p_1), \quad A_{11}V_1 = \exp(\mathrm{i}k\hat{\boldsymbol{b}}\cdot\boldsymbol{p}_1),$$

$$A_{22}U_2 = 2u_{\text{inc}}(p_2), \quad A_{22}V_2 = \exp(-\mathrm{i}k\hat{\boldsymbol{b}}\cdot\boldsymbol{p}_2);$$

these correspond to solving two problems for the scatterer S_1 in isolation, and two for S_2. Then, linearity gives

$$u_1 = U_1 - \mathcal{A}_1 V_1 \quad \text{and} \quad u_2 = U_2 - \mathcal{A}_2 V_2.$$

To find $\mathcal{A}_1$ and $\mathcal{A}_2$, substitute for u_1 and u_2 in the right-hand sides of (8.90) and (8.89), respectively, giving two algebraic equations in $\mathcal{A}_1$ and $\mathcal{A}_2$. Solving these equations gives Karp's wide-spacing approximation.

We can give a physical interpretation of (8.91) and (8.92). The first of these says that S_1 is excited by the incident field, u_{inc}, and a *plane* wave coming from S_2. Similarly, (8.92) says that S_2 is excited by u_{inc} and a plane wave coming from S_1. The amplitudes of the plane waves, $\mathcal{A}_1$ and $\mathcal{A}_2$, are found by solving a 2×2 system; this ensures self-consistency.

Evidently, Karp's method will generalise to N scatterers; there would be $N(N-1)$ plane-wave amplitudes to determine, in general. For a description in the context of electromagnetic waves, see [564, §8.37].

Karp's method was applied in [589] to the problem of scattering by a wide gap between two coplanar semi-infinite screens. Subsequently, more refined approximations were developed by Zitron & Karp [1382, 1383]; see also [161]. For applications to scattering by circular cylinders, see [511, 996].

8.5.4 Three-dimensional water waves

Consider two immersed three-dimensional bodies, S_1 and S_2; a prototype is a pair of fixed, half-immersed spheres. Choose Cartesian coordinates $Oxyz$ with the mean free surface in the xy-plane, z increasing with depth. For simplicity, suppose that the water is deep. Take the incident wave as

$$\phi_{\text{inc}} = \mathrm{e}^{-Kz}\,\mathrm{e}^{\mathrm{i}K(x\cos\alpha+y\sin\alpha)} = \mathrm{e}^{-Kz+\mathrm{i}Kr\cos(\theta-\alpha)}, \tag{8.93}$$

where $K = \omega^2/g$ is the wavenumber, α is the angle of incidence and (r, θ) are plane polar coordinates in the xy-plane. Choose an origin O_j at $(x, y) = (\xi_j, \eta_j)$ in the vicinity of S_j, and plane polar coordinates (r_j, θ_j) at O_j, so that $x = \xi_j + r_j\cos\theta_j$ and $y = \eta_j + r_j\sin\theta_j$, $j = 1, 2$.

Imagine the incident wave encountering the first body, S_1; a diagrammatic representation of the scattering process is given in Fig. 8.1 (a similar process is initiated by the incident wave encountering S_2). The incident wave will induce a scattered wave, spreading out in all directions: it behaves like

$$r_1^{-1/2}\,\mathrm{e}^{\mathrm{i}Kr_1} \quad \text{as } r_1 \to \infty, \tag{8.94}$$

on the free surface, and so is attenuated. In addition, there is a local field near the body. This field decays like r_1^{-2} along the free surface. So, if the second body is sufficiently far away, the local field can be ignored, leaving only the attenuated scattered wave from the first body. This wave will be partly scattered to infinity ($\boxed{1}$ in Fig. 8.1) and partly scattered by the second body, again with an associated local field; this is the *second order of scattering*, $\boxed{2}$. When this wave reaches the first body, it will be of order b^{-1} compared with the incident field; scattering it off the first body gives the *third order of scattering*, $\boxed{3}$. We can make one more step, to fourth order $\boxed{4}$; subsequent steps cannot be justified as they give rise to contributions that are comparable to the neglected local fields.

There is an additional contribution at fourth order. It arises from the next term in (8.94); specifically, when ϕ_{inc} is scattered by S_1 at O_1, in isolation, the scattered field is given by

$$\phi_{\text{sc}}^1 \simeq I_1\,\frac{\mathrm{e}^{\mathrm{i}Kr_1}}{\sqrt{r_1}}\left\{f_1(\theta_1;\alpha) + \frac{1}{r_1}\,g_1(\theta_1;\alpha)\right\} \quad \text{as } r_1 \to \infty,$$

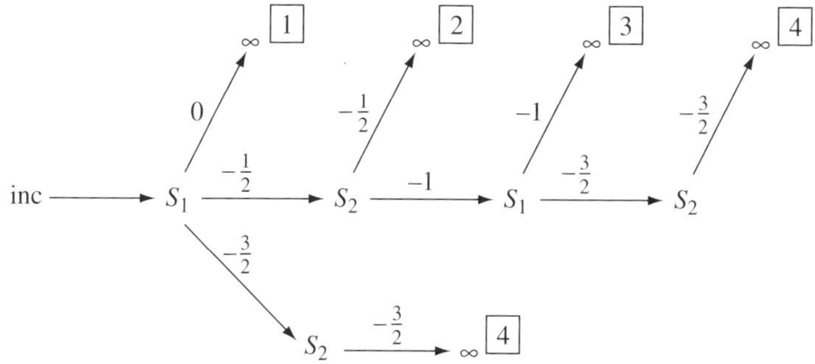

Fig. 8.1. Scattering of water waves by two widely-spaced bodies in three dimensions. The number γ above any arrow signifies a contribution of size b^γ, where b is the spacing. Contributions of $O(b^{-2})$ are omitted.

on the free surface, where $I_1 = \exp\{\mathrm{i}K(\xi_1 \cos\alpha + \eta_1 \sin\alpha)\}$ is a known phase factor, and f_1 (the *far-field pattern*) and g_1 are assumed known. The term involving g_1 gives a contribution of order $b^{-3/2}$ to the second order of scattering (indicated by the arrows pointing down in Fig. 8.1), which is comparable to the leading contribution to the fourth order of scattering: both contributions are marked by [4] in Fig. 8.1.

The quantities f_1 and g_1 come from solving the single-body problem (in fact, g_1 can be expressed in terms of f_1; see (8.96)). $f_1(\theta; \alpha)$ gives the far-field amplitude in the direction θ for the incident wave (8.93) when S_1 is located at O. We assume that $f_1(\theta; \alpha)$ is known (or computable) for any choices of θ and α.

The approach described above has been used by Greenhow [436] for a number of hemispheres. For two hemispheres, he calculated the second order of scattering [2], but he was 'anxious to avoid' calculating to third order [436, p. 299]. In fact, it is not too difficult to calculate *all* the contributions in Fig. 8.1. The key to making these calculations tractable was given by Zitron & Karp [1382] in their 1961 paper on the multiple scattering of acoustic waves in two dimensions.

Let $(\xi_2 - \xi_1, \eta_2 - \eta_1) = b(\cos\delta, \sin\delta)$. Expand ϕ^1_{sc} in a neighbourhood of O_2, for large b. Straightforward calculations show that

$$\phi^1_{\mathrm{sc}} \simeq I_1 \, \mathrm{e}^{\mathrm{i}K(b+X)} \, b^{-1/2} \left\{ f_1(\delta; \alpha) + b^{-1} \, \Lambda(X, Y) \right\}, \tag{8.95}$$

$$\Lambda(X, Y) = \frac{1}{2}(\mathrm{i}KY^2 - X) f_1(\delta; \alpha) + Y f_1'(\delta; \alpha) + g_1(\delta; \alpha),$$

where $X = r_2 \cos(\theta_2 - \delta)$, $Y = r_2 \sin(\theta_2 - \delta)$ and

$$f_1'(\delta; \alpha) = (\partial/\partial\theta) f_1(\theta; \alpha) \quad \text{evaluated at } \theta = \delta;$$

g_1 is given exactly by

$$2\mathrm{i}K g_1(\delta; \alpha) = \frac{1}{4} f_1(\delta; \alpha) + f_1''(\delta; \alpha). \tag{8.96}$$

The expression (8.95) shows that, to leading order in b, the field scattered by S_1 is approximately a plane wave at S_2, propagating in the direction from O_1 to O_2 and with a known amplitude. Consequently, the scattering of this field by S_2 can be calculated.

The term involving $\Lambda(X, Y)$ in (8.95) is not a plane wave, but it can be written as a linear combination of certain derivatives of plane waves with respect to the angle of incidence. Thus, if

$$v(\varphi) = \exp\{\mathrm{i}Kr_2 \cos(\theta_2 - \varphi)\},$$

then $v'(\delta) = \mathrm{i}KY\,\mathrm{e}^{\mathrm{i}KX}$ and $v''(\delta) = \mathrm{i}K(\mathrm{i}KY^2 - X)\,\mathrm{e}^{\mathrm{i}KX}$, which are precisely the combinations occurring in (8.95).

The leading-order plane-wave form of (8.95) is as expected from Section 8.5.3. It was exploited in Simon's plane-wave approximation [1104], refined in [833], and used in [827].

8.6 Random arrangements of small scatterers; suspensions

There are three main classes of theory which may be applied to the propagation of ultrasound in suspensions. Phenomenological theories attempt to derive average properties for suspensions using just the relative amounts of the constituents, ignoring the details of their spatial arrangement. In hydrodynamic theories, the controlling factor is the viscous drag between the suspended particles and the surrounding fluid. In scattering theories, the particles are treated as fixed scatterers which perturb the ultrasonic field.

(Harker & Temple [479, p. 20])

We are interested here in the third class of methods, as described in the review by Harker & Temple [479]. (For later reviews and applications to suspensions, see [679, 820, 985, 1134, 300, 1133].) Such methods are intended to model the propagation of waves through a random arrangement of identical spherical scatterers, for example. We shall not discuss statistical aspects in detail, but it is of interest to see how multiple-scattering theories enter into the story. For a review, see [368, §VI].

We start with a simplified version of Foldy's model [354]: N small, identical scatterers, located at the points $\boldsymbol{r}_1, \ldots, \boldsymbol{r}_N$. Denote the configuration of points by $\Lambda_N = \{\boldsymbol{r}_1, \boldsymbol{r}_2, \ldots, \boldsymbol{r}_N\}$. The deterministic model for multiple scattering by the configuration Λ_N is summarised here; the total field is given by

$$u(\boldsymbol{r}|\Lambda_N) = u_{\mathrm{inc}}(\boldsymbol{r}) + g \sum_{j=1}^{N} u_{\mathrm{ex}}(\boldsymbol{r}_j; \boldsymbol{r}_j|\Lambda_N)\, G(\boldsymbol{r} - \boldsymbol{r}_j), \tag{8.97}$$

where g is the scattering coefficient, the exciting field u_{ex} is given by

$$u_{\mathrm{ex}}(\boldsymbol{r}; \boldsymbol{r}_n|\Lambda_N) = u_{\mathrm{inc}}(\boldsymbol{r}) + g \sum_{\substack{j=1 \\ j\neq n}}^{N} u_{\mathrm{ex}}(\boldsymbol{r}_j; \boldsymbol{r}_j|\Lambda_N)\, G(\boldsymbol{r} - \boldsymbol{r}_j) \tag{8.98}$$

and $u_{ex}(\boldsymbol{r}_n; \boldsymbol{r}_n|\Lambda_N)$ is to be determined by solving the linear system obtained by evaluating (8.98) at $\boldsymbol{r}=\boldsymbol{r}_n$. Note that $u_{ex}(\boldsymbol{r}; \boldsymbol{r}_n|\Lambda_N)$ was denoted by $u_n(\boldsymbol{r})$ in Section 8.3.

We are interested in average properties of u, with respect to changes in Λ_N. Thus, we need some elements of probability theory.

8.6.1 Some probability theory

Assume that, on average, there is a constant number of scatterers per unit volume, denoted by n_0. Let B_N be a region of volume $|B_N| = N/n_0$, so that, on average, B_N contains N scatterers.

Let $p(\boldsymbol{r}_1, \boldsymbol{r}_2, \ldots, \boldsymbol{r}_N)\,\mathrm{d}V_1\,\mathrm{d}V_2 \cdots \mathrm{d}V_N$ be the probability of finding the scatterers in a configuration in which the first scatterer is in the volume element $\mathrm{d}V_1$ about the point $\boldsymbol{r}_1$, the second scatterer is in the volume element $\mathrm{d}V_2$ about the point $\boldsymbol{r}_2$, and so on, up to $\boldsymbol{r}_N$. We assume that the integral of the joint probability distribution $p(\boldsymbol{r}_1, \boldsymbol{r}_2, \ldots, \boldsymbol{r}_N)$ over all configurations is normalised to unity. We also assume that the N scatterers are indistinguishable, so that the order of the arguments of p is irrelevant; in particular, if the scatterers are statistically independent (their locations are uncorrelated), we have

$$p(\boldsymbol{r}_1, \boldsymbol{r}_2, \ldots, \boldsymbol{r}_N) = p(\boldsymbol{r}_1)\, p(\boldsymbol{r}_2) \cdots p(\boldsymbol{r}_N).$$

In the absence of any other information, the probability of finding a particular scatterer in $\mathrm{d}V_1$ at $\boldsymbol{r}_1$ is

$$p(\boldsymbol{r}_1)\,\mathrm{d}V_1 = \mathrm{d}V_1 \int \cdots \int p(\boldsymbol{r}_1, \boldsymbol{r}_2, \ldots, \boldsymbol{r}_N)\,\mathrm{d}V_2 \cdots \mathrm{d}V_N,$$

where each volume integration is over B_N. As each of the N scatterers is equally likely to occupy $\mathrm{d}V_1$, the density of scatterers at $\boldsymbol{r}_1$ is

$$Np(\boldsymbol{r}_1) = n_0. \tag{8.99}$$

The joint probability of finding the first scatterer in $\mathrm{d}V_1$ and the second in $\mathrm{d}V_2$ is given by

$$p(\boldsymbol{r}_1, \boldsymbol{r}_2)\,\mathrm{d}V_1\,\mathrm{d}V_2 = \mathrm{d}V_1\,\mathrm{d}V_2 \int \cdots \int p(\boldsymbol{r}_1, \boldsymbol{r}_2, \ldots, \boldsymbol{r}_N)\,\mathrm{d}V_3 \cdots \mathrm{d}V_N.$$

We have

$$p(\boldsymbol{r}_1, \boldsymbol{r}_2) = p(\boldsymbol{r}_1)\, p(\boldsymbol{r}_2|\boldsymbol{r}_1),$$

which defines the conditional probability $p(\boldsymbol{r}_2|\boldsymbol{r}_1)$. Combining (8.99) with the normalisation condition gives

$$1 = \iint p(\boldsymbol{r}_1, \boldsymbol{r}_2)\,\mathrm{d}V_1\,\mathrm{d}V_2 = \frac{n_0}{N} \iint p(\boldsymbol{r}_2|\boldsymbol{r}_1)\,\mathrm{d}V_1\,\mathrm{d}V_2. \tag{8.100}$$

The configurational (or ensemble) average of a quantity $f(\boldsymbol{r}|\Lambda_N)$ is defined by

$$\langle f(\boldsymbol{r})\rangle_N = \int\cdots\int p(\boldsymbol{r}_1,\boldsymbol{r}_2,\ldots,\boldsymbol{r}_N)\, f(\boldsymbol{r}|\Lambda_N)\,\mathrm{d}V_1\cdots\mathrm{d}V_N, \tag{8.101}$$

where the subscript N on the left-hand side indicates that the integration is over N copies of the volume B_N. Similarly, the partial average with one or more scatterers held fixed is obtained by averaging over the appropriate conditional probability: thus, $\langle f(\boldsymbol{r}|\boldsymbol{r}_1)\rangle_{N-1}$, the average of $f(\boldsymbol{r}|\Lambda_N)$ over all configurations for which the first scatterer is fixed at $\boldsymbol{r}_1$, is given by

$$\langle f(\boldsymbol{r}|\boldsymbol{r}_1)\rangle_{N-1} = \int\cdots\int p(\boldsymbol{r}_2,\ldots,\boldsymbol{r}_N|\boldsymbol{r}_1)\, f(\boldsymbol{r}|\Lambda_N)\,\mathrm{d}V_2\cdots\mathrm{d}V_N. \tag{8.102}$$

8.6.2 The ensemble average of *u*

Let us try to compute the ensemble average of u, $\langle u\rangle_N$. From (8.97), we multiply by $p(\boldsymbol{r}_1,\boldsymbol{r}_2,\ldots,\boldsymbol{r}_N)$ and integrate (according to (8.101)), giving

$$\begin{aligned}\langle u(\boldsymbol{r})\rangle_N = u_{\mathrm{inc}}(\boldsymbol{r}) + g\sum_{j=1}^{N}\int\cdots\int p(\boldsymbol{r}_1,\boldsymbol{r}_2,\ldots,\boldsymbol{r}_N)\\ \times u_{\mathrm{ex}}(\boldsymbol{r}_j;\boldsymbol{r}_j|\Lambda_N)\,G(\boldsymbol{r}-\boldsymbol{r}_j)\,\mathrm{d}V_1\cdots\mathrm{d}V_N.\end{aligned} \tag{8.103}$$

But $p(\boldsymbol{r}_1,\boldsymbol{r}_2,\ldots,\boldsymbol{r}_N) = p(\boldsymbol{r}_j)\,p(\boldsymbol{r}_1,\ldots,\boldsymbol{r}_{j-1},\boldsymbol{r}_{j+1},\ldots,\boldsymbol{r}_N|\boldsymbol{r}_j)$ and

$$\begin{aligned}\langle u_{\mathrm{ex}}(\boldsymbol{r};\boldsymbol{r}_j|\boldsymbol{r}_j)\rangle_{N-1} = \int\cdots\int p(\boldsymbol{r}_1,\ldots,\boldsymbol{r}_{j-1},\boldsymbol{r}_{j+1},\ldots,\boldsymbol{r}_N|\boldsymbol{r}_j)\\ \times u_{\mathrm{ex}}(\boldsymbol{r};\boldsymbol{r}_j|\Lambda_N)\,\mathrm{d}V_1\cdots\mathrm{d}V_{j-1}\mathrm{d}V_{j+1}\cdots\mathrm{d}V_N,\end{aligned}$$

whence (8.103) becomes

$$\begin{aligned}\langle u(\boldsymbol{r})\rangle_N &= u_{\mathrm{inc}}(\boldsymbol{r}) + g\sum_{j=1}^{N}\int \langle u_{\mathrm{ex}}(\boldsymbol{r}_j;\boldsymbol{r}_j|\boldsymbol{r}_j)\rangle_{N-1}\, p(\boldsymbol{r}_j)\,G(\boldsymbol{r}-\boldsymbol{r}_j)\,\mathrm{d}V_j\\ &= u_{\mathrm{inc}}(\boldsymbol{r}) + g n_0\int \langle u_{\mathrm{ex}}(\boldsymbol{r}';\boldsymbol{r}'|\boldsymbol{r}')\rangle_{N-1}\,G(\boldsymbol{r}-\boldsymbol{r}')\,\mathrm{d}V(\boldsymbol{r}'),\end{aligned} \tag{8.104}$$

where we have used (8.99) and the indistinguishability of the scatterers.

Having examined the right-hand side of (8.104), we next try to compute the partial average $\langle u_{\mathrm{ex}}(\boldsymbol{r};\boldsymbol{r}_n|\boldsymbol{r}_n)\rangle_{N-1}$. From (8.98), we multiply by

$$\begin{aligned}&p(\boldsymbol{r}_1,\ldots,\boldsymbol{r}_{n-1},\boldsymbol{r}_{n+1},\ldots,\boldsymbol{r}_N|\boldsymbol{r}_n)\\ &\quad= p(\boldsymbol{r}_j|\boldsymbol{r}_n)\,p(\boldsymbol{r}_1,\ldots,\boldsymbol{r}_{j-1},\boldsymbol{r}_{j+1},\ldots,\boldsymbol{r}_{n-1},\boldsymbol{r}_{n+1},\ldots,\boldsymbol{r}_N|\boldsymbol{r}_j,\boldsymbol{r}_n)\end{aligned}$$

and integrate (according to (8.102)), giving

$$\langle u_{\text{ex}}(\boldsymbol{r};\boldsymbol{r}_n|\boldsymbol{r}_n)\rangle_{N-1} = u_{\text{inc}}(\boldsymbol{r}) + g\sum_{\substack{j=1\\ j\neq n}}^{N}\int p(\boldsymbol{r}_j|\boldsymbol{r}_n)$$
$$\times\langle u_{\text{ex}}(\boldsymbol{r}_j;\boldsymbol{r}_j|\boldsymbol{r}_j,\boldsymbol{r}_n)\rangle_{N-2}\,G(\boldsymbol{r}-\boldsymbol{r}_j)\,\mathrm{d}V_j$$
$$= u_{\text{inc}}(\boldsymbol{r}) + g(N-1)\int p(\boldsymbol{r}'|\boldsymbol{r}_n)\,\langle u_{\text{ex}}(\boldsymbol{r}';\boldsymbol{r}'|\boldsymbol{r}',\boldsymbol{r}_n)\rangle_{N-2}\,G(\boldsymbol{r}-\boldsymbol{r}')\,\mathrm{d}V(\boldsymbol{r}'). \tag{8.105}$$

In (8.105), $\langle u_{\text{ex}}(\boldsymbol{r};\boldsymbol{r}_j|\boldsymbol{r}_j,\boldsymbol{r}_n)\rangle_{N-2}$ is the partial average of $u_{\text{ex}}(\boldsymbol{r};\boldsymbol{r}_j|\Lambda_N)$ over all configurations with both $\boldsymbol{r}_j$ and $\boldsymbol{r}_n$ held fixed. Thus,

> *the fact that the exciting field with one scatterer fixed is given in terms of the field with two scatterers fixed, is the basic difficulty encountered in the implicit approach to multiple scattering.*
>
> (Waterman & Truell [1297, p. 515])

Repeating the argument above, keeping more and more scatterers fixed, leads to a hierarchy of equations. If N is large, it would be far too expensive to try to solve this hierarchy exactly. Instead, it is usual to break the hierarchy in some way.

8.6.3 Foldy's approximation

Foldy [354, p. 111] argued that the average exciting field near a scatterer held fixed at $\boldsymbol{r}_j$ is approximately equal to the average total field near $\boldsymbol{r}_j$, provided that N is large:

$$\langle u_{\text{ex}}(\boldsymbol{r};\boldsymbol{r}_j|\boldsymbol{r}_j)\rangle_{N-1} \simeq \langle u(\boldsymbol{r})\rangle_N \quad \text{for } \boldsymbol{r} \text{ near } \boldsymbol{r}_j. \tag{8.106}$$

When this approximation is used in (8.104), we obtain

$$\langle u(\boldsymbol{r})\rangle_N = u_{\text{inc}}(\boldsymbol{r}) + g n_0\int_{B_N}\langle u(\boldsymbol{r}')\rangle_N\,G(\boldsymbol{r}-\boldsymbol{r}')\,\mathrm{d}V(\boldsymbol{r}'), \tag{8.107}$$

which is an integral equation for $\langle u\rangle_N$.

If we apply the operator (∇^2+k^2) to (8.107), recalling the definition (8.45) and the property (H.1), we find that

$$(\nabla^2+k_{\text{eff}}^2)\langle u\rangle_N = 0, \tag{8.108}$$

where

$$k_{\text{eff}}^2 = k^2 - 4\pi \mathrm{i} g n_0/k. \tag{8.109}$$

We see that k_{eff} can be interpreted as the effective complex propagation constant for the medium: the real part gives the effective wavelength and the imaginary part gives the attenuation.

The two-dimensional version of (8.109) is

$$k_{\text{eff}}^2 = k^2 - 4\mathrm{i}g\hat{n}_0, \tag{8.110}$$

where $\hat{n}_0$ is the number of scatterers per unit area. The formula (8.110) can be found in [1200, 40], for example.

The result (8.109) was obtained by Foldy [354]. In fact, he allowed for scatterers with a variable density $n(\boldsymbol{r})$, whereas we took $n(\boldsymbol{r}) = n_0$ for simplicity. For further discussion of Foldy's analysis, see [884, pp. 1499–1501], [541, Chapter 14], [1297, 492, 493] and some papers by Angel *et al.* [32, 33, 12, 40]. In particular, Foldy's approximation (8.106) has been studied in detail, but only for one-dimensional problems: it can be justified in certain circumstances. See the pair of papers by Haacke & Foldy [460] and references therein for more information.

For applications to elastic waves, see [768, 264, 265, 266, 603, 602]. For an application of (8.110) to the ultrasonic inspection of cancellous bones, see [751].

Foldy's theory has been used to understand the propagation of sound through bubbly liquids; see, for example, [1242, §4]. For more information on this problem, see [393, 291, 227, 1053, 1054, 1355, 492, 570].

8.6.4 Lax's approximation

Lax [688] suggested breaking the hierarchy obtained in Section 8.6.2 at the next level. For finite N, this approximation, which is known as the *quasi-crystalline approximation* (QCA), is

$$\langle u_{\text{ex}}(\boldsymbol{r};\boldsymbol{r}_j|\boldsymbol{r}_j,\boldsymbol{r}_n)\rangle_{N-2} \simeq \langle u_{\text{ex}}(\boldsymbol{r};\boldsymbol{r}_j|\boldsymbol{r}_j)\rangle_{N-1}. \tag{8.111}$$

Thus, writing $v_N(\boldsymbol{r}) = \langle u_{\text{ex}}(\boldsymbol{r};\boldsymbol{r}|\boldsymbol{r})\rangle_{N-1}$ in (8.104) and (8.105), we obtain

$$\langle u(\boldsymbol{r})\rangle_N = u_{\text{inc}}(\boldsymbol{r}) + g n_0 \int_{B_N} v_N(\boldsymbol{r}')\, G(\boldsymbol{r}-\boldsymbol{r}')\,\mathrm{d}V(\boldsymbol{r}'), \tag{8.112}$$

$$v_N(\boldsymbol{r}) = u_{\text{inc}}(\boldsymbol{r}) + g(N-1)\int_{B_N} p(\boldsymbol{r}'|\boldsymbol{r})\, v_N(\boldsymbol{r}')\, G(\boldsymbol{r}-\boldsymbol{r}')\,\mathrm{d}V(\boldsymbol{r}'). \tag{8.113}$$

Equation (8.113) is an integral equation of the second kind for v_N; once solved, the average (coherent) field is given by (8.112).

In order to use (8.113), we have to specify $p(\boldsymbol{r}_2|\boldsymbol{r}_1)$, consistent with (8.100). Also, we want to ensure that scatterers do not overlap. For scatterers of diameter $2a$, a simple choice (involving a parameter b) is

$$p(\boldsymbol{r}_2|\boldsymbol{r}_1) = p_0\, H(R_{12} - b) \quad \text{with} \quad b \geq 2a, \tag{8.114}$$

where $H(x)$ is the Heaviside unit function, $R_{12} = |\boldsymbol{r}_1 - \boldsymbol{r}_2|$ and p_0 is a constant determined by (8.100). In three dimensions, we obtain

$$p_0 = \left\{ |B_N| - \frac{4}{3}\pi b^3 \right\}^{-1} \simeq n_0/N, \tag{8.115}$$

assuming that $b^3 n_0/N \ll 1$. (The same approximation holds in two dimensions.) The equality in (8.115) assumes that the 'hole' at $\boldsymbol{r}_1$ of radius b does not cut the boundary of B_N. Evidently, taking this possibility into account would not change the approximation $p_0 \simeq n_0/N$.

8.6.5 A scattering problem

The scattering problem leading to (8.107) or (8.113) is complicated. However, if we are mainly interested in obtaining approximations to k_{eff}, we can simplify the geometry. Thus, let B_N be a rectangular box, defined by $|x| < \ell_N$, $|y| < \ell_N$ and $0 < z < \ell_N$, with $\ell_N = [N/(4n_0)]^{1/3}$; eventually, we let $N \to \infty$. Then, we can consider the following scattering problem. The semi-infinite region $z < 0$ is filled with a homogeneous compressible fluid. The semi-infinite region $z > 0$ contains the same fluid and many scatterers; for definiteness, we suppose that the scatterers are identical spheres of radius a. A plane wave, defined by (8.6), is incident from $z = -\infty$; thus

$$u_{\text{inc}} = \exp(\mathrm{i}k\boldsymbol{r} \cdot \hat{\boldsymbol{\alpha}}),$$

with $\hat{\boldsymbol{\alpha}} \cdot \hat{\mathbf{z}} > 0$ and $\hat{\mathbf{z}} = (0, 0, 1)$. There will be a reflected wave field and a transmitted wave field. Within the inhomogeneous half-space, away from the 'interface' at $z = 0$, we expect to see a plane wave, on average, propagating towards $z = +\infty$. This wave should have the form

$$\exp(\mathrm{i}k_{\text{eff}}\boldsymbol{r} \cdot \hat{\boldsymbol{\beta}}),$$

where the unit vector $\hat{\boldsymbol{\beta}}$ gives the direction of propagation, $\hat{\boldsymbol{\beta}} \cdot \hat{\mathbf{z}} \geq 0$ and $\operatorname{Im} k_{\text{eff}} > 0$: the wave decays (attenuates) as $z \to \infty$. The goal is to calculate k_{eff}. We should expect that k_{eff} depends on the various parameters of the problem *except the angle of incidence*, θ_{inc}, defined by $\hat{\boldsymbol{\beta}} \cdot \hat{\mathbf{z}} = \cos\theta_{\text{inc}}$. If k_{eff} were to depend on θ_{inc}, then we could not regard k_{eff} as being the effective wavenumber: we hope to replace the complicated scattering medium by an 'equivalent' homogeneous, isotropic medium, at least if we are not too close to $z = 0$.

One way to tackle the scattering problem described above is to start with an exact formulation for scattering by N obstacles, such as the multipole method of Section 4.8 for scattering by N spheres; then perform ensemble averaging, invoking one of the closure approximations, such as the QCA; then let $N \to \infty$; and then derive some equations which determine k_{eff}. Clearly, there are many choices to be made, and some of these are open to criticism. Nevertheless, progress has been made, as described next.

8.6.6 Finite-size effects

As before, we restrict ourselves here to spherical scatterers of radius a. It is common to express k_{eff} in terms of the volume fraction, ϕ; for spheres,

$$\phi = \frac{4}{3}\pi a^3 n_0,$$

and then Foldy's estimate (8.109) becomes

$$(k_{\text{eff}}/k)^2 = 1 - 3\mathrm{i}\phi(ka)^{-3}g. \tag{8.116}$$

Many papers have been written with the aim of generalising this formula to non-isotropic scattering, to scatterers of finite size, and to scatterers with correlated locations. Some of these are mentioned below.

We can solve the problem of scattering by one sphere, exactly, using the method of separation of variables; see Section 4.6. Specifically, for an incident plane wave, we have

$$u_{\text{inc}} = \mathrm{e}^{\mathrm{i}kz} = \sum_{n=0}^{\infty}(2n+1)\,\mathrm{i}^n j_n(kr)\,P_n(\cos\theta),$$

and then we can write the scattered field as

$$u_{\text{sc}} = \sum_{n=0}^{\infty}(2n+1)\,\mathrm{i}^n B_n\,h_n(kr)\,P_n(\cos\theta).$$

(The coefficients B_n are determined by the boundary conditions on the sphere.) In particular, the far-field pattern is defined by (4.44) and (4.45) as

$$f(\theta) = \sum_{n=0}^{\infty}(2n+1)\,B_n\,P_n(\cos\theta).$$

Thus, the *forward-scattering amplitude* is

$$f(0) = \sum_{n=0}^{\infty}(2n+1)\,B_n$$

and the *back-scattering amplitude* is

$$f(\pi) = \sum_{n=0}^{\infty}(2n+1)\,(-1)^n\,B_n.$$

Waterman & Truell [1297] began with the scattering problem of Section 8.6.5, for normal incidence ($\theta_{\text{inc}} = 0$). They used Foldy's closure assumption (8.106) for an uncorrelated distribution of finite-sized scatterers and obtained the formula

$$\left(\frac{k_{\text{eff}}}{k}\right)^2 = \left[1 - \frac{3\mathrm{i}\phi}{2(ka)^3}f(0)\right]^2 + \left[\frac{3\phi}{2(ka)^3}f(\pi)\right]^2 \tag{8.117}$$

$$= 1 - 3\mathrm{i}\phi(ka)^{-3}f(0) + \frac{9}{4}\phi^2(ka)^{-6}Q[f], \tag{8.118}$$

where

$$Q[f] = \{f(\pi)\}^2 - \{f(0)\}^2. \tag{8.119}$$

Twersky [1200] considered oblique incidence, and obtained (8.118) with

$$Q[f] = \left[\{f(\pi - 2\theta_{\text{inc}})\}^2 - \{f(0)\}^2\right] \sec^2 \theta_{\text{inc}}. \tag{8.120}$$

Javanaud & Thomas [554] considered normal incidence and obtained the formula $Q[f] = \{f(\pi) - f(0)\}f(0)$. Ye & Ding [1355] obtained $Q[f] = -2\{f(0)\}^3$; this approximation is discussed further in [492, 493, 586].

If the volume fraction is small, $\phi \ll 1$, we may discard the last term in (8.118), and so Foldy's approximation (8.116) is recovered provided we identify $f(0)$ with the scattering coefficient g; indeed, this identification is correct for small sound-soft spheres: compare (8.9) and (8.60). In fact, the formula (8.118) with $Q = 0$ was given by Fermi in his lectures on nuclear physics [342, p. 202]. Moreover, the formula (8.117) was obtained by Urick & Ament in 1949 [1221], using a different argument.

Lloyd & Berry [742] showed that the correct expression for Q is

$$Q[f] = -\{f(\pi)\}^2 + \{f(0)\}^2 + \int_0^\pi \frac{1}{\sin(\theta/2)} \frac{\mathrm{d}}{\mathrm{d}\theta} \left\{[f(\theta)]^2\right\} \mathrm{d}\theta. \tag{8.121}$$

(The integral can be evaluated explicitly in terms of the coefficients B_n, using the linearisation formula for Legendre polynomials (B.10) and the formula

$$\int_{-1}^{1} \frac{1}{\sqrt{1-x}} P_n'(x)\,\mathrm{d}x = \begin{cases} n\sqrt{2}, & \text{when } n \text{ is even,} \\ (n+1)\sqrt{2}, & \text{when } n \text{ is odd.)} \end{cases}$$

The method of Waterman & Truell [1297] was extended to electromagnetic waves by Mathur & Yeh [811] and subsequently criticised and corrected by Lloyd & Berry [742].

Applications of (8.118) can be found in [1200, 1065, 478, 822, 599, 11], in [1185, §34] and in [985, Chapter 4].

The two-dimensional formula for $(k_{\text{eff}}/k)^2$, correct to $O(\hat{n}_0^2)$, is

$$k_{\text{eff}}^2 = k^2 - 4\mathrm{i}\hat{n}_0 f(0) + \frac{8\hat{n}_0^2}{\pi k^2} \int_0^\pi \cot(\theta/2) \frac{\mathrm{d}}{\mathrm{d}\theta} [f(\theta)]^2 \,\mathrm{d}\theta. \tag{8.122}$$

This formula, which is reminiscent of the Lloyd–Berry formula [742], is derived in [730].

8.6.7 Effective field methods and effective medium methods

The methods described above, starting with Foldy [354] and Lax [688], are realisations of a class of self-consistent methods known as *effective field methods*. In fact,

> *it is possible to point out two main underlying ideas of self-consistency: the* effective field *and the* effective medium *methods. These methods are based on different assumptions and, generally speaking, give different results when applied*

> *to a given composite medium.... It is necessary to emphasize that the hypotheses of the methods are introduced in a heuristic manner and the resulting models and theories are* only *approximations. That is why it is difficult to evaluate* ad hoc *the region of their validity. Only the comparison of the theoretical predictions with experimental data or with exact solutions (when and if such exist) allows us to estimate the regions where the methods yield reliable results.*
>
> (Kanaun [581, p. 243])

The basic hypotheses of *effective field methods* are as follows [581, p. 246].

EFM1. Each scatterer in the heterogeneous medium behaves as if isolated and immersed in the original homogeneous medium (outside the scatterers).

EFM2. The effective (exciting) field which acts on each scatterer represents a plane wave, one and the same for every scatterer.

Higher-order statistics, describing the spatial distribution of the scatterers, can be incorporated into effective field methods. For example, one could use Lax's QCA, (8.111), and then make specific assumptions concerning the form of the conditional probability remaining, $p(\boldsymbol{r}_2|\boldsymbol{r}_1)$, depending on what is known about how the scatterers are correlated.

For applications of effective field methods in acoustics, see [346, 1192, 1245, 755, 1109] in three dimensions and [123, 1251, 1243, 122, 1108, 736, 737] in two dimensions. For applications in elastodynamics, see [767, 1324, 1246, 1247]. For applications in electromagnetics, see [1244, 1257, 140, 581, 1189, 951, 952]. Many of the papers by the Varadans and their colleagues make use of T-matrix methods to accommodate non-spherical scatterers; for reviews, see [756, 1254].

Twersky has used his multiple-scattering theory (outlined in Section 6.13) to analyse scattering by random arrangements of scatterers. See, for example, [1200, 1204, 1211, 1212, 1214] for acoustics and [1213] for electromagnetics. His theory has been implemented numerically in [754, 1250]. For elastic waves, see [972].

The basic hypotheses of *effective medium methods* are as follows [581, p. 247].

EMM1. Each scatterer in the heterogeneous medium behaves as if isolated and immersed in a homogeneous medium that has the (unknown) effective properties of the actual heterogeneous medium.

EMM2. The average wavefield in the heterogeneous medium coincides with the wavefield propagating in the homogeneous effective medium.

Several variants of these methods exist; see [581, pp. 247–250] and [583] for short surveys, and the book by Choy [209]. In general, these variants have been introduced because the results obtained using EMM1 and EMM2 are not always in good agreement with experiments:

> *Multiple scattering is indeed the culprit that hinders effective medium theory and it needs to be better understood...*
>
> (Choy [209, p. 73])

For a heterogeneous medium containing a distribution of spheres of radius a, we consider the following acoustic 'coated-sphere problem': surround a typical sphere $0 \le r < a$ by a concentric coating, $a < r < b$; the actual boundary condition of the problem (such as $u = 0$ for soft spheres) is imposed on $r = a$; within the coating, the wavenumber is k; outside the coating, $r > b$, the wavenumber is k_{eff} (unknown); transmission conditions are imposed across $r = b$; and the incident field is a plane wave with wavenumber k_{eff}. This leads to what Twersky called a 'two-space' or 'schizoid' problem: the scattered wave travels in k-space whereas the incident wave travels in k_{eff}-space; see his 1962 paper [1201], and [1204] for comparisons with [1200]. A T-matrix method for the one-body two-space problem is given in [1284].

The coated-sphere problem is one possibility for EMM1. Given b and k_{eff}, it can be solved exactly by separation of variables. Common choices for the radius b are $b = a$ (no coating) or such that $(a/b)^3 = \phi$, the volume fraction occupied by the spheres.

For EMM2, we could replace k by k_{eff} in (8.118), giving

$$1 = 1 - 3\text{i}\phi(k_{\text{eff}}a)^{-3} f_{\text{eff}}(0) + \frac{9}{4}\phi^2(k_{\text{eff}}a)^{-6}Q[f_{\text{eff}}], \tag{8.123}$$

where $f_{\text{eff}}(\theta)$ is the far-field pattern due to the incident wave $\exp(\text{i}k_{\text{eff}}z)$. Equation (8.123) simplifies to

$$f_{\text{eff}}(0) = -\frac{3}{4}\text{i}\phi(k_{\text{eff}}a)^{-3}Q[f_{\text{eff}}], \tag{8.124}$$

and is evidently a non-linear equation for k_{eff}. A two-dimensional version of (8.123), with Q given by (8.119), has been used by Yang & Mal [1351] in the context of fibre-reinforced composites; see also [582]. For extensions to elastic waves, see [1351, 1352, 1059] for plane-strain problems and [1350, 605] for spherical inclusions.

Equation (8.124) is a realisation of EMM2. More simply, one could require

$$f_{\text{eff}}(0) = 0 : \tag{8.125}$$

'the effective medium should be chosen so the mean forward amplitude of the field, scattered [by] a coated inclusion embedded in the effective medium, vanishes' [581, p. 249]. For some applications of (8.125), see [1156, 912].

The condition (8.125) is equivalent to the *coherent potential approximation* (CPA) of solid-state physics [1379, §9.4], [422], [209, §5.2], [424, §5.4]; for the mentioned equivalence, see, for example, [1379, p. 424]. In addition, the KKR method (see Section 1.1.5) is often combined with the CPA; see, for example, [422, Chapter XVIII] or [423, Chapter 12].

In formulating EMM1, we considered scattering by a single coated sphere. This is not the only option. We could choose a (representative) cluster of N scatterers. In particular, we could choose the entire heterogeneous medium, leading to methods that combine the QCA and the CPA; for detailed expositions, see [281, 1186, 1189].

An interesting application of the CPA was made by Jing *et al.* [558]. They used clusters of two, four and ten coated spheres to study wave propagation through suspensions. They found that, at low frequencies, just one (longitudinal) mode can propagate through the suspension, but additional modes can propagate at higher frequencies.

Many other self-consistent schemes have been developed. We mention the papers by Berryman [91, 92], Bobeth & Diener [107], Kerr [599, 600] and Sabina *et al.* [1042, 1041, 1120, 1121, 583] on elastic inclusions, and those by Schwartz *et al.* [1085, 1086, 271] on suspensions. A two-dimensional (plane-strain) version of [1042] is developed in [167]. Earlier than these is the paper by O'Connell & Budiansky [921] on effective wave-speeds in cracked solids; see also [144, 922, 521]. Chatterjee *et al.* [184] compared the effective medium method of [921] with effective field methods and found that they agreed to order ϕ, but not to order ϕ^2.

As the last part of the quotation at the beginning of this section suggests [581, p. 243], it is difficult to justify any particular effective field or effective medium method. Consequently, it is not surprising that controversies have arisen. For some comparisons and criticisms, see [1085, 1086, 271, 36, 394, 395, 821, 599, 600, 37, 94, 503]. For example, [1085, footnote 15] 'in [281, 1186, 1187, 1188, 1192], an analog of the CPA is developed. However, the proposed equations are identical to those formulated by Korringa and Mills [649] and have been shown to be seriously deficient [1105]'. For comparisons with Monte Carlo simulations, see [1194, 1386].

To conclude, we recall that Keller [596] has famously classified methods for solving problems involving random arrangements of scatterers (and other problems of wave propagation through random media). Thus

> *there are two kinds of methods . . . called "honest" and "dishonest" methods respectively. In an honest method the solution $u(\mathbf{r}|\Lambda_N)$ is first determined for each Λ_N. . . . The second step . . . is to compute the mean value of $u(\mathbf{r}|\Lambda_N)$, as well as its variance and other statistics. . . . In a dishonest method randomness is utilized before $u(\mathbf{r}|\Lambda_N)$ is determined. For example, if the mean value $\langle u\rangle$ is sought, the original equations for u may be "averaged" to yield equations for $\langle u\rangle$. In such cases the averaged equations also involve terms like $\langle u^2\rangle$ or $\langle nu\rangle$. Dishonesty enters through the assumption that $\langle u^2\rangle = \langle u\rangle^2$ or $\langle nu\rangle = \langle n\rangle\langle u\rangle$ or some similar unproved assumption. . . . Dishonest methods have the advantage over honest ones that they simplify the problem to be solved. . . . many of the significant and non-trivial results in the theory of wave propagation in random media have been obtained by dishonest methods.*
>
> (Keller [596, pp. 228–229])

Subsequently, at the end of his well-known 1968 review, Frisch wrote as follows.

> *Even if* [dishonest methods] *have produced many interesting results in the past, they are now obsolete, for using dishonest methods is like gambling: one does not know in advance whether the result will be valid or not.*
>
> (Frisch [368, p. 188])

As it has turned out, dishonest methods continue to be used widely.

Appendices

A Legendre functions

In this appendix, we collect together various formulae concerning the associated Legendre function P_n^m. We use the following definition (Ferrers's definition) of this special function.

Definition A.1 *For* $m = 0, 1, \ldots, n$, $n = 0, 1, 2, \ldots$ *and* $-1 \le t \le 1$,

$$P_n^m(t) = (1-t^2)^{m/2} \frac{\mathrm{d}^m}{\mathrm{d}t^m} P_n(t) = \frac{(1-t^2)^{m/2}}{2^n n!} \frac{\mathrm{d}^{m+n}}{\mathrm{d}t^{m+n}} (t^2-1)^n, \tag{A.1}$$

where P_n *is a Legendre polynomial. The integers n and m are called the* degree *and the* order, *respectively.*

The Legendre polynomial $P_n \equiv P_n^0$ is defined by

$$P_n(t) = \frac{1}{2^n n!} \frac{\mathrm{d}^n}{\mathrm{d}t^n} (t^2-1)^n$$

for $n \ge 0$, giving $P_0(t) = 1$ and $P_1(t) = t$; P_n also satisfies the recurrence relation

$$(n+1)P_{n+1}(t) - (2n+1)tP_n(t) + nP_{n-1}(t) = 0, \quad n = 1, 2, \ldots.$$

Evaluating the right-hand side of (A.1) gives

$$P_n^m(t) = (1-t^2)^{m/2} \sum_{l=0}^{[(n-m)/2]} B_l^{n,m} t^{n-m-2l} \tag{A.2}$$

for $0 \le m \le n$, where $B_l^{n,m}$ is defined by (3.27). Here, $[n]$ is the integer part of n, so that the series terminates with a term in t^1 or t^0. In particular, we have

$$P_n(t) = \frac{1}{2^n} \sum_{l=0}^{[n/2]} \frac{(-1)^l (2n-2l)!}{l!\,(n-l)!\,(n-2l)!} t^{n-2l}. \tag{A.3}$$

P_n^m can also be expressed in terms of the hypergeometric function, F. A convenient representation is (GR 8.751(1))[†]

$$P_n^m(t) = \frac{(n+m)!}{(n-m)!}\,\frac{(1-t^2)^{m/2}}{2^m m!}\,F\left(m-n, m+n+1; m+1; \frac{1}{2}(1-t)\right). \tag{A.4}$$

Definition A.1 can also be used for negative integer orders. This is equivalent to the following definition (GR 8.752(2)):

$$P_n^{-m}(t) = (-1)^m \frac{(n-m)!}{(n+m)!}\,P_n^m(t), \quad m = 0, 1, \ldots, n \text{ and } n \geq 0. \tag{A.5}$$

For other integer orders, we have $P_n^m(t) = 0$ for $|m| > n$.

A consequence of (A.2) and (A.5) is

$$P_n^m(-t) = (-1)^{n+m} P_n^m(t) \quad \text{for } 0 \leq |m| \leq n, \tag{A.6}$$

which shows that

$$P_n^m(0) = 0 \text{ when } n+m \text{ is odd.} \tag{A.7}$$

Two special values of note are

$$P_n^m(1) = \delta_{0m} \quad \text{and} \quad P_n^m(-1) = (-1)^n\,\delta_{0m}. \tag{A.8}$$

The associated Legendre functions can be defined for complex order, complex degree and complex argument. However, we are only interested in positive integer degrees ($n \geq 0$), integer orders (m) and real arguments with $-1 \leq t \leq 1$.

From GR 8.733(1), we have

$$(1-t^2)\frac{\mathrm{d}}{\mathrm{d}t}P_n^m(t) = (n+1)t\,P_n^m(t) - (n-m+1)\,P_{n+1}^m(t) \tag{A.9}$$

$$= -nt\,P_n^m(t) + (n+m)\,P_{n-1}^m(t) \tag{A.10}$$

$$= \sqrt{1-t^2}\,P_n^{m+1}(t) - mt\,P_n^m(t) \tag{A.11}$$

$$= -(n-m+1)(n+m)\sqrt{1-t^2}\,P_n^{m-1}(t) + mt\,P_n^m(t) \tag{A.12}$$

for $1 \leq |m| \leq n$. From (A.9) and (A.10), we obtain GR 8.733(2):

$$(2n+1)t\,P_n^m(t) = (n+m)\,P_{n-1}^m(t) + (n-m+1)\,P_{n+1}^m(t). \tag{A.13}$$

[†] This shorthand notation means equation 1 of §8.751 in the book by Gradshteyn & Ryzhik [427]. Recall that their definition of P_n^m differs by a factor of $(-1)^m$ from (A.1).

Eliminating tP_n^m between (A.9) and (A.10) gives

$$(2n+1)(1-t^2)\frac{\mathrm{d}}{\mathrm{d}t}P_n^m(t) = (n+1)(n+m)\,P_{n-1}^m(t) - n(n-m+1)\,P_{n+1}^m(t). \quad \text{(A.14)}$$

From (A.11), we have

$$\sqrt{1-t^2}\,\frac{\mathrm{d}}{\mathrm{d}t}P_n^m(t) = P_n^{m+1}(t) - \frac{mt}{\sqrt{1-t^2}}\,P_n^m(t), \quad \text{(A.15)}$$

whereas (A.11) and (A.12) give

$$\frac{2mt}{\sqrt{1-t^2}}\,P_n^m(t) = P_n^{m+1}(t) + (n-m+1)(n+m)\,P_n^{m-1}(t); \quad \text{(A.16)}$$

eliminating P_n^m between (A.15) and (A.16) then gives

$$2\sqrt{1-t^2}\,\frac{\mathrm{d}}{\mathrm{d}t}P_n^m(t) = P_n^{m+1}(t) - (n-m+1)(n+m)\,P_n^{m-1}(t). \quad \text{(A.17)}$$

Also, multiplying (A.16) by t gives

$$\frac{2m}{\sqrt{1-t^2}}\,P_n^m(t) = t\left[P_n^{m+1}(t) + (n-m+1)(n+m)\,P_n^{m-1}(t)\right] + 2m\sqrt{1-t^2}\,P_n^m(t). \quad \text{(A.18)}$$

From (A.10) and (A.12), we obtain

$$(n-m+1)\sqrt{1-t^2}\,P_n^{m-1}(t) = t\,P_n^m(t) - P_{n-1}^m(t), \quad \text{(A.19)}$$

after cancelling a factor of $(n+m)$. Eliminating $tP_n^m(t)$, using (A.13), then gives GR 8.733(4):

$$(2n+1)\sqrt{1-t^2}\,P_n^m(t) = P_{n+1}^{m+1}(t) - P_{n-1}^{m+1}(t). \quad \text{(A.20)}$$

Similarly, (A.9) and (A.11) give (replacing m by $m-1$)

$$\sqrt{1-t^2}\,P_n^m(t) = (n+m)t\,P_n^{m-1}(t) - (n-m+2)\,P_{n+1}^{m-1}(t), \quad \text{(A.21)}$$

whereas subtracting (A.10) from (A.11) gives

$$\sqrt{1-t^2}\,P_n^{m+1}(t) = (m-n)t\,P_n^m(t) + (n+m)\,P_{n-1}^m(t). \quad \text{(A.22)}$$

B Integrating a product of three spherical harmonics; Gaunt coefficients

Consider the integral

$$I_{n\nu N}^{m\mu M} = \int_\Omega Y_n^m\,Y_\nu^\mu\,Y_N^M\,\mathrm{d}\Omega, \quad \text{(B.1)}$$

where n, ν and N are non-negative. This has been studied extensively in the quantum-mechanics literature; for an early discussion, see [229, Chapter VI, §8]. Edmonds [305, p. 63] proves that

$$I_{n\nu N}^{m\mu M} = \sqrt{\frac{(2n+1)(2\nu+1)(2N+1)}{4\pi}} \begin{pmatrix} n & \nu & N \\ 0 & 0 & 0 \end{pmatrix} \begin{pmatrix} n & \nu & N \\ m & \mu & M \end{pmatrix}, \tag{B.2}$$

where

$$\begin{pmatrix} j_1 & j_2 & j_3 \\ m_1 & m_2 & m_3 \end{pmatrix} \tag{B.3}$$

is called the *Wigner* 3-j *symbol*. It is known that the 3-j symbol (B.3) vanishes unless $m_1 + m_2 + m_3 = 0$; this is consistent with (B.1), which vanishes unless $m + \mu + M = 0$ due to the orthogonality of $\{e^{im\phi}\}$. The 3-j symbol (B.3) also vanishes unless the 'triangle inequalities' $|j_1 - j_2| \le j_3 \le j_1 + j_2$ hold. Thus, (B.2) vanishes unless there exists a (possibly degenerate) triangle with sides of lengths n, ν and N.

The 3-j symbol was introduced by the physicist E.P. Wigner in a famous widely-circulated typescript written in 1940 but published much later [1319]. Properties of the 3-j symbol can be found in the books by Wigner [1318, Chapter 24], Messiah [846, Appendix C.I], Edmonds [305, §3.7], Zare [1371, Chapter 2], Brink & Satchler [141, Appendix I], Dahlen & Tromp [250, §C.7.3] and Mishchenko *et al.* [870, Appendix D].

Wigner's 3-j symbol is related to the *Clebsch–Gordan* (or *vector-coupling*) coefficient $\mathcal{C}$ by

$$\begin{pmatrix} n & \nu & N \\ m & \mu & M \end{pmatrix} = \frac{(-1)^{n+\nu+M}}{\sqrt{2N+1}} \mathcal{C}(n, \nu, N; m, \mu, -M); \tag{B.4}$$

see [846, eqn (C.12)], [305, eqn (3.7.3)] or [141, eqn (3.3)]. The notation for Clebsch–Gordan coefficients varies; see Table B.1.

An explicit formula for $\mathcal{C}$ was obtained by Racah [993]; it contains a finite sum of terms, each involving many factorials. See [846, eqn (C.21)], [1, §27.9], [305, eqn (3.6.10)] or [141, eqn (2.34)].

When $m_1 = m_2 = m_3 = 0$, the 3-j symbol (B.3) vanishes unless $j_1 + j_2 + j_3$ is even. Then

$$\begin{pmatrix} n & \nu & N \\ 0 & 0 & 0 \end{pmatrix} = \frac{(-1)^s\, s!\, \sqrt{(2s-2n)!\,(2s-2\nu)!\,(2s-2N)!}}{(s-n)!\,(s-\nu)!\,(s-N)!\,\sqrt{(2s+1)!}}, \tag{B.5}$$

where $n + \nu + N = 2s$; see [846, eqn (C.23*b*)], [305, eqn (3.7.17)] or [250, eqn (C.220)]. In particular, we have

$$\begin{pmatrix} n & \nu & n+\nu \\ 0 & 0 & 0 \end{pmatrix} = \frac{(-1)^{n+\nu}\,(n+\nu)!}{n!\,\nu!} \sqrt{\frac{(2n)!\,(2\nu)!}{(2n+2\nu+1)!}}.$$

Table B.1. *Some notations for the Clebsch–Gordan coefficients. For others, see* [305, Table 3.1], [99, p. 150] *and* [141, Appendix I].

This book	$\mathcal{C}(j_1, j_2, j_3; m_1, m_2, m_3)$
Condon & Shortley [229], also [1]	$(j_1 j_2 m_1 m_2 \vert j_1 j_2 j_3 m_3)$
Rose [1028, 1029]	$C(j_1 j_2 j_3; m_1 m_2 m_3)$
Messiah [846], Brink & Satchler [141]	$\langle j_1 j_2 m_1 m_2 \vert j_3 m_3 \rangle$
Goldberger & Watson [416]	$\langle j_1, j_2; m_1, m_2 \vert j_3, m_3 \rangle$
Edmonds [305]	$(j_1\, m_1\, j_2\, m_2 \vert j_1\, j_2\, j_3\, m_3)$
Biedenharn & Louck [99]	$C^{j_1 j_2 j_3}_{m_1 m_2 m_3}$
Zare [1371]	$\langle j_1 m_1, j_2 m_2 \vert j_3 m_3 \rangle$

Another useful explicit formula is

$$\begin{pmatrix} n & \nu & n+\nu \\ m & \mu & -m-\mu \end{pmatrix} = (-1)^{n+\nu+m+\mu} \sqrt{\frac{(2n)!\,(2\nu)!}{(2n+2\nu+1)!}} \times \sqrt{\frac{(n+\nu-m-\mu)!\,(n+\nu+m+\mu)!}{(n+m)!\,(n-m)!\,(\nu+\mu)!\,(\nu-\mu)!}}; \tag{B.6}$$

it is given in [305, eqn (3.7.10)] and in [141, p. 138].

If we combine the known explicit formula for $\mathcal{C}$ with (B.4), we obtain

$$I^{m\mu M}_{n\nu N} = A_1 A_2 A_3 A_4\, \delta_{m+\mu+M,0}, \tag{B.7}$$

where

$$A_1 = \sqrt{\frac{(2n+1)(2\nu+1)(2N+1)}{4\pi}},$$

$$A_2 = \frac{(-1)^{s+N+M}\, s!\,(2s-2n)!\,(2s-2\nu)!}{(2s+1)!\,(s-n)!\,(s-\nu)!\,(s-N)!},$$

$$A_3 = \sqrt{\frac{(n+m)!\,(\nu-\mu)!}{(n-m)!\,(\nu+\mu)!}}\, \frac{\sqrt{(N+M)!\,(N-M)!}}{(N-\nu+m)!\,(N-n-\mu)!},$$

$$A_4 = \sum_{k=0} \frac{(-1)^k\,(2s-2N)!\,(n-m)!\,(\nu+\mu)!}{k!\,(2s-2N-k)!\,(n-m-k)!\,(\nu+\mu-k)!} \times \frac{(N-\nu+m)!\,(N-n-\mu)!}{(N-\nu+m+k)!\,(N-n-\mu+k)!},$$

and $2s = n+\nu+N$. The summation is over those k for which the factorials have non-negative arguments. (If desired, an explicit formula for the general 3-j symbol can be obtained by combining (B.2), (B.5) and (B.7).)

Now, the *generalised hypergeometric function* ${}_3F_2$ is defined by [52]

$$ {}_3F_2\left[\begin{matrix} a_1,\ a_2,\ a_3;\ z \\ b_1,\ b_2 \end{matrix}\right] = \sum_{k=0}^{\infty} \frac{(a_1)_k (a_2)_k (a_3)_k}{(b_1)_k (b_2)_k} \frac{z^k}{k!}, $$

where $(a)_k = a(a+1)(a+2)\cdots(a+k-1)$ is *Pochhammer's symbol*; by definition, $(a)_0 = 1$. Note that

$$ (n)_k = \frac{(n+k-1)!}{(n-1)!} \quad \text{and} \quad (-n)_k = (-1)^k \frac{n!}{(n-k)!}, $$

where n is a positive integer and $k \geq 0$. Hence

$$ A_4 = {}_3F_2\left[\begin{matrix} N-n-\nu, & m-n, & -\nu-\mu; & 1 \\ & N-\nu+m+1, & N-n-\mu+1 & \end{matrix}\right]. $$

Rose [1028, eqn (B.2)] gave a similar formula for $\mathcal{C}$. These formulae can be transformed into many equivalent forms [52, Chapter 3]. For a detailed discussion and further references, see [100, Chapter 5, Topic 11].

Another way of viewing (B.1) is via the formula

$$ P_n^m(x)\, P_\nu^\mu(x) = \sum_q \tilde{\mathcal{G}}(n, m; \nu, \mu; q)\, P_q^{m+\mu}(x), \tag{B.8} $$

which is the linearisation formula for products of associated Legendre functions. The summation is finite; it runs from $q = q_0(n, m; \nu, \mu)$ to $q = (n+\nu)$ in steps of 2, where q_0 is defined by (3.74). Use of (3.10) gives

$$ \begin{aligned} \tilde{\mathcal{G}}(n, m; \nu, \mu; q) &= \frac{2q+1}{2} \frac{(q-m-\mu)!}{(q+m+\mu)!} \\ &\quad \times \int_{-1}^{1} P_n^m(x)\, P_\nu^\mu(x)\, P_q^{m+\mu}(x)\,\mathrm{d}x \\ &= (q+\tfrac{1}{2})(-1)^{m+\mu} \int_{-1}^{1} P_n^m(x)\, P_\nu^\mu(x)\, P_q^{-m-\mu}(x)\,\mathrm{d}x. \end{aligned} \tag{B.9} $$

The coefficient $\tilde{\mathcal{G}}$ is related to the Gaunt coefficient $\mathcal{G}$ (defined by (3.71)) by (3.72). It follows that

$$ \begin{aligned} I_{n\nu N}^{m\mu M} &= \delta_{m+\mu+M,0}\,(-1)^M \sqrt{\frac{(2n+1)(2\nu+1)}{4\pi(2N+1)}} \\ &\quad \times \sqrt{\frac{(n-m)!\,(\nu-\mu)!\,(N-M)!}{(n+m)!\,(\nu+\mu)!\,(N+M)!}}\, \tilde{\mathcal{G}}(n, m; \nu, \mu; N) \\ &= \delta_{m+\mu+M,0}\,(-1)^M\, \mathcal{G}(n, m; \nu, \mu; N). \end{aligned} $$

Also, in terms of the Clebsch–Gordan coefficient $\mathcal{C}$, we have

$$\mathcal{G}(n, m; \nu, \mu; N) = \sqrt{\frac{(2n+1)(2\nu+1)}{4\pi(2N+1)}} \times \mathcal{C}(n, \nu, N; 0, 0, 0)\, \mathcal{C}(n, \nu, N; m, \mu, m+\mu).$$

The special case of (B.8) with $\mu = -m$ has been discussed by Bruning & Lo [145]; they give recursion relations for $\tilde{\mathcal{G}}(n, m; \nu, -m; q)$. Another special case is given next; see also [31, Corollary 6.8.3].

Example B.1 (Linearisation formula for Legendre polynomials) If we set $m = \mu = 0$ in (B.8), we obtain

$$P_n(x)\, P_\nu(x) = \sum_q (2q+1) \begin{pmatrix} n & \nu & q \\ 0 & 0 & 0 \end{pmatrix}^2 P_q(x), \tag{B.10}$$

where the summation is from $q = |n-\nu|$ to $q = n+\nu$ in steps of 2, and the Wigner 3-j symbol is given explicitly by (B.5).

It is common to refer to $\tilde{\mathcal{G}}$ as a Gaunt coefficient, because Gaunt [396] published a detailed investigation of the integral on the right-hand side of (B.9) in 1929. (We shall always refer to $\mathcal{G}$ as a Gaunt coefficient.) Gaunt [396] gave various expressions for $\tilde{\mathcal{G}}$, including some special cases where he wrote $\tilde{\mathcal{G}}$ as a generalised hypergeometric function ${}_3F_2$ with argument 1, which he evaluated using some results of Ramanujan as described by Hardy [477]; see also [88, Chapter 10]. Gaunt [396] also cited a paper by J.C. Adams [8] from 1878 on the special case $m = \mu = 0$, where the integrand in (B.9) reduces to the product of three Legendre polynomials. (Adams is famous for his calculations in 1845 predicting the existence of the planet Neptune.) In fact, Adams [9, pp. 398–400] evaluated $\tilde{\mathcal{G}}(n, m; \nu, \mu; q)$ in general, although his results are not widely known. They were published posthumously, and edited by his brother, W.G. Adams:

> [It] *has been no easy task to piece together into this connected whole the detached portions of the work which were delivered into my hands, without any explanation as to any part of them or as to their connection with one another…*
>
> (Adams [9, p. xxviii])

C Rotation matrices

> *So, round and around and around we go.*
> *Where the world's headed, nobody knows.*
> *Oh, great googalooga, can't you hear me talking to you?*
> *Just a ball of confusion.*
>
> ('Ball of confusion', N. Whitfield & B. Strong, 1970)

Suppose that a set of Cartesian axes $Oxyz$ is rotated into axes $Ox'y'z'$. This can be specified in various ways (using direction cosines, for example), but the most economical way is to use three Eulerian angles, α, β and γ. To effect the rotation, one does three successive finite rotations, as follows.

(i) Rotate anti-clockwise by α about the z-axis. This takes the y-axis into a new line called the *line of nodes*.
(ii) Rotate anti-clockwise by β about the line of nodes. This takes the z-axis into the z'-axis.
(iii) Rotate anti-clockwise by γ about the z'-axis. This takes the line of nodes into the y'-axis.

This process is described in books on mechanics, such as [1316, §10] and [417, §4-4], and in books on angular momentum in the context of quantum mechanics; see, for example, [1029, §13], [305, §1.3], [1371, §3.2] or [141, §2.4].

The prescription 'anti-clockwise' needs a little more precision. Both sets of axes, $Oxyz$ and $Ox'y'z'$, are right-handed. Draw the axes Oxy on paper, with the z-axis pointing up, out of the paper. Then, rotating Oxy anti-clockwise in the plane of the paper defines an anti-clockwise rotation about the z-axis. Similarly, for a rotation about the x-axis (y-axis), draw Oyz (Ozx) on paper with the x-axis (y-axis) pointing out of the paper.

We are especially interested in the effect of rotations on spherical harmonics. Specifically, we have

$$Y_n^m(\hat{\boldsymbol{r}}') = \sum_{l=-n}^{n} D_n^{ml}(\alpha, \beta, \gamma)\, Y_n^l(\hat{\boldsymbol{r}}), \tag{C.1}$$

where the *rotation matrix* $\mathbf{D}(\boldsymbol{\omega})$ with entries D_n^{ml} is to be found. We use the notation $\boldsymbol{\omega} = (\alpha, \beta, \gamma)$ where convenient.

The form of this expansion is clear: if we multiply by $r'^n = r^n$, we see that we are expanding a harmonic polynomial of degree n on the left in terms of a basis of such polynomials on the right; see Section 3.4.

An explicit formula for $\mathbf{D}$ can be given. Derivations are given in [1029, Appendix II] and [555], and in the appendix to Chapter 7 of [243]; this latter appendix was prepared by W. Magnus based on notes of G. Herglotz. Other derivations in the context of quantum mechanics are also available; see, for example, [305, §4.1] or [1371, §3.5].

Before giving formulae for $\mathbf{D}$, consider rotations in the plane. Evidently, if axes Oxy are rotated into $Ox'y'$ by an angle α, and if the corresponding polar angles are ϕ and ϕ', we have $\phi = \phi' + \alpha$ so that

$$e^{il\phi'} = e^{-il\alpha} e^{il\phi}.$$

As the first rotation is about the z-axis, it follows that

$$D_n^{ml}(\alpha, \beta, \gamma) = e^{-il\alpha}\, \widetilde{D}_n^{ml}(\beta, \gamma),$$

say. Similarly, the third rotation is about the z'-axis, so that we can write

$$D_n^{ml}(\alpha, \beta, \gamma) = \mathrm{e}^{-\mathrm{i}l\alpha}\, d_n^{ml}(\beta)\, \mathrm{e}^{-\mathrm{i}m\gamma}, \tag{C.2}$$

say, where d_n^{ml} are the entries in a reduced matrix $\mathbf{d}(\beta)$.

We can deduce some properties of $\mathbf{D}(\boldsymbol{\omega})$ directly. Thus, if we replace m by $-m$ and l by $-l$ in (C.1), and use (3.7), we find that

$$D_n^{ml}(\boldsymbol{\omega}) = (-1)^{m+l}\,\overline{D_n^{-m,-l}(\boldsymbol{\omega})},$$

whence (C.2) gives

$$d_n^{ml}(\beta) = (-1)^{m+l}\,\overline{d_n^{-m,-l}(\beta)}. \tag{C.3}$$

As $\mathbf{D}$ *is* a rotation matrix, it has two general properties. First, a rotation can be reversed by reversing the steps, so that

$$\left[D_n^{ml}(\alpha, \beta, \gamma)\right]^{-1} = D_n^{ml}(-\gamma, -\beta, -\alpha).$$

Second, the rotation matrix must be a unitary matrix, so that $\mathbf{D}^{-1} = \overline{\mathbf{D}^{T}}$. Comparing these two gives

$$\overline{D_n^{lm}(\alpha, \beta, \gamma)} = D_n^{ml}(-\gamma, -\beta, -\alpha), \tag{C.4}$$

whence (C.2) gives

$$\overline{d_n^{lm}(\beta)} = d_n^{ml}(-\beta). \tag{C.5}$$

Equations (C.3) and (C.5) are often called 'symmetry relations'. The first of these means that, without loss of generality, we can assume that $m + l \geq 0$.

Courant & Hilbert [243, p. 545] prove that

$$P_n^m(\cos\theta')\,\mathrm{e}^{\mathrm{i}m\phi'} = \sum_{l=-n}^{n} \frac{(n-l)!}{(n+l)!}\, S_n^{(m,l)}\, P_n^l(\cos\theta)\,\mathrm{e}^{\mathrm{i}l\phi}, \tag{C.6}$$

where

$$S_n^{(m,l)} = U_n^{(m,l)}(\tau)\, \exp\{\mathrm{i}(m-l)\sigma - \mathrm{i}(m+l)\rho\}$$

and σ, ρ and τ are related to the Eulerian angles by

$$\alpha = \rho + \sigma, \quad \beta = 2\tau \quad \text{and} \quad \gamma = \rho - \sigma.$$

(These relations may be proved by comparing corresponding expressions for the 3×3 orthogonal matrices that rotate one coordinate system into the other.) Once the definition of Y_n^m (Definition 3.2) has been used in (C.6), we obtain

$$D_n^{ml} = (-1)^{m+l} S_n^{(m,l)} \sqrt{\frac{(n+l)!\,(n-l)!}{(n+m)!\,(n-m)!}},$$

whence

$$d_n^{ml}(\beta) = (-1)^{m+l} U_n^{(m,l)}(\tfrac{1}{2}\beta) \sqrt{\frac{(n+l)!\,(n-l)!}{(n+m)!\,(n-m)!}}. \tag{C.7}$$

Courant & Hilbert [243, pp. 544–545] give two expressions for $U_n^{(m,l)}$, namely

$$\begin{aligned} U_n^{(m,l)}(\tau) &= \frac{(\cos\tau)^{l+m}(\sin\tau)^{l-m}}{(n+l)!} \frac{\mathrm{d}^{n+l}}{\mathrm{d}t^{n+l}} \left\{ t^{n-m}(t-1)^{n+m} \right\} \Big|_{t=\cos^2\tau} \\ &= (-1)^{n+l} \binom{n+m}{n-l} (\cos\tau)^{l+m}(\sin\tau)^{l-m} \\ &\quad \times F(l-n,\, n+l+1;\, l+m+1;\, \cos^2\tau) \end{aligned} \tag{C.8}$$

for $l+m \geq 0$. As $l-n \leq 0$, the hypergeometric function F is a polynomial in $\cos^2\tau$. In fact, it can be written as a Jacobi polynomial $P_N^{(a,b)}(X)$. Thus, from equations 8.960 and 8.962(1) in [427], we have

$$\begin{aligned} P_N^{(a,b)}(X) &= \frac{(-1)^N\,\Gamma(N+b+1)}{N!\,\Gamma(b+1)} F(-N, N+a+b+1; b+1; X_0) \\ &= \frac{1}{2^N} \sum_{k=0}^{N} \binom{N+a}{N-k} \binom{N+b}{k} (X-1)^k (X+1)^{N-k}, \end{aligned}$$

with $X_0 = (1+X)/2$. Comparing with (C.8), we obtain $N = n-l$, $a = l-m$, $b = l+m$ and $X = \cos 2\tau$, whence

$$\begin{aligned} U_n^{(m,l)}(\tau) &= (\cos\tau)^{l+m}(\sin\tau)^{l-m}\, P_{n-l}^{(l-m,l+m)}(\cos 2\tau) \\ &= \cos^{2n}\tau \sum_{k=0}^{n-l} (-1)^k \binom{n-m}{n-l-k} \binom{n+m}{k} (\tan\tau)^{2k+l-m}. \end{aligned} \tag{C.9}$$

(The formula (C.9) can also be found in [133, §6.1] and [31, §9.14].) It follows from (C.7) that

$$d_n^{ml}(\beta) = \sum_{k=0}^{n-l} (-1)^{k+l+m} \Lambda_k(n, m, l) \left(\cos\frac{\beta}{2}\right)^{2n} \left(\tan\frac{\beta}{2}\right)^{2k+l-m},$$

with

$$\Lambda_k(n, m, l) = \frac{\sqrt{(n+l)!\,(n-l)!\,(n+m)!\,(n-m)!}}{(n-l-k)!\,(n+m-k)!\,(k+l-m)!\,k!}.$$

This result agrees with [1371, eqn (3.57)]. It holds for $l+m \geq 0$; for other values of l and m, we can use (C.3). Note also that d_n^{ml} is real so that the overbars can be removed from (C.3) and (C.5). In particular, (C.5), (C.7) and (C.9) give

$$d_n^{lm}(\beta) = U_n^{(m,l)}(\tfrac{1}{2}\beta) \sqrt{\frac{(n+l)!\,(n-l)!}{(n+m)!\,(n-m)!}},$$

which agrees with $d_{lm}^{(n)}$ in [305, eqn (4.1.23)]. Guseinov [454] advocates writing

$$\Lambda_k(n, m, l) = \left[\binom{2n}{n+l} \Big/ \binom{2n}{n+m} \right]^{1/2} \binom{n-1}{k} \binom{n+1}{k+1-m}$$

and then computing the binomial coefficients directly.

As a special case of (C.9), we have

$$U_n^{(0,l)}\left(\frac{1}{2}\beta\right) = \left(\frac{1}{2}\sin\beta\right)^l P_{n-l}^{(l,l)}(\cos\beta) = \frac{n!}{(n+l)!} P_n^l(\cos\beta),$$

whence

$$D_n^{0l} = (-1)^l \sqrt{\frac{(n-l)!}{(n+l)!}} P_n^l(\cos\beta)\, \mathrm{e}^{-il\alpha} = \sqrt{\frac{4\pi}{2n+1}}\, \overline{Y_n^l(\beta, \alpha)}.$$

Although we have explicit formulae for d_n^{ml} and D_n^{ml}, they can also be computed using recursion relations; see [1326, §VI], [418, 763, 208, 448] and [250, §C.8.5].

D One-dimensional finite-part integrals

Is it possible to obtain the required result without having recourse to this finally useless mediation?

I thought it worth while to attain this, though we cannot do so without introducing a rather paradoxical notion which I shall now speak of.

(Hadamard [466, p. 133])

A numerical value may be assigned to a divergent integral in various ways. For example, we have

$$\int_{-1}^{1} \frac{\mathrm{d}t}{t^2} = -2 \tag{D.1}$$

if we interpret the integral in the finite-part sense. Evidently, the integral on the left-hand side of (D.1) is divergent, the integrand is positive and the assigned numerical value is negative! Thus, intuition is not reliable when using non-standard definitions of 'integral'.

We shall mainly discuss integrals of the form

$$I_n(x) = \int_a^b \frac{f(t)}{(t-x)^n}\, \mathrm{d}t, \quad n = 1, 2,$$

where $a < x < b$ and $f(t)$ is a given function. When the integrals are properly defined, I_1 is a Cauchy principal-value (CPV) integral and I_2 is a Hadamard finite-part integral; I_1 is also called a *singular* integral and I_2 is also called a *hypersingular* integral.

To be sure that $I_n(x)$ exists, $f(t)$ must have certain smoothness or continuity properties. Classically, these are usually expressed in terms of Hölder continuity and the function spaces $C^{m,\alpha}$. These spaces are described next.

- f is Hölder continuous, $f \in C^{0,\alpha}$. This means that we can find positive constants A and α so that

$$|f(t_1) - f(t_2)| < A|t_1 - t_2|^\alpha \quad \text{with } 0 < \alpha \le 1$$

  for all t_1 and t_2 in $a < t < b$. Roughly speaking, functions in $C^{0,\alpha}$ are smoother than merely continuous functions, but they need not be differentiable (even if $\alpha = 1$). For example, $f(t) = |t|$ is in $C^{0,1}$ but it is not differentiable at $t = 0$.
- f has a Hölder-continuous first derivative, $f' \in C^{0,\alpha}$ or $f \in C^{1,\alpha}$. Roughly speaking, functions in $C^{1,\alpha}$ are smoother than merely differentiable functions, but they need not have two continuous derivatives.

D.1 Cauchy principal-value integrals

Definition D.1 *The Cauchy principal-value* (CPV) *integral of a function f is defined, for $a < x < b$, by*

$$I_1(x) = ⨍_a^b \frac{f(t)}{t-x}\,\mathrm{d}t = \lim_{\varepsilon\to 0}\left\{\int_a^{x-\varepsilon} \frac{f(t)}{t-x}\,\mathrm{d}t + \int_{x+\varepsilon}^b \frac{f(t)}{t-x}\,\mathrm{d}t\right\}. \tag{D.2}$$

This definition was given by Cauchy in 1822; see [1119, §3.8].

It is well known [898] that a sufficient condition for the existence of I_1 is that f be Hölder continuous, $f \in C^{0,\alpha}$. In particular, when $f \equiv 1$, we obtain

$$⨍_a^b \frac{\mathrm{d}t}{t-x} = \log\frac{b-x}{x-a}, \qquad a < x < b. \tag{D.3}$$

The CPV integral can be regularised: for $f \in C^1$, we have

$$⨍_a^b \frac{f(t)}{t-x}\,\mathrm{d}t = f(b)\log(b-x) - f(a)\log(x-a) - \int_a^b f'(t)\log|t-x|\,\mathrm{d}t. \tag{D.4}$$

This formula, which can be obtained by combining Definition D.1 with integration by parts, shows that a CPV integral can be replaced by an ordinary improper integral, but only if f is sufficiently smooth.

Again, if f is sufficiently smooth, $I_1(x)$ can be differentiated. Thus, if $f \in C^{1,\alpha}$, we can differentiate (D.4) to obtain

$$\frac{\mathrm{d}}{\mathrm{d}x}⨍_a^b \frac{f(t)}{t-x}\,\mathrm{d}t = -\frac{f(b)}{b-x} - \frac{f(a)}{x-a} + ⨍_a^b \frac{f'(t)}{t-x}\,\mathrm{d}t; \tag{D.5}$$

for a proof, see [853, Chapter II, Lemma 6.1].

Much is known about the properties of the singular integral I_1 and how to solve singular integral equations. See, for example, [898] or the encyclopaedic book by Mikhlin & Prössdorf [853].

Smithies [1119, §3.8] has noted that Cauchy actually considered a generalisation of (D.2), namely

$$\lim_{\varepsilon\to 0}\left\{\int_a^{x-\kappa'\varepsilon} \frac{f(t)}{t-x}\,\mathrm{d}t + \int_{x+\kappa''\varepsilon}^{b} \frac{f(t)}{t-x}\,\mathrm{d}t\right\}, \qquad a < x < b, \tag{D.6}$$

where κ' and κ'' are given positive constants. Cauchy showed that the value of (D.6) is

$$⨍_a^b \frac{f(t)}{t-x}\,\mathrm{d}t + f(x)\log\frac{\kappa'}{\kappa''},$$

which he calls the 'general value' of the integral.

> *No serious use of this notion of the "general value" of an improper integral seems to have been made by Cauchy or anyone else; on the other hand, the concept of "principal value" has survived and has proved to be useful in numerous parts of analysis.*
>
> (Smithies [1119, pp. 74–75])

In fact, these generalisations have been discussed in, for example, [534, 878], and they have some relevance to changes of variable, especially in the context of integrals over curves or surfaces.

D.2 Hadamard finite-part integrals

> *In June 1985 Sobolev told us "...* [Hadamard's] *use of the finite part of a divergent integral is just a great foresight. I realized its meaning only later..."*
>
> (Maz'ya & Shaposhnikova [818, pp. 216–217])

Definition D.2 *The Hadamard finite-part integral of a function f is defined, for $a < x < b$, by*

$$\begin{aligned} I_2(x) &= ⨎_a^b \frac{f(t)}{(t-x)^2}\,\mathrm{d}t \\ &= \lim_{\varepsilon\to 0}\left\{\int_a^{x-\varepsilon} \frac{f(t)\,\mathrm{d}t}{(t-x)^2} + \int_{x+\varepsilon}^{b} \frac{f(t)\,\mathrm{d}t}{(t-x)^2} - \frac{2f(x)}{\varepsilon}\right\}. \end{aligned} \tag{D.7}$$

This definition assumes implicitly that f is continuous at x. In fact, it is well known [800] that a sufficient condition for the existence of I_2 is that f' be Hölder continuous, $f \in C^{1,\alpha}$. In particular, when $f \equiv 1$, we obtain

$$⨎_a^b \frac{\mathrm{d}t}{(t-x)^2} = \frac{b-a}{(a-x)(b-x)}, \qquad a < x < b; \tag{D.8}$$

this value is always negative, and it reduces to (D.1) when $a = -1$, $b = 1$ and $x = 0$.

Equation (D.7) defines what might be called a *two-sided finite part integral of order* 2; it is 'two-sided' because the singularity of the integrand is not at an end-point and it is 'of order 2' because of the exponent in the denominator. In fact, Hadamard [466] is mainly concerned with one-sided finite-part integrals of non-integer order.

Definition D.3 *Let μ be a real number with $1 < \mu < 2$. One-sided Hadamard finite-part integrals of order μ are defined by*

$$\fint_a^x \frac{f(t)}{(x-t)^\mu}\,\mathrm{d}t = \lim_{\varepsilon\to 0}\left\{\int_a^{x-\varepsilon} \frac{f(t)}{(x-t)^\mu}\,\mathrm{d}t - \frac{f(x)}{(\mu-1)\varepsilon^{\mu-1}}\right\}, \tag{D.9}$$

$$\fint_x^b \frac{f(t)}{(t-x)^\mu}\,\mathrm{d}t = \lim_{\varepsilon\to 0}\left\{\int_{x+\varepsilon}^{b} \frac{f(t)}{(t-x)^\mu}\,\mathrm{d}t - \frac{f(x)}{(\mu-1)\varepsilon^{\mu-1}}\right\}. \tag{D.10}$$

These are the definitions originally introduced by Hadamard [466, §§80–87], in his investigations into linear hyperbolic partial differential equations; see [818, §15.4]. Later, hyperbolic equations were solved using finite-part integrals in the context of supersonic aerodynamics; see, for example, [1285, §3.4]. Bureau [160] gives many applications and a good historical survey.

To motivate these definitions, consider

$$\begin{aligned}\int_a^{x-\varepsilon} \frac{f(t)}{(x-t)^\mu}\,\mathrm{d}t &= \int_a^{x-\varepsilon} \frac{f(t)-f(x)}{(x-t)^\mu}\,\mathrm{d}t + f(x)\int_a^{x-\varepsilon}\frac{\mathrm{d}t}{(x-t)^\mu}\\ &= \int_a^{x-\varepsilon} \frac{f(t)-f(x)}{(x-t)^\mu}\,\mathrm{d}t - \frac{f(x)}{(\mu-1)(x-a)^{\mu-1}}\\ &\quad + \frac{f(x)}{(\mu-1)\varepsilon^{\mu-1}}.\end{aligned} \tag{D.11}$$

By adding and subtracting one term of the Taylor series of $f(t)$ about the singular point, we obtain an ordinary improper integral on the right-hand side of (D.11), together with a bounded term and a term proportional to $\varepsilon^{1-\mu}$. This last term is unbounded as $\varepsilon \to 0$ because $\mu > 1$. So, if we put this term on the left-hand side, and then let $\varepsilon \to 0$, we obtain a finite result. Thus, 'we give a meaning to our integral by *removing fractional infinities*' at x [466, p. 136].

The definitions (D.9) and (D.10) are easily extended to larger non-integer values of μ, provided that f is sufficiently smooth. One simply adds and subtracts more terms from the Taylor series of f about the singular point, and then 'removes the fractional infinities'. The same ideas work when $\mu = n$, an integer, although the resulting one-sided definitions involve $\log\varepsilon$. Hadamard [466, §83] considers these briefly, and observes that they lead to definitions that are not independent of simple changes of variables. The same ideas also arise in the theory of generalised functions; see, for example, [1084, Chapter II, §2], [399, Chapter I, §§1.7 & 4] and [185, §§3.1 & 3.2].

We can also use Taylor expansions to regularise I_1 and I_2. Thus

$$\fint_a^b \frac{f(t)}{t-x}\,\mathrm{d}t = \int_a^b \frac{f(t)-f(x)}{t-x}\,\mathrm{d}t + f(x)\log\frac{b-x}{x-a}, \quad a < x < b, \tag{D.12}$$

where we have used (D.3). Note that the integral on the right-hand side is an ordinary improper integral if $f \in C^{0,\alpha}$. Thus, we have two regularisations of a CPV integral, namely (D.4) and (D.12).

Similarly, using (D.3) and (D.8),

$$\times\!\!\!\!\!\!\int_a^b \frac{f(t)}{(t-x)^2}\,\mathrm{d}t = \int_a^b \frac{f(t)-f(x)-(t-x)f'(x)}{(t-x)^2}\,\mathrm{d}t + \frac{f(x)\,(b-a)}{(a-x)(b-x)} + f'(x)\,\log\frac{b-x}{x-a}, \tag{D.13}$$

for $a < x < b$; again, the remaining integral is merely improper if $f \in C^{1,\alpha}$. As an alternative regularisation, we have

$$\times\!\!\!\!\!\!\int_a^b \frac{f(t)}{(t-x)^2}\,\mathrm{d}t = -\frac{f(b)}{b-x} - \frac{f(a)}{x-a} + \fint_a^b \frac{f'(t)}{t-x}\,\mathrm{d}t. \tag{D.14}$$

If we compare this formula with (D.5), we obtain

$$\frac{\mathrm{d}}{\mathrm{d}x}\fint_a^b \frac{f(t)}{t-x}\,\mathrm{d}t = \times\!\!\!\!\!\!\int_a^b \frac{f(t)}{(t-x)^2}\,\mathrm{d}t, \tag{D.15}$$

assuming that $f \in C^{1,\alpha}$. This important formula suggests why finite-part integrals might be *useful*: one can differentiate a CPV integral by simply differentiating the integrand, *provided the resulting integral is interpreted correctly*. It is precisely this direct, natural result that Hadamard wanted, without the 'useless mediation' via improper integrals mentioned in the quotation at the beginning of this appendix. In fact, we note that (D.15) is sometimes taken as the *definition* of a two-sided finite-part integral of order 2.

D.3 Integrals over curves

When two-dimensional boundary-value problems are reduced to one-dimensional boundary integral equations, integrals over curves are encountered. We have seen (in Section 5.4) that such integrals can be reduced to integrals over intervals, by parametrising the curves. However, we can define finite-part integrals over curves directly.

Before doing so, let us digress to consider the classical definition of a singular integral [853, Chapter II, Definition 2].

Definition D.4 *Let $\psi(z)$ be a function of a complex variable z, and let Γ be a smooth contour in the z-plane. For sufficiently small $\varepsilon > 0$, and for a fixed point $z_0 \in \Gamma$, put $\Gamma_\varepsilon = \{z \in \Gamma : |z - z_0| \geq \varepsilon\}$. Then*

$$\fint_\Gamma \frac{\psi(z)}{z-z_0}\,\mathrm{d}z = \lim_{\varepsilon\to 0}\int_{\Gamma_\varepsilon} \frac{\psi(z)}{z-z_0}\,\mathrm{d}z.$$

Thus, one removes a piece of Γ by drawing a circle of radius ε, centred at z_0. If Γ is actually a segment of the real axis, this definition reduces to Definition D.1.

Now, consider finite-part integrals over curves, defined as follows.

Definition D.5 *Let μ be a function defined on a smooth curve S. For sufficiently small $\varepsilon > 0$, and for a fixed point $p \in S$, put*

$$S_\varepsilon = \{q \in S : \mathcal{R} \equiv |\boldsymbol{r}_p - \boldsymbol{r}_q| \geq \varepsilon\}.$$

Then, the Hadamard finite-part integral of μ is defined by

$$⨎_S \frac{\mu(q)}{\mathcal{R}^2}\,\mathrm{d}s_q = \lim_{\varepsilon\to 0}\left\{\int_{S_\varepsilon} \frac{\mu(q)}{\mathcal{R}^2}\,\mathrm{d}s_q - \frac{\mu(p)}{\varepsilon}\right\}, \quad p \in S. \tag{D.16}$$

In discussing this definition, we should check that it supplies a finite value. Parametrising S, the end-points of S_ε are given by

$$\varepsilon^2 = \mathcal{R}^2 = [x(t) - x(s)]^2 + [y(t) - y(s)]^2 \simeq (t-s)^2 v^2,$$

where $v(t) = \{[x'(t)]^2 + [y'(t)]^2\}^{1/2}$ and $v \equiv v(s)$. Thus, the end-points are at $t = s \pm \varepsilon/v$. So, in a neighbourhood of p, the expression inside braces on the right-hand side of (D.16) becomes

$$\int^{s-\varepsilon/v} \frac{\varphi(t)\,v(t)}{[(t-s)v]^2}\,\mathrm{d}t + \int_{s+\varepsilon/v} \frac{\varphi(t)\,v(t)}{[(t-s)v]^2}\,\mathrm{d}t - \frac{2\varphi(s)}{\varepsilon}$$

$$\simeq \frac{\varphi(s)}{v}\int^{s-\varepsilon/v} \frac{\mathrm{d}t}{(t-s)^2} + \frac{\varphi(s)}{v}\int_{s+\varepsilon/v} \frac{\mathrm{d}t}{(t-s)^2} - \frac{2\varphi(s)}{\varepsilon},$$

where $\varphi(t) = \mu(q(t))$. Evaluating the two integrals, one finds that all the terms in ε^{-1} cancel, leaving a finite value as $\varepsilon \to 0$.

A form of (D.16) can be found in [113, eqn (12)].

D.4 Literature

There is a considerable literature on one-dimensional finite-part integrals. This covers analytical properties, definite integrals and numerical evaluation. Much of the literature concerns methods for the analysis and numerical treatment of hypersingular integral equations.

For a good table of finite-part integrals, see [594]. Fox [359] gave an early discussion of two-sided finite-part integrals of integer order n with $n > 2$.

Quadrature rules of various kinds have been developed in [674, 938, 732, 136, 535, 98, 277, 309, 483]. See also the review [878].

Methods for regularising finite-part integrals have been discussed in [1168, 802, 24, 803, 187]. See also the collection [1113].

The basic one-dimensional (two-sided) hypersingular integral equation can be written as (6.88); see Section 6.7.1. Many applications of such integral equations are described in [789]. We also mention a paper by Erdélyi [320] in which the (one-sided) integral equation

$$\oint_a^x \frac{u(t)}{(x^2-t^2)^{\mu/2}} P_\nu^\mu(x/t)\,\mathrm{d}t = f(x), \quad 0 < a \le x < b,$$

is solved; here, P_ν^μ is an associated Legendre function and the parameters μ and ν are unrestricted. Hypersingular integral equations involving contour integrals have been discussed in [791, 723].

An important special case of (6.88) is that with $K \equiv 0$. Then, after some simple scaling, we obtain the *dominant equation*,

$$\frac{1}{\pi} \oint_{-1}^1 \frac{u(t)}{(t-x)^2}\,\mathrm{d}t = f(x), \quad -1 < x < 1.$$

The general solution of this equation is

$$u(x) = \frac{1}{\pi} \int_{-1}^1 f(t) \log\left(\frac{|x-t|}{1-xt+\sqrt{(1-x^2)(1-t^2)}} \right) \mathrm{d}t + \frac{A+Bx}{\sqrt{1-x^2}},$$

where A and B are arbitrary constants; if $u(-1) = u(1) = 0$, it follows that $A = B = 0$. This solution can be found in [1073, 1032, 419, 791].

E Proof of Theorem 5.4

For $N\mu$, we begin with the acoustic double-layer potential, defined by (5.8) for $P = (X, Y)$ in B. Then, parametrising S, we obtain

$$(D\mu)(P) = \int_0^1 \varphi(t)\,\Lambda(t, P)\,\mathcal{H}(\mathcal{R}_1)\,\mathrm{d}t,$$

where $\mathcal{R}_1 = |\boldsymbol{r}_P - \boldsymbol{r}_q| = \left\{[x(t)-X]^2 + [y(t)-Y]^2\right\}^{1/2}$,

$$\Lambda(t, P) = [x(t)-X]y'(t) - [y(t)-Y]x'(t)$$

and $\mathcal{H}(\mathcal{R}_1) = \frac{1}{2}\mathrm{i}(k/\mathcal{R}_1)\,H_1^{(1)}(k\mathcal{R}_1)$. Choose X and Y so that P is close to the boundary, along the normal at $p = (x(s), y(s))$; thus, we set

$$X = x(s) + (h/v)y'(s) \quad \text{and} \quad Y = y(s) - (h/v)x'(s), \quad 0 < s < 1,$$

where h is the distance along the normal at p and $v \equiv v(s) = \{[x'(s)]^2 + [y'(s)]^2\}^{1/2}$. Then, we are interested in

$$(N\mu)(p) = \lim_{h\to 0} \frac{\partial}{\partial h}(D\mu)(P).$$

The calculation of this limit is complicated, so we begin with a simple explicit example (for another, see Example 5.1).

Example E.1 Take $k=0$, $x(t)=t$ and $y(t)=0$, whence $v=1$, $X=s$, $Y=\Lambda=-h$, $\mathcal{H}=\pi^{-1}\mathcal{R}_1^{-2}$, $\mathcal{R}_1^2=(t-s)^2+h^2$ and

$$(D\mu)(P)=-\frac{h}{\pi}\int_0^1\frac{\varphi(t)}{(t-s)^2+h^2}\,\mathrm{d}t=\mathcal{D}_0(h), \tag{E.1}$$

say, where $0<s<1$. Next, suppose that φ is linear, so that we can write $\varphi(t)=\pi\{\varphi_0+(t-s)\varphi_1\}$, where φ_0 and φ_1 are constants. Hence

$$\mathcal{D}_0(h)=-\varphi_0\left\{\tan^{-1}\frac{1-s}{h}+\tan^{-1}\frac{s}{h}\right\}-\frac{h\varphi_1}{2}\log\frac{(1-s)^2+h^2}{s^2+h^2}$$

and so, for $0<s<1$,

$$(N\mu)(p)=\lim_{h\to 0}\frac{\mathrm{d}}{\mathrm{d}h}\mathcal{D}_0(h)=\frac{\varphi_0}{s(1-s)}-\varphi_1\log\frac{1-s}{s}. \tag{E.2}$$

Now, return to (E.1), and proceed formally: differentiate with respect to h and then set $h=0$ to give

$$\mathcal{D}_0'(0)=-\frac{1}{\pi}\int_0^1\frac{\varphi(t)}{(t-s)^2}\,\mathrm{d}t=-\varphi_0\int_0^1\frac{\mathrm{d}t}{(t-s)^2}-\varphi_1\int_0^1\frac{\mathrm{d}t}{t-s}.$$

These integrals are divergent. However, if we interpret the first integral as a Hadamard finite-part integral and the second as a Cauchy principal-value integral (see Appendix D for definitions and properties), we obtain

$$\mathcal{D}_0'(0)=-\varphi_0 ⨎_0^1\frac{\mathrm{d}t}{(t-s)^2}-\varphi_1 ⨍_0^1\frac{\mathrm{d}t}{t-s};$$

evaluating these, using (D.3) and (D.8), we obtain (E.2), as before.

In this example, we obtained the correct result by differentiating under the integral sign and then interpreting the resulting divergent integral as a finite-part integral.

The proof of Theorem 5.4 proceeds by reducing the problem to a small neighbourhood of $p=q$ $(t=s)$, in which approximations can be made, leading to integrals similar to (E.1).

To isolate the small neighbourhood of the singularity at $t=s$, we write $(D\mu)(P)=D_0(h)+D_1(h)$, where

$$D_0(h)=\int_{s-\delta}^{s+\delta}\varphi(t)\,\Lambda(t,P)\,\mathcal{H}(\mathcal{R}_1)\,\mathrm{d}t, \tag{E.3}$$

δ is small and positive and $D_1(h)=(D\mu)(P)-D_0(h)$. For $D_1(h)$, we can simply differentiate with respect to h under the integral sign and then let $h\to 0$, because the singularity at $t=s$ has been excluded. Thus

$$D_1'(0)=\lim_{h\to 0}D_1'(h)=\int_{T_\delta}Q(s,t)\,\varphi(t)\,\mathrm{d}t,$$

where $T_\delta=(0,s-\delta)\cup(s+\delta,1)$ and $Q(s,t)$ is defined by (5.25).

Next, we consider $D_0(h)$, defined by (E.3). In this formula, we can replace $\mathcal{H}(\mathcal{R}_1)$ by its asymptotic approximation for small $\mathcal{R}_1$, namely

$$\pi^{-1}\left(\mathcal{R}_1^{-2} - \frac{1}{2}k^2 \log \mathcal{R}_1\right), \tag{E.4}$$

as the difference leads to non-singular integrals; indeed, the difference is continuously differentiable at $h = 0$. Also, for small $|t - s|$, we have $\Lambda(t, P) \sim -hv$. Therefore, the main contribution coming from the logarithmic term is

$$\frac{k^2 v}{2\pi} h\, \mathcal{S}(h) \quad \text{with} \quad \mathcal{S}(h) = \int_{s-\delta}^{s+\delta} \varphi(t) \log \mathcal{R}_1 \, \mathrm{d}t,$$

with derivative $\frac{1}{2}k^2(v/\pi)\{\mathcal{S}(h) + h\mathcal{S}'(h)\}$. The integral $\mathcal{S}(h)$ is recognised as a classical harmonic single-layer potential. It is well known that $\mathcal{S}(h)$ is continuous at $h = 0$, whereas $h\mathcal{S}'(h) \to 0$ as $h \to 0$ (because $\mathcal{S}'(h)$ has a finite limit as $h \to 0$). Thus, we conclude that the term involving $\log \mathcal{R}_1$ is continuously differentiable in h at $h = 0$.

Turning to the most singular term in (E.4), we have

$$\mathcal{R}_1^2 \simeq h^2 + (t - s)^2 v^2$$

for small $|t - s|$ and h. Thus, proceeding as before, we consider

$$-\frac{hv}{\pi} \int_{s-\delta}^{s+\delta} \frac{\varphi(t)}{h^2 + (t - s)^2 v^2} \mathrm{d}t = \mathcal{D}(h), \tag{E.5}$$

say. Again, the difference is found to be continuously differentiable at $h = 0$; in other words,

$$\lim_{h \to 0} \frac{\partial}{\partial h} \left\{ \frac{\Lambda(t, P)}{\mathcal{R}_1^2} + \frac{hv}{h^2 + (t - s)^2 v^2} \right\}$$

exists, and is finite for all t, including $t = s$.

So far, we have only assumed that $\varphi(t)$ is continuous. We now assume more, and suppose that φ has a Hölder-continuous first derivative, $\varphi \in C^{1,\alpha}$; see Appendix D. This means that

$$|\varphi'(t) - \varphi'(s)| \le A|t - s|^\alpha \tag{E.6}$$

for some A and α, with $0 < \alpha \le 1$. This condition cannot be weakened (pointwise), as the following example shows.

Example E.2 Take $\varphi(t) = |t - s| \in C^{0,1}$. Then, from (E.5),

$$\mathcal{D}(h) = -\frac{2hv}{\pi} \int_0^\delta \frac{x\, \mathrm{d}x}{h^2 + v^2 x^2} = -\frac{h}{v\pi} \log\left(1 + \frac{v^2\delta^2}{h^2}\right)$$

and so

$$\mathcal{D}'(h) = -\frac{1}{v\pi} \log\left(1 + \frac{v^2\delta^2}{h^2}\right) + \frac{2v\delta^2}{\pi(h^2 + v^2\delta^2)}.$$

The first term on the right-hand side is unbounded as $h \to 0$.

Let us return to $\mathcal{D}(h)$, defined by (E.5). As $\varphi \in C^{1,\alpha}$, we can integrate by parts to give

$$\mathcal{D}(h) = -\frac{1}{\pi}\{\varphi(s+\delta)+\varphi(s-\delta)\}\tan^{-1}\frac{v\delta}{h} + \mathcal{D}_1(h),$$

where

$$\begin{aligned}\mathcal{D}_1(h) &= \frac{1}{\pi}\int_{s-\delta}^{s+\delta}\varphi'(t)\tan^{-1}\frac{v(t-s)}{h}\,\mathrm{d}t \\ &= \frac{1}{\pi}\int_{s-\delta}^{s+\delta}\{\varphi'(t)-\varphi'(s)\}\tan^{-1}\frac{v(t-s)}{h}\,\mathrm{d}t;\end{aligned}$$

the second equality follows by noting that $\tan^{-1}x$ is an odd function of x. Hence

$$\mathcal{D}'(h) = \{\varphi(s+\delta)+\varphi(s-\delta)\}\frac{v\delta}{\pi(h^2+v^2\delta^2)} + \mathcal{D}_1'(h), \tag{E.7}$$

where

$$\mathcal{D}_1'(h) = -\frac{v}{\pi}\int_{s-\delta}^{s+\delta}\{\varphi'(t)-\varphi'(s)\}\frac{t-s}{h^2+v^2(t-s)^2}\,\mathrm{d}t.$$

The first term in (E.7) is continuous at $h=0$, whereas

$$\mathcal{D}_1'(0) = -\frac{1}{v\pi}\int_{s-\delta}^{s+\delta}\frac{\varphi'(t)-\varphi'(s)}{t-s}\,\mathrm{d}t.$$

The following lemma proves that $|\mathcal{D}_1'(h)-\mathcal{D}_1'(0)| \le Bh^\beta$ for some positive β, whence $\mathcal{D}'(h)$ is continuous at $h=0$ with

$$\mathcal{D}'(0) = \frac{1}{\pi v\delta}\{\varphi(s+\delta)+\varphi(s-\delta)\} - \frac{1}{v\pi}\int_{s-\delta}^{s+\delta}\frac{\varphi'(t)-\varphi'(s)}{t-s}\,\mathrm{d}t. \tag{E.8}$$

Lemma E.3 *For $\varphi \in C^{1,\alpha}$, we have*

$$|\mathcal{D}_1'(h)-\mathcal{D}_1'(0)| \le Bh^\beta, \tag{E.9}$$

where $0<\beta<\alpha$ and B is a constant.

Proof We have

$$\mathcal{D}_1'(h)-\mathcal{D}_1'(0) = \frac{h^2}{v\pi}\int_{s-\delta}^{s+\delta}\frac{\varphi'(t)-\varphi'(s)}{(t-s)^{\gamma+1}}\frac{(t-s)^\gamma}{h^2+v^2(t-s)^2}\,\mathrm{d}t$$

for any γ. Making use of Hölder's inequality for integrals, we obtain

$$|\mathcal{D}_1'(h)-\mathcal{D}_1'(0)| \le \frac{h^2}{v\pi}(I_1)^{1/p}(I_2)^{1/q}, \tag{E.10}$$

where $p>1$, $p^{-1}+q^{-1}=1$,

$$I_1 = \int_{s-\delta}^{s+\delta}\left|\frac{\varphi'(t)-\varphi'(s)}{(t-s)^{\gamma+1}}\right|^p\mathrm{d}t, \quad I_2 = \int_{s-\delta}^{s+\delta}\left|\frac{(t-s)^\gamma}{h^2+v^2(t-s)^2}\right|^q\mathrm{d}t$$

and γ is now restricted so that I_1 and I_2 exist. For I_1, we use (E.6) to obtain

$$I_1 \le 2A^p \int_s^{s+\delta} (t-s)^{(\alpha-\gamma-1)p} \mathrm{d}t = \frac{2A^p \delta^{(\alpha-\gamma-1)p+1}}{(\alpha-\gamma-1)p+1},$$

assuming that the denominator is positive.

For I_2, we choose γ and q so that $\gamma q = 1$, whence

$$\begin{aligned} I_2 &= 2\int_s^{s+\delta} \frac{t-s}{[h^2+v^2(t-s)^2]^q}\,\mathrm{d}t \\ &\le \int_0^\infty \frac{2x\,\mathrm{d}x}{(h^2+v^2x^2)^q} = \frac{h^{2(1-q)}}{v^2(q-1)}, \end{aligned}$$

as $q > 1$. Hence, (E.10) gives (E.9) with $\beta = 2 + 2(1-q)/q = 2/q$.

Now, p, q and γ must satisfy

$$\gamma q = 1, \quad p^{-1} + q^{-1} = 1, \quad p > 1 \quad \text{and} \quad \alpha - \gamma - 1 + p^{-1} > 0.$$

With $\beta = 2/q$, these conditions become

$$\gamma = \frac{1}{2}\beta, \quad p = 2/(2-\beta), \quad \beta > 0 \quad \text{and} \quad \alpha - \beta > 0,$$

respectively, and so the lemma is proved. □

Return now to $\mathcal{D}(h)$, defined by (E.5). Differentiate with respect to h and formally set $h = 0$ to give

$$\mathcal{D}'(0) = -\frac{1}{v\pi} \int_{s-\delta}^{s+\delta} \frac{\varphi(t)}{(t-s)^2}\,\mathrm{d}t.$$

Of course, this integral does not exist. However, from (D.14), we have

$$-\frac{1}{v\pi} \fint_{s-\delta}^{s+\delta} \frac{\varphi(t)}{(t-s)^2}\,\mathrm{d}t = \frac{1}{\pi v \delta}\{\varphi(s+\delta) + \varphi(s-\delta)\} - \frac{1}{v\pi} \fint_{s-\delta}^{s+\delta} \frac{\varphi'(t)}{t-s}\,\mathrm{d}t. \quad \text{(E.11)}$$

As the range of integration is symmetric about $t = s$, we see that the right-hand sides of (E.8) and (E.11) are equal. Therefore, combining all of the above results, we obtain Theorem 5.4.

F Two-dimensional finite-part integrals

In this appendix, we consider integrals of the form

$$I(p) = \int_{\mathcal{P}} \mu(q)\, \mathcal{K}(p,q)\,\mathrm{d}s_q, \quad p \in \mathcal{P},$$

where $\mathcal{K}(p,q) \to \infty$ as $q \to p$. Parametrising the surface patch $\mathcal{P}$ as in Section 5.5, we obtain

$$I(p) = \int_{\mathcal{T}} K(s_1, s_2; t_1, t_2)\, \varphi(t_1, t_2)\,\mathrm{d}t_1\,\mathrm{d}t_2, \quad (s_1, s_2) \in \mathcal{T},$$

where $\mathcal{T}$ is flat. Introduce plane polar coordinates (r, θ) at the *pole* (s_1, s_2), so that

$$t_1 = s_1 + r\cos\theta \quad \text{and} \quad t_2 = s_2 + r\sin\theta. \tag{F.1}$$

Then, with $\tilde{\varphi}(r, \theta) = \varphi(t_1, t_2)$, we can write

$$I(p) = \int_{\mathcal{T}} K(s_1, s_2; r, \theta)\, \tilde{\varphi}(r, \theta)\, r\, \mathrm{d}r\, \mathrm{d}\theta, \tag{F.2}$$

with a slight abuse of notation. By assumption, $K \to \infty$ as $r \to 0$; to make this explicit, we extract the dominant behaviour and write

$$K(s_1, s_2; r, \theta) = r^{-m} K_m(s_1, s_2; r, \theta),$$

where $K_m \to$ a constant as $r \to 0$ and $m > 0$. In practice, $K - r^{-m}K_m$ may be unbounded as $r \to 0$, implying that K also has milder singularities at $r = 0$. Such singularities can be subtracted by repeating the above procedure. Hence, we consider kernels of the form

$$K(s_1, s_2; r, \theta) = r^{-m} f(\theta; s_1, s_2),$$

where f is called the *characteristic* of the corresponding operator.

If the parameter m satisfies $0 < m < 2$, the corresponding kernel is said to be *weakly singular*: the integral exists as an ordinary improper integral. Note that, as in (F.2), $\mathrm{d}t_1\, \mathrm{d}t_2 = r\, \mathrm{d}r\, \mathrm{d}\theta$, and the Jacobian r dilutes the singularity in K. In particular, the integrand vanishes at $r = 0$ if $0 < m < 1$. Integral operators with weakly-singular kernels occur frequently in potential-theoretic problems involving smooth surfaces; see, for example, [597] or [223, §2.3]. However, kernels with stronger singularities also occur.

F.1 Cauchy principal-value integrals

Suppose that $m \geq 2$. Then, $I(p)$ does not exist as an ordinary improper integral. We proceed by breaking $\mathcal{T}$ into three pieces,

$$\mathcal{T} = \mathcal{D}_\varepsilon \cup \mathcal{A}_\varepsilon \cup \mathcal{O}_a,$$

where $\mathcal{D}_\varepsilon$ is a small circular disc with $0 \leq r < \varepsilon$, $\mathcal{A}_\varepsilon$ is an annulus with $\varepsilon < r < a$ (where $a > 0$ is fixed) and $\mathcal{O}_a$ is the remaining 'outer' part of $\mathcal{T}$. Clearly, the contribution from integrating over $\mathcal{O}_a$ is finite, and independent of ε. We cut out the small circular disc $\mathcal{D}_\varepsilon$ of radius ε, centred at the pole, and consider

$$\int_{\mathcal{A}_\varepsilon} K(s_1, s_2; r, \theta)\, \tilde{\varphi}(r, \theta)\, r\, \mathrm{d}r\, \mathrm{d}\theta = \int_\varepsilon^a \int_0^{2\pi} r^{1-m} f(\theta; s_1, s_2)\, \tilde{\varphi}(r, \theta)\, \mathrm{d}\theta\, \mathrm{d}r. \tag{F.3}$$

Next, we introduce the Taylor expansion of φ about the pole,

$$\tilde{\varphi}(r, \theta) = \varphi(t_1, t_2) = \varphi(s_1, s_2) + r\left\{\cos\theta\, \frac{\partial\varphi}{\partial t_1} + \sin\theta\, \frac{\partial\varphi}{\partial t_2}\right\} + \cdots, \tag{F.4}$$

where the partial derivatives are evaluated at $(t_1, t_2) = (s_1, s_2)$. The leading term in (F.4) contributes

$$f_0(s_1, s_2)\, \varphi(s_1, s_2) \int_\varepsilon^a r^{1-m} \mathrm{d}r \tag{F.5}$$

to (F.3), where

$$f_0(s_1, s_2) = \int_0^{2\pi} f(\theta; s_1, s_2)\, \mathrm{d}\theta. \tag{F.6}$$

Clearly, (F.5) is unbounded as $\varepsilon \to 0$ unless the characteristic is such that

$$f_0(s_1, s_2) = 0. \tag{F.7}$$

Moreover, if $2 \le m < 3$, the contribution from the first-derivative terms in the Taylor expansion of φ, (F.4), will exist as a weakly-singular integral. For $m = 2$, these considerations lead to the definition of two-dimensional Cauchy principal-value integrals.

Definition F.1 *Let $\mathcal{T}$ be a flat two-dimensional region, and let φ be a function defined on $\mathcal{T}$. For sufficiently small $\varepsilon > 0$, and for a fixed point $(s_1, s_2) \in \mathcal{T}$, put*

$$\mathcal{T}_\varepsilon = \left\{ (t_1, t_2) \in \mathcal{T} : r \equiv \{(t_1 - s_1)^2 + (t_2 - s_2)^2\}^{1/2} \ge \varepsilon \right\}. \tag{F.8}$$

Then, the Cauchy principal-value (CPV) *integral of φ is defined by*

$$⨍_{\mathcal{T}} \frac{f(\theta; s_1, s_2)}{r^2}\, \varphi(t_1, t_2)\, \mathrm{d}t_1\, \mathrm{d}t_2 = \lim_{\varepsilon \to 0} \int_{\mathcal{T}_\varepsilon} \frac{f(\theta; s_1, s_2)}{r^2}\, \varphi(t_1, t_2)\, \mathrm{d}t_1\, \mathrm{d}t_2, \tag{F.9}$$

where f is a given characteristic and θ is defined by (F.1).

For the CPV integral (F.9) to exist, it is necessary and sufficient that the characteristic f satisfies (F.7). This result was obtained by Tricomi in 1928. (Note that (F.7) is *not* satisfied if $f \equiv 1$.) Much is known about properties of multi-dimensional singular integrals, such as (F.9). See, for example, the books by Mikhlin [852] and Mikhlin & Prössdorf [853]. Three-dimensional singular integrals are discussed in Section H.2.

F.2 Finite-part integrals

If (F.7) is not satisfied, the CPV integral is not defined, but a finite value may be assigned by extracting a finite part; see, for example, [878, Definition 3.3]. However, such integrals seldom arise in practice.

In fact, we can define finite-part integrals for all $m \ge 2$, provided φ is sufficiently smooth. For our purposes, the most useful has $m = 3$; this is the two-dimensional analogue of Definition D.2.

Definition F.2 *Let $\mathcal{T}$ be a flat two-dimensional region, and let φ be a function defined on $\mathcal{T}$. The finite-part integral of φ is defined by*

$$\oint_{\mathcal{T}} \frac{f(\theta; s_1, s_2)}{r^3} \varphi(t_1, t_2) \, \mathrm{d}t_1 \, \mathrm{d}t_2 = \lim_{\varepsilon \to 0} \left\{ \int_{\mathcal{T}_\varepsilon} \frac{f(\theta; s_1, s_2)}{r^3} \varphi \, \mathrm{d}t_1 \, \mathrm{d}t_2 \right.$$
$$\left. - \frac{f_0}{\varepsilon} \varphi(s_1, s_2) + \left[f_c \frac{\partial \varphi}{\partial t_1} + f_s \frac{\partial \varphi}{\partial t_2} \right] \log \varepsilon \right\}, \quad \text{(F.10)}$$

for $(s_1, s_2) \in \mathcal{T}$, where f is a given characteristic, $\mathcal{T}_\varepsilon$ is defined by (F.8), *the partial derivatives are evaluated at $(t_1, t_2) = (s_1, s_2)$,*

$$f_c(s_1, s_2) = \int_0^{2\pi} f(\theta; s_1, s_2) \cos\theta \, \mathrm{d}\theta,$$

$$f_s(s_1, s_2) = \int_0^{2\pi} f(\theta; s_1, s_2) \sin\theta \, \mathrm{d}\theta$$

and $f_0(s_1, s_2)$ is defined by (F.6),

In applications, it is common to find that the characteristic is independent of θ, whence Definition F.2 simplifies as follows.

Definition F.3 *Let $\mathcal{T}$ be a flat two-dimensional region, and let φ be a function defined on $\mathcal{T}$. The finite-part integral of φ is defined by*

$$\oint_{\mathcal{T}} \frac{\varphi(t_1, t_2)}{r^3} \mathrm{d}t_1 \, \mathrm{d}t_2 = \lim_{\varepsilon \to 0} \left\{ \int_{\mathcal{T}_\varepsilon} \frac{\varphi(t_1, t_2)}{r^3} \mathrm{d}t_1 \, \mathrm{d}t_2 - \frac{2\pi}{\varepsilon} \varphi(s_1, s_2) \right\} \quad \text{(F.11)}$$

for $(s_1, s_2) \in \mathcal{T}$, where $\mathcal{T}_\varepsilon$ is defined by (F.8).

F.3 Integrals over surfaces

All three definitions given above assume that a small circular disc $\mathcal{D}_\varepsilon$ has been cut out from the flat region $\mathcal{T}$. If the cut-out region is not circular, these definitions must be modified, in general. For more details, see, for example, [853, Chapter IX, §1.3] or [878, §3.1]. Specifically, suppose that the boundary of the cut-out region is given by

$$r = \varepsilon\rho(\theta; \varepsilon), \quad \text{where} \quad \lim_{\varepsilon \to 0} \rho(\theta; \varepsilon) = \rho_0(\theta). \quad \text{(F.12)}$$

Then, clearly, if $\rho_0(\theta)$ is actually independent of θ, the definitions remain unchanged: in effect, the cut-out region becomes circular as $\varepsilon \to 0$. If ρ_0 does depend on θ, it may still happen that the definitions remain unchanged and, in fact, this usually happens when applications to boundary integral equations are being considered.

These remarks on non-circular cut-out regions become apposite when we define finite-part integrals over surfaces. Thus, for example, we generalise Definition F.3 as follows.

Definition F.4 *Let μ be a function defined on a smooth surface patch $\mathcal{P}$. For sufficiently small $\varepsilon > 0$, and for a fixed point $p \in \mathcal{P}$, put*

$$\mathcal{P}_\varepsilon = \left\{ q \in \mathcal{P} : \mathcal{R} = |\boldsymbol{r}_p - \boldsymbol{r}_q| \geq \varepsilon \right\}.$$

Then, the finite-part integral of μ is defined by

$$⨎_{\mathcal{P}} \frac{\mu(q)}{\mathcal{R}^3} \, \mathrm{d}s_q = \lim_{\varepsilon \to 0} \left\{ \int_{\mathcal{P}_\varepsilon} \frac{\mu(q)}{\mathcal{R}^3} \, \mathrm{d}s_q - \frac{2\pi}{\varepsilon} \mu(p) \right\}, \quad p \in \mathcal{P}. \tag{F.13}$$

Let us check that this definition does give a finite value. To do this, we have to estimate the integral for small ε. Thus, near the pole, we have

$$\begin{aligned} x(t_1, t_2) \simeq x(s_1, s_2) &+ (t_1 - s_1)\frac{\partial x}{\partial t_1} + (t_2 - s_2)\frac{\partial x}{\partial t_2} + \frac{1}{2}(t_1 - s_1)^2 \frac{\partial^2 x}{\partial {t_1}^2} \\ &+ (t_1 - s_1)(t_2 - s_2)\frac{\partial^2 x}{\partial t_1 \, \partial t_2} + \frac{1}{2}(t_2 - s_2)^2 \frac{\partial^2 x}{\partial {t_2}^2}, \end{aligned}$$

with similar Taylor approximations for $y(t_1, t_2)$ and $z(t_1, t_2)$. Hence

$$\begin{aligned} \mathcal{R}^2 &= |\boldsymbol{r}_p - \boldsymbol{r}_q|^2 = [x(t_1, t_2) - x(s_1, s_2)]^2 \\ &\quad + [y(t_1, t_2) - y(s_1, s_2)]^2 + [z(t_1, t_2) - z(s_1, s_2)]^2 \\ &= r^2 \{\ell_2(\theta) + r\ell_3(\theta)\} + O(r^4) \end{aligned} \tag{F.14}$$

as $r \to 0$, where we have used (F.1),

$$\ell_2(\theta) = E\cos^2\theta + F\sin 2\theta + G\sin^2\theta, \tag{F.15}$$

$$\ell_3(\theta) = E_1\cos^3\theta + F_1\cos^2\theta\,\sin\theta + F_2\cos\theta\,\sin^2\theta + G_1\sin^3\theta, \tag{F.16}$$

$$E = \boldsymbol{X}_1 \cdot \boldsymbol{X}_1, \quad F = \boldsymbol{X}_1 \cdot \boldsymbol{X}_2, \quad G = \boldsymbol{X}_2 \cdot \boldsymbol{X}_2,$$

$$E_1 = \boldsymbol{X}_1 \cdot \boldsymbol{X}_{11}, \quad F_1 = 2\boldsymbol{X}_1 \cdot \boldsymbol{X}_{12} + \boldsymbol{X}_2 \cdot \boldsymbol{X}_{11},$$

$$G_1 = \boldsymbol{X}_2 \cdot \boldsymbol{X}_{22}, \quad F_2 = 2\boldsymbol{X}_2 \cdot \boldsymbol{X}_{12} + \boldsymbol{X}_1 \cdot \boldsymbol{X}_{22},$$

$$\boldsymbol{X}_i = \frac{\partial \boldsymbol{r}_q}{\partial t_i} \quad \text{and} \quad \boldsymbol{X}_{ij} = \frac{\partial^2 \boldsymbol{r}_q}{\partial t_i \, \partial t_j} \quad \text{evaluated at } (t_1, t_2) = (s_1, s_2).$$

The quantities E, F and G occur in the *first fundamental form* of differential geometry ([1146, Chapter IV, §5], [185, §6.4]). Note also that the normal vector $\boldsymbol{N}(p)$, defined by (5.26), is given by $\boldsymbol{N}(p) = \boldsymbol{X}_1 \times \boldsymbol{X}_2$, whence $|\boldsymbol{N}(p)| = \sqrt{EG - F^2}$.

Now, for small ε, the boundary of $\mathcal{P}_\varepsilon$ is given by $\varepsilon^2 = \mathcal{R}^2 \simeq r^2\ell_2(\theta)$, so that (F.12) gives

$$\rho_0(\theta) = [\ell_2(\theta)]^{-1/2}. \tag{F.17}$$

We also have

$$\frac{1}{\mathcal{R}^3} = \frac{1}{r^3[\ell_2(\theta)]^{3/2}} - \frac{3}{2r^2}\frac{\ell_3(\theta)}{[\ell_2(\theta)]^{5/2}} + O(r^{-1}) \tag{F.18}$$

as $r \to 0$, using (F.14) and the binomial theorem.

The function $\varphi(t_1, t_2) = \mu(q)$ has a Taylor expansion about the pole, given by (F.4). Similarly, the element of surface area is given by

$$\mathrm{d}s_q = |\boldsymbol{N}(q)|\,\mathrm{d}t_1\,\mathrm{d}t_2 = \sqrt{EG-F^2}\,\mathrm{d}t_1\,\mathrm{d}t_2,$$

where, in this formula only, the square-root term is evaluated at (t_1, t_2); this also has a Taylor expansion about the pole. Combining these two expansions gives

$$\mu(q)\,\mathrm{d}s_q \simeq \Big\{\sqrt{EG-F^2}\,\varphi(s_1, s_2) \\ + r\,[\Lambda_c(s_1, s_2)\cos\theta + \Lambda_s(s_1, s_2)\sin\theta]\Big\}\, r\,\mathrm{d}r\,\mathrm{d}\theta, \tag{F.19}$$

where explicit expressions for Λ_c and Λ_s could be given but are not required here.

Next, we combine (F.18) with (F.19) to give

$$\int_{\mathcal{P}_\varepsilon} \frac{\mu(q)}{\mathcal{R}^3}\,\mathrm{d}s_q \simeq \sqrt{EG-F^2}\left\{\varphi(s_1, s_2)\int_0^{2\pi}\frac{1}{[\ell_2(\theta)]^{3/2}}\int_{\varepsilon\rho_0}\frac{\mathrm{d}r}{r^2}\,\mathrm{d}\theta \right. \\ \left. + \int_0^{2\pi}\mathcal{L}(\theta)\int_{\varepsilon\rho_0}\frac{\mathrm{d}r}{r}\,\mathrm{d}\theta\right\}, \tag{F.20}$$

where $\mathcal{L}(\theta) = \ell_2^{-3/2}\{\Lambda_c\cos\theta + \Lambda_s\sin\theta\} - \frac{3}{2}\ell_3\ell_2^{-5/2}\varphi(s_1, s_2)$. The last inner integral in (F.20) gives

$$\int_{\varepsilon\rho_0}\frac{\mathrm{d}r}{r} \simeq -\log(\varepsilon\rho_0) \sim -\log\varepsilon$$

as $\varepsilon \to 0$, and so the contribution involving $\log\varepsilon$ is proportional to

$$\int_0^{2\pi}\mathcal{L}(\theta)\,\mathrm{d}\theta = \int_0^{\pi}\{\mathcal{L}(\theta) + \mathcal{L}(\pi+\theta)\}\,\mathrm{d}\theta = 0,$$

because $\mathcal{L}(\pi+\theta) = -\mathcal{L}(\theta)$; this follows from (F.15), (F.16), $\ell_2(\pi+\theta) = \ell_2(\theta)$ and $\ell_3(\pi+\theta) = -\ell_3(\theta)$. Hence, the term in $\log\varepsilon$ is absent.

Next, consider the first term in (F.20). The inner integral is

$$\int_{\varepsilon\rho_0}\frac{\mathrm{d}r}{r^2} \simeq \frac{1}{\varepsilon\rho_0} \simeq \frac{[\ell_2(\theta)]^{1/2}}{\varepsilon},$$

using (F.17), whence

$$\int_{\mathcal{P}_\varepsilon}\frac{\mu(q)}{\mathcal{R}^3}\,\mathrm{d}s_q = \frac{\mu(p)}{\varepsilon}\sqrt{EG-F^2}\,I + O(1)$$

as $\varepsilon \to 0$, where

$$I = \int_0^{2\pi} \frac{d\theta}{\ell_2(\theta)} = \int_0^{2\pi} \frac{d\theta}{E\cos^2\theta + F\sin 2\theta + G\sin^2\theta} = \frac{2\pi}{\sqrt{EG - F^2}};$$

this integral can be evaluated easily using the calculus of residues, with an integral around the unit circle in the complex z-plane, where $z = e^{i\theta}$. This leaves a term in ε^{-1} that is exactly cancelled by the second term inside the braces on the right-hand side of (F.13).

The calculation described above for $\mathcal{R}^{-3}$ generalises to other kernels. Details are given in the paper by Schwab & Wendland [1083]; see also [361] for non-smooth surfaces. The calculations involved can be performed using symbolic-manipulation software, as demonstrated in [601]. Moreover, it turns out that typical operators arising from the reduction of elliptic boundary-value problems to boundary integral equations are such that transforming from a surface patch $\mathcal{P}$ to a flat region $\mathcal{T}$ does not introduce any modifications into the definitions of the various finite-part integrals, provided that the mapping between $\mathcal{P}$ and $\mathcal{T}$ is sufficiently smooth.

F.4 Literature

There is a considerable literature on Cauchy principal-value integrals, defined over flat regions and over surfaces; the standard reference is [853]. Much of this is concerned with properties of so-called *singular integral operators* [210, 173]. As Mikhlin [852, p. 2] remarks: 'The first important work on multidimensional singular equations is due to Tricomi' in 1928. He considered integrals of the form (F.9) (with f independent of s_1 and s_2), established the condition (F.7), and examined the effects of changing the shape of the cut-out region.

If we suppose that the Tricomi condition (F.7) is *not* satisfied, then we enter the realm of finite-part integrals. Thus, following Schwartz [1084], we may define

$$⨎_{\mathcal{T}} \frac{\varphi(t_1, t_2)}{r^2}\, dt_1\, dt_2 = \lim_{\varepsilon\to 0}\left\{\int_{\mathcal{T}_\varepsilon} \frac{\varphi(t_1, t_2)}{r^2}\, dt_1\, dt_2 + 2\pi\,\varphi(s_1, s_2)\log\varepsilon\right\}; \qquad \text{(F.21)}$$

this definition is given in [175, p. 17], together with some regularised versions and some discussion of Definition F.3. Indeed, both (F.11) and (F.21) are special cases of equation (II, 3; 5) on p. 44 of Schwartz's book [1084].

Several physical problems involving thin flat screens, plates or cracks can be reduced to a two-dimensional hypersingular integral equation of the form (6.89); see Section 6.7.2. For information on regularisation methods and applications to boundary element methods, see the corresponding references in Section D.4. Simple finite-part integrals over flat regions $\mathcal{T}$ can sometimes be evaluated analytically. Examples are linear variation of φ in (F.11) over triangles [533], and integrals of $t_1^\alpha t_2^\beta r^{-\gamma}$ over convex regions, where α, β and γ are integers [1264].

G Maue's formula

We sketch a proof of Maue's formula (5.18). Let $\mathcal{S}$ by an open surface with boundary $\partial\mathcal{S}$. For a vector field $\boldsymbol{F}$, Stokes's theorem gives

$$\int_{\mathcal{S}} n_\ell \epsilon_{\ell mi} \frac{\partial F_i}{\partial x_m} \, \mathrm{d}s(\boldsymbol{x}) = \int_{\partial\mathcal{S}} F_i \, \mathrm{d}x_i,$$

where ϵ_{ijk} is the alternating tensor, and $\mathcal{S}$ has been oriented in the standard way. We apply Stokes's theorem to the vector field

$$F_i = g\, \epsilon_{ijk} \frac{\partial}{\partial x_j} G(P, Q) = -g\, \epsilon_{ijk} \frac{\partial}{\partial x'_j} G(P, Q),$$

where Q is at (x_1, x_2, x_3), P is at (x'_1, x'_2, x'_3) and $g(\boldsymbol{x})$ is a smooth scalar field. We suppose to begin with that P is not on $\mathcal{S}$. Hence

$$\begin{aligned}\int_{\partial\mathcal{S}} F_i \, \mathrm{d}x_i &= \int_{\mathcal{S}} \epsilon_{\ell mi} \epsilon_{ijk} n_\ell \frac{\partial}{\partial x_m}\left(g \frac{\partial G}{\partial x_j}\right) \mathrm{d}s(\boldsymbol{x}) \\ &= \int_{\mathcal{S}} \left\{ n_j \frac{\partial}{\partial x_k}\left(g \frac{\partial G}{\partial x_j}\right) - n_k \frac{\partial}{\partial x_j}\left(g \frac{\partial G}{\partial x_j}\right)\right\} \mathrm{d}s(\boldsymbol{x}),\end{aligned}$$

using $\epsilon_{\ell mi} = \epsilon_{i\ell m}$ and $\epsilon_{ijk}\epsilon_{i\ell m} = \delta_{j\ell}\delta_{km} - \delta_{jm}\delta_{k\ell}$. As G satisfies the Helmholtz equation, we obtain

$$\begin{aligned}\int_{\partial\mathcal{S}} F_i \, \mathrm{d}x_i = k^2 \int_{\mathcal{S}} g n_k G \, \mathrm{d}s(\boldsymbol{x}) &- \frac{\partial}{\partial x'_k} \int_{\mathcal{S}} g \frac{\partial G}{\partial n_q} \, \mathrm{d}s(\boldsymbol{x}) \\ &+ \int_{\mathcal{S}} \left(n_k \frac{\partial g}{\partial x_j} - n_j \frac{\partial g}{\partial x_k}\right) \frac{\partial G}{\partial x'_j} \, \mathrm{d}s(\boldsymbol{x}).\end{aligned}$$

Next, put $g = \nu$ and let $P \to p \in \mathcal{S}$ to give

$$\begin{aligned}\lim_{P\to p} \frac{\partial}{\partial x'_k} \int_{\mathcal{S}} \nu \frac{\partial G}{\partial n_q} \, \mathrm{d}s(\boldsymbol{x}) = \lim_{P\to p} \Bigg\{ & k^2 \int_{\mathcal{S}} \nu n_k G \, \mathrm{d}s(\boldsymbol{x}) - \int_{\partial\mathcal{S}} F_i \, \mathrm{d}x_i \\ & + \frac{\partial}{\partial x'_j} \int_{\mathcal{S}} \mu_{jk}(q)\, G \, \mathrm{d}s(\boldsymbol{x}) \Bigg\},\end{aligned}$$

where $\mu_{jk} = n_k \, \partial\nu/\partial x_j - n_j \, \partial\nu/\partial x_k$. The first two terms on the right-hand side are continuous as $P \to p$. The last term is the gradient of a single-layer potential; from [223, Theorem 2.17], its limiting value is

$$\pm n_j(p)\mu_{jk}(p) + ⨍_{\mathcal{S}} \mu_{jk}(q) \frac{\partial G}{\partial x'_j} \, \mathrm{d}s(\boldsymbol{x}),$$

the sign being + (−) when $\boldsymbol{n}(p)$ points towards (away from) P. Hence, multiplying by $n_k(p)$, we obtain

$$
\begin{aligned}
&\frac{\partial}{\partial n_p}\int_{\mathcal{S}} \nu(q)\,\frac{\partial G}{\partial n_q}\,\mathrm{d}s_q \\
&\quad = k^2\int_{\mathcal{S}} n_k(p)\,n_k(q)\,G\nu\,\mathrm{d}s_q - n_k(p)\int_{\partial\mathcal{S}} \nu\,\epsilon_{ijk}\,\frac{\partial G}{\partial x_j}\,\mathrm{d}x_i \\
&\qquad + n_k(p) \,⨍_{\mathcal{S}} \left(n_k\frac{\partial\nu}{\partial x_j} - n_j\frac{\partial\nu}{\partial x_k}\right)\frac{\partial G}{\partial x_j'}\,\mathrm{d}s(\boldsymbol{x}) \\
&\quad = k^2\int_{\mathcal{S}} \boldsymbol{n}(p)\cdot\boldsymbol{n}(q)\,G(p,q)\,\nu(q)\,\mathrm{d}s_q \\
&\qquad + \int_{\partial\mathcal{S}} \nu(q)\,\boldsymbol{n}(p)\cdot\big(d\boldsymbol{r}\times\operatorname{grad}_p G(p,q)\big) \\
&\qquad + ⨍_{\mathcal{S}} \big(\boldsymbol{n}(q)\times\operatorname{grad}_q\nu\big)\cdot\big(\boldsymbol{n}(p)\times\operatorname{grad}_p G(p,q)\big)\,\mathrm{d}s_q,
\end{aligned}
$$

where we have noted that $n_j n_k \mu_{jk} = 0$. The line integral around $\partial\mathcal{S}$ will vanish if $\mathcal{S}$ is a closed surface or if $\nu = 0$ on $\partial\mathcal{S}$, and then we obtain Maue's formula.

The proof given above is modelled on a paper by Epton [317]. For other proofs, see [1320, 876, 848, 1170], [223, Theorem 2.23] and [976, §10.2]. Related regularisations for Laplace's equation are given in [142].

H Volume potentials

Define a *volume potential* $W(P)$ by

$$W(P) = \int_B \varphi(Q)\,G(P,Q)\,\mathrm{d}V_Q,$$

where $G(P,Q) = -\exp(\mathrm{i}kR)/(2\pi R)$. The properties of such potentials are similar to those of *Newtonian potentials* for which $k=0$: thus, define

$$W_0(P) = -\frac{1}{2\pi}\int_B \frac{\varphi(Q)}{R}\,\mathrm{d}V_Q.$$

From [225, §8.2], we have

$$(\nabla^2 + k^2)W = \begin{cases} 0, & P\in B_\mathrm{e}, \\ 2\varphi(P), & P\in B, \end{cases} \tag{H.1}$$

where B_e is the region exterior to B. Also, if φ is piecewise continuous, then $W(P)$ and its first partial derivatives are continuous everywhere in three-dimensional space;

see, for example, [597, Chapter VI, §3], [244, Chapter IV, §1.2], [1231, §3.9], [225, Theorem 8.1] and [405, §4.2].

H.1 Discontinuity properties

In addition to the classical results given above, we also require the behaviour of the second derivatives of W near the boundary of B, S. As Kellogg remarks [597, p. 156], 'in general, the derivatives of second order will not exist. It is clear that they cannot all be continuous, for as we pass from an exterior to an interior point through the boundary where φ is not 0, $\nabla^2 W_0$ experiences a break of 2φ.' It is this discontinuous behaviour that interests us here.

The appropriate discontinuities can be found in [762, p. 175] and [778, p. 125]. Here, we sketch a derivation.

It is sufficient to examine the behaviour of $W_0(P)$ when $\varphi(Q) \equiv 1$ and B has a simple shape. To see this, we write

$$W(P) = \int_B \varphi(Q) \left\{ G(P, Q) + \frac{1}{2\pi R} \right\} \mathrm{d}V_Q - \int_{B \setminus B_1} \frac{\varphi(Q)}{2\pi R} \mathrm{d}V_Q$$

$$- \int_{B_1} \frac{\varphi(Q) - \varphi(p)}{2\pi R} \mathrm{d}V_Q + \varphi(p)\, W_1(P), \tag{H.2}$$

where $B_1 \subset B$, $p \in \partial B_1 \cap S$, ∂B_1 is the boundary of B_1, and

$$W_1(P) = -\frac{1}{2\pi} \int_{B_1} \frac{1}{R} \mathrm{d}V_Q.$$

We are interested in the discontinuity in certain second derivatives of $W(P)$ as P crosses S at p. The first three terms on the right-hand side of (H.2) have no such discontinuities: the first term reduces the analysis to the static case, the second term reduces the analysis to a (small) volume B_1, and the third term reduces the analysis to constant φ. For B_1, we take a truncated circular cylinder,

$$0 \le r \le a, \quad 0 \le \theta < 2\pi, \quad 0 \le z \le h,$$

where (r, θ, z) are cylindrical polar coordinates. The base of the cylinder, $z = 0$, is a tangent plane to S at p, which is itself at the origin, $r = 0$, $z = 0$. Normal differentiation is in the negative z direction. It is also convenient to introduce local Cartesian coordinates (x, y, z), with $x = r\cos\theta$ and $y = r\sin\theta$. Then, suppose that P is at (X, Y, Z) with respect to these coordinates, and consider $(\partial/\partial X) W_1(P)$ at a point $(0, 0, Z)$ on the z-axis; we obtain

$$2\pi \frac{\partial W_1}{\partial X} = \int_{B_1} \frac{\partial}{\partial x}\left(\frac{1}{R}\right) \mathrm{d}V_Q = -\int_{B_1} \frac{x}{R^3} \mathrm{d}V_Q = 0,$$

by symmetry. Here, the integration point Q is at (x, y, z). Similarly, $(\partial/\partial Y)W_1(P)=0$ at $(0,0,Z)$. Also

$$\frac{\partial W_1}{\partial Z} = \frac{1}{2\pi}\int_{B_1}\frac{\partial}{\partial z}\left(\frac{1}{R}\right)\mathrm{d}V_Q = \int_0^a \left[\frac{1}{R}\right]_{z=0}^{z-h} r\,dr$$
$$= \left[\sqrt{r^2+(h-Z)^2}-\sqrt{r^2+Z^2}\right]_{r=0}^{r=a}$$
$$= |Z|-|h-Z|+\sqrt{a^2+(h-Z)^2}-\sqrt{a^2+Z^2},$$

whence

$$\lim_{Z\to 0+}\frac{\partial^2 W_1}{\partial Z^2} - \lim_{Z\to 0-}\frac{\partial^2 W_1}{\partial Z^2} = 2. \tag{H.3}$$

Now, consider

$$\left[\frac{\partial}{\partial n}\mathcal{L}v\right] = \lim_{Z\to 0+}\frac{\partial}{\partial Z}\mathcal{L}v - \lim_{Z\to 0-}\frac{\partial}{\partial Z}\mathcal{L}v, \tag{H.4}$$

where $[\cdot]$ is defined by (6.22) and $\mathcal{L}v$ is defined by (6.19). The second term in (6.19) is a volume potential: it does not contribute as its first derivatives are continuous across S. The calculations above show that the only part of the first term in (6.19) that contributes comes from

$$-\frac{\partial}{\partial Z}\int_B\left(\frac{\rho_e}{\rho_0}-1\right)\frac{\partial v}{\partial z}\left(\frac{-1}{4\pi R}\right)\mathrm{d}V_Q \simeq \varphi(p)\,\frac{\partial W_1}{\partial Z},$$

where

$$\varphi(p) = \frac{1}{2}\left(\frac{\rho_e}{\rho_0}-1\right)\frac{\partial v}{\partial n_p}.$$

Hence, (H.3) and (H.4) give (6.23).

H.2 Three-dimensional Cauchy principal-value integrals

In Section 6.4.5, we met volume integrals of the form

$$\int_B \varphi(Q)\,\frac{\partial^2}{\partial x_i^Q\,\partial x_j^Q}G(P,Q)\,\mathrm{d}V_Q, \quad P\in B, \tag{H.5}$$

where $P\equiv(x_1^P, x_2^P, x_3^P)$ and $Q\equiv(x_1^Q, x_2^Q, x_3^Q)$. These integrals exist as three-dimensional Cauchy principal-value (CPV) integrals. To see this, it is sufficient to examine the corresponding expressions for Newtonian potentials with $\varphi(Q)\equiv -2\pi$, namely

$$\int_B \frac{\partial^2}{\partial x_i^Q\,\partial x_j^Q}\left(\frac{1}{R}\right)\mathrm{d}V_Q, \quad P\in B; \tag{H.6}$$

the difference between (H.5) and (H.6) is an ordinary improper integral, provided that φ is Hölder-continuous.

Introduce spherical polar coordinates (r, θ, ϕ) at the pole P, so that

$$x_1^Q = x_1^P + r\sin\theta\cos\phi, \quad x_2^Q = x_2^P + r\sin\theta\sin\phi, \quad x_3^Q = x_3^P + r\cos\theta. \tag{H.7}$$

Then, as in Section F.1, break B into three pieces, $B = B_\varepsilon \cup \mathcal{A}_\varepsilon \cup \mathcal{O}_a$, where B_ε is a small ball with $0 \le r < \varepsilon$, $\mathcal{A}_\varepsilon$ is a spherical shell with $\varepsilon < r < a$ (where $a > 0$ is fixed) and $\mathcal{O}_a$ is the remaining 'outer' part of B. Clearly, the contribution from integrating over $\mathcal{O}_a$ is finite and independent of ε. We remove the ball B_ε of radius ε, centred at the pole, and consider

$$\int_{\mathcal{A}_\varepsilon} \frac{\partial^2}{\partial x_i^Q\,\partial x_j^Q}\left(\frac{1}{R}\right) \mathrm{d}V_Q = \int_\varepsilon^a \int_0^\pi \int_{-\pi}^{\pi} f_{ij}(\theta, \phi)\,\frac{\sin\theta}{r}\,\mathrm{d}\phi\,\mathrm{d}\theta\,\mathrm{d}r, \tag{H.8}$$

where f_{ij} is defined by

$$\frac{\partial^2}{\partial x_i^Q\,\partial x_j^Q}\left(\frac{1}{R}\right) = -\frac{1}{R^3}\delta_{ij} + \frac{3}{R^5}(x_i^Q - x_i^P)(x_j^Q - x_j^P) = \frac{1}{r^3} f_{ij}(\theta, \phi).$$

Explicitly, we have

$$f_{11} = -1 + 3\sin^2\theta\cos^2\phi, \qquad f_{22} = -1 + 3\sin^2\theta\sin^2\phi,$$

$$f_{33} = -1 + 3\cos^2\theta, \qquad f_{12} = 3\sin^2\theta\sin\phi\cos\phi,$$

$$f_{23} = 3\sin\theta\cos\theta\sin\phi, \qquad f_{31} = 3\sin\theta\cos\theta\cos\phi$$

and $f_{ij} = f_{ji}$. Elementary calculations show that

$$\int_0^\pi \int_{-\pi}^{\pi} f_{ij}(\theta, \phi)\sin\theta\,\mathrm{d}\phi\,\mathrm{d}\theta = 0, \quad i, j = 1, 2, 3, \tag{H.9}$$

and so the integral on the left-hand side of (H.8) vanishes for all $\varepsilon > 0$. Thus, the integrals in (H.5) exist as CPV integrals. These are defined as follows (cf. Definition F.1).

Definition H.1 *Let $\mathcal{B}$ be a three-dimensional region, and let φ be a function defined on $\mathcal{B}$. For sufficiently small $\varepsilon > 0$, and for a fixed point $P \in \mathcal{B}$, put*

$$\mathcal{B}_\varepsilon = \left\{ Q \in \mathcal{B} : r \equiv \{(x_1^P - x_1^Q)^2 + (x_2^P - x_2^Q)^2 + (x_3^P - x_3^Q)^2\}^{1/2} \ge \varepsilon \right\}.$$

Then, the Cauchy principal value (CPV) *integral of φ is defined by*

$$⨍_{\mathcal{B}} \frac{f(\theta, \phi; P)}{r^3}\,\varphi(Q)\,\mathrm{d}V_Q = \lim_{\varepsilon\to 0}\int_{\mathcal{B}_\varepsilon} \frac{f(\theta, \phi; P)}{r^3}\,\varphi(Q)\,\mathrm{d}V_Q, \tag{H.10}$$

where f is a given characteristic and θ, ϕ are defined by (H.7).

For the CPV integral (H.10) to exist, it is necessary and sufficient that f satisfies (H.9), so that the integral of f over the unit sphere must vanish [853, Chapter IX, Theorem 1.1].

I Boundary integral equations for G^{E}

Fix points P and Q in B_{e}, and then apply Green's theorem to $G^{\mathrm{E}}(L; P)$ and $G(L, Q)$ with respect to the point L; as both of these are singular solutions, the result is

$$2G^{\mathrm{E}}(Q; P) - 2G(P, Q) = -\int_S G^{\mathrm{E}}(l; P)\frac{\partial}{\partial n_l}G(l, Q)\,\mathrm{d}s_l. \tag{I.1}$$

If we apply the same argument, but with $G(L, Q)$ replaced by $G^{\mathrm{E}}(L; Q)$, we deduce that

$$G^{\mathrm{E}}(P; Q) = G^{\mathrm{E}}(Q; P) \quad \text{for all } P \text{ and } Q \text{ in } B_{\mathrm{e}},\ P \neq Q. \tag{I.2}$$

Then, combining this symmetry property with (I.1) gives

$$2G^{\mathrm{E}}(Q; P) - 2G(P, Q) = -\int_S G^{\mathrm{E}}(l; Q)\frac{\partial}{\partial n_l}G(l, P)\,\mathrm{d}s_l. \tag{I.3}$$

We can obtain integral equations from the two different integral representations for G^{E}. First, let $Q \to q \in S$ in (I.1), giving

$$G^{\mathrm{E}}(q; P) + \int_S G^{\mathrm{E}}(l; P)\frac{\partial}{\partial n_l}G(l, q)\,\mathrm{d}s_l = 2G(q, P), \tag{I.4}$$

which is a boundary integral equation for $G^{\mathrm{E}}(q; P)$; note that, for fixed P, the operator on the left-hand side is precisely $(I + \overline{K^*})$. In particular, letting $P \to p \in S$, we obtain (6.135).

Second, interchange P and Q in (I.3), and then let $Q \to q \in S$; the result is

$$G^{\mathrm{E}}(P; q) + \int_S G^{\mathrm{E}}(l; P)\frac{\partial}{\partial n_l}G(l, q)\,\mathrm{d}s_l = 2G(q, P). \tag{I.5}$$

Subtracting this equation from (I.4) shows that

$$G^{\mathrm{E}}(q; P) = G^{\mathrm{E}}(P; q) \quad \text{for all } P \in B_{\mathrm{e}} \text{ and } q \in S. \tag{I.6}$$

Third, let $Q \to q \in S$ in (I.3), giving

$$2G^{\mathrm{E}}(q; P) = 2G(q, P) - \int_S G^{\mathrm{E}}(l; q)\frac{\partial}{\partial n_l}G(l, P)\,\mathrm{d}s_l \tag{I.7}$$

(there is no jump as $Q \to q$), which is not an integral equation. However, if we let $P \to p \in S$, we obtain

$$G^{\mathrm{E}}(q; p) + \int_S G^{\mathrm{E}}(l; q)\frac{\partial}{\partial n_l}G(l, p)\,\mathrm{d}s_l = 2G(q, p). \tag{I.8}$$

If we interchange p and q in this equation and then subtract the result from (5.58), we deduce that

$$G^{\mathrm{E}}(q; p) = G^{\mathrm{E}}(p; q) \quad \text{for all } p \text{ and } q \text{ on } S, \tag{I.9}$$

whence (I.8) becomes

$$G^{\mathrm{E}}(q;\, p) + \int_S G^{\mathrm{E}}(q;\, l)\frac{\partial}{\partial n_l}G(l,\, p)\,\mathrm{d}s_l = 2G(q,\, p). \tag{I.10}$$

Multiplying this equation by $f(q)$ and then integrating over $q \in S$ gives $(I+\overline{K^*})S^{\mathrm{E}} f = 2Sf$.

Combining (I.2), (I.6) and (I.9), we see that G^{E} is symmetric for all locations of the source and field points, (6.131).

Finally, we note that we can derive integral equations giving the gradient of G^{E} at any point P not on S, $\mathrm{grad}_P G(q;\, P)$. Thus, taking the gradient of (I.4) gives

$$\mathrm{grad}_P G^{\mathrm{E}}(q;\, P) + \int_S \mathrm{grad}_P G^{\mathrm{E}}(l;\, P)\frac{\partial}{\partial n_l}G(l,\, q)\,\mathrm{d}s_l = 2\,\mathrm{grad}_P G(q,\, P),$$

which is an integral equation for $\mathrm{grad}_P G^{\mathrm{E}}(q;\, P)$. Then, (I.1) gives

$$2\,\mathrm{grad}_P G^{\mathrm{E}}(Q;\, P) = 2\,\mathrm{grad}_P G(P,\, Q) - \int_S \mathrm{grad}_P G^{\mathrm{E}}(l;\, P)\frac{\partial}{\partial n_l}G(l,\, Q)\,\mathrm{d}s_l$$

for any point Q not on S.

References

Comments on the references

G.H. Hardy used to advise the referees of papers that they should ask three questions: "Is it new? Is it true? Is it interesting?"

(Boas [105, p. 434])

This book contains an extensive list of references. We have only cited books, reports and articles that we have actually seen. All items in the reference list have been cited in the book; the places of citation are indicated by 'C:' at the end of each item. There is an alphabetical listing of all authors in the references beginning on p. 421.

Many journals are cited frequently, and so we have used abbreviations for their titles, as defined in the following table.

ACM	*Advances in Computational Mathematics*
AMR	*Applied Mechanics Reviews*
AO	*Applied Optics*
AOR	*Applied Ocean Research*
ARFM	*Annual Review of Fluid Mechanics*
ARMA	*Archive for Rational Mechanics and Analysis*
BSSA	*Bulletin of the Seismological Society of America*
CJP	*Canadian Journal of Physics*
CMAME	*Computer Methods in Applied Mechanics and Engineering*
CPAM	*Communications on Pure and Applied Mathematics*
EABE	*Engineering Analysis with Boundary Elements*
EESD	*Earthquake Engineering and Structural Dynamics*
EFM	*Engineering Fracture Mechanics*
GJI	*Geophysical Journal International*
GJRAS	*Geophysical Journal of the Royal Astronomical Society*
IEAP	*IEEE Transactions on Antennas and Propagation*
IEAPM	*IEEE Antennas and Propagation Magazine*
IEE-H	*IEE Proc. Part H: Microwaves, Antennas & Propagation*
IEGRS	*IEEE Transactions on Geoscience and Remote Sensing*
IEMT	*IEEE Transactions on Microwave Theory and Techniques*
IJES	*International Journal of Engineering Science*
IJNME	*International Journal for Numerical Methods in Engineering*
IJNMF	*International Journal for Numerical Methods in Fluids*
IJSS	*International Journal of Solids and Structures*
IMAP	*IMA Journal of Applied Mathematics*
JAM	*Journal of Applied Mechanics*

JAP	*Journal of Applied Physics*
JASA	*Journal of the Acoustical Society of America*
JCAM	*Journal of Computational and Applied Mathematics*
JCP	*Journal of Computational Physics*
JEM	*Journal of Engineering Mathematics*
JEWA	*Journal of Electromagnetic Waves and Applications*
JFM	*Journal of Fluid Mechanics*
JGR	*Journal of Geophysical Research*
JMAA	*Journal of Mathematical Analysis and Applications*
JMP	*Journal of Mathematical Physics*
JMPS	*Journal of the Mechanics and Physics of Solids*
JOSA	*Journal of the Optical Society of America*
JPDAP	*Journal of Physics* D*: Applied Physics*
JQSRT	*Journal of Quantitative Spectroscopy & Radiative Transfer*
JSR	*Journal of Ship Research*
JSV	*Journal of Sound and Vibration*
JWPCOE	*Journal of Waterway, Port, Coastal, & Ocean Engineering*
MM	*Mechanics of Materials*
MMAS	*Mathematical Methods in the Applied Sciences*
MOTL	*Microwave and Optical Technology Letters*
MPCPS	*Mathematical Proceedings of the Cambridge Philosophical Society*
NM	*Numerische Mathematik*
OC	*Optics Communications*
OE	*Ocean Engineering*
OL	*Optics Letters*
PCPS	*Proceedings of the Cambridge Philosophical Society*
PR	*Physical Review*
PRSA	*Proceedings of the Royal Society* A
PRSE	*Proceedings of the Royal Society of Edinburgh*
PTRSA	*Philosophical Transactions of the Royal Society* A
QAM	*Quarterly of Applied Mathematics*
QJMAM	*Quarterly Journal of Mechanics and Applied Mathematics*
RMP	*Reviews of Modern Physics*
RS	*Radio Science*
SDEE	*Soil Dynamics & Earthquake Engineering*
SIAP	*SIAM Journal on Applied Mathematics*
SIMA	*SIAM Journal on Mathematical Analysis*
SINUM	*SIAM Journal on Numerical Analysis*
SISC	*SIAM Journal on Scientific Computing*
SPA	*Soviet Physics – Acoustics*
SPD	*Soviet Physics – Doklady*
WM	*Wave Motion*
ZAMM	*Zeitschrift für angewandte Mathematik und Mechanik*

1. M. Abramowitz & I.A. Stegun (eds.), *Handbook of Mathematical Functions*. New York: Dover, 1965. C: pp. 28, 34, 35, 38, 39, 47, 49, 63, 91, 103, 108, 132, 150, 261, 326 & Table B.1 (p. 327)
2. J.D. Achenbach, *Wave Propagation in Elastic Solids*. Amsterdam: North-Holland, 1973. C: pp. 20, 22
3. J.D. Achenbach, *Reciprocity in Elastodynamics*. Cambridge: Cambridge University Press, 2003. C: p. 26

4. J.D. Achenbach, G.E. Kechter & Y.-L. Xu, Off-boundary approach to the boundary element method. *CMAME* **70** (1988) 191–201. C: p. 244
5. J.D. Achenbach & M. Kitahara, Reflection and transmission of an obliquely incident wave by an array of spherical cavities. *JASA* **80** (1986) 1209–1214. C: p. 203
6. J.D. Achenbach & M. Kitahara, Harmonic waves in a solid with a periodic distribution of spherical cavities. *JASA* **81** (1987) 595–598. C: p. 203
7. J.D. Achenbach, Y.-C. Lu & M. Kitahara, 3-D reflection and transmission of sound by an array of rods. *JSV* **125** (1988) 463–476. C: p. 178
8. J.C. Adams, On the expression of the product of any two Legendre's coefficients by means of a series of Legendre's coefficients. *Proc. Roy. Soc.* **27** (1878) 63–71. C: p. 329
9. J.C. Adams, *The Scientific Papers of John Couch Adams*, vol. II. Cambridge: Cambridge University Press, 1900. C: p. 329
10. R.J. Adams, Combined field integral equation formulations for electromagnetic scattering from convex geometries. *IEAP* **52** (2004) 1294–1303. C: p. 219
11. D.G. Aggelis, D. Polyzos & T.P. Philippidis, Wave dispersion and attenuation in fresh mortar: theoretical predictions vs. experimental results. *JMPS* **53** (2005) 857–883. C: p. 319
12. A.R. Aguiar & Y.C. Angel, Antiplane coherent scattering from a slab containing a random distribution of cavities. *PRSA* **456** (2000) 2883–2909. C: p. 316
13. J.F. Ahner, The exterior time harmonic elasticity problem with prescribed displacement vector on the boundary. *Archiv der Mathematik* **27** (1976) 106–111. C: p. 219
14. J.F. Ahner, Scattering by an inhomogeneous medium. *J. Inst. Maths. Applics.* **19** (1977) 425–439. C: pp. 15, 135, 186
15. J.F. Ahner, A scattering trinity: the reproducing kernel, null- field equations and modified Green's functions. *QJMAM* **39** (1986) 153–162. C: p. 271
16. K.A. Al-Badwaihy & J.L. Yen, Extended boundary condition integral equations for perfectly conducting and dielectric bodies: formulation and uniqueness. *IEAP* **23** (1975) 546–551. C: p. 247
17. G.E. Albert & J.L. Synge, The general problem of antenna radiation and the fundamental integral equation, with application to an antenna of revolution – part I. *QAM* **6** (1948) 117–131. C: p. 244
18. J.R. Allegra & S.A. Hawley, Attenuation of sound in suspensions and emulsions: theory and experiments. *JASA* **51** (1972) 1545–1564. C: p. 143
19. A. Alvarez, C.-C. Wang & Z. Ye, A numerical algorithm of the multiple scattering from an ensemble of arbitrary scatterers. *JCP* **154** (1999) 231–236. C: p. 304
20. C.J.S. Alves, Numerical simulations on resonance poles – the trapping case of two plane cracks. *JCAM* **111** (1999) 267–279. C: p. 209
21. C.J.S. Alves, Acoustic scattering by multicracks using a boundary finite element method. In: *Advances in Boundary Element Techniques* V (eds. V.M.A. Leitão & M.H. Aliabadi). Eastleigh: EC Ltd, 2004, pp. 161–168. C: p. 208
22. S. Amini, An iterative method for the boundary element solution of the exterior acoustic problem. *JCAM* **20** (1987) 109–117. C: p. 219
23. S. Amini, Boundary integral solution of the exterior Helmholtz problem. *Computational Mech.* **13** (1993) 2–11. C: p. 219
24. S. Amini & N.D. Maines, Regularization of strongly singular integrals in boundary integral equations. *Comm. Numer. Meth. Eng.* **12** (1996) 787–793. C: p. 338
25. S. Amini & N.D. Maines, Preconditioned Krylov subspace methods for boundary element solution of the Helmholtz equation. *IJNME* **41** (1998) 875–898. C: p. 219
26. S. Amini & A.T.J. Profit, Multi-level fast multipole solution of the scattering problem. *EABE* **27** (2003) 547–564. C: p. 240

27. S. Amini & D.T. Wilton, An investigation of boundary element methods for the exterior acoustic problem. *CMAME* **54** (1986) 49–65. C: p. 219
28. P. Andersen & H. Wuzhou, On the calculation of two-dimensional added mass and damping coefficients by simple Green's function technique. *OE* **12** (1985) 425–451. C: p. 206
29. V.C. Anderson, Sound scattering from a fluid sphere. *JASA* **22** (1950) 426–431. C: p. 135
30. M.G. Andreasen, Scattering from parallel metallic cylinders with arbitrary cross sections. *IEAP* **12** (1964) 746–754. C: p. 178
31. G.E. Andrews, R. Askey & R. Roy, *Special Functions*. Cambridge: Cambridge University Press, 1999. C: pp. 36, 65, 72, 72, 83, 109, 329, 332
32. Y.C. Angel & A. Bolshakov, Oblique coherent waves inside and outside a randomly cracked elastic solid. *JASA* **102** (1997) 3290–3299. C: p. 316
33. Y.C. Angel & Y.K. Koba, Complex-valued wavenumber, reflection and transmission in an elastic solid containing a cracked slab region. *IJSS* **35** (1998) 573–592. C: p. 316
34. J.H. Ansell, Legendre functions, the Hilbert transform and surface waves on a sphere. *GJRAS* **32** (1973) 95–117. C: p. 142
35. J.H. Ansell, On the scattering of *SH* waves from a point source by a sphere. *GJRAS* **54** (1978) 349–387. C: p. 142
36. L.W. Anson & R.C. Chivers, Ultrasonic propagation in suspensions – A comparison of a multiple scattering and an effective medium approach. *JASA* **85** (1989) 535–540. C: p. 322
37. L.W. Anson & R.C. Chivers, Ultrasonic velocity in suspensions of solids in solids – a comparison of theory and experiment *JPDAP* **26** (1993) 1566–1575. C: p. 322
38. J. Antonio & A. Tadeu, 3D scattering by multiple cylindrical cavities buried in an elastic formation. *European J. Mech. A/Solids* **20** (2001) 367–383. C: p. 202
39. E. Argyropoulos, K. Kiriaki & G.F. Roach, A modified Green's-function technique for the exterior Dirichlet problem in linear elasticity. *QJMAM* **52** (1998) 275–295. C: p. 217
40. C. Aristégui & Y.C. Angel, New results for isotropic point scatterers: Foldy revisited. *WM* **36** (2002) 383–399. C: p. 316
41. F.M. Arscott & A. Darai, Curvilinear co-ordinate systems in which the Helmholtz equation separates. *IMAP* **27** (1981) 33–70. C: pp. 122, 148, 150
42. E. Arvas & J.R. Mautz, On the non-uniqueness of the surface EFIE applied to multiple conducting and/or dielectric bodies. *Archiv für Elektronik und Übertragungstechnik* **42** (1988) 364–369. C: p. 195
43. E. Arvas, S.M. Rao & T.K. Sarkar, *E*-field solution of TM-scattering from multiple perfectly conducting and lossy dielectric cylinders of arbitrary cross-section. *IEE-H* **133** (1986) 115–121. C: p. 195
44. S. Asano & G. Yamamoto, Light scattering by a spheroidal particle. *AO* **14** (1975) 29–49. C: p. 150
45. R. Askey, Linearization of the product of orthogonal polynomials. In: *Problems in Analysis* (ed. R.C. Gunning) pp. 131–138. Princeton: Princeton University Press, 1970. C: p. 83
46. R. Askey, *Orthogonal Polynomials and Special Functions*. Philadelphia: SIAM, 1975. C: pp. 29, 62, 83
47. C. Athanasiadis, P.A. Martin & I.G. Stratis, Electromagnetic scattering by a homogeneous chiral obstacle: boundary integral equations and low-chirality approximations. *SIAP* **59** (1999) 1745–1762. C: p. 18
48. C. Athanasiadis, P.A. Martin & I.G. Stratis, Electromagnetic scattering by a homogeneous chiral obstacle: scattering relations and the far-field operator. *MMAS* **22** (1999) 1175–1188. C: p. 18

49. K.E. Atkinson, *The Numerical Solution of Integral Equations of the Second Kind*. Cambridge: Cambridge University Press, 1997. C: pp. 162, 168, 169, 177, 263, 266
50. J.-C. Auger & B. Stout, A recursive centered T-matrix algorithm to solve the multiple scattering equation: numerical validation. *JQSRT* **79-80** (2003) 533–547. C: p. 283
51. K. Aydin & A. Hizal, On the completeness of the spherical vector wave functions. *JMAA* **117** (1986) 428–440. C: p. 257
52. W.N. Bailey, *Generalized Hypergeometric Series*. London: Cambridge University Press, 1935. C: p. 328
53. B.B. Baker & E.T. Copson, *The Mathematical Theory of Huygens' Principle*, third edition. New York: Chelsea, 1987. C: pp. 11, 13, 152, 156, 157, 165, 165
54. M. Balabane, Boundary decomposition for Helmholtz and Maxwell equations 1: disjoint sub-scatterers. *Asymptotic Analysis* **38** (2004) 1–10. C: p. 228
55. M. Balabane & V. Tirel, Décomposition de domaine pour un calcul hybride de l'équation de Helmholtz. *C. R. Acad. Sci. Paris*, Série I, **324** (1997) 281–286. C: p. 228
56. T.F. Balsa, Low frequency two dimensional flows through a sparse array of bodies. *JSV* **82** (1982) 489–504. C: p. 297
57. R.P. Banaugh, Application of integral representations of displacement potentials in elastodynamics. *BSSA* **54** (1964) 1073–1086. C: p. 201
58. R.P. Banaugh & W. Goldsmith, Diffraction of steady acoustic waves by surfaces of arbitrary shape. *JASA* **35** (1963) 1590–1601. C: p. 178
59. R.P. Banaugh & W. Goldsmith, Diffraction of steady elastic waves by surfaces of arbitrary shape. *JAM* **30** (1963) 589–597. C: p. 201
60. P.K. Banerjee, *The Boundary Element Methods in Engineering*. London: McGraw-Hill, 1994. C: p. 11
61. R. Barakat, Diffraction of plane waves by an elliptic cylinder. *JASA* **35** (1963) 1990–1996. C: p. 149
62. R.G. Barantsev, The method of separation of variables in the problem of scattering by a body of arbitrary shape. *SPD* **7** (1963) 974–975. C: p. 248
63. P. Barber & C. Yeh, Scattering of electromagnetic waves by arbitrarily shaped dielectric bodies. *AO* **14** (1975) 2864–2872. C: p. 251
64. P.W. Barber, Electromagnetic power deposition in prolate spheroidal models of man and animals at resonance. *IEEE Trans. on Biomedical Engng.* **24** (1977) 513–521. C: p. 251
65. P.W. Barber, Resonance electromagnetic absorption by nonspherical dielectric objects. *IEMT* **25** (1977) 373–381. C: p. 251
66. P.W. Barber, Scattering and absorption efficiencies for nonspherical dielectric objects – biological models. *IEEE Trans. on Biomedical Engng.* **25** (1978) 155–159. C: p. 251
67. D.M. Barnett, Bulk, surface, and interfacial waves in anisotropic linear elastic solids. *IJSS* **37** (2000) 45–54. C: p. 153
68. M.L. Baron, H.H. Bleich & A.T. Matthews, Forced vibrations of an elastic circular cylindrical body of finite length submerged in an acoustic fluid. *IJSS* **1** (1965) 3–22. C: p. 172
69. P.J. Barratt, Scattering of plane harmonic waves in an infinite elastic solid by an arbitrary configuration of two-dimensional obstacles. *J. Inst. Maths Applics.* **4** (1968) 233–249. C: pp. 22, 235
70. P.J. Barratt, Multiple scattering of plane harmonic elastic waves in an infinite solid by an arbitrary configuration of obstacles. *PCPS* **66** (1969) 469–480. C: p. 235
71. F. Bass & M. Fix, Influence of the shape of small scatterers upon their resonance features. *PR* E **56** (1997) 7235–7239. C: p. 305

72. F. Bass, V.D. Freilikher & V.V. Prosentsov, Electromagnetic wave scattering from small scatterers of arbitrary shape. *JEWA* **14** (2000) 269–283. C: p. 305
73. H. Bateman, *Electrical and Optical Wave-Motion*. Cambridge: Cambridge University Press, 1915. Reprinted by Dover, New York, 1955. C: pp. 63, 82, 140, 223
74. H. Bateman, *Partial Differential Equations of Mathematical Physics*. Cambridge: Cambridge University Press, 1944. C: pp. 68, 223
75. R.H.T. Bates, Modal expansions for electromagnetic scattering from perfectly conducting cylinders of arbitrary cross-section. *Proc. IEE* **115** (1968) 1443–1445. C: p. 250
76. R.H.T. Bates, J.R. James, I.N.L. Gallett & R.F. Millar, An overview of point matching. *The Radio & Electronic Engineer* **43** (1973) 193–200. C: p. 137
77. R.H.T. Bates & D.J.N. Wall, Null field approach to scalar diffraction I. General method. *PTRSA* **287** (1977) 45–78. C: pp. 247, 258, 265, 274, 280
78. L. Bencheikh, Modified fundamental solutions for the scattering of elastic waves by a cavity. *QJMAM* **43** (1990) 57–73. C: p. 217
79. L. Bencheikh, Modified fundamental solutions for the scattering of elastic waves by a cavity: numerical results. *IJNME* **36** (1993) 3283–3302. C: p. 217
80. R. Benites, K. Aki & K. Yomogida, Multiple scattering of *SH* waves in 2-D media with many cavities. *Pure & Applied Geophysics* **138** (1992) 353–390. C: p. 246
81. R.A. Benites, P.M. Roberts, K. Yomogida & M. Fehler, Scattering of elastic waves in 2-D composite media I. Theory and test. *Physics of the Earth & Planetary Interiors* **104** (1997) 161–173. C: p. 246
82. A. Ben-Menahem, An operational representation of the addition theorems for spherical waves. *J. Maths & Phys.* **41** (1962) 201–204. C: p. 88
83. A. Ben-Menahem & S.J. Singh, *Seismic Waves and Sources*. New York: Springer, 1981. C: pp. 20, 35, 71, 107, 141, 142
84. S.B. Bennett, Scattering of a plane elastic wave from objects near an interface. *JAM* **39** (1972) 1019–1026. C: p. 235
85. W. Benthien & [H.]A. Schenck, Nonexistence and nonuniqueness problems associated with integral equation methods in acoustics. *Computers & Structures* **65** (1997) 295–305. C: pp. 209, 221
86. S. Bergman & M. Schiffer, *Kernel Functions and Elliptic Differential Equations in Mathematical Physics*. New York: Academic Press, 1953. C: p. 223
87. P.G. Bergmann, The wave equation in a medium with a variable index of refraction. *JASA* **17** (1946) 329–333. C: pp. 14, 16
88. B.C. Berndt, *Ramanujan's Notebooks*, Part II. New York: Springer, 1989. C: p. 329
89. M. Berry, Paul Dirac: the purest soul in physics. *Physics World* **11** (Feb. 1998) 36–40. C: p. 284
90. M.V. Berry, Quantizing a classically ergodic system: Sinai's billiard and the KKR method. *Annals of Physics* **131** (1981) 163–216. C: p. 9
91. J.G. Berryman, Long-wavelength propagation in composite elastic media I. Spherical inclusions. *JASA* **68** (1980) 1809–1819. C: p. 322
92. J.G. Berryman, Long-wavelength propagation in composite elastic media II. Ellipsoidal inclusions. *JASA* **68** (1980) 1820–1831. C: p. 322
93. J.G. Berryman, Scattering by a spherical inhomogeneity in a fluid-saturated porous medium. *JMP* **26** (1985) 1408–1419. C: p. 143
94. J.G. Berryman & P.A. Berge, Critique of two explicit schemes for estimating elastic properties of multiphase composites. *MM* **22** (1996) 149–164. C: p. 322
95. D.E. Beskos, Boundary element methods in dynamic analysis. *AMR* **40** (1987) 1–23. C: p. 203

96. D.E. Beskos, Boundary element methods in dynamic analysis: part II (1986–1996). *AMR* **50** (1997) 149–197. C: p. 203
97. S.J. Bever & J.P. Allebach, Multiple scattering by a planar array of parallel dielectric cylinders. *AO* **31** (1992) 3524–3532. C: p. 129
98. B. Bialecki, Sinc quadrature rule for Hadamard finite-part integrals. *NM* **57** (1990) 263–269. C: p. 338
99. L.C. Biedenharn & J.D. Louck, *Angular Momentum in Quantum Physics*. Reading: Addison-Wesley, 1981. C: Table B.1 (p. 327)
100. L.C. Biedenharn & J.D. Louck, *The Racah–Wigner Algebra in Quantum Theory*. Reading: Addison-Wesley, 1981. C: p. 328
101. G.D. Birkhoff, The mathematical work of Oliver Dimon Kellogg. *Bull. Amer. Math. Soc.* **39** (1933) 171–177. C: p. 154
102. R.E.D. Bishop & W.G. Price (eds.), *The Dynamics of Marine Vehicles and Structures in Waves*. London: Institution of Mechanical Engineers, London, 1975. C: p. 23
103. S. Biwa, S. Yamamoto, F. Kobayashi & N. Ohno, Computational multiple scattering analysis for shear wave propagation in uni directional composites. *IJSS* **41** (2004) 435–457. C: p. 138
104. J. Björkberg & G. Kristensson, Electromagnetic scattering by a perfectly conducting elliptic disk. *CJP* **65** (1987) 723–734. C: p. 274
105. I.E. Block & R.F. Drenick (eds.), A symposium on the publication of mathematical literature. *SIAM Review* **6** (1964) 431–454. C: p. 357
106. J. Board & K. Schulten, The fast multipole algorithm. *IEEE Computing in Science & Engng.* **2**, 1 (2000) 76–79. C: p. 235
107. M. Bobeth & G. Diener, Effective scattering and damping of elastic waves in multicomponent mixtures. *physica status solidi* (b) **145** (1988) 123–138. C: p. 322
108. T.J. Bogdan & D.C. Fox, Multiple scattering of acoustic waves by a pair of uniformly magnetized flux tubes. *Astrophysical J.* **379** (1991) 758–775. C: p. 129
109. C.F. Bohren, *Clouds in a Glass of Beer*. New York: Wiley, 1987. C: p. 2
110. C.F. Bohren, Multiple scattering of light and some of its observable consequences. *Amer. J. Phys.* **55** (1987) 524–533. C: p. 2
111. C.F. Bohren & D.R. Huffman, *Absorption and Scattering of Light by Small Particles*. New York: Wiley, 1983. C: pp. 2, 140
112. B.A. Boley, A method for the construction of Green's functions. *QAM* **14** (1956) 249–257. C: p. 223
113. J.-C. Bolomey & W. Tabbara, Numerical aspects on coupling between complementary boundary value problems. *IEAP* **21** (1973) 356–363. C: pp. 218, 250, 338
114. J.C. Bolomey & A. Wirgin, Numerical comparison of the Green's function and the Waterman and Rayleigh theories of scattering from a cylinder with arbitrary cross-section. *Proc. IEE* **121** (1974) 794–804. C: p. 250
115. O. Bolza, *Lectures on Integral Equations*. Handwritten notes prepared by W.N. Lovitt, University of Chicago, 1913. C: p. 160
116. M. Bonnet, *Boundary Integral Equation Methods for Solids and Fluids*. Chichester: Wiley, 1999. C: pp. 11, 203
117. F. Borghese, P. Denti & R. Saija, *Scattering from Model Nonspherical Particles*. Berlin: Springer, 2003. C: pp. 140, 141
118. F. Borghese, P. Denti, R. Saija, G. Toscano & O.I. Sindoni, Use of group theory for the description of electromagnetic scattering from molecular systems. *JOSA* A **1** (1984) 183–191. C: p. 137

119. F. Borghese, P. Denti, G. Toscano & O.I. Sindoni, Electromagnetic scattering by a cluster of spheres. *AO* **18** (1979) 116–120. C: p. 141
120. F. Borghese, P. Denti, G. Toscano & O.I. Sindoni, An addition theorem for vector Helmholtz harmonics. *JMP* **21** (1980) 2754–2755. C: p. 107
121. M. Born & E. Wolf, *Principles of Optics*, sixth (corrected) edition. Cambridge: Cambridge University Press, 1997. C: pp. 140, 247
122. S.K. Bose, Ultrasonic plane SH wave reflection from a uni-directional fibrous composite slab. *JSV* **193** (1996) 1069–1078. C: p. 320
123. S.K. Bose & A.K. Mal, Longitudinal shear waves in a fiber-reinforced composite. *IJSS* **9** (1973) 1075–1085. C: p. 320
124. M.G. Bostock, Surface wave scattering from 3-D obstacles. *GJI* **104** (1991) 351–370. C: pp. 274, 281
125. M.G. Bostock & B.L.N. Kennett, Multiple scattering of surface waves from discrete obstacles. *GJI* **108** (1992) 52–70. C: p. 281
126. A. Boström, Multiple scattering of elastic waves by bounded obstacles. *JASA* **67** (1980) 399–413. C: p. 281
127. A. Boström, Scattering by a smooth elastic obstacle. *JASA* **67** (1980) 1904–1913. C: p. 251
128. A. Boström, Scattering of stationary acoustic waves by an elastic obstacle immersed in a fluid. *JASA* **67** (1980) 390–398. C: pp. 264, 274
129. A. Boström, Scattering of acoustic waves by a layered elastic obstacle in a fluid – an improved null field approach. *JASA* **76** (1984) 588–593. C: p. 263
130. A. Boström, Review of hypersingular integral equation method for crack scattering and application to modeling of ultrasonic nondestructive evaluation. *AMR* **56** (2003) 383–405. C: p. 207
131. A. Boström & A.S. Eriksson, Scattering by two penny-shaped cracks with spring boundary conditions. *PRSA* **443** (1993) 183–201. C: pp. 209, 274
132. A. Boström & G. Kristensson, Elastic wave scattering by a three-dimensional inhomogeneity in an elastic half space. *WM* **2** (1980) 335–353. C: p. 251
133. A. Boström, G. Kristensson & S. Ström, Transformation properties of plane, spherical and cylindrical scalar and vector wave functions. In: *Field Representations and Introduction to Scattering* (eds. V.V. Varadan, A. Lakhtakia & V.K. Varadan) pp. 165–210. Amsterdam: Elsevier, 1991. C: pp. 89, 90, 107, 110, 332
134. C.J. Bouwkamp, Diffraction theory. *Reports on Progress in Physics* **17** (1954) 35–100. C: pp. 156, 157
135. J.J. Bowman, T.B.A. Senior & P.L.E. Uslenghi (eds.), *Electromagnetic and Acoustic Scattering by Simple Shapes*. Amsterdam: North-Holland, 1969. Revised printing, Hemisphere, Levittown, Pennsylvania, 1987. C: pp. 11, 124, 134, 140, 149, 150, 172, 293, 295, 296
136. M.P. Brandão, Improper integrals in theoretical aerodynamics: the problem revisited. *AIAA J.* **25** (1987) 1258–1260. C: p. 338
137. D.W. Brandt, C. Eftimiu & P.L. Huddleston, Electromagnetic scattering by closed conducting bodies: the problem of internal resonances. *Proc. 4th Int. Conf. on Antennas & Propagation* (ICAP 85), Conference Publication No. 248, pp. 434–437. London: IEE, 1985. C: p. 217
138. L.M. Brekhovskikh, *Waves in Layered Media*. New York: Academic, 1960. C: pp. 16, 307
139. D. Brill & G. Gaunaurd, Resonance theory of elastic waves ultrasonically scattered from an elastic sphere. *JASA* **81** (1987) 1–21. C: p. 142

140. V.N. Bringi, V.V. Varadan & V.K. Varadan, The effects on pair correlation function of coherent wave attenuation in discrete random media. *IEAP* **30** (1982) 805–808. C: p. 320
141. D.M. Brink & G.R. Satchler, *Angular Momentum*, third edition. Oxford: Oxford University Press, 1993. C: pp. 326, 327, 330 & Table B.1 (p. 327)
142. T.E. Brockett, M.-H. Kim & J.-H. Park, Limiting forms for surface singularity distributions when the field point is on the surface. *JEM* **23** (1989) 53–79. C: p. 351
143. G.B. Brundrit, A solution to the problem of scalar scattering from a smooth, bounded obstacle using integral equations. *QJMAM* **18** (1965) 473–489. C: p. 172
144. W.M. Bruner, Comment on 'Seismic velocities in dry and saturated cracked solids' by Richard J. O'Connell and Bernard Budiansky [reference [921]], *JGR* **81** (1976) 2573–2576. C: p. 322
145. J.H. Bruning & Y.T. Lo, Multiple scattering of EM waves by spheres, part I – Multipole expansion and ray-optical solutions. *IEAP* **19** (1971) 378–390. C: pp. 93, 140, 245, 329
146. J.H. Bruning & Y.T. Lo, Multiple scattering of EM waves by spheres, part II – Numerical and experimental results. *IEAP* **19** (1971) 391–400. C: pp. 140, 245
147. O.P. Bruno, C.A. Geuzaine, J.A. Monroe Jr & F. Reitich, Prescribed error tolerances within fixed computational times for scattering problems of arbitrarily high frequency: the convex case. *PTRSA* **362** (2004) 629–645. C: p. 219
148. O.P. Bruno & E.M. Hyde, An efficient, preconditioned, high- order solver for scattering by two-dimensional inhomogeneous media. *JCP* **200** (2004) 670–694. C: p. 186
149. O.P. Bruno & L.A. Kunyansky, A fast, high-order algorithm for the solution of surface scattering problems: basic implementation, tests, and applications. *JCP* **169** (2001) 80–110. C: pp. 162, 218
150. O.P. Bruno & L.A. Kunyansky, Surface scattering in three dimensions: an accelerated high-order solver. *PRSA* **457** (2001) 2921–2934. C: pp. 162, 218
151. O.P. Bruno & A. Sei, A fast high-order solver for EM scattering from complex penetrable bodies: TE case. *IEAP* **48** (2000) 1862–1864. C: p. 186
152. O.P. Bruno & A. Sei, A fast high-order solver for problems of scattering by heterogeneous bodies. *IEAP* **51** (2003) 3142–3154. C: p. 186
153. G.R. Buchanan & G.R. Ramirez, A note on the vibration of transversely isotropic solid spheres. *JSV* **253** (2002) 724–732. C: p. 141
154. K. Budal, Theory for absorption of wave power by a system of interacting bodies. *JSR* **21** (1977) 248–253. C: p. 145
155. R.J. Buehler & J.O. Hirschfelder, Bipolar expansion of Coulombic potentials. *PR* **83** (1951) 628–633. Addenda: **85** (1952) 149. C: p. 96
156. A. Buffa, M. Costabel & C. Schwab, Boundary element methods for Maxwell's equations on non-smooth domains. *NM* **92** (2002) 679–710. C: p. 195
157. A. Buffa & R. Hiptmair, A coercive combined field integral equation for electromagnetic scattering. *SINUM* **42** (2004) 621–640. C: p. 219
158. A. Buffa & R. Hiptmair, Regularized combined field integral equations. *NM* **100** (2005) 1–19. C: pp. 218, 218
159. T.V. Burchuladze, N.P. Vekua, T.G. Gegelia, L.D. Kudryavtsev, S.M. Nikol'skii, S.L. Sobolev & A.N. Tikhonov, Viktor Dmitrievich Kupradze (obituary). *Russ. Math. Surveys* **41** (1986) 177–179. C: p. 171
160. F.J. Bureau, Divergent integrals and partial differential equations. *CPAM* **8** (1955) 143–202. C: p. 336
161. J.E. Burke, D. Censor & V. Twersky, Exact inverse-separation series for multiple scattering in two dimensions. *JASA* **37** (1965) 5–13. C: pp. 273, 310

162. J.E. Burke & V. Twersky, On scattering of waves by an elliptic cylinder and by a semielliptic protuberance on a ground plane. *JOSA* **54** (1964) 732–744. C: pp. 47, 149
163. J.E. Burke & V. Twersky, On scattering of waves by many bodies. *J. Res. National Bureau of Standards* **68D** (1964) 500–510. C: pp. 10, 122
164. J.E. Burke & V. Twersky, On scattering of waves by the infinite grating of elliptic cylinders. *IEAP* **14** (1966) 465–480. C: p. 235
165. A.J. Burton, Numerical solution of scalar diffraction problems. In: *Numerical Solution of Integral Equations* (eds. L.M. Delves & J. Walsh) pp. 275–290. Oxford: Oxford University Press, 1974. C: p. 219
166. A.J. Burton & G.F. Miller, The application of integral equation methods to the numerical solution of some exterior boundary-value problems. *PRSA* **323** (1971) 201–210. C: pp. 209, 219, 240 & §6.10.2 (p. 218)
167. P.G.J. Bussink, P.L. Iske, J. Oortwijn & G.L.M.M. Verbist, Self-consistent analysis of elastic wave propagation in two-dimensional matrix-inclusion composites. *JMPS* **43** (1995) 1673–1690. C: p. 322
168. W.H. Butler, P.H. Dederichs, A. Gonis & R.L. Weaver (eds.), *Applications of Multiple Scattering Theory to Materials Science*. Pittsburgh: Materials Research Society, 1992. C: p. 1
169. L.-W. Cai & J.H. Williams Jr, Large-scale multiple scattering problems. *Ultrasonics* **37** (1999) 453–462. C: p. 282
170. L.-W. Cai & J.H. Williams Jr, Full-scale simulations of elastic wave scattering in fiber-reinforced composites. *Ultrasonics* **37** (1999) 463–482. C: p. 282
171. L.-W. Cai & J.H. Williams Jr, NDE via stop band formation in fiber reinforced composites having square fiber arrangements. *Ultrasonics* **37** (1999) 483–492. C: p. 282
172. F.X. Canning, Robust use of supplementary conditions for moment method solution near internal resonances. *IEAP* **43** (1995) 264–269. C: p. 220
173. A. Carbery, Harmonic analysis of the Calderón–Zygmund school, 1970–1993. *Bull. London Math. Soc.* **30** (1998) 11–23. C: p. 349
174. B.C. Carlson & G.S. Rushbrooke, On the expansion of a Coulomb potential in spherical harmonics. *PCPS* **46** (1950) 626–633. C: p. 96
175. A. Caron, Approximations of finite parts and numerical solutions of singular integral equations. *Calcolo* **19** (1982) 15–34. C: p. 349
176. A. Caron, M. Dupuy & Ch. Pichot, Numerical solution of singular integral equations. Application to the flanged plane waveguide. *JCP* **38** (1980) 357–377. C: p. 207
177. G.F. Carrier, M. Krook & C.E. Pearson, *Functions of a Complex Variable*. New York: McGraw-Hill, 1966. C: p. 230
178. F. Cassot, Contribution à l'étude de la diffraction par un écran mince. *Acustica* **34** (1975) 64–71. C: p. 208
179. T. Cebeci (ed.), *Legacy of a Gentle Giant. The Life of A.M.O. Smith*. Long Beach, California: Horizons Publishing, 1999. C: pp. 169, 172
180. S.K. Chakrabarti, Wave forces on multiple vertical cylinders. *J. Waterway, Port, Coastal and Ocean Div., Proc. ASCE* **104** (1978) 147–161. C: p. 129
181. S.K. Chakrabarti, Wave interaction with multiple horizontal cylinders. *AOR* **1** (1979) 213–216. C: p. 205
182. S.K. Chakrabarti, Hydrodynamic interaction forces on multi-moduled structures. *OE* **27** (2000) 1037–1063. C: p. 283
183. A. Charalambopoulos, G. Dassios & M. Hadjinicolaou, An analytic solution for low-frequency scattering by two soft spheres. *SIAP* **58** (1998) 370–386. C: p. 296

184. A.K. Chatterjee, A.K. Mal & L. Knopoff, Elastic moduli of two-component systems. *JGR* **83** (1978) 1785–1792. C: p. 322
185. G. Chen & J. Zhou, *Boundary Element Methods*. London: Academic Press, 1992. C: pp. 160, 218, 336, 347
186. I.L. Chen, J.T. Chen & M.T. Liang, Analytical study and numerical experiments for radiation and scattering problems using the CHIEF method. *JSV* **248** (2001) 809–828. C: p. 221
187. J.T. Chen & H.-K. Hong, Review of dual boundary element methods with emphasis on hypersingular integrals and divergent series. *AMR* **52** (1999) 17–33. C: p. 338
188. L.H. Chen & D.G. Schweikert, Sound radiation from an arbitrary body. *JASA* **35** (1963) 1626–1632. C: p. 171
189. Y. Chen, A fast, direct algorithm for the Lippmann–Schwinger integral equation in two dimensions. *ACM* **16** (2002) 175–190. C: p. 186
190. Y.-Y. Chen & Z. Ye, Acoustic attenuation by two-dimensional arrays of rigid cylinders. *Phys. Rev. Lett.* **87** (2001) paper 184301. C: p. 129
191. A.H.-D. Cheng & D.T. Cheng, Heritage and early history of the boundary element method. *EABE* **29** (2005) 268–302. C: p. 168
192. S.L. Cheng, Multiple scattering of elastic waves by parallel cylinders. *JAM* **36** (1969) 523–527. C: p. 142
193. S.L. Cheng, Dynamic stresses in a plate with circular holes. *JAM* **39** (1972) 129–132. C: p. 142
194. G. Chertock, Sound radiation from vibrating surfaces. *JASA* **36** (1964) 1305–1313. C: p. 178
195. W.C. Chew, An N^2 algorithm for the multiple scattering solution of N scatterers. *MOTL* **2** (1989) 380–383. C: p. 282
196. W.C. Chew, *Waves and Fields in Inhomogeneous Media*. New York: Van Nostrand, 1990. C: pp. 18, 182, 186, 195, 282
197. W.C. Chew, Recurrence relations for three-dimensional scalar addition theorem. *JEWA* **6** (1992) 133–142. C: pp. 86, 94
198. W.C. Chew, Computational electromagnetics: the physics of smooth versus oscillatory fields. *PTRSA* **362** (2004) 579–602. C: p. 154
199. W.C. Chew, J.A. Friedrich & R. Geiger, A multiple scattering solution for the effective permittivity of a sphere mixture. *IEGRS* **28** (1990) 207–214. C: p. 282
200. W.C. Chew, L. Gürel, Y.-M. Wang, G. Otto, R.L. Wagner & Q.H. Liu, A generalized recursive algorithm for wave-scattering solutions in two dimensions. *IEMT* **40** (1992) 716–723. C: p. 282
201. W.C. Chew, S. Koc, J.M. Song, C.C. Lu & E. Michielssen, A succinct way to diagonalize the translation matrix in three dimensions. *MOTL* **15** (1997) 144–147. C: pp. 94, 98
202. W.C. Chew & C.-C. Lu, The use of Huygens' equivalence principle for solving the volume integral equation of scattering. *IEAP* **41** (1993) 897–904. C: p. 186
203. W.C. Chew & C.C. Lu, The recursive aggregate interaction matrix algorithm for multiple scatterers. *IEAP* **43** (1995) 1483–1486. C: p. 282
204. W.C. Chew, C.C. Lu & Y.M. Wang, Efficient computation of three-dimensional scattering of vector electromagnetic waves. *JOSA* A **11** (1994) 1528–1537. C: p. 282
205. W.C. Chew & Y.M. Wang, Efficient ways to compute the vector addition theorem. *JEWA* **7** (1993) 651–665. C: p. 107
206. C.C. Chien, H. Rajiyah & S.N. Atluri, An effective method for solving the hypersingular integral equations in 3-D acoustics. *JASA* **88** (1990) 918–937. C: p. 219
207. S.K. Cho, *Electromagnetic Scattering*. New York: Springer, 1990. C: pp. 195, 218

208. C.H. Choi, J. Ivanic, M.S. Gordon & K. Ruedenberg, Rapid and stable determination of rotation matrices between spherical harmonics by direct recursion. *J. Chem. Phys.* **111** (1999) 8825–8831. C: p. 333
209. T.C. Choy, *Effective Medium Theory*. Oxford: Oxford University Press, 1999. C: pp. 281, 282, 320, 321
210. M. Christ, *Lectures on Singular Integral Operators*. Providence: American Mathematical Society, 1990. C: p. 349
211. S. Christiansen, Condition number of matrices derived from two classes of integral equations. *MMAS* **3** (1981) 364–392. C: p. 244
212. S. Christiansen & E.B. Hansen, Numerical solution of boundary value problems through integral equations. *ZAMM* **58** (1978) T11–T25. C: pp. 168, 177
213. S. Christiansen & R.E. Kleinman, On a misconception involving point collocation and the Rayleigh hypothesis. *IEAP* **44** (1996) 1309–1316. C: p. 137
214. Y.K. Chung & M.I. Coleman, Hydrodynamic forces and moments for oscillating cylinders. *Proc. Civil Engineering in the Oceans* III. New York: ASCE, 1975, pp. 899–913. C: p. 205
215. I.R. Ciric & [M.]F.R. Cooray, Benchmark solutions for electromagnetic scattering by systems of randomly oriented spheroids. *JQSRT* **63** (1999) 131–148. C: p. 150
216. R.D. Ciskowski & C.A. Brebbia (eds.), *Boundary Element Methods in Acoustics*. Southampton: Computational Mechanics Publications, 1991. C: p. 168
217. H.J.H. Clercx & P.P.J.M. Schram, Retarded hydrodynamic interactions in suspensions. *Physica* A **174** (1991) 325–354. C: p. 86
218. H.J.H. Clercx & P.P.J.M. Schram, An alternative expression for the addition theorems of spherical wave solutions of the Helmholtz equation. *JMP* **34** (1993) 5292–5302. C: pp. 71, 86
219. R. Coifman, V. Rokhlin & S. Wandzura, The fast multipole method for the wave equation: a pedestrian prescription. *IEAPM* **35**, 3 (1993) 7–12. C: pp. 80, 240
220. D. Colton, The scattering of acoustic waves by a spherically stratified medium and an obstacle. *SIMA* **9** (1978) 935–942. C: p. 135
221. D. Colton & R. Kress, The construction of solutions to acoustic scattering problems in a spherically stratified medium. *QJMAM* **31** (1978) 9–17. C: p. 135
222. D. Colton & R. Kress, The construction of solutions to acoustic scattering problems in a spherically stratified medium. II. *QJMAM* **32** (1979) 53–62. C: p. 135
223. D. Colton & R. Kress, *Integral Equation Methods in Scattering Theory*. New York: Wiley, 1983. C: pp. 12, 13, 14, 17, 152, 154, 156, 162, 163, 167, 168, 170, 172, 180, 188, 189, 190, 193, 195, 209, 211, 212, 218, 218, 232, 344, 350, 351
224. D. Colton & R. Kress, The unique solvability of the null field equations of acoustics. *QJMAM* **36** (1983) 87–95. C: p. 257
225. D. Colton & R. Kress, *Inverse Acoustic and Electromagnetic Scattering Theory*, second edition. Berlin: Springer, 1998. C: pp. 15, 16, 18, 18, 63, 64, 65, 82, 186, 232, 244, 273, 351
226. U. Comberg & T. Wriedt, Comparison of scattering calculations for aggregated particles based on different models. *JQSRT* **63** (1999) 149–162. C: p. 245
227. K.W. Commander & A. Prosperetti, Linear pressure waves in bubbly liquids: comparison between theory and experiments. *JASA* **85** (1989) 732–746. C: p. 316
228. C.A. Condat & T.R. Kirkpatrick, Resonant scattering and Anderson localization of acoustic waves. *PR* B **36** (1987) 6782–6793. C: p. 11
229. E.U. Condon & G.H. Shortley, *The Theory of Atomic Spectra*. Cambridge: Cambridge University Press, 1935. C: p. 326 & Table B.1 (p. 327)

230. C. Constanda, *Direct and Indirect Boundary Integral Equation Methods*. Boca Raton: Chapman & Hall/CRC, 2000. C: p. 168
231. H. Contopanagos, B. Dembart, M. Epton, J.J. Ottusch, V. Rokhlin, J.L. Visher & S.M. Wandzura, Well-conditioned boundary integral equations for three-dimensional electromagnetic scattering. *IEAP* **50** (2002) 1824–1830. C: p. 219
232. M.F.R. Cooray & I.R. Ciric, Electromagnetic wave scattering by a system of two spheroids of arbitrary orientation. *IEAP* **37** (1989) 608–618. C: p. 150
233. M.F.R. Cooray & I.R. Ciric, Scattering of electromagnetic waves by a system of two dielectric spheroids of arbitrary orientation. *IEAP* **39** (1991) 680–684. C: p. 150
234. M.F.R. Cooray, I.R. Ciric & B.P. Sinha, Electromagnetic scattering by a system of two parallel dielectric prolate spheroids. *CJP* **68** (1990) 376–384. C: p. 150
235. L.G. Copley, Integral equation method for radiation from vibrating bodies. *JASA* **41** (1967) 807–816. C: p. 244
236. F.J. Corbató & A.C. Switendick, Integrals for diatomic molecular calculations. In: *Methods in Computational Physics* vol. 2 (eds. B. Alder, S. Fernbach & M. Rotenberg) pp. 155–179. New York: Academic Press, 1963. C: p. 93
237. L.M. Correia, A comparison of integral equations with unique solution in the resonance region for scattering by conducting bodies. *IEAP* **41** (1993) 52–58. C: p. 219
238. M. Costabel, Boundary integral operators on Lipschitz domains: elementary results. *SIMA* **19** (1988) 613–626. C: p. 169
239. M. Costabel & E. Stephan, A direct boundary integral equation method for transmission problems. *JMAA* **106** (1985) 367–413. C: p. 180
240. M. Costabel & E. Stephan, Integral equations for transmission problems in linear elasticity. *J. Integ. Eqns & Applics.* **2** (1990) 211–223. C: p. 202
241. B. Count (ed.), *Power from Sea Waves*. London: Academic Press, 1980. C: p. 145
242. B.M. Count & E.R. Jefferys, Wave power, the primary interface. *Proc. 13th Symposium on Naval Hydrodynamics*, 1980, pp. 817–826. Discussion, pp. 827–828. C: p. 205
243. R. Courant & D. Hilbert, *Methods of Mathematical Physics*, vol. I. New York: Interscience, 1953. C: pp. 64, 170, 253, 330, 331
244. R. Courant & D. Hilbert, *Methods of Mathematical Physics*, vol. II. New York: Interscience, 1962. C: pp. 166, 210, 352
245. O. Coutant, Numerical study of the diffraction of elastic waves by fluid-filled cracks. *JGR* **94** (1989) 17805–17818. C: p. 208
246. L. Cruz, L.F. Fonseca & M. Gómez, T-matrix approach for the calculation of local fields in the neighborhood of small clusters in the electrodynamic regime. *PR* B **40** (1989) 7491–7500. C: p. 281
247. O.R. Cruzan, Translational addition theorems for spherical vector wave functions. *QAM* **20** (1962) 33–40. C: p. 107
248. C.W. Cryer, *Numerical Functional Analysis*. Oxford: Oxford University Press, 1982. C: pp. 169, 253, 254
249. D.K. Dacol & D.G. Roy, Wave scattering in waveguides. *JMP* **44** (2003) 2133–2148. C: p. 272
250. F.A. Dahlen & J. Tromp, *Theoretical Global Seismology*. Princeton: Princeton University Press, 1998. C: pp. 20, 63, 93, 106, 141, 326, 333
251. A.G. Dallas, Basis properties of traces and normal derivatives of spherical-separable solutions of the Helmholtz equation. Technical Report No. 2000-6, Dept. of Mathematical Sciences, Univ. of Delaware, 2000. C: pp. 253, 266 & §7.7.2 (p. 254)

252. J. Dalmas & R. Deleuil, Multiple scattering of electromagnetic waves from two infinitely conducting prolate spheroids which are centered in a plane perpendicular to their axes of rotation. *RS* **20** (1985) 575–581. C: p. 150
253. J. Dalmas, R. Deleuil & R.H. MacPhie, Rotational-translational addition theorems for spheroidal vector wave functions. *QAM* **47** (1989) 351–364. C: p. 150
254. M. Danos & L.C. Maximon, Multipole matrix elements of the translation operator. *JMP* **6** (1965) 766–778. C: pp. 90, 98, 107
255. E. Darve, The fast multipole method I: error analysis and asymptotic complexity. *SINUM* **38** (2000) 98–128. C: pp. 80, 240
256. E. Darve, The fast multipole method: numerical implementation. *JCP* **160** (2000) 195–240. C: p. 240
257. E. Darve & P. Havé, A fast multipole method for Maxwell equations stable at all frequencies. *PTRSA* **362** (2004) 603–628. C: p. 241
258. E. Darve & P. Havé, Efficient fast multipole method for low- frequency scattering. *JCP* **197** (2004) 341–363. C: p. 241
259. S. Das & L. Debnath, Diffraction of SH-waves by collinear Griffith cracks in an infinite transversely isotropic medium. *Can. Appl. Math. Quart.* **8** (2000) 17–26. C: p. 208
260. G. Dassios, The Atkinson–Wilcox theorem in ellipsoidal geometry. *JMAA* **274** (2002) 828–845. C: p. 150
261. G. Dassios, K. Kiriaki & D. Polyzos, On the scattering amplitudes for elastic waves. *J. Appl. Math. & Phys.* (ZAMP) **38** (1987) 856–873. C: p. 199
262. G. Dassios & R. Kleinman, *Low Frequency Scattering*. Oxford: Oxford University Press, 2000. C: pp. 14, 18, 150, 150, 195, 272, 285, 286, 287, 288, 288, 296
263. G. Dassios & Z. Rigou, Elastic Herglotz functions. *SIAP* **55** (1995) 1345–1361. C: pp. 106, 107
264. S.K. Datta, Propagation of SH-waves through a fiber-reinforced composite – elliptic cylindrical fibers. *JAM* **42** (1975) 165–170. C: p. 316
265. S.K. Datta, Scattering of elastic waves by a distribution of inclusions. *Archives of Mechanics* **28** (1976) 317–324. C: p. 316
266. S.K. Datta, A self-consistent approach to multiple scattering by elastic ellipsoidal inclusions. *JAM* **44** (1977) 657–662. C: p. 316
267. M. David, On a Dirichlet–Neumann–third mixed boundary value problem for the Helmholtz equation. *PRSE* **94A** (1983) 221–233. C: p. 171
268. P.J. Davies, D.B. Duncan & S.A. Funken, Accurate and efficient algorithms for frequency domain scattering from a thin wire. *JCP* **168** (2001) 155–183. C: p. 189
269. C.A. Davis, V.W. Lee & J.P. Bardet, Transverse response of underground cavities and pipes to incident SV waves. *EESD* **30** (2001) 383–410. C: pp. 22, 143
270. E.J. Davis & G. Schweiger, *The Airborne Microparticle*. Berlin: Springer, 2002. C: p. 140
271. V.A. Davis & L. Schwartz, Electromagnetic propagation in close-packed disordered suspensions. *PR* B **31** (1985) 5155–5165. C: p. 322
272. W.A. Davis & R. Mittra, A new approach to the thin scatterer problem using the hybrid equations. *IEAP* **25** (1977) 402–406. C: p. 207
273. Y. Decanini, A. Folacci, E. Fournier & P. Gabrielli, Exact *S*-matrix for *N*-disc systems and various boundary conditions: I. Generalization of the Korringa–Kohn–Rostoker–Berry method. *J. Phys. A: Math. Gen.* **31** (1998) 7865–7889. C: pp. 129, 137
274. Y. Decanini, A. Folacci, E. Fournier & P. Gabrielli, Exact *S*-matrix for *N*-disc systems and various boundary conditions: II. Determination and partial classification of resonances. *J. Phys. A: Math. Gen.* **31** (1998) 7891–7900. C: pp. 129, 137

275. Y. Decanini, A. Folacci, P. Gabrielli & J.-L. Rossi, Algebraic aspects of multiple scattering by two parallel cylinders: classification and physical interpretation of scattering resonances. *JSV* **221** (1999) 785–804. C: p. 137
276. D. Deirmendjian, *Electromagnetic Scattering on Spherical Polydispersions*. New York: Elsevier, 1969. C: p. 3
277. D. Delbourgo & D. Elliott, On the approximate evaluation of Hadamard finite-part integrals. *IMA J. Numer. Anal.* **14** (1994) 485–500. C: p. 338
278. J.A. DeSanto, *Scalar Wave Theory*. Berlin: Springer, 1992. C: p. 100
279. J.A. DeSanto, Scattering by rough surfaces. In: [977], pp. 15–36. C: p. 11
280. J.A. DeSanto & G.S. Brown, Analytical techniques for multiple scattering from rough surfaces. In: *Progress in Optics*, vol. 23 (ed. E. Wolf) pp. 1–62. Amsterdam: North-Holland, 1986. C: p. 11
281. A.J. Devaney, Multiple scattering theory for discrete, elastic, random media. *JMP* **21** (1980) 2603–2611. C: p. 321
282. P. de Vries, D.V. van Coevorden & A. Lagendijk, Point scatterers for classical waves. *RMP* **70** (1998) 447–466. C: pp. 305, 306
283. A. Doicu, Yu.A. Eremin & T. Wriedt, Convergence of the T-matrix method for light scattering from a particle on or near a surface. *OC* **159** (1999) 266–277. C: p. 273
284. A. Doicu, Yu.A. Eremin & T. Wriedt, Projection schemes in the null field method. *JQSRT* **63** (1999) 175–189. C: p. 266
285. A. Doicu, Yu. Eremin & T. Wriedt, *Acoustic and Electromagnetic Scattering Analysis using Discrete Sources*. London: Academic Press, 2000. C: pp. 90, 244, 245, 251, 253, 254, 256, 257, 265, 266
286. A. Doicu & T. Wriedt, T-matrix method for electromagnetic scattering from scatterers with complex structure. *JQSRT* **70** (2001) 663–673. C: p. 282
287. E. Dokumaci & A.S. Sarigül, Analysis of the near field acoustic radiation characteristics of two radially vibrating spheres by the Helmholtz integral equation formulation and a critical study of the efficacy of the "CHIEF" overdetermination method in two-body problems. *JSV* **187** (1995) 781–798. C: pp. 178, 221
288. C.L. Dolph, The integral equation method in scattering theory. In: *Problems in Analysis* (ed. R.C. Gunning) pp. 201–227. Princeton: Princeton University Press, 1970. C: p. 171
289. E. Domany & O. Entin-Wohlman, Application of multiple scattering theory to subsurface defects. *JAP* **56** (1984) 137–142. C: p. 227
290. E. Domany, O. Entin-Wohlman & L. Mizrachi, Multiple scattering formalism: application to scattering by two spheres. *JAP* **56** (1984) 132–136. C: pp. 226, 227
291. S.N. Domenico, Acoustic wave propagation in air-bubble curtains in water – Part I: history and theory. *Geophysics* **47** (1982) 345–353. C: p. 316
292. T.E. Doyle, Iterative simulation of elastic wave scattering in particle-filled materials. In: *Review of Progress in Quantitative Nondestructive Evaluation*, vol. 24 (eds. D.O. Thompson & D.E. Chimenti) pp. 1121–1128. New York: American Institute of Physics, 2005. C: p. 143
293. M. Dravinski, Amplification of P, SV, and Rayleigh waves by two alluvial valleys. *SDEE* **2** (1983) 66–77. C: p. 246
294. M. Dravinski, Scattering of elastic waves by a general anisotropic basin. Part 2: a 3D model. *EESD* **32** (2003) 653–670. C: p. 246
295. M. Dravinski & Y. Niu, Three-dimensional time-harmonic Green's functions for a triclinic full-space using a symbolic computation system. *IJNME* **53** (2002) 455–472. C: p. 202

296. M. Dravinski & M.S. Wilson, Scattering of elastic waves by a general anisotropic basin. Part 1: a 2D model. *EESD* **30** (2001) 675–689. C: p. 246
297. M. Dravinski & T. Zheng, Numerical evaluation of three- dimensional time-harmonic Green's functions for a nonisotropic full-space. *WM* **32** (2000) 141–151. C: p. 202
298. G. Duclos & A.H. Clément, Wave propagation through arrays of unevenly spaced vertical piles. *OE* **31** (2004) 1655–1668. C: p. 129
299. R.J. Duffin & J.H. McWhirter, An integral equation formulation of Maxwell's equations. *J. Franklin Institute* **298** (1974) 385–394. C: p. 179
300. A.S. Dukhin & P.J. Goetz, Acoustic and electroacoustic spectroscopy for characterizing concentrated dispersions and emulsions. *Advances in Colloid & Interface Science* **92** (2001) 73–132. C: p. 312
301. J.H. Duncan, R.A. Barr & Y.Z. Liu, Computations of the coupled response of two bodies in a seaway. In: *International Workshop on Ship and Platform Motions* (ed. R.W. Yeung), University of California, Berkeley, 1983, pp. 491–511. C: p. 205
302. J.H. Duncan & C.E. Brown, Development of a numerical method for the calculation of power absorption by arrays of similar arbitrarily shaped bodies in a seaway. *JSR* **26** (1982) 38–44. C: p. 205
303. R. Eatock Taylor & F.P. Chau, Wave diffraction theory – some developments in linear and nonlinear theory. *J. Offshore Mechanics & Arctic Engng.* **114** (1992) 185–194. C: p. 205
304. R. Eatock Taylor & J. Zietsman, Hydrodynamic loading on multi- component bodies. In: *Behaviour of Off-Shore Structures* (BOSS82), vol. 1 (eds. C. Chryssostomidis & J.J. Connor) pp. 424–443. Washington, DC: Hemisphere, 1983. C: p. 206
305. A.R. Edmonds, *Angular Momentum in Quantum Mechanics*, second edition, third printing with corrections. Princeton: Princeton University Press, 1974. C: pp. 63, 64, 84, 93, 326, 327, 330, 330, 333 & Table B.1 (p. 327)
306. N.G. Einspruch & C.A. Barlow, Jr, Scattering of a compressional wave by a prolate spheroid. *QAM* **19** (1961) 253–258. C: p. 150
307. N.G. Einspruch & R. Truell, Scattering of a plane longitudinal wave by a spherical fluid obstacle in an elastic medium. *JASA* **32** (1960) 214–220. C: p. 141
308. N.G. Einspruch, E.J. Witterholt & R. Truell, Scattering of a plane transverse wave by a spherical obstacle in an elastic medium. *JAP* **31** (1960) 806–818. C: p. 142
309. D. Elliott & E. Venturino, Sigmoidal transformations and the Euler–Maclaurin expansion for evaluating certain Hadamard finite-part integrals. *NM* **77** (1997) 453–465. C: p. 338
310. M. El-Shenawee, Scattering from multiple objects buried beneath two-dimensional random rough surface using the steepest descent fast multipole method. *IEAP* **51** (2003) 802–809. C: p. 241
311. A.Z. Elsherbeni, A comparative study of two-dimensional multiple scattering techniques. *RS* **29** (1994) 1023–1033. C: p. 129
312. A.Z. Elsherbeni & M. Hamid, Scattering by parallel conducting circular cylinders. *IEAP* **35** (1987) 355–358. C: p. 172
313. T.F.W. Embleton, Mutual interaction between two spheres in a plane sound field. *JASA* **34** (1962) 1714–1720. C: pp. 137, 145
314. N. Engheta, W.D. Murphy, V. Rokhlin & M.S. Vassiliou, The fast multipole method (FMM) for electromagnetic scattering problems. *IEAP* **40** (1992) 634–641. C: p. 240
315. R. England, F.J. Sabina & I. Herrera, Scattering of SH waves by surface cavities of arbitrary shape using boundary methods. *Physics of the Earth & Planetary Interiors* **21** (1980) 148–157. C: p. 263

316. P.S. Epstein & R.R. Carhart, The absorption of sound in suspensions and emulsions. I. Water fog in air. *JASA* **25** (1953) 553–565. C: p. 143
317. M.A. Epton, Integration-by-parts formulas for boundary-element methods. *AIAA J.* **30** (1992) 496–504. C: p. 351
318. M.A. Epton & B. Dembart, Multipole translation theory for the three-dimensional Laplace and Helmholtz equations. *SISC* **16** (1995) 865–897. C: pp. 80, 89, 91, 95, 240
319. A. Erdélyi, Zur Theorie der Kugelwellen. *Physica* **4** (1937) 107–120. C: p. 71
320. A. Erdélyi, Some integral equations involving finite parts of divergent integrals. *Glasgow Math. J.* **8** (1967) 50–54. C: p. 339
321. A. Erdélyi, W. Magnus, F. Oberhettinger & F.G. Tricomi, *Higher Transcendental Functions*, volumes I and II. New York: McGraw-Hill, 1953. C: pp. 63, 110, 230
322. A. Erdélyi, W. Magnus, F. Oberhettinger & F.G. Tricomi, *Higher Transcendental Functions*, vol. III. New York: McGraw-Hill, 1955. C: pp. 48, 49, 150, 150
323. Ju.A. Eremin, N.V. Orlov & V.I. Rozenberg, Multiple electromagnetic scattering by a linear array of electrified raindrops. *J. Atmospheric & Terrestrial Physics* **57** (1995) 311–319. C: p. 245
324. E. Erez & Y. Leviatan, Analysis of scattering from structures containing a variety of length scales using a source-model technique. *JASA* **93** (1993) 3027–3031. C: p. 245
325. E. Erez & Y. Leviatan, Current-model analysis of electromagnetic scattering from objects containing a variety of length-scales. *JOSA* A **11** (1994) 1500–1504. C: p. 245
326. V.J. Ervin & E.P. Stephan, Collocation with Chebyshev polynomials for a hypersingular integral equation on an interval. *JCAM* **43** (1992) 221–229. C: p. 208
327. D.V. Evans, Diffraction of water waves by a submerged vertical plate. *JFM* **40** (1970) 433–451. C: p. 144
328. D.V. Evans, Power from water waves. *ARFM* **13** (1981) 157– 187. C: p. 145
329. D.V. Evans, The wide-spacing approximation applied to multiple scattering and sloshing problems. *JFM* **210** (1990) 647–658. C: p. 308
330. D.V. Evans & R. Porter, Near-trapping of waves by circular arrays of vertical cylinders. *AOR* **19** (1997) 83–99. C: p. 129
331. H.W. Eves, *In Mathematical Circles, Quadrants III and IV.* Boston: Prindle, Weber & Schmidt, 1969. C: p. 11
332. L. Eyges, Some nonseparable boundary value problems and the many- body problem. *Annals of Physics* **2** (1957) 101–128. C: p. 137
333. C. Fadiman, *Fantasia Mathematica.* New York: Copernicus, Springer, 1997. C: p. 249
334. G. Fairweather & A. Karageorghis, The method of fundamental solutions for elliptic boundary value problems. *ACM* **9** (1998) 69–95. C: p. 245
335. G. Fairweather, A. Karageorghis & P.A. Martin, The method of fundamental solutions for scattering and radiation problems. *EABE* **27** (2003) 759–769. C: p. 245
336. J. Falnes, Radiation impedance matrix and optimum power absorption for interacting oscillators in surface waves. *AOR* **2** (1980) 75–80. C: p. 145
337. J. Falnes, *Ocean Waves and Oscillating Systems.* Cambridge: Cambridge University Press, 2002. C: p. 145
338. O.M. Faltinsen & F.C. Michelsen, Motions of large structures in waves at zero Froude number. In: [102], pp. 91–106. C: p. 205
339. J.J. Faran, Jr, Sound scattering by solid cylinders and spheres. *JASA* **23** (1951) 405–418. C: p. 141
340. D. Felbacq, G. Tayeb & D. Maystre, Scattering by a random set of parallel cylinders. *JOSA* A **11** (1994) 2526–2538. C: p. 282

341. B.U. Felderhof & R.B. Jones, Addition theorems for spherical wave solutions of the vector Helmholtz equation. *JMP* **28** (1987) 836–839. C: p. 107
342. E. Fermi, *Nuclear Physics*, revised edition. Chicago: University of Chicago Press, 1950. C: p. 319
343. N.M. Ferrers, *An Elementary Treatise on Spherical Harmonics and Subjects Connected to This*. London: Macmillan & Co., London, 1877. C: p. 63
344. R.P. Feynman, *The Pleasure of Finding Things Out*. Cambridge: Perseus, 1999. C: p. 73
345. J.G. Fikioris, Multiple scattering from inhomogeneous media. *IEAP* **14** (1966) 202–207. C: p. 299
346. J.G. Fikioris & P.C. Waterman, Multiple scattering of waves. II. "Hole corrections" in the scalar case. *JMP* **5** (1964) 1413–1420. C: p. 320
347. P. Filippi & G. Dumery, Etude théorique et numérique de la diffraction par un écran mince. *Acustica* **21** (1969) 343–350. C: pp. 207, 208
348. P. Filippi, D. Habault, J.-P. Lefebvre & A. Bergassoli, *Acoustics: Basic Physics, Theory and Methods*. London: Academic Press, 1999. C: p. 218
349. P.J.T. Filippi, Layer potentials and acoustic diffraction. *JSV* **54** (1977) 473–500. C: p. 159
350. C. Flammer, *Spheroidal Wave Functions*. Stanford: Stanford University Press, 1957. C: p. 150
351. P.J. Flatau, K.A. Fuller & D.W. Mackowski, Scattering by two spheres in contact: comparisons between discrete-dipole approximation and modal analysis. *AO* **32** (1993) 3302–3305. C: p. 195
352. L. Flax, L.R. Dragonette, V.K. Varadan & V.V. Varadan, Analysis and computation of the acoustic scattering by an elastic prolate spheroid obtained from the T-matrix formulation. *JASA* **71** (1982) 1077–1082. C: p. 274
353. J.T. Fokkema & P.M. van den Berg, *Seismic Applications of Acoustic Reciprocity*. Amsterdam: Elsevier, 1993. C: p. 263
354. L.L. Foldy, The multiple scattering of waves. I. General theory of isotropic scattering by randomly distributed scatterers. *PR* **67** (1945) 107–119. C: pp. 5, 10, 312, 319, §8.3 (p. 297) & §8.6.3 (p. 315)
355. L.L. Foldy, Theory of passive linear electroacoustic transducers with fixed velocity distribution. *JASA* **21** (1949) 595–604. C: p. 223
356. L.L. Foldy & H. Primakoff, A general theory of passive linear electroacoustic transducers and the electroacoustic reciprocity theorem. I. *JASA* **17** (1945) 109–120. C: p. 223
357. A.R. Forsyth, Old Tripos days at Cambridge. *Mathematical Gazette* **19** (1935) 162–179. C: p. 12
358. J. Fourier, *The Analytical Theory of Heat*. New York: Dover, 1955. C: p. 124
359. C. Fox, A generalization of the Cauchy principal value. *Can. J. Math.* **9** (1957) 110–117. C: p. 338
360. D.T.I. Francis, A gradient formulation of the Helmholtz integral equation for acoustic radiation and scattering. *JASA* **93** (1993) 1700–1709. C: p. 219
361. A. Frangi & M. Guiggiani, Free terms and compatibility conditions for 3D hypersingular boundary integral equations. *ZAMM* **81** (2001) 651–664. C: p. 349
362. W. Freeden, T. Gervens & M. Schreiner, *Constructive Approximation on the Sphere*. Oxford: Oxford University Press, 1998. C: pp. 68, 72
363. A. Frenkel, A Chebyshev expansion of singular integrodifferential equations with a $\partial^2 \log|s-t|/\partial s\,\partial t$ kernel. *JCP* **51** (1983) 335–342. C: pp. 161, 208
364. D. Frenkel & R. Portugal, Algebraic methods to compute Mathieu functions. *J. Phys. A: Math. Gen.* **34** (2001) 3541–3551. C: p. 49
365. F.G. Friedlander, *Sound Pulses*. Cambridge: Cambridge University Press, 1958. C: p. 11

366. B. Friedman & J. Russek, Addition theorems for spherical waves. *QAM* **12** (1954) 13–23. C: pp. 87, 89, 90
367. M.B. Friedman & R. Shaw, Diffraction of pulses by cylindrical obstacles of arbitrary cross section. *JAM* **29** (1962) 40–46. C: p. 11
368. U. Frisch, Wave propagation in random media. In: *Probabilistic Methods in Applied Mathematics*, vol. 1 (ed. A.T. Bharucha-Reid) pp. 75–198. New York: Academic Press, 1968. C: pp. 312, 322
369. G.V. Frisk & J.A. DeSanto, Scattering by spherically symmetric inhomogeneities. *JASA* **47** (1970) 172–180. C: p. 135
370. V. Fritzsche, A new spherical-wave approximation for photoelectron diffraction, EXAFS and MEED. *J. Phys.: Condensed Matter* **2** (1990) 1413–1424. C: pp. 104, 105
371. V. Fritzsche, Shortcomings of separable representations of the Green function in multiple-scattering theory for photoelectron diffraction and extended X-ray absorption fine structure. *J. Electron Spectroscopy & Related Phenomena* **58** (1992) 299–314. C: p. 105
372. H. Fujiwara, The fast multipole method for integral equations of seismic scattering problems. *GJI* **133** (1998) 773–782. C: p. 241
373. H. Fujiwara, The fast multipole method for solving integral equations of three-dimensional topography and basin problems. *GJI* **140** (2000) 198–210. C: p. 241
374. K.A. Fuller, Optical resonances and two-sphere systems. *AO* **30** (1991) 4716–4731. C: p. 140
375. K.A. Fuller & G.W. Kattawar, Consummate solution to the problem of classical electromagnetic scattering by an ensemble of spheres. I: linear chains. *OL* **13** (1988) 90–92. C: p. 141
376. K.A. Fuller & G.W. Kattawar, Consummate solution to the problem of classical electromagnetic scattering by an ensemble of spheres. II: clusters of arbitrary configuration. *OL* **13** (1988) 1063–1065. C: p. 141
377. P. Gabrielli & M. Mercier-Finidori, Acoustic scattering by two spheres: multiple scattering and symmetry considerations. *JSV* **241** (2001) 423–439. C: p. 137
378. M. Ganesh & I.G. Graham, A high-order algorithm for obstacle scattering in three dimensions. *JCP* **198** (2004) 211–242. C: p. 162
379. P.R. Garabedian, *Partial Differential Equations*. New York: Wiley, 1964. C: p. 223
380. F.J. García de Abajo, Multiple scattering of radiation in clusters of dielectrics. *PR* B **60** (1999) 6086–6102. C: p. 282
381. C.J.R. Garrett, Wave forces on a circular dock. *JFM* **46** (1971) 129–139. C: p. 148
382. F. Garrido & A. Salazar, Thermal wave scattering by spheres. *JAP* **95** (2004) 140–149. C: p. 139
383. C.J. Garrison, Hydrodynamics of large objects in the sea, Part I — Hydrodynamic analysis. *J. Hydronautics* **8** (1974) 5–12. C: p. 205
384. C.J. Garrison, Hydrodynamic loading of large offshore structures: three-dimensional source distribution methods. In: *Numerical Methods in Offshore Engineering* (eds. O.C. Zienkiewicz, R.W. Lewis & K.G. Stagg) pp. 87–140. Chichester: Wiley, 1978. C: p. 205
385. C.J. Garrison & P.Y. Chow, Wave forces on submerged bodies. *J. Waterways, Harbors & Coastal Engng. Div., Proc. ASCE* **98** (1972) 375–392. C: p. 205
386. C.J. Garrison & V. Seetharama Rao, Interaction of waves with submerged objects. *J. Waterways, Harbors & Coastal Engng. Div., Proc. ASCE* **97** (1971) 259–277. C: p. 205
387. P. Gaspard & S.A. Rice, Exact quantization of the scattering from a classically chaotic repellor. *J. Chem. Phys.* **90** (1989) 2255–2262. C: p. 129

388. G.C. Gaunaurd & H. Huang, Acoustic scattering by a spherical body near a plane boundary. *JASA* **96** (1994) 2526–2536. C: p. 137
389. G.C. Gaunaurd & H. Huang, Acoustic scattering by an air-bubble near the sea surface. *IEEE J. Oceanic Engng.* **20** (1995) 285–292. C: p. 137
390. G.C. Gaunaurd & H. Huang, Sound scattering by a spherical object near a hard flat bottom. *IEEE Trans. on Ultrasonics, Ferroelectrics, & Frequency Control* **43** (1996) 690–700. C: p. 137
391. G.C. Gaunaurd & H. Huang, Sound scattering by bubble clouds near the sea surface. *JASA* **107** (2000) 95–102. C: p. 137
392. G.C. Gaunaurd, H. Huang & H.C. Strifors, Acoustic scattering by a pair of spheres. *JASA* **98** (1995) 495–507. Addenda and corrigenda: **101** (1997) 2983–2985. C: p. 137
393. G.C. Gaunaurd & H. Überall, Resonance theory of bubbly liquids. *JASA* **69** (1981) 362–370. C: p. 316
394. G.C. Gaunaurd & W. Wertman, Comparison of effective medium theories for inhomogeneous continua. *JASA* **85** (1989) 541– 554. C: p. 322
395. G.C. Gaunaurd & W. Wertman, Comparison of effective medium and multiple-scattering theories of predicting the ultrasonic properties of dispersions: a reexamination of results. *JASA* **87** (1990) 2246–2248. C: p. 322
396. J.A. Gaunt, The triplets of helium. *PTRSA* **228** (1929) 151–196. C: p. 329
397. R. Gayen & B.N. Mandal, Water wave scattering by two thin symmetric inclined plates submerged in finite depth water. *Int. J. Applied Mechanics & Engineering* **8** (2003) 589–601. C: p. 208
398. J. Geer, The scattering of a scalar wave by a slender body of revolution. *SIAP* **34** (1978) 348–370. C: p. 244
399. I.M. Gel'fand & G.E. Shilov, *Generalized Functions*, vol. 1. New York: Academic Press, 1964. C: p. 336
400. J.M. Gérardy & M. Ausloos, Absorption spectrum of clusters of spheres from the general solution of Maxwell's equations. II. Optical properties of aggregated metal spheres. *PR* B **25** (1982) 4204–4229. C: pp. 94, 141
401. E. Gerjuoy & D.S. Saxon, Variational principles for the acoustic field. *PR* **94** (1954) 1445–1458. C: p. 185
402. K. Germey, Die Beugung einer ebenen elektromagnetischen Welle an zwei parallelen unendlich langen idealleitenden Zylindern von elliptischem Querschnitt. *Annalen der Physik*, 7 Folge, **13** (1964) 237–251. C: p. 149
403. O.A. Germogenova, The scattering of a plane electromagnetic wave by two spheres. *Bulletin (Izvestiya) of the Academy of Sciences, USSR, Geophysics Series* (1963) 403–405. C: p. 140
404. C. Geuzaine, O. Bruno & F. Reitich, On the $\mathcal{O}(1)$ solution of multiple-scattering problems. *IEEE Trans. Magnetics* **41** (2005) 1488–1491. C: pp. 219, 227
405. D. Gilbarg & N.S. Trudinger, *Elliptic Partial Differential Equations of Second Order*, second edition. Berlin: Springer, 1998. C: pp. 286, 352
406. D. Gintides & K. Kiriaki, Multiple scattering of elastic waves by sphere configurations. In: *Mathematical Methods in Scattering Theory and Biomedical Technology* (eds. G. Dassios, D.I. Fotiadis, K. Kiriaki & C.V. Massalas), Pitman Research Notes in Math. 390, pp. 201–214. Harlow: Addison Wesley, Longman, 1998. C: p. 143
407. J. Giroire, Integral equation methods for the Helmholtz equation. *Integral Equations & Operator Theory* **5** (1982) 506– 517. C: p. 172
408. J. Giroire & J.C. Nedelec, Numerical solution of an exterior Neumann problem using a double layer potential. *Mathematics of Computation* **32** (1978) 973–990. C: p. 172

409. V.E. Glazanov, Diffraction of a plane longitudinal wave by a grating of cylindrical cavities in an elastic medium. *SPA* **13** (1968) 303–308. C: p. 142
410. V.E. Glazanov, Diffraction of a wave radiated by a cylinder at a grating of acoustically compliant cylinders. *SPA* **14** (1969) 447–451. C: p. 129
411. A.W. Glisson, An integral equation for electromagnetic scattering from homogeneous dielectric bodies. *IEAP* **32** (1984) 173–175. C: p. 195
412. M. Golberg & H. Bowman, Optimal convergence rates for some discrete projection methods. *Appl. Math. & Comp.* **96** (1998) 237–271. C: p. 208
413. M.A. Golberg, The convergence of several algorithms for solving integral equations with finite-part integrals. *J. Integ. Eqns.* **5** (1983) 329–340. C: p. 208
414. M.A. Golberg, The convergence of several algorithms for solving integral equations with finite part integrals. II. *J. Integ. Eqns.* **9** (1985) 267–275. C: p. 208
415. M.A. Golberg & C.S. Chen, *Discrete Projection Methods for Integral Equations*. Southampton: Computational Mechanics Publications, 1997. C: p. 168
416. M.L. Goldberger & K.M. Watson, *Collision Theory*. New York: Wiley, 1964. C: p. 84 & Table B.1 (p. 327)
417. H. Goldstein, *Classical Mechanics*. Reading: Addison-Wesley, 1950. C: p. 330
418. J.D. Goldstein, The effect of coordinate system rotations on spherical harmonic expansions: a numerical method. *JGR* **89** (1984) 4413–4418. C: p. 333
419. J.J. Golecki, An elementary approach to two-dimensional elastostatic crack theory. *SM Archives* **1** (1981) 1–29. C: p. 339
420. V.T. Golovchan, Solution of fundamental boundary-value problems for the wave equation in a half-space with spherical cavities. *SPA* **17** (1971) 198–201. C: p. 137
421. V.T. Golovchan & A.N. Guz', Diffraction of elastic waves by an infinite row of circular cylinders. *SPD* **14** (1969) 503–505. C: p. 142
422. A. Gonis, *Green Functions for Ordered and Disordered Systems*. Amsterdam: North-Holland, 1992. C: pp. 1, 82, 96, 321
423. A. Gonis, *Theoretical Materials Science*. Warrendale: Materials Research Society, 2000. C: pp. 1, 9
424. A. Gonis & W.H. Butler, *Multiple Scattering in Solids*. New York: Springer, 2000. C: pp. 1, 9, 90, 321
425. I.R. Gonsalves, D.J. Shippy & F.J. Rizzo, The direct boundary integral equation method for the three-dimensional elastodynamic transmission problem. *Mathl Comput. Modelling* **15** (1991) 155–164. C: pp. 202, 221
426. A.G. Gorshkov & D.V. Tarlakovsky, *Transient Aerohydroelasticity of Spherical Bodies*. Berlin: Springer, 2001. C: p. 142
427. I.S. Gradshteyn & I.M. Ryzhik, *Table of Integrals, Series, and Products*, fourth edition. New York: Academic Press, 1980. C: pp. 35, 49, 57, 63, 109, 324, 332
428. J.H. Graf, Ueber die Addition und Subtraction der Argumente bei Bessel'schen Functionen nebst einer Anwendung. *Mathematische Annalen* **43** (1893) 136–144. C: p. 38
429. K.F. Graff, *Wave Motion in Elastic Solids*. Oxford: Oxford University Press, 1975. Reprinted by Dover, New York, 1991. C: pp. 20, 141
430. I.G. Graham & I.H. Sloan, Fully discrete spectral boundary integral methods on smooth closed surfaces in R^3. *NM* **92** (2002) 289–323. C: pp. 162, 263
431. W.T. Grandy, Jr, *Scattering of Waves from Large Spheres*. Cambridge: Cambridge University Press, 2000. C: p. 140
432. E.P. Gray, Scattering of a surface wave by a submerged sphere. *JEM* **12** (1978) 15–41. C: p. 206

433. L.J. Gray, Hypersingular integrals and the Helmholtz equation in two dimensions. In: *Advances in Boundary Element Methods in Japan and USA* (eds. M. Tanaka, C.A. Brebbia & R. Shaw) pp. 379–388. Southampton: Computational Mechanics Publications, 1990. C: p. 208
434. G. Green, *Mathematical Papers of George Green* (ed. N.M. Ferrers). New York: Chelsea, 1970. First published in London, 1871. C: p. 222
435. L. Greengard, J. Huang, V. Rokhlin & S. Wandzura, Accelerating fast multipole methods for the Helmholtz equation at low frequencies. *IEEE Computational Sci. & Engng.* **5**, 3 (1998) 32–38. C: p. 241
436. M.J.L. Greenhow, The hydrodynamic interactions of spherical wave power devices in surface waves. In: **[241]**, pp. 287–343. C: pp. 145, 311
437. D. Greenspan & P. Werner, A numerical method for the exterior Dirichlet problem for the reduced wave equation. *ARMA* **23** (1966) 288–316. C: p. 218
438. R.D. Gregory, An expansion theorem applicable to problems of wave propagation in an elastic half-space containing a cavity. *PCPS* **63** (1967) 1341–1367. C: p. 53
439. R.D. Gregory, The propagation of waves in an elastic half-space containing a cylindrical cavity. *PCPS* **67** (1970) 689–710. C: p. 143
440. R.D. Gregory, The non-existence of standing modes in certain problems of linear elasticity. *MPCPS* **77** (1974) 385–404. C: p. 23
441. R.D. Gregory & D.M. Austin, Scattering of waves by a semicylindrical groove in the surface of an elastic half-space. *QJMAM* **43** (1990) 293–315. C: p. 143
442. J. Groenenboom & R. Snieder, Attenuation, dispersion, and anisotropy by multiple scattering of transmitted waves through distributions of scatterers. *JASA* **98** (1995) 3482–3492. C: p. 299
443. D. Gross & Ch. Zhang, Diffraction of SH waves by a system of cracks: solution by an integral equation method. *IJSS* **24** (1988) 41–49. C: p. 208
444. M.J. Grote & C. Kirsch, Dirichlet-to-Neumann boundary conditions for multiple scattering problems. *JCP* **201** (2004) 630–650. C: p. 129
445. J.E. Gubernatis, E. Domany & J.A. Krumhansl, Formal aspects of the theory of the scattering of ultrasound by flaws in elastic materials. *JAP* **48** (1977) 2804–2811. C: p. 201
446. J.E. Gubernatis, E. Domany, J.A. Krumhansl & M. Huberman, The Born approximation in the theory of the scattering of elastic waves by flaws. *JAP* **48** (1977) 2812–2819. C: p. 142
447. N.A. Gumerov & R. Duraiswami, Computation of scattering from N spheres using multipole reexpansion. *JASA* **112** (2002) 2688–2701. C: pp. 137, 139
448. N.A. Gumerov & R. Duraiswami, Recursions for the computation of multipole translation and rotation coefficients for the 3-d Helmholtz equation. *SISC* **25** (2003) 1344–1381. C: pp. 94, 333
449. N.A. Gumerov & R. Duraiswami, *Fast Multipole Methods for the Helmholtz Equation in Three Dimensions*. Amsterdam: Elsevier, 2004. C: pp. 137, 235
450. N.A. Gumerov & R. Duraiswami, Computation of scattering from clusters of spheres using the fast multipole method. *JASA* **117** (2005) 1744–1761. C: p. 282
451. L. Gurel & W.C. Chew, On the connection of T matrices and integral equations. *International IEEE Antennas and Propagation Society Symposium Digest*, London, Ontario, June 1991, pp. 1624–1627. New York: IEEE. C: p. 271
452. L. Gürel & W.C. Chew, A recursive T-matrix algorithm for strips and patches. *RS* **27** (1992) 387–401. C: p. 271
453. M.E. Gurtin, The linear theory of elasticity. In: *Handbuch der Physik*, vol. VIa/2 (ed. C. Truesdell) pp. 1–295. Berlin: Springer, 1972. C: p. 198

454. I.I. Guseinov, On the evaluation of rotation coefficients for spherical harmonics using binomial coefficients. *J. Molecular Structure* (*Theochem*) **366** (1996) 119–121. C: p. 333
455. I.I. Guseinov, A. Özmen, Ü. Atav & H. Yüksel, Computation of Clebsch–Gordan and Gaunt coefficients using binomial coefficients. *JCP* **122** (1995) 343–347. C: p. 93
456. H.L. Gustafson, III & P.R. Stepanishen, Effects of multiple scattering and acoustic interactions among two parallel hydrophones and a planar baffle. *JASA* **69** (1981) 380–389. C: p. 129
457. K. Gustafson & T. Abe, The third boundary condition – was it Robin's? *Mathematical Intelligencer* **20**, 1 (1998) 63–71. C: p. 13
458. A.N. Guz & V.T. Golovchan, *Diffraction of Elastic Waves in Multiply Connected Bodies.* Report FTD-MT-24-418-73, Wright-Patterson Air Force Base, Ohio, 1973. Available from U.S. National Technical Information Service, Accession No. AD 7678931. C: p. 142
459. A.N. Guz', V.D. Kubenko & M.A. Cherevko, Diffraction of elastic waves. *Soviet Applied Mechanics* **18** (1978) 789–798. C: p. 142
460. E.M. Haacke & L.L. Foldy, Multiple scattering of scalar waves by point scatterers in one dimension. I & II. *PR* C **23** 1320–1329 & 1330–1339. C: p. 316
461. R.H. Hackman, The transition matrix for acoustic and elastic wave scattering in prolate spheroidal coordinates. *JASA* **75** (1984) 35–45. C: pp. 265, 274
462. R.H. Hackman, Acoustic scattering from elastic solids. In: *Physical Acoustics*, vol. 22 (eds. A.D. Pierce & R.N. Thurston) pp. 1–194. San Diego: Academic Press, 1993. C: p. 141
463. R.H. Hackman & R. Lim, Development and application of the spheroidal coordinate based *T* matrix solution to elastic wave scattering. *RS* **29** (1994) 1035–1049. C: pp. 265, 274
464. R.H. Hackman, G.S. Sammelmann, K.L. Williams & D.H. Trivett, A reanalysis of the acoustic scattering from elastic spheroids. *JASA* **83** (1988) 1255–1266. C: p. 274
465. R.H. Hackman & D.G. Todoroff, An application of the spheroidal-coordinate based transition matrix: The acoustic scattering from high aspect ratio solids. *JASA* **78** (1985) 1058–1071. C: p. 274
466. J. Hadamard, *Lectures on Cauchy's Problem in Linear Partial Differential Equations.* New Haven: Yale University Press, 1923. C: pp. 158, 333, 336, 336
467. J. Hadamard, *The Psychology of Invention in the Mathematical Field.* New York: Dover, 1954. C: p. 168
468. T. Ha Duong, La méthode de Schenck pour la résolution numérique du problème de radiation acoustique. *Electricité de France, Bulletin de la Direction des Etudes et Recherches, Série C, Mathematiques, Informatique* No. 2 (1979) 15–49. C: p. 220
469. C. Hafner, *Post-modern Electromagnetics.* Chichester: Wiley, 1999. C: p. 245
470. M.A. Haider, S.P. Shipman & S. Venakides, Boundary-integral calculations of two-dimensional electromagnetic scattering in infinite photonic crystal slabs: channel defects and resonances. *SIAP* **62** (2002) 2129–2148. C: p. 180
471. W.S. Hall & X.Q. Mao, Boundary element method calculation for coherent electromagnetic scattering from two and three dielectric spheres. *EABE* **15** (1995) 313–320. C: p. 195
472. A.-K. Hamid, Iterative solution to the TM scattering by two infinitely long lossy dielectric elliptic cylinders. *JEWA* **18** (2004) 529–546. C: p. 149
473. A.-K. Hamid, I.R. Ciric & M. Hamid, Electromagnetic scattering by an arbitrary configuration of dielectric spheres. *CJP* **68** (1990) 1419–1428. C: p. 141
474. A.K. Hamid & M.I. Hussein, Iterative solution to the electromagnetic plane wave scattering by two parallel conducting elliptic cylinders. *JEWA* **17** (2003) 813–828. C: p. 149

475. T.B. Hansen & A.D. Yaghjian, Low-frequency scattering from two-dimensional perfect conductors. *IEAP* **40** (1992) 1389–1402. C: pp. 293, 295, 296
476. W.W. Hansen, A new type of expansion in radiation problems. *PR* **47** (1935) 139–143. C: p. 106
477. G.H. Hardy, A chapter from Ramanujan's note-book. *PCPS* **21** (1923) 492–503. C: p. 329
478. A.H. Harker & J.A.G. Temple, Velocity and attenuation of ultrasound in suspensions of particles in fluids. *JPDAP* **21** (1988) 1576–1588. C: p. 319
479. A.H. Harker & J.A.G. Temple, Propagation and attenuation of ultrasound in suspensions. In: *Developments in Acoustics and Ultrasonics* (eds. M.J.W. Povey & D.J. McClements) pp. 19–31. Bristol: Institute of Physics Publishing, 1992. C: p. 312
480. R. Harrington, Origin and development of the method of moments for field computation. *IEAPM* **32**, 3 (1990) 31–36. C: p. 195
481. R.F. Harrington, *Field Computation by Moment Methods*. New York: Macmillan, 1968. C: p. 195
482. R.F. Harrington, Boundary integral formulations for homogeneous material bodies. *JEWA* **3** (1989) 1–15. C: pp. 18, 195
483. T. Hasegawa, Uniform approximations to finite Hilbert transform and its derivative. *JCAM* **163** (2004) 127–138. C: p. 338
484. M. Hasheminejad & T.L. Geers, Modal impedances for two spheres in a thermoviscous fluid. *JASA* **94** (1993) 2205–2214. C: p. 143
485. S.M. Hasheminejad & S.A. Badsar, Acoustic scattering by a pair of poroelastic spheres. *QJMAM* **57** (2004) 95–113. C: p. 143
486. S.M. Hasheminejad & S.A. Badsar, Elastic wave scattering by two spherical inclusions in a poroelastic medium. *J. Engng. Mech.* **131** (2005) 953–965. C: p. 143
487. M.L. Hastriter, S. Ohnuki & W.C. Chew, Error control of the translation operator in 3D MLFMA. *MOTL* **37** (2003) 184–188. C: p. 80
488. Sir T. Havelock, Waves due to a floating sphere making periodic heaving oscillations. *PRSA* **231** (1955) 1–7. C: pp. 110, 143, 145
489. O. Heaviside, *Electromagnetic Theory*, vol. 1. London: "The Electrician" Printing & Publishing Co., 1893. C: p. 1
490. B.K. Henderson, V.K. Kinra & K.I. Maslov, Characterization of resonances of a periodic linear array of particles. *Nondestructive Testing & Evaluation* **15** (1999) 265–278. C: p. 143
491. M. Henseler, A. Wirzba & T. Guhr, Quantization of hyperbolic N-sphere scattering systems in three dimensions. *Annals of Physics* **258** (1997) 286–319. C: p. 139
492. F.S. Henyey, Corrections to Foldy's effective medium theory for propagation in bubble clouds and other collections of very small scatterers. *JASA* **105** (1999) 2149–2154. C: pp. 316, 316, 319
493. F.S. Henyey, Reply to "Comment on 'Corrections to Foldy's effective medium theory for propagation in bubble clouds and other collections of very small scatterers' (reference [492])" [J. Acoust. Soc. Am. 111, 1552–1555 (2002)]. *JASA* **111** (2002) 1556–1559. C: pp. 316, 319
494. K.F. Herzfeld, The scattering of sound-waves by small elastic spheres. *Phil. Mag.*, Series 7, **9** (1930) 741–751. C: p. 141
495. J.L. Hess, Review of the historical development of surface source methods. In: *Computational Methods in Potential Aerodynamics* (ed. L. Morino) pp. 21–37. Berlin: Springer, 1985. C: p. 172

496. J.L. Hess, Development and application of panel methods. In: *Advanced Boundary Element Methods* (ed. T.A. Cruse) pp. 165–177. Berlin: Springer, 1988, pp. 165–177. C: p. 172
497. J.L. Hess, Panel methods in computational fluid dynamics. *ARFM* **22** (1990) 255–274. C: p. 172
498. J.L. Hess & A.M.O. Smith, Calculation of nonlifting potential flow about arbitrary three-dimensional bodies. *JSR* **8**, 2 (1964) 22–44. C: p. 172
499. R. Hickling & N.M. Wang, Scattering of sound by a rigid movable sphere. *JASA* **39** (1966) 276–279. C: p. 134
500. J. Hildenbrand & G. Kuhn, Non-linear co-ordinate transformation of finite-part integrals in two-dimensional boundary element analysis. *IJNME* **36** (1993) 2939–2954. C: p. 208
501. N.L. Hills, Semi-infinite diffraction gratings. II. Inward resonance. *CPAM* **18** (1965) 389–395. C: p. 10
502. N.L. Hills & S.N. Karp, Semi-infinite diffraction gratings – I. *CPAM* **18** (1965) 203–233. C: p. 10
503. A.K. Hipp, G. Storti & M. Morbidelli, On multiple-particle effects in the acoustic characterization of colloidal dispersions. *JPDAP* **32** (1999) 568–576. C: p. 322
504. R. Hiptmair, Coercive combined field integral equations. *J. Numerical Math.* **11** (2003) 115–134. C: p. 218
505. J.O. Hirschfelder, C.F. Curtiss & R.B. Bird, *Molecular Theory of Gases and Liquids*. New York: Wiley, 1954. C: p. 96
506. R.-T. Ho & A. Harten, Green's function techniques for solutions of floating body problems. *Proc. Civil Engineering in the Oceans* III. New York: ASCE 1975, pp. 939–958. C: p. 206
507. E.W. Hobson, *The Theory of Spherical and Ellipsoidal Harmonics*. Cambridge: Cambridge University Press, 1931. C: pp. 31, 33, 63, 65, 68, 68
508. C.H. Hodges, Confinement of vibration by structural irregularity. *JSV* **82** (1982) 411–424. C: pp. 11, 307
509. N. Hogben & R.G. Standing, Wave loads on large bodies. In: [102], pp. 258–277. C: p. 205
510. S.Y. Hong, J.H. Kim, S.K. Cho, Y.R. Choi & Y.S. Kim, Numerical and experimental study on hydrodynamic interaction of side-by-side moored multiple vessels. *OE* **32** (2005) 783–801. C: p. 205
511. K. Hongo, Multiple scattering by two conducting circular cylinders. *IEAP* **26** (1978) 748–751. C: pp. 2, 310
512. H. Hönl, A.W. Maue & K. Westpfahl, Theorie der Beugung. In: *Handbuch der Physik*, vol. 25/1 (ed. S. Flügge) pp. 218–573. Berlin: Springer, 1961. C: p. 157
513. B.A. Howarth, Multiple-scattering resonances between parallel conducting cylinders. *CJP* **51** (1973) 2415–2427. C: p. 129
514. L. Howarth, One hundred years of Lamb's Hydrodynamics. *JFM* **90** (1979) 202–207. C: p. 12
515. M. Hron & M. Razavy, Integral equation approach to the problem of scattering by a number of fixed scatterers. *CJP* **55** (1977) 1442–1452. C: p. 178
516. G. Hsiao & R.C. MacCamy, Solution of boundary value problems by integral equations of the first kind. *SIAM Review* **15** (1973) 687–705. C: p. 172
517. H. Huang & G.C. Gaunaurd, Acoustic scattering of a plane wave by two spherical elastic shells above the coincidence frequency. *JASA* **101** (1997) 2659–2668. C: p. 143
518. H. Huang & G.C. Gaunaurd, Scattering of a plane acoustic wave by a spherical elastic shell near a free surface. *IJSS* **34** (1997) 591–602. C: p. 143

519. J.Y. Huang & H. So, Diffraction of P waves by two cracks at arbitrary position in an elastic medium. *EFM* **29** (1988) 335–347. C: p. 208
520. J.A. Hudson, *The Excitation and Propagation of Elastic Waves*. Cambridge: Cambridge University Press, 1980. C: pp. 20, 22
521. J.A. Hudson, Overall properties of heterogeneous material. *GJI* **107** (1991) 505–511. C: p. 322
522. A. Hulme, The wave forces acting on a floating hemisphere undergoing forced periodic oscillations. *JFM* **121** (1982) 443–463. C: p. 145
523. A. Hulme, A ring-source/integral-equation method for the calculation of hydrodynamic forces exerted on floating bodies of revolution. *JFM* **128** (1983) 387–412. C: p. 206
524. J.F. Hunka & K.K. Mei, Electromagnetic scattering by two bodies of revolution. *Electromagnetics* **1** (1981) 329–347. C: p. 273
525. E.M. Hyde & O.P. Bruno, A fast, higher-order solver for scattering by penetrable bodies in three dimensions. *JCP* **202** (2005) 236–261. C: p. 186
526. F. Ihlenburg, *Finite Element Analysis of Acoustic Scattering*. New York: Springer, 1998. C: p. 11
527. T. Ijima, C.R. Chou & A. Yoshida, Method of analyses for two-dimensional water wave problems. *Proc. 15th Coastal Engineering Conf.*, 1976, pp. 2717–2736. ASCE: New York, 1976. C: p. 206
528. H. Ikuno & K. Yasuura, Improved point-matching method with application to scattering from a periodic surface. *IEAP* **21** (1973) 657–662. C: p. 263
529. M.G. Imhof, Multiple multipole expansions for acoustic scattering. *JASA* **97** (1995) 754–763. C: p. 245
530. M.G. Imhof, Multiple multipole expansions for elastic scattering. *JASA* **100** (1996) 2969–2979. C: p. 245
531. M.G. Imhof, Computing the elastic scattering from inclusions using the multiple multipoles method in three dimensions. *GJI* **156** (2004) 287–296. C: p. 245
532. M.G. Imhof & M.N. Toksöz, Acousto-elastic multiple scattering: a comparison of ultrasonic experiments with multiple multipole expansions. *JASA* **101** (1997) 1836–1846. C: p. 245
533. N.I. Ioakimidis, Exact expression for a two-dimensional finite-part integral appearing during the numerical solution of crack problems in three-dimensional elasticity. *Comm. in Appl. Numer. Methods* **1** (1985) 183–189. C: p. 349
534. N.I. Ioakimidis, A new interpretation of Cauchy type singular integrals with an application to singular integral equations. *JCAM* **14** (1986) 271–278. C: p. 335
535. N.I. Ioakimidis & M.S. Pitta, Remarks on the Gaussian quadrature rule for finite-part integrals with a second-order singularity. *CMAME* **69** (1988) 325–343. C: p. 338
536. M.P. Ioannidou, N.C. Skaropoulos & D.P. Chrissoulidis, Study of interactive scattering by clusters of spheres. *JOSA A* **12** (1995) 1782–1789. C: p. 141
537. G. Iovane, I.K. Lifanov & M.A. Sumbatyan, On direct numerical treatment of hypersingular integral equations arising in mechanics and acoustics. *Acta Mechanica* **162** (2003) 99–110. C: p. 208
538. M. de St.Q. Isaacson, Vertical cylinders of arbitrary section in waves. *J. Waterway, Port, Coastal and Ocean Div., Proc. ASCE* **104** (1978) 309–324. C: p. 172
539. M. de St.Q. Isaacson, Interference effects between large cylinders in waves. *J. Petroleum Technology* **31** (1979) 505–512. C: p. 172
540. M. de St.Q. Isaacson, Fixed and floating axisymmetric structures in waves. *J. Waterway, Port, Coastal and Ocean Div., Proc. ASCE* **108** (1982) 180–199. C: p. 206

541. A. Ishimaru, *Wave Propagation and Scattering in Random Media*, vol. 2. Boston: Academic Press, 1978. C: p. 316
542. M. Iskandarani & P.L.-F. Liu, Multiple scattering of surface water-waves and wave forces on cylinder arrays. *AOR* **10** (1988) 170–180. C: p. 148
543. M.F. Iskander, P.W. Barber, C.H. Durney & H. Massoudi, Irradiation of prolate spheroidal models of humans in the near field of a short electric dipole. *IEMT* **28** (1980) 801–807. C: p. 251
544. M.F. Iskander, A. Lakhtakia & C.H. Durney, A new procedure for improving the solution stability and extending the frequency range of the EBCM. *IEAP* **31** (1983) 317–324. C: p. 265
545. S. Itou, Dynamic stress concentration around two coplanar Griffith cracks in an infinite elastic medium. *JAM* **45** (1978) 803–806. C: p. 208
546. S. Itou, 3D dynamic stress intensity factors at three rectangular cracks in an infinite elastic medium subjected to a time-harmonic stress wave. *Archive of Applied Mechanics* **69** (1999) 286–298. C: p. 209
547. S. Itou, Three-dimensional dynamic stress intensity factors around two parallel square cracks in an infinite elastic medium subjected to a time-harmonic stress wave. *Acta Mechanica* **143** (2000) 79–90. C: p. 209
548. S. Itou & H. Haliding, Dynamic stress intensity factors around three cracks in an infinite elastic plane subjected to time-harmonic stress waves. *Int. J. Fracture* **83** (1997) 379–391. C: p. 208
549. J.D. Jackson, *Classical Electrodynamics*, second edition. New York: Wiley, 1975. C: pp. 19, 63, 107, 116, 140
550. E. Jahnke, F. Emde & F. Lösch, *Tables of Higher Functions*, sixth edition. Stuttgart: Teubner, 1960. C: p. 73
551. R.W. James, The Adams and Elsasser dynamo integrals. *PRSA* **331** (1973) 469–478. C: p. 93
552. M.A. Jaswon, Integral equation methods in potential theory. I. *PRSA* **275** (1963) 23–32. C: p. 177
553. M.A. Jaswon & G.T. Symm, *Integral Equation Methods in Potential Theory and Elastostatics*. London: Academic Press, 1977. C: p. 177
554. C. Javanaud & A. Thomas, Multiple scattering using the Foldy–Twersky integral equation. *Ultrasonics* **26** (1988) 341–343. C: p. 319
555. B. Jeffreys, Transformation of tesseral harmonics under rotation. *GJRAS* **10** (1965) 141–145. C: p. 330
556. H. Jeffreys & B. Jeffreys, *Methods of Mathematical Physics*, third edition. Cambridge: Cambridge University Press, 1956. C: p. 63
557. Y.-Q. Jin & X. Huang, Numerical simulations of T-matrix solution for polarized bistatic scattering from a cluster of scatterers. *OC* **124** (1996) 27–32. C: p. 282
558. X. Jing, P. Sheng & M. Zhou, Acoustic and electromagnetic quasimodes in dispersed random media. *PR* A **46** (1992) 6513–6534. C: p. 321
559. F. John, On the motion of floating bodies II. Simple harmonic motions. *CPAM* **3** (1950) 45–101. C: pp. 26, 56, 114, 146, 204, 204
560. G. Johnson & R. Truell, Numerical computations of elastic scattering cross sections. *JAP* **36** (1965) 3466–3475. C: p. 141
561. R.K. Johnson, Sound scattering from a fluid sphere revisited. *JASA* **61** (1977) 375–377. Erratum: **63** (1978) 626. C: p. 135
562. A.R. Jones, Electromagnetic wave scattering by assemblies of particles in the Rayleigh approximation. *PRSA* **366** (1979) 111–127. Correction: **375** (1981) 453–454. C: p. 299

563. A.R. Jones, Scattering efficiency factors for agglomerates of small spheres. *JPDAP* **12** (1979) 1661–1672. C: p. 299
564. D.S. Jones, *The Theory of Electromagnetism*. Oxford: Pergamon, 1964. C: pp. 13, 19, 63, 107, 124, 140, 157, 165, 165, 177, 189, 244, 310
565. D.S. Jones, Integral equations for the exterior acoustic problem. *QJMAM* **27** (1974) 129–142. C: pp. 210, 221
566. D.S. Jones, Numerical methods for antenna problems. *Proc. IEE* **121** (1974) 573–582. C: p. 221
567. D.S. Jones, Note on the integral equation for a straight wire antenna. *IEE-H* **128** (1981) 114–116. C: p. 189
568. D.S. Jones, An exterior problem in elastodynamics. *MPCPS* **96** (1984) 173–182. C: p. 217
569. D.S. Jones, Boundary integrals in elastodynamics. *IMAP* **34** (1985) 83–97. C: p. 219
570. P.M. Jordan & C. Feuillade, On the propagation of harmonic acoustic waves in bubbly liquids. *IJES* **42** (2004) 1119–1128. C: p. 316
571. K. Jörgens, *Linear Integral Operators*. London: Pitman, 1982. C: p. 162
572. G. Jost, Integral equations with modified fundamental solution in time-harmonic electromagnetic scattering. *IMAP* **40** (1988) 129–143. C: p. 217
573. P. Juhl, An axisymmetric integral equation formulation for free space non-axisymmetric radiation and scattering of a known incident wave. *JSV* **163** (1993) 397–406. C: p. 178
574. M. Kafesaki, R.S. Penciu & E.N. Economou, Air bubbles in water: a strongly multiple scattering medium for acoustic waves. *Phys. Rev. Lett.* **84** (2000) 6050–6053. C: p. 139
575. H. Kagemoto, M. Murai, M. Saito, B. Molin & Š. Malenica, Experimental and theoretical analysis of the wave decay along a long array of vertical cylinders. *JFM* **456** (2002) 113–135. C: p. 283
576. H. Kagemoto & D.K.P. Yue, Interactions among multiple three-dimensional bodies in water waves: an exact algebraic method. *JFM* **166** (1986) 189–209. C: p. 283
577. F.M. Kahnert, J.J. Stamnes & K. Stamnes, Application of the extended boundary condition method to homogeneous particles with point-group symmetries. *AO* **40** (2001) 3110–3123. C: p. 251
578. F.M. Kahnert, J.J. Stamnes & K. Stamnes, Surface-integral formulation for electromagnetic scattering in spheroidal coordinates. *JQSRT* **77** (2003) 61–78. C: p. 265
579. L. Kai & P. Massoli, Scattering of electromagnetic-plane waves by radially inhomogeneous spheres: a finely stratified sphere model. *AO* **33** (1994) 501–511. C: p. 140
580. D.I. Kaklamani & H.T. Anastassiu, Aspects of the method of auxiliary sources (MAS) in computational electromagnetics. *IEAPM* **44**, 3 (2002) 48–64. C: p. 244
581. S.K. Kanaun, Self-consistent methods in the problem of wave propagation through heterogeneous media. In: *Heterogeneous Media: Micromechanics Modeling Methods and Simulations* (eds. K. Markov & L. Preziosi) pp. 214–319. Boston: Birkhäuser, 2000. C: pp. 320, 320, 321
582. S.K. Kanaun & V.M. Levin, Effective medium method in the problem of axial elastic shear wave propagation through fiber composites. *IJSS* **40** (2003) 4859–4878. C: p. 321
583. S.K. Kanaun, V.M. Levin & F.J. Sabina, Propagation of elastic waves in composites with random set of spherical inclusions (effective medium approach). *WM* **40** (2004) 69–88. C: pp. 320, 322
584. G. Kapodistrias & P.H. Dahl, Effects of interaction between two bubble scatterers. *JASA* **107** (2000) 3006–3017. C: p. 137
585. D.L. Karabalis & D.E. Beskos, Dynamic soil–structure interaction. In: *Boundary Element Methods in Mechanics* (ed. D.E. Beskos) pp. 492–562. Amsterdam: North-Holland, 1987. C: p. 22

586. S.G. Kargl, Effective medium approach to linear acoustics in bubbly liquids. *JASA* **111** (2002) 168–173. C: p. 319
587. S.N. Karp, Diffraction by combinations of obstacles. *Proc. McGill Symposium on Microwave Optics*, McGill University, Montreal, June 1953. Published by U.S. Air Force, Cambridge Research Center, Bedford, MA, as AFCRC-TR-59-118, part II, 1959, pp. 196–201. C: p. 307 & §8.5.3 (p. 308)
588. S.N. Karp, A convergent 'farfield' expansion for two-dimensional radiation functions. *CPAM* **14** (1961) 427–434. C: p. 273
589. S.N. Karp & A. Russek, Diffraction by a wide slit. *JAP* **27** (1956) 886–894. C: p. 310
590. M. Kashiwagi, Hydrodynamic interactions among a great number of columns supporting a very large flexible structure. *J. Fluids & Structures* **14** (2000) 1013–1034. C: p. 283
591. M. Kashiwagi, K. Endo & H. Yamaguchi, Wave drift forces and moments on two ships arranged side by side in waves. *OE* **32** (2005) 529–555. C: p. 205
592. M. Katory, A.A. Lacey & P.K.Y. Tam, Application of theoretical hydrodynamics to the design of arbitrarily shaped and multibodied marine structures. *Trans. Royal Institution of Naval Architects* **122** (1980) 253–263. Discussion, 263–267. C: p. 205
593. S.E. Kattis, D. Polyzos & D.E. Beskos, Vibration isolation by a row of piles using a 3-D frequency domain BEM. *IJNME* **46** (1999) 713–728. C: p. 202
594. A.C. Kaya & F. Erdogan, On the solution of integral equations with strongly singular kernels. *QAM* **45** (1987) 105–122. C: pp. 208, 338
595. A.J. Keane, P. Temarel, X.-J. Wu & Y. Wu, Hydroelasticity of non-beamlike ships in waves. *PTRSA* **334** (1991) 339–353. Discussion, 353–355. C: p. 205
596. J.B. Keller, Wave propagation in random media. In: *Hydrodynamic Instability* (eds. G. Birkhoff, R. Bellman & C.C. Lin) pp. 227–246. New York: American Mathematical Society, 1962. C: p. 322
597. O.D. Kellogg, *Foundations of Potential Theory*. Berlin: Springer, 1929. C: pp. 116, 154, 163, 168, 223, 286, 344, 352,
598. M. Kerker (ed.), *Selected Papers on Light Scattering*. SPIE vol. 951, part 1. Bellingham: SPIE – The Int. Soc. for Optical Engng., 1988. C: p. 140
599. F.H. Kerr, The scattering of a plane elastic wave by spherical elastic inclusions. *IJES* **30** (1992) 169–186. C: pp. 319, 322
600. F.H. Kerr, An effective medium approach to the study of plane wave propagation in an elastic matrix with spherical elastic inclusions. *IJES* **30** (1992) 187–198. C: p. 322
601. R. Kieser, C. Schwab & W.L. Wendland, Numerical evaluation of singular and finite-part integrals on curved surfaces using symbolic manipulation. *Computing* **49** (1992) 279–301. C: p. 349
602. M. Kikuchi, Dispersion and attenuation of elastic waves due to multiple scattering from cracks. *Physics of the Earth & Planetary Interiors* **27** (1981) 100–105. C: p. 316
603. M. Kikuchi, Dispersion and attenuation of elastic waves due to multiple scattering from inclusions. *Physics of the Earth & Planetary Interiors* **25** (1981) 159–162. C: p. 316
604. C.H. Kim, The hydrodynamic interaction between two cylindrical bodies floating in beam seas. Rept. SIT-OE-72-10, Ocean Engineering Dept., Stevens Institute of Technology, Hoboken, New Jersey, 1972. C: p. 205
605. J.-Y. Kim, On the generalized self-consistent model for elastic wave propagation in composite materials. *IJSS* **41** (2004) 4349–4360. C: p. 321
606. K.T. Kim, The translation formula for vector multipole fields and the recurrence relations of the translation coefficients of scalar and vector multipole fields. *IEAP* **44** (1996) 1482–1487. C: pp. 86, 94, 107

607. M.-H. Kim, Interaction of waves with *N* vertical circular cylinders. *JWPCOE* **119** (1993) 671–689. C: pp. 129, 283
608. W.D. Kim, On the harmonic oscillations of a rigid body on a free surface. *JFM* **21** (1965) 427–451. C: p. 205
609. B.J. King & A.L. Van Buren, A general addition theorem for spheroidal wave functions. *SIMA* **4** (1973) 149–160. C: p. 150
610. L.V. King, On the acoustic radiation pressure on spheres. *PRSA* **147** (1934) 212–240. C: p. 134
611. R.W.P. King, R.B. Mack & S.S. Sandler, *Arrays of Cylindrical Dipoles*. Cambridge: Cambridge University Press 1968. C: p. 189
612. V.K. Kinra, N.A. Day, K. Maslov, B.K. Henderson & G. Diderich, The transmission of a longitudinal wave through a layer of spherical inclusions with a random or periodic arrangement. *JMPS* **46** (1998) 153–165. C: p. 143
613. V.K. Kinra, B.K. Henderson & K.I. Maslov, Elastodynamic response of layers of spherical particles in hexagonal and square periodic arrangements. *JMPS* **47** (1999) 2147–2170. C: p. 143
614. K. Kiriaki, A uniquely solvable integral equation for the Neumann problem in linear elasticity. *Applicable Analysis* **73** (1999) 379–392. C: p. 219
615. T.R. Kirkpatrick, Localization of acoustic waves. *PR* B **31** (1985) 5746–5755. C: p. 11
616. A. Kirsch, *An Introduction to the Mathematical Theory of Inverse Problems*. New York: Springer, 1996. C: p. 266
617. A.A. Kishk & L. Shafai, Different formulations for numerical solution of single or multibodies of revolution with mixed boundary conditions. *IEAP* **34** (1986) 666–673. C: p. 195
618. H.S. Kit, M.V. Khaj & V.V. Mykhas'kiv, Analysis of dynamic stress concentration in an infinite body with parallel penny-shaped cracks by BIEM. *EFM* **55** (1996) 191–207. C: p. 209
619. M. Kitahara, *Boundary Integral Equation Methods in Eigenvalue Problems of Elastodynamics and Thin Plates*. Amsterdam: Elsevier, 1985. C: p. 203
620. M. Kitahara & J.D. Achenbach, BIE method to analyze wave motion in solids with periodically distributed cavities. *CMAME* **64** (1987) 523–536. C: p. 203
621. M. Kitahara & J.D. Achenbach, BIE calculations for harmonic waves in a solid with periodically distributed inhomogeneities. In: *Advanced Boundary Element Methods* (ed. T.A. Cruse) pp. 197–205. Berlin: Springer, 1988. C: p. 203
622. M. Kitahara & K. Nakagawa, Boundary integral equation methods in three dimensional elastodynamics. In: *Boundary Elements VII* (eds. C.A. Brebbia & G. Maier) pp. 6-27–6-36. Berlin: Computational Mechanics Publications, Springer, 1985. C: p. 202
623. M. Kitahara, K. Nakagawa & J.D. Achenbach, Boundary-integral equation method for elastodynamic scattering by a compact inhomogeneity. *Computational Mech.* **5** (1989) 129–144. C: p. 202
624. R. Kittappa & R.E. Kleinman, Acoustic scattering by penetrable homogeneous objects. *JMP* **16** (1975) 421–432. C: p. 179
625. A.I. Kleev & A.B. Manenkov, The convergence of point-matching techniques. *IEAP* **37** (1989) 50–54. C: p. 137
626. R. Kleinman & B. Vainberg, Full low-frequency asymptotic expansion for second-order elliptic equations in two dimensions. *MMAS* **17** (1994) 989–1004. C: pp. 292, 294, 296 & §8.2.4 (p. 291)
627. R.E. Kleinman & P.A. Martin, On single integral equations for the transmission problem of acoustics. *SIAP* **48** (1988) 307–325. C: pp. 14, 180

628. R.E. Kleinman & G.F. Roach, Boundary integral equations for the three-dimensional Helmholtz equation. *SIAM Review* **16** (1974) 214–236. C: pp. 152, 154, 170, 172
629. R.E. Kleinman & G.F. Roach, On modified Green functions in exterior problems for the Helmholtz equation. *PRSA* **383** (1982) 313–332. C: p. 210
630. R.E. Kleinman & G.F. Roach, Operators of minimal norm via modified Green's functions. *PRSE* **94A** (1983) 163–178. C: p. 210
631. R.E. Kleinman, G.F. Roach & S.E.G. Ström, The null field method and modified Green functions. *PRSA* **394** (1984) 121–136. C: pp. 254, 270
632. R.E. Kleinman & T.B.A. Senior, Rayleigh scattering. In: *Low and High Frequency Asymptotics* (eds. V.K. Varadan & V.V. Varadan) pp. 1–70. Amsterdam: Elsevier, 1986. C: pp. 288, 296
633. R.E. Kleinman & W.L. Wendland, On Neumann's method for the exterior Neumann problem for the Helmholtz equation. *JMAA* **57** (1977) 170–202. C: p. 169
634. A.A. Kleshchev, Sound scattering by spheroidal bodies near an interface. *SPA* **23** (1977) 225–228. C: p. 150
635. L.F. Knockaert & D. De Zutter, Integral equation for the fields inside a dielectric cylinder immersed in an incident E-wave. *IEAP* **34** (1986) 1065–1067. C: p. 180
636. L. Knopoff, Scattering of compression waves by spherical obstacles. *Geophysics* **24** (1959) 30–39. C: p. 141
637. L. Knopoff, Scattering of shear waves by spherical obstacles. *Geophysics* **24** (1959) 209–219. C: p. 142
638. S. Kobayashi, Elastodynamics. In: *Boundary Element Methods in Mechanics* (ed. D.E. Beskos) pp. 191–255. Amsterdam: North-Holland, 1987. C: p. 203
639. S. Kobayashi & N. Nishimura, On the indeterminacy of BIE solutions for the exterior problems of time-harmonic elastodynamics and incompressible elastostatics. In: *Boundary Elements in Engineering* (ed. C.A. Brebbia) pp. 282–296. Berlin: Springer, 1982. C: p. 217
640. S. Koc & W.C. Chew, Calculation of acoustical scattering from a cluster of scatterers. *JASA* **103** (1998) 721–734. C: pp. 98, 282
641. S. Koc, J. Song & W.C. Chew, Error analysis for the numerical evaluation of diagonal forms of the scalar spherical addition theorem. *SINUM* **36** (1999) 906–921. C: p. 98
642. W. Kohn & N. Rostoker, Solution of the Schrödinger equation in periodic lattices with an application to metallic lithium. *PR* **94** (1954) 1111–1120. C: p. 87
643. Y.-L. Kok, General solution to the multiple-metallic-grooves scattering problem: the fast-polarization case. *AO* **32** (1993) 2573–2581. C: p. 148
644. G.C. Kokkorakis & J.G. Fikioris, Acoustic field induced in spheres with inhomogeneous density by external sources. *JASA* **115** (2004) 478–487. C: p. 135
645. G.C. Kokkorakis, J.G. Fikioris & G. Fikioris, Field induced in inhomogeneous spheres by external sources. I. The scalar case. *JASA* **112** (2002) 1297–1306. C: p. 135
646. G.H. Koopmann & H. Benner, Method for computing the sound power of machines based on the Helmholtz integral. *JASA* **71** (1982) 78–89. C: p. 221
647. V.A. Korneev & L.R. Johnson, Scattering of elastic waves by a spherical inclusion–I. Theory and numerical results. *GJI* **115** (1993) 230–250. C: p. 142
648. J. Korringa, Early history of multiple scattering theory for ordered systems. *Physics Reports* **238** (1994) 341–360. C: pp. 7, 9
649. J. Korringa & R.L. Mills, Coherent-potential approximation for random systems with short-range correlations. *PR* B **5** (1972) 1654–1655. C: p. 322
650. D.W. Kraft & M.C. Franzblau, Scattering of elastic waves from a spherical cavity in a solid medium. *JAP* **42** (1971) 3019– 3024. C: p. 142

651. M.A. Krasnosel'skii, G.M. Vainikko, P.P. Zabreiko, Ya.B. Rutitskii & V.Ya. Stetsenko, *Approximate Solution of Operator Equations*. Groningen: Wolters-Noordhoff, 1972. C: p. 267
652. V.V. Kravtsov, On the non-basicity of a system of functions. *USSR Comput. Maths. Math. Phys.* **21**, 6 (1981) 229–231. C: p. 245
653. R. Kress, A Nyström method for boundary integral equations in domains with corners. *NM* **58** (1990) 145–161. C: p. 169
654. R. Kress, Boundary integral equations in time-harmonic acoustic scattering. *Mathl Comput. Modelling* **15** (1991) 229– 243. C: pp. 161, 169
655. R. Kress, *Linear Integral Equations*, second edition. New York: Springer, 1999. C: pp. 160, 168, 169, 177, 254
656. R. Kress & A. Mohsen, On the simulation source technique for exterior problems in acoustics. *MMAS* **8** (1986) 585–597. C: p. 244
657. R. Kress & G.F. Roach, On mixed boundary value problems for the Helmholtz equation. *PRSE* **77A** (1977) 65–77. C: pp. 14, 171
658. R. Kress & G.F. Roach, Transmission problems for the Helmholtz equation. *JMP* **19** (1978) 1433–1437. C: pp. 180, 240
659. R. Kress & W.T. Spassov, On the condition number of boundary integral operators for the exterior Dirichlet problem for the Helmholtz equation. *NM* **42** (1983) 77–95. C: p. 218
660. G.A. Kriegsmann & E.L. Reiss, Low frequency scattering by local inhomogeneities. *SIAP* **43** (1983) 923–934. C: p. 15
661. G. Krishnasamy, F.J. Rizzo & T.J. Rudolphi, Hypersingular boundary integral equations: their occurrence, interpretation, regularization and computation. In: *Developments in Boundary Element Methods* (eds. P.K. Banerjee & S. Kobayashi) pp. 207–252. Amsterdam: Elsevier, 1992. C: p. 208
662. G. Kristensson, Electromagnetic scattering from buried inhomogeneities – a general three-dimensional formalism. *JAP* **51** (1980) 3486–3500. C: pp. 251, 273
663. G. Kristensson, Acoustic scattering by a soft elliptic disk. *JSV* **103** (1985) 487–498. C: p. 274
664. G. Kristensson, A.G. Ramm & S. Ström, Convergence of the T-matrix approach in scattering theory. II. *JMP* **24** (1983) 2619–2631. C: pp. 254, 257, 266, 266
665. G. Kristensson & S. Ström, Scattering from buried inhomogeneities – a general three-dimensional formalism. *JASA* **64** (1978) 917–936. C: p. 250
666. G. Kristensson & S. Ström, Electromagnetic scattering from geophysical targets by means of the T matrix approach: a review of some recent results. *RS* **17** (1982) 903–912. C: p. 273
667. G. Kristensson & P.C. Waterman, The T matrix for acoustic and electromagnetic scattering by circular disks. *JASA* **72** (1982) 1612–1625. C: p. 274
668. U.R. Kristiansen & T.F. Johansen, The horn loudspeaker as a screen-diffraction problem. *JSV* **133** (1989) 449–456. C: p. 208
669. P.A. Krutitskii, A new approach to reduction of the Neumann problem in acoustic scattering to a non-hypersingular integral equation. *IMAP* **64** (2000) 259–269. C: p. 221
670. W.D. Kupradse, *Randwertaufgaben der Schwingungstheorie und Integralgleichungen*. Berlin: VEB Deutscher Verlag der Wissenschaften, 1956. C: p. 171
671. V.D. Kupradze, *Dynamical Problems in Elasticity*. Progress in Solid Mechanics (eds. I. N. Sneddon & R. Hill), vol. III. Amsterdam: North-Holland, 1963. C: pp. 20, 201, 243
672. V.D. Kupradze, *Potential Methods in the Theory of Elasticity*. Jerusalem: Israel Program for Scientific Translations, 1965. C: p. 243

673. V.D. Kupradze, T.G. Gegelia, M.O. Basheleishvili & T.V. Burchuladze, *Three-dimensional Problems of the Mathematical Theory of Elasticity and Thermoelasticity*. Amsterdam: North-Holland, 1979. C: pp. 20, 197, 197, 201, 202
674. H.R. Kutt, The numerical evaluation of principal value integrals by finite-part integration. *NM* **24** (1975) 205–210. C: p. 338
675. N. Kuznetsov, V. Maz'ya & B. Vainberg, *Linear Water Waves*. Cambridge: Cambridge University Press 2002. C: pp. 23, 25, 27, 56, 114, 204, 204
676. N. Kuznetsov, P. McIver & C.M. Linton, On uniqueness and trapped modes in the water-wave problem for vertical barriers. *WM* **33** (2001) 283–307. C: p. 208
677. S.O. Kvyatkovskii, Possible ways of simplifying the computational process in the T-matrix method on the basis of an analysis of surface integrals. *SPA* **33** (1987) 74–76. C: p. 262
678. Å. Kyllingstad, A low-scattering approximation for the hydrodynamic interactions of small wave-power devices. *AOR* **6** (1984) 132–139. C: p. 145
679. H.K. Kytömaa, Theory of sound propagation in suspensions: a guide to particle size and concentration characterization. *Powder Technology* **82** (1995) 115–121. C: p. 312
680. A. Lakhtakia, V.K. Varadan & V.V. Varadan, Iterative extended boundary condition method for scattering by objects of high aspect ratios. *JASA* **76** (1984) 906–912. C: p. 265
681. A. Lakhtakia, V.V. Varadan & V.K. Varadan, Scattering of ultrasonic waves by oblate spheroidal voids of high aspect ratios. *JAP* **58** (1985) 4525–4530. C: p. 269
682. H. Lamb, *Hydrodynamics*, sixth edition. Cambridge: Cambridge University Press 1932. C: pp. 12, 23, 63, 124, 134, 165
683. K.J. Langenberg, A thorough look at the nonuniqueness of the electromagnetic scattering integral equation solutions as compared to the scalar acoustic ones. *RS* **38** (2003) 8021, doi: 10.1029/2001RS002558. C: p. 191
684. T. Larsen, A survey of the theory of wire grids. *IRE Trans. on Microwave Theory & Techniques* **10** (1962) 191–201. C: p. 10
685. P. Latimer & P. Barber, Scattering by ellipsoids of revolution: a comparison of theoretical methods. *J. Colloid & Interface Sci.* **63** (1978) 310–316. C: p. 251
686. S.M. Lau & G.E. Hearn, Suppression of irregular frequency effects in fluid-structure interaction problems using a combined integral equation method. *IJNMF* **9** (1989) 763–782. C: p. 221
687. M. Lax, Multiple scattering of waves. *RMP* **23** (1951) 287–310. C: pp. 4, 5, 10
688. M. Lax, Multiple scattering of waves. II. The effective field in dense systems. *PR* **85** (1952) 621–629. C: pp. 10, 319 & §8.6.4 (p. 316)
689. M. Lax, Localized perturbations. *PR* **94** (1954) 1391–1392. C: p. 305
690. M. Lax & H. Feshbach, Absorption and scattering for impedance boundary conditions on spheres and circular cylinders. *JASA* **20** (1948) 108–124. C: pp. 124, 134
691. N.N. Lebedev, *Special Functions and their Applications*. Englewood Cliffs: Prentice-Hall, 1965. C: pp. 36, 63, 147
692. C.-H. Lee, WAMIT theory manual. Rept. 95-2, Dept. of Ocean Engineering, Massachusetts Institute of Technology, 1995. C: p. 205
693. C.-H. Lee & J.N. Newman, Computation of wave effects using the panel method. In: *Numerical Models in Fluid-Structure Interaction* (ed. S.K. Chakrabarti) pp. 211–251. Southampton: WIT Press, 2005. C: p. 205
694. C.-H. Lee, J.N. Newman & X. Zhu, An extended boundary integral equation method for the removal of irregular frequency effects. *IJNMF* **23** (1996) 637–660. C: p. 221
695. C.-H. Lee & P.D. Sclavounos, Removing the irregular frequencies from integral equations in wave-body interactions. *JFM* **207** (1989) 393–418. C: p. 219

696. C.M. Lee, H.D. Jones & R.M. Curphey, Prediction of motion and hydrodynamic loads of catamarans. *Marine Technology* **10** (1973) 392–405. C: p. 205
697. J. Lee & A.K. Mal, A volume integral equation technique for multiple scattering problems in elastodynamics. *Appl. Math. & Comp.* **67** (1995) 135–159. C: p. 202
698. J. Lee, H. Lee & A. Mal, A mixed volume and boundary integral equation technique for elastic wave field calculations in heterogeneous materials. *WM* **39** (2004) 1–19. C: p. 202
699. P.A. Lee & J.B. Pendry, Theory of the extended x-ray absorption fine structure. *PR* B **11** (1975) 2795–2811. C: p. 105
700. S.-C. Lee, Dependent scattering of an obliquely incident plane wave by a collection of parallel cylinders. *JAP* **68** (1990) 4952–4957. C: p. 129
701. A.G. Leiko & V.I. Mayatskii, Scattered field of an array of acoustically compliant elliptical cylinders. *SPA* **19** (1973) 246–249. C: p. 149
702. A.G. Leiko & B.A. Omel'chenko, Diffraction of a plane sound wave by a circular array of acoustically rigid elliptical-cylindrical scatterers. *SPA* **22** (1976) 171–173. C: p. 149
703. A.G. Leiko & B.A. Omel'chenko, Scattered field of a circular array of acoustically compliant elliptical-cylindrical scatterers. *SPA* **22** (1976) 140–143. C: p. 149
704. H. Levine, Scattering of surface waves by a submerged circular cylinder. *JMP* **6** (1965) 1231–1243. C: p. 206
705. H. Levine, *Partial Differential Equations*. Providence: American Mathematical Society, 1997. C: p. 124
706. H. Levine & J. Schwinger, On the theory of diffraction by an aperture in an infinite plane screen. I. *PR* **74** (1948) 958–974. C: p. 156
707. S. Levine & G.O. Olaofe, Scattering of electromagnetic waves by two equal spherical particles. *J. Colloid & Interface Sci.* **27** (1968) 442–457. C: p. 140
708. L. Li, J. He, Z. Liu, X. Dong & L. Carin, MLFMA analysis of scattering from multiple targets in the presence of a half-space. *IEAP* **51** (2003) 810–819. C: p. 241
709. L. Li, Z. Liu, X. Dong, J.A. Thompson & L. Carin, Scalable multilevel fast multipole method for multiple targets in the vicinity of a half-space. *IEGRS* **41** (2003) 791–802. C: p. 241
710. L.W. Li, M.S. Leong, P.S. Kooi & T.S. Yeo, Spheroidal vector wave eigenfunction expansion of dyadic Green's functions for a dielectric spheroid. *IEAP* **49** (2001) 645–659. C: p. 150
711. Y.L. Li, C.H. Liu & S.J. Franke, Three-dimensional Green's function for wave propagation in a linearly inhomogeneous medium – the exact analytic solution. *JASA* **87** (1990) 2285–2291. C: p. 15
712. C. Liang & Y.T. Lo, Scattering by two spheres. *RS* **2** (1967) 1481–1495. C: p. 140
713. S. Liapis, A method for suppressing the irregular frequencies from integral equations in water wave-structure interaction problems. *Computational Mech.* **12** (1993) 59–68. C: p. 221
714. J. Lighthill, *Waves in Fluids*. Cambridge: Cambridge University Press, 1978. C: p. 12
715. P. Lilienfeld, Gustav Mie: the person. *AO* **30** (1991) 4696–4698. C: p. 140
716. R. Lim, Multiple scattering by many bounded obstacles in a multilayered acoustic medium. *JASA* **92** (1992) 1593–1612. C: p. 282
717. R. Lim & R.H. Hackman, A formulation of multiple scattering by many bounded obstacles using a multicentered, *T* supermatrix. *JASA* **91** (1992) 613–638. C: p. 282
718. N. Limić, The exterior Neumann problem for the Helmholtz equation. *Glasnik Matematički* **16** (1981) 51–64. C: p. 244
719. T.-C. Lin, A proof for the Burton and Miller integral equation approach for the Helmholtz equation. *JMAA* **103** (1984) 565–574. C: p. 218

720. T.-C. Lin, The numerical solution of Helmholtz's equation for the exterior Dirichlet problem in three dimensions. *SINUM* **22** (1985) 670–686. C: p. 218

721. W.H. Lin & S.S. Chen, On the added mass and radiation damping of rod bundles oscillating in compressible fluids. *JSV* **74** (1981) 441–453. C: p. 129

722. W.H. Lin & A.C. Raptis, Sound scattering by a group of oscillatory cylinders. *JASA* **77** (1985) 15–28. C: p. 129

723. A.M. Linkov & S.G. Mogilevskaya, Complex hypersingular integrals and integral equations in plane elasticity. *Acta Mechanica* **105** (1994) 189–205. C: p. 339

724. C.M. Linton, Radiation and diffraction of water waves by a submerged sphere in finite depth. *OE* **18** (1991) 61–74. C: p. 145

725. C.M. Linton, The finite dock problem. *J. Appl. Math. & Phys.* (ZAMP) **52** (2001) 640–656. C: p. 205

726. C.M. Linton & D.V. Evans, The interaction of waves with arrays of vertical circular cylinders. *JFM* **215** (1990) 549–569. Corrigendum: **218** (1990) 663. C: pp. 129, 130, 132, 178

727. C.M. Linton & D.V. Evans, The interaction of waves with a row of circular cylinders. *JFM* **251** (1993) 687–708. C: pp. 9, 129

728. C.M. Linton & N.G. Kuznetsov, Non-uniqueness in two-dimensional water wave problems: numerical evidence and geometrical restrictions. *PRSA* **453** (1997) 2437–2460. C: p. 208

729. C.M. Linton & P.A. Martin, Semi-infinite arrays of isotropic point scatterers. A unified approach. *SIAP* **64** (2004) 1035–1056. C: p. 10

730. C.M. Linton & P.A. Martin, Multiple scattering by random configurations of circular cylinders: second-order corrections for the effective wavenumber. *JASA* **117** (2005) 3413–3423. C: p. 319

731. C.M. Linton & P. McIver, *Handbook of Mathematical Techniques for Wave/Structure Interactions*. Boca Raton: Chapman & Hall/CRC, 2001. C: pp. 23, 25, 55, 110, 129, 132, 144, 145, 145, 148, 203, 205, 283, 308

732. P. Linz, On the approximate computation of certain strongly singular integrals. *Computing* **35** (1985) 345–353. C: p. 338

733. B.A. Lippmann & J. Schwinger, Variational principles for scattering processes. I. *PR* **79** (1950) 469–480. C: p. 186

734. J.E. Littlewood, *Littlewood's Miscellany* (ed. B. Bollobás). Cambridge: Cambridge University Press, 1986. Revised version of *A Mathematician's Miscellany*, London: Methuen, 1953. C: pp. 67, 164

735. E. Liu & Z. Zhang; Numerical study of elastic wave scattering by cracks or inclusions using the boundary integral equation method. *J. Computational Acoustics* **9** (2001) 1039–1054. C: pp. 172, 201

736. W. Liu & R.D. Kriz, Shear waves in fiber-reinforced composites with interfacial cracks. *IJSS* **35** (1998) 1425–1449. C: p. 320

737. W. Liu & R.D. Kriz, Axial shear waves in fiber-reinforced composites with multiple interfacial layers between fiber core and matrix. *MM* **31** (1999) 117–129. C: p. 320

738. Y. Liu & S. Chen, A new form of the hypersingular boundary integral equation for 3-D acoustics and its implementation with C^0 boundary elements. *CMAME* **173** (1999) 375–386. C: p. 219

739. Y. Liu & F.J. Rizzo, A weakly singular form of the hypersingular boundary integral equation applied to 3-D acoustic wave problems. *CMAME* **96** (1992) 271–287. C: p. 219

740. Y. Liu & F.J. Rizzo, Hypersingular boundary integral equations for radiation and scattering of elastic waves in three dimensions. *CMAME* **107** (1993) 131–144. C: pp. 202, 219
741. Z. Liu, C.T. Chan, P. Sheng, A.L. Goertzen & J.H. Page, Elastic wave scattering by periodic structures of spherical objects: theory and experiment. *PR* B **62** (2000) 2446–2457. C: p. 143
742. P. Lloyd & M.V. Berry, Wave propagation through an assembly of spheres IV. Relations between different multiple scattering theories. *Proc. Physical Society* **91** (1967) 678–688. C: p. 319
743. N.A. Logan, Survey of some early studies of the scattering of plane waves by a sphere. *Proc. IEEE* **53** (1965) 773– 785. C: pp. 140, 141
744. A.T. Lonseth, Sources and applications of integral equations. *SIAM Review* **19** (1977) 241–278. C: p. 168
745. A.E.H. Love, *A Treatise on the Mathematical Theory of Elasticity*, fourth edition. Cambridge: Cambridge University Press, 1927. C: p. 141
746. W.N. Lovitt, *Linear Integral Equations*. New York: McGraw-Hill, 1924. Reprinted by Dover, New York, 1950. C: p. 160
747. C.C. Lu & W.C. Chew, Fast algorithm for solving hybrid integral equations. *IEE-H* **140** (1993) 455–460. C: p. 240
748. J.E. Luco & L. Contesse, Dynamic structure-soil-structure interaction. *BSSA* **63** (1973) 1289–1303. C: p. 129
749. A.C. Ludwig, Scattering by two and three spheres computed by the generalized multipole technique. *IEAP* **39** (1991) 703–705. C: p. 245
750. C.J. Luke & P.A. Martin, Fluid–solid interaction: acoustic scattering by a smooth elastic obstacle. *SIAP* **55** (1995) 904–922. C: p. 21
751. F. Luppé, J.-M. Conoir & H. Franklin, Multiple scattering in a trabecular bone: influence of the marrow viscosity on the effective properties. *JASA* **113** (2003) 2889–2892. C: p. 316
752. F. Luzón, F.J. Sánchez-Sesma, J.L. Rodríguez-Zúñiga, A.M. Posadas, J.M. García, J. Martín, M.D. Romacho & M. Navarro, Diffraction of *P*, *S* and Rayleigh waves by three-dimensional topographies. *GJI* **129** (1997) 571–578. C: p. 202
753. A.E. Løken, Hydrodynamic interaction between several floating bodies of arbitrary form in waves. *Proc. Hydrodynamics in Ocean Engineering*, pp. 745–779. Trondheim: Norwegian Institute of Technology, 1981. C: p. 24
754. Y. Ma, V.K. Varadan & V.V. Varadan, Application of Twersky's multiple scattering formalism to a dense suspension of elastic particles in water. *JASA* **75** (1984) 335–339. C: p. 320
755. Y. Ma, V.K. Varadan & V.V. Varadan, Comments on ultrasonic propagation in suspensions. *JASA* **87** (1990) 2779–2782. C: p. 320
756. Y. Ma, V.V. Varadan & V.K. Varadan, Multiple scattering theory for wave propagation in discrete random media. *IJES* **22** (1984) 1139–1148. C: p. 320
757. C. Macaskill & E.O. Tuck, Evaluation of the acoustic impedance of a screen. *J. Australian Math. Soc.*, Series B, **20** (1977) 46–61. C: p. 207
758. R.C. MacCamy, On the heaving motion of cylinders of shallow draft. *JSR* **5**, 3 (1961) 34–43. C: p. 205
759. R.C. MacCamy, On the scattering of water waves by a circular disk. *ARMA* **8** (1961) 120–138. C: p. 205
760. R.C. MacCamy, The motion of cylinders of shallow draft. *JSR* **7**, 3 (1964) 1–11. C: p. 205
761. R.C. MacCamy, Low frequency acoustic oscillations. *QAM* **23** (1965) 247–256. C: pp. 291, 296

762. W.D. MacMillan, *The Theory of the Potential*. New York: Dover, 1958. C: p. 352
763. D.W. Mackowski, Analysis of radiative scattering for multiple sphere configurations. *PRSA* **433** (1991) 599–614. C: pp. 94, 141, 333
764. D.W. Mackowski, Calculation of total cross sections of multiple-sphere clusters. *JOSA* A **11** (1994) 2851–2861. C: p. 141
765. R.H. MacPhie, J. Dalmas & R. Deleuil, Rotational-translational addition theorems for scalar spheroidal wave functions. *QAM* **44** (1987) 737–749. C: p. 150
766. H. Maeda, Hydrodynamical forces on a cross-section of a stationary structure. In: [102], pp. 80–90. C: p. 205
767. A.K. Mal & S.K. Bose, Dynamic elastic moduli of a suspension of imperfectly bonded spheres. *PCPS* **76** (1974) 587–600. C: pp. 143, 320
768. A.K. Mal & L. Knopoff, Elastic wave velocities in two-component systems. *J. Inst. Maths. Applics.* **3** (1967) 376–387. C: pp. 200, 316
769. F. Manar & Ch. Brouder, A fast free Green function calculation in multiple-scattering theory. *Physica* B **208–209** (1995) 79–80. C: p. 94
770. B.N. Mandal (ed.), *Mathematical Techniques for Water Waves*. Southampton: Computational Mechanics Publications, 1997. C: p. 144
771. B.N. Mandal & R. Gayen, Water-wave scattering by two symmetric circular-arc-shaped thin plates. *JEM* **44** (2002) 297–309. C: p. 208
772. K.W. Mangler, Improper integrals in theoretical aerodynamics. Aeronautical Research Council Current Paper No. 94, 1952. C: p. 158
773. G.D. Manolis, A comparative study on three boundary element method approaches to problems in elastodynamics. *IJNME* **19** (1983) 73–91. C: p. 11
774. G.D. Manolis & D.E. Beskos, *Boundary Element Methods in Elastodynamics*. London: Unwin Hyman, 1988. C: p. 203
775. P.M. Marcus, A simple derivation of the layer-KKR theory of LEED. In: [168], pp. 459–469. C: p. 9
776. L.A. Marnevskaya, Diffraction of a plane scalar wave by two spheres. *SPA* **14** (1969) 356–360. C: p. 137
777. L.A. Marnevskaya, Plane wave scattering by two acoustically-rigid spheres. *SPA* **15** (1970) 499–502. C: p. 137
778. E. Martensen, *Potentialtheorie*. Stuttgart: Teubner, 1968. C: p. 352
779. P.A. Martin, On the null-field equations for the exterior problems of acoustics. *QJMAM* **33** (1980) 385–396. C: p. 249
780. P.A. Martin, On the null-field equations for water-wave radiation problems. *JFM* **113** (1981) 315–332. C: pp. 114, 217, 251
781. P.A. Martin, Acoustic scattering and radiation problems, and the null-field method. *WM* **4** (1982) 391–408. C: p. 249
782. P.A. Martin, On the null-field equations for water-wave scattering problems. *IMAP* **33** (1984) 55–69. C: p. 251
783. P.A. Martin, Multiple scattering of surface water waves. In: *Classical Scattering* (ed. G.F. Roach) pp. 110–123. Nantwich: Shiva Publishing, 1984. C: p. 267
784. P.A. Martin, Multiple scattering of surface water waves and the null-field method. *Proc. 15th Symposium on Naval Hydrodynamics*, pp. 119–132. Washington, DC: National Academy Press, 1984. C: pp. 274, 281, 308
785. P.A. Martin, Integral-equation methods for multiple-scattering problems I. Acoustics. *QJMAM* **38** (1985) 105–118. C: p. 217
786. P.A. Martin, Integral-equation methods for multiple-scattering problems II. Water waves. *QJMAM* **38** (1985) 119–133. C: pp. 61, 217

787. P.A. Martin, On the T-matrix for water-wave scattering problems. *WM* **7** (1985) 177–193. C: pp. 25, 274
788. P.A. Martin, On the scattering of elastic waves by an elastic inclusion in two dimensions. *QJMAM* **43** (1990) 275–291. C: p. 202
789. P.A. Martin, End-point behaviour of solutions to hypersingular integral equations. *PRSA* **432** (1991) 301–320. C: p. 339
790. P.A. Martin, Boundary integral equations for scattering of elastic waves by elastic inclusions with thin interface layers. *J. Nondestructive Evaluation* **11** (1992) 167–174. C: p. 21
791. P.A. Martin, Exact solution of a simple hypersingular integral equation. *J. Integ. Eqns & Applics.* **4** (1992) 197–204. Addendum: **5** (1993) 297. C: pp. 339, 339
792. P.A. Martin, Multiple scattering and the Rehr–Albers–Fritzsche formula for the propagator matrix. *J. Phys. A: Math. Gen.* **31** (1998) 8923–8932. C: pp. 102, 104
793. P.A. Martin, Acoustic scattering by inhomogeneous spheres. *JASA* **111** (2002) 2013–2018. C: pp. 14, 135
794. P.A. Martin, Multiple scattering and modified Green's functions. *JMAA* **275** (2002) 642–656. C: p. 217
795. P.A. Martin, On connections between boundary integral equations and T-matrix methods. *EABE* **27** (2003) 771–777. C: pp. 267, 271
796. P.A. Martin, Acoustic scattering by inhomogeneous obstacles. *SIAP* **64** (2003) 297–308. C: pp. 181, 182
797. P.A. Martin & A.G. Dixon, The scattering of regular surface waves by a fixed, half-immersed, circular cylinder. *AOR* **5** (1983) 13–23. C: p. 144
798. P.A. Martin & L. Farina, Radiation of water waves by a heaving submerged horizontal disc. *JFM* **337** (1997) 365–379. C: p. 205
799. P.A. Martin & P. Ola, Boundary integral equations for the scattering of electromagnetic waves by a homogeneous dielectric obstacle. *PRSE* **123A** (1993) 185–208. C: pp. 18, 188, 195, 195
800. P.A. Martin & F.J. Rizzo, On boundary integral equations for crack problems. *PRSA* **421** (1989) 341–355. C: pp. 161, 208, 335
801. P.A. Martin & F.J. Rizzo, Partitioning, boundary integral equations, and exact Green's functions. *IJNME* **38** (1995) 3483–3495. C: p. 226
802. P.A. Martin & F.J. Rizzo, Hypersingular integrals: how smooth must the density be? *IJNME* **39** (1996) 687–704. C: p. 338
803. P.A. Martin, F.J. Rizzo & T.A. Cruse, Smoothness-relaxation strategies for singular and hypersingular integral equations. *IJNME* **42** (1998) 885–906. C: p. 338
804. P.A. Martin & G.R. Wickham, Diffraction of elastic waves by a penny-shaped crack: analytical and numerical results. *PRSA* **390** (1983) 91–129. C: p. 207
805. E. Marx, Single integral equation for wave scattering. *JMP* **23** (1982) 1057–1065. C: pp. 180, 195
806. E. Marx, Integral equation for scattering by a dielectric. *IEAP* **32** (1984) 166–172. C: pp. 180, 195
807. K. Maslov, V.K. Kinra & B.K. Henderson, Lattice resonances of a planar array of spherical inclusions: an experimental study. *MM* **31** (1999) 175–186. C: p. 143
808. K. Maslov, V.K. Kinra & B.K. Henderson, Elastodynamic response of a coplanar periodic layer of elastic spherical inclusions. *MM* **32** (2000) 785–795. C: p. 143
809. J.C. Mason, Chebyshev polynomials of the second, third and fourth kinds in approximation, indefinite integration, and integral transforms. *JCAM* **49** (1993) 169–178. C: p. 208

810. G. Masters & K. Richards-Dinger, On the efficient calculation of ordinary and generalized spherical harmonics. *GJI* **135** (1998) 307–309. C: p. 65
811. N.C. Mathur & K.C. Yeh, Multiple scattering of electromagnetic waves by random scatterers of finite size. *JMP* **5** (1964) 1619–1628. C: p. 319
812. T. Matsui & T. Tamaki, Hydrodynamic interaction between groups of vertical axisymmetric bodies floating in waves. *Proc. Hydrodynamics in Ocean Engineering*, pp. 817–836. Trondheim: Norwegian Institute of Technology, 1981. C: p. 205
813. A.W. Maue, Zur Formulierung eines allgemeinen Beugungsproblems durch eine Integralgleichung. *Zeitschrift für Physik* **126** (1949) 601–618. Translated as: Toward formulation of a general diffraction problem via an integral equation. In: *Computational Electromagnetics: Frequency-Domain Method of Moments* (eds. E.K. Miller, L. Medgyesi-Mitschang & E.H. Newman) pp. 7–14. New York: IEEE Press, 1992. C: pp. 157, 177, 195
814. J.R. Mautz, A stable integral equation for electromagnetic scattering from homogeneous dielectric bodies. *IEAP* **37** (1989) 1070–1071. C: p. 195
815. S.A. Mavrakos & P. Koumoutsakos, Hydrodynamic interaction among vertical axisymmetric bodies restrained in waves. *AOR* **9** (1987) 128–140. C: p. 148
816. S.A. Mavrakos & P. McIver, Comparison of methods for computing hydrodynamic characteristics of arrays of wave power devices. *AOR* **19** (1997) 283–291. C: p. 23
817. D. Maystre & P. Vincent, Diffraction d'une onde electromagnetique plane par un objet cylindrique non infiniment conducteur de section arbitraire. *OC* **5** (1972) 327–330. C: p. 180
818. V. Maz'ya & T. Shaposhnikova, *Jacques Hadamard, a Universal Mathematician*. Providence: American Mathematical Society, 1998. C: pp. 335, 336
819. R.J. McBride & D.W. Kraft, Scattering of a transverse elastic wave by an elastic sphere in a solid medium. *JAP* **43** (1972) 4853–4861. C: p. 142
820. D.J. McClements, Principles of ultrasonic droplet size determination in emulsions. *Langmuir* **12** (1996) 3454– 3461. C: p. 312
821. D.J. McClements, P. Fairley & M.J.W. Povey, Comparison of effective medium and multiple-scattering theories of predicting the ultrasonic properties of dispersions. *JASA* **87** (1990) 2244–2246. C: p. 322
822. D.J. McClements & M.J.W. Povey, Scattering of ultrasound by emulsions. *JPDAP* **22** (1989) 38–47. C: p. 319
823. J.M. McCormick & M.L. Baron, Sound radiation from submerged cylindrical shells of finite length. *J. Engineering for Industry* **87** (1965) 393–405. C: p. 172
824. M.E. McCormick, *Ocean Wave Energy Conversion*. New York: Wiley, 1981. C: p. 145
825. W.G. McDougal, A.N. Williams & K. Furukawa, Multiple-pit breakwaters. *JWPCOE* **122** (1996) 27–33. C: p. 180
826. M. McIver, An example of non-uniqueness in the two-dimensional linear water wave problem. *JFM* **315** (1996) 257–266. C: p. 27
827. P. McIver, Wave forces on arrays of floating bodies. *JEM* **18** (1984) 273–285. C: p. 312
828. P. McIver, Wave forces on adjacent floating bridges. *AOR* **8** (1986) 67–75. C: p. 148
829. P. McIver, Mean drift forces on arrays of bodies due to incident long waves. *JFM* **185** (1987) 469–482. C: p. 297
830. P. McIver, The scattering of long water waves by a group of submerged, horizontal cylinders. *QJMAM* **43** (1990) 499–515. C: p. 297
831. P. McIver, Low frequency asymptotics of hydrodynamic forces on fixed and floating structures. In: *Ocean Waves Engineering* (ed. M. Rahman) pp. 1–49. Southampton: Computational Mechanics Publications, 1994. C: p. 296
832. P. McIver, Wave interaction with arrays of structures. *AOR* **24** (2002) 121–126. C: p. 23

833. P. McIver & D.V. Evans, Approximation of wave forces on cylinder arrays. *AOR* **6** (1984) 101–107. C: p. 312
834. N.W. McLachlan, Historical note on Heaviside's operational method. *Mathematical Gazette* **22** (1938) 255–260. C: p. 62
835. N.W. McLachlan, *Theory and Application of Mathieu Functions*. New York: Dover, 1964. C: pp. 48, 48, 149
836. W. McLean, *Strongly Elliptic Systems and Boundary Integral Equations*. Cambridge: Cambridge University Press, 2000. C: pp. 11, 158, 159, 168, 168, 172
837. K. McTaggart, D. Cumming, C.C. Hsiung & L. Li, Seakeeping of two ships in close proximity. *OE* **30** (2003) 1051–1063. C: p. 205
838. L.N. Medgyesi-Mitschang, J.M. Putnam & M.B. Gedera, Generalized method of moments for three-dimensional penetrable scatterers. *JOSA* A **11** (1994) 1383–1398. C: p. 195
839. W.C. Meecham, Variational method for the calculation of the distribution of energy reflected from a periodic surface. I. *JAP* **27** (1956) 361–367. C: p. 263
840. E. Mehlum, A circular cylinder in waves. *AOR* **2** (1980) 171–177. C: p. 145
841. C.C. Mei, Numerical methods in water-wave diffraction and radiation. *ARFM* **10** (1978) 393–416. C: pp. 206, 206
842. C.C. Mei, *The Applied Dynamics of Ocean Surface Waves*, 2nd printing with corrections. Singapore: World Scientific, 1994. C: pp. 23, 25, 56, 144, 147
843. J. Mei, Z. Liu, J. Shi & D. Tian, Theory for elastic wave scattering by a two-dimensional periodical array of cylinders: an ideal approach for band-structure calculations. *PR* B **67** (2003) 245107. C: p. 142
844. K.K. Mei & J.G. Van Bladel, Scattering by perfectly-conducting rectangular cylinders. *IEAP* **11** (1963) 185–192. C: p. 178
845. J. Meixner & F.W. Schäfke, *Mathieusche Funktionen und Sphäroidfunktionen*. Berlin: Springer, 1954. C: pp. 48, 49, 53, 150
846. A. Messiah, *Quantum Mechanics*, vol. II. Amsterdam: North- Holland, 1962. C: pp. 63, 84, 326 & Table B.1 (p. 327)
847. W.L. Meyer, W.A. Bell, M.P. Stallybrass & B.T. Zinn, Prediction of the sound field radiated from axisymmetric surfaces. *JASA* **65** (1979) 631–638. C: p. 219
848. W.L. Meyer, W.A. Bell, B.T. Zinn & M.P. Stallybrass, Boundary integral solutions of three dimensional acoustic radiation problems. *JSV* **59** (1978) 245–262. C: pp. 219, 351
849. G. Mie, Beiträge zur Optik trüber Medien, speziell kolloidaler Metallösungen. *Annalen der Physik*, 4 Folge, **25** (1908) 377–445. C: p. 140
850. S.G. Mikhlin, *Integral Equations*, second edition. Oxford: Pergamon Press, 1964. C: p. 171
851. S.G. Mikhlin, *Variational Methods in Mathematical Physics*. Oxford: Pergamon Press, 1964. C: p. 263
852. S.G. Mikhlin, *Multidimensional Singular Integrals and Integral Equations*. Oxford: Pergamon Press, 1965. C: pp. 345, 349
853. S.G. Mikhlin & S. Prössdorf, *Singular Integral Operators*. Berlin: Springer, 1986. C: pp. 162, 201, 334, 337, 345, 346, 349, 354
854. J.W. Miles, Motion of a rigid cylinder due to a plane elastic wave. *JASA* **32** (1960) 1656–1659. C: p. 141
855. J.H. Milgram & J.E. Halkyard, Wave forces on large objects in the sea. *JSR* **15** (1971) 115–124. C: pp. 205, 206
856. R.F. Millar, The scattering of a plane wave by a row of small cylinders. *CJP* **38** (1960) 272–289. C: pp. 177, 223

857. R.F. Millar, Plane wave spectra in grating theory I. Scattering by a finite number of bodies. *CJP* **41** (1963) 2106–2134. C: pp. 234, 282
858. R.F. Millar, Plane wave spectra in grating theory II. Scattering by an infinite grating of identical cylinders. *CJP* **41** (1963) 2135–2154. C: p. 235
859. R.F. Millar, Plane wave spectra in grating theory III. Scattering by a semiinfinite grating of identical cylinders. *CJP* **42** (1964) 1149–1184. C: pp. 10, 235
860. R.F. Millar, Plane wave spectra in grating theory IV. Scattering by a finite grating of identical cylinders. *CJP* **42** (1964) 2395–2410. C: p. 235
861. R.F. Millar, Plane wave spectra in grating theory V. Scattering by a semi-infinite grating of isotropic scatterers. *CJP* **44** (1966) 2839–2874. C: p. 10
862. R.F. Millar, The Rayleigh hypothesis and a related least-squares solution to scattering problems for periodic surfaces and other scatterers. *RS* **8** (1973) 785–796. C: pp. 244, 249, 257, 263
863. R.F. Millar, On the completeness of sets of solutions to the Helmholtz equation. *IMAP* **30** (1983) 27–37. C: p. 257
864. S. Mingde & P. Yu, Shallow water diffraction around multiple large cylinders. *AOR* **9** (1987) 31–36. C: p. 129
865. M.I. Mishchenko, Coherent backscattering by two-sphere clusters. *OL* **21** (1996) 623–625. C: p. 141
866. M.I. Mishchenko, J.W. Hovenier & L.D. Travis (eds.), *Light Scattering by Nonspherical Particles*. San Diego: Academic Press, 2000. C: pp. 2, 2, 140, 150, 242, 265, 269, 273, 282
867. M.I. Mishchenko & D.W. Mackowski, Electromagnetic scattering by randomly oriented bispheres: comparison of theory and experiment and benchmark calculations. *JQSRT* **55** (1996) 683–694. C: p. 281
868. M.I. Mishchenko, D.W. Mackowski & L.D. Travis, Scattering of light by bispheres with touching and separated components. *AO* **34** (1995) 4589–4599. C: pp. 2, 141, 281
869. M.I. Mishchenko & L.D. Travis, T-matrix computations of light scattering by large spheroidal particles. *OC* **109** (1994) 16–21. C: pp. 265, 273
870. M.I. Mishchenko, L.D. Travis & A.A. Lacis, *Scattering, Absorption, and Emission of Light by Small Particles*. Cambridge: Cambridge University Press, 2002. C: pp. 10, 141, 150, 195, 242, 251, 265, 273, 282, 326
871. M.I. Mishchenko, L.D. Travis & D.W. Mackowski, T-matrix computations of light scattering by nonspherical particles: a review. *JQSRT* **55** (1996) 535–575. C: pp. 242, 265, 273, 282
872. M.I. Mishchenko & G. Videen, Single-expansion EBCM computations for osculating spheres. *JQSRT* **63** (1999) 231– 236. C: p. 251
873. M.I. Mishchenko, G. Videen, V.A. Babenko, N.G. Khlebtsov & T. Wriedt, T-matrix theory of electromagnetic scattering by particles and its applications: a comprehensive reference database. *JQSRT* **88** (2004) 357–406. C: pp. 242, 273
874. D. Mitrea, M. Mitrea & J. Pipher, Vector potential theory on nonsmooth domains in R^3 and applications to electromagnetic scattering. *J. Fourier Analysis & Applications* **3** (1997) 131–192. C: p. 195
875. M. Mitrea, The method of layer potentials in electromagnetic scattering theory on non-smooth domains. *Duke Mathematical J.* **77** (1995) 111–133. C: p. 195
876. K.M. Mitzner, Acoustic scattering from an interface between media of greatly different density. *JMP* **7** (1966) 2053–2060. C: p. 351
877. F.R. Molino & P.C. Sabatier, Elastic waves in discontinuous media: three-dimensional scattering. *JMP* **35** (1994) 4594– 4636. C: p. 202

878. G. Monegato, Numerical evaluation of hypersingular integrals. *JCAM* **50** (1994) 9–31. C: pp. 335, 338, 345, 346
879. P. Monk, *Finite Element Methods for Maxwell's Equations*. Oxford: Oxford University Press, 2003. C: p. 11
880. C.L. Morfey, J. Garcia-Bonito & A.T. Peplow, Acoustic radiation from a pair of spheres. *Acustica – acta acustica* **84** (1998) 701–711. C: pp. 137, 178
881. N. Morita, Another method of extending the boundary condition for the problem of scattering by dielectric cylinders. *IEAP* **27** (1979) 97–99. C: p. 251
882. N. Morita, Resonant solutions involved in the integral equations approach to scattering from conducting and dielectric cylinders. *IEAP* **27** (1979) 869–871. C: p. 180
883. J.A. Morrison, Integral equations for electromagnetic scattering by perfect conductors with two-dimensional geometry. *Bell System Technical J.* **58** (1979) 409–425. C: p. 161
884. P.M. Morse & H. Feshbach, *Methods of Theoretical Physics*. New York: McGraw-Hill, 1953. C: pp. 36, 46, 48, 49, 63, 107, 124, 134, 140, 150, 253, 296, 316
885. P.M. Morse & K.U. Ingard, *Theoretical Acoustics*. New York: McGraw-Hill, 1968. Reprinted by Princeton University Press, Princeton 1986. C: pp. 12, 14, 124, 134, 181, 185, 187
886. P.M. Morse & P.J. Rubenstein, The diffraction of waves by ribbons and by slits. *PR* **54** (1938) 895–898. C: pp. 48, 149
887. M. Moshinsky, Transformation brackets for harmonic oscillator functions. *Nuclear Physics* **13** (1959) 104–116. C: p. 89
888. A.B. Movchan, N.V. Movchan & C.G. Poulton, *Asymptotic Models of Fields in Dilute and Densely Packed Composites*. London: Imperial College Press, 2002. C: pp. 7, 9
889. A. Mugnai & W.J. Wiscombe, Scattering from nonspherical Chebyshev particles. 1: cross sections, single-scattering albedo, asymmetry factor, and backscattering fraction. *AO* **25** (1986) 1235–1244. C: p. 251
890. A. Mugnai & W.J. Wiscombe, Scattering from nonspherical Chebyshev particles. 3: variability in angular scattering patterns. *AO* **28** (1989) 3061–3073. C: p. 251
891. S. Mukherjee, CPV and HFP integrals and their applications in the boundary element method. *IJSS* **37** (2000) 6623–6634. C: p. 208
892. C. Müller, *Spherical Harmonics* (*Lect. Notes in Math.* vol. 17). Berlin: Springer, 1966. C: p. 72
893. C. Müller, *Analysis of Spherical Symmetries in Euclidean Spaces*. New York: Springer, 1998. C: pp. 34, 72, 82
894. C. Müller & H. Kersten, Zwei Klassen vollständiger Funktionensysteme zur Behandlung der Randwertaufgaben der Schwingungsgleichung $\triangle U + k^2 U = 0$. *MMAS* **2** (1980) 48–67. C: pp. 254, 257
895. M. Murai, H. Kagemoto & M. Fujino, On the hydroelastic responses of a very large floating structure in waves. *J. Marine Science & Technology* **4** (1999) 123–153. C: p. 283
896. W.D. Murphy, V. Rokhlin & M.S. Vassiliou, Solving electro magnetic scattering problems at resonance frequencies. *JAP* **67** (1990) 6061–6065. C: pp. 210, 240
897. F.H. Murray, Conductors in an electromagnetic field ($E^0 e^{pt}$, $H^0 e^{pt}$). *Amer. J. Math.* **53** (1931) 275–288. C: p. 195
898. N.I. Muskhelishvili, *Singular Integral Equations*. Groningen: Noordhoff, 1953. C: pp. 334, 334
899. M. Nagase, Diffraction of elastic waves by a spherical surface. *J. Physical Soc. of Japan* **11** (1956) 279–301. C: p. 142
900. D. Natroshvili, Boundary integral equation method in the steady state oscillation problems for anisotropic bodies. *MMAS* **20** (1997) 95–119. C: p. 202

901. G. Neave, A uniquely solvable integral equation for the exterior electromagnetic scattering problem. *QJMAM* **40** (1987) 57–70. C: p. 217
902. J.C. Nedelec, Integral equations with non integrable kernels. *Integral Equations & Operator Theory* **5** (1982) 562–572. C: pp. 158, 172
903. J.-C. Nédélec, *Acoustic and Electromagnetic Equations: Integral Representations for Harmonic Problems*. New York: Springer, 2001. C: pp. 11, 63, 65, 158, 172, 218
904. R. New & T.J. Eisler, Acoustic radiation from multiple spheres. *JSV* **22** (1972) 1–17. C: p. 137
905. J.N. Newman, Propagation of water waves past long two-dimensional obstacles. *JFM* **23** (1965) 23–29. C: p. 308
906. J.N. Newman, Diffraction of water waves. *AMR* **25** (1972) 1–7. C: p. 24
907. J.N. Newman, Water wave singularities in a prolate spheroidal coordinate system. *JSR* **16** (1972) 41–46. C: p. 110
908. J.N. Newman, Three-dimensional wave interactions with ships and platforms. In: *International Workshop on Ship and Platform Motions* (ed. R.W. Yeung) pp. 418–442. Berkeley: University of California, 1983. C: pp. 143, 205
909. J.N. Newman, Wave effects on multiple bodies. In: *Hydrodynamics in Ship and Ocean Engineering* (ed. M. Kashiwagi) pp. 3–26. Kyushu: Research Institute for Applied Mechanics, Kyushu University, 2001. C: p. 23
910. J.N. Newman & C.-H. Lee, Boundary-element methods in offshore structure analysis. *J. Offshore Mechanics & Arctic Engng.* **124** (2002) 81–89. C: p. 205
911. R.G. Newton, *Scattering Theory of Waves and Particles*, second edition. New York: Springer, 1982. C: pp. 15, 16, 186
912. G.A. Niklasson & C.G. Granqvist, Optical properties and solar selectivity of coevaporated Co-Al_2O_3 composite films. *JAP* **55** (1984) 3382–3410. C: p. 321
913. N. Nishimura, Fast multipole accelerated boundary integral equation methods. *AMR* **55** (2002) 299–324. C: pp. 235, 238, 241
914. Y. Niu & M. Dravinski, Direct 3D BEM for scattering of elastic waves in a homogeneous anisotropic half-space. *WM* **38** (2003) 165–175. C: p. 202
915. Y. Niu & M. Dravinski, Three-dimensional BEM for scattering of elastic waves in general anisotropic media. *IJNME* **58** (2003) 979–998. C: p. 202
916. B. Noble, Integral equation perturbation methods in low-frequency diffraction. In: *Electromagnetic Waves* (ed. R.E. Langer) pp. 323–360. Madison: University of Wisconsin Press, 1962. C: pp. 177, 293, 296
917. N. Nordenstrøm, O. Faltinsen & B. Pedersen, Prediction of wave-induced motions and loads for catamarans. *Proc. Third Offshore Technology Conference*, vol. II, Paper OTC 1418, 1971, pp. 13–58. C: p. 205
918. A.N. Norris, Scattering of elastic waves by spherical inclusions with applications to low frequency wave propagation in composites. *IJES* **24** (1986) 1271–1282. C: p. 142
919. R. Nozawa, Bipolar expansion of screened Coulomb potentials, Helmholtz' solid harmonics, and their addition theorems. *JMP* **7** (1966) 1841–1860. C: pp. 71, 90
920. M. Ochmann, The source simulation technique for acoustic radiation problems. *Acustica* **81** (1995) 512–527. C: p. 244
921. R.J. O'Connell & B. Budiansky, Seismic velocities in dry and saturated cracked solids. *JGR* **79** (1974) 5412–5426. C: p. 322
922. R.J. O'Connell & B. Budiansky, Reply [to reference [144]]. *JGR* **81** (1976) 2577–2578. C: p. 322
923. J.A. Ogilvy, *Theory of Wave Scattering from Random Rough Surfaces*. Bristol: Adam Hilger, 1991. C: p. 11

924. T. Oguchi, Scattering from hydrometeors: a survey. *RS* **16** (1981) 691–730. C: p. 10
925. M. Ohkusu, On the heaving motion of two circular cylinders on the surface of a fluid. *Reports of Research Institute for Applied Mechanics*, Kyushu University **17**, No. 58 (1969) 167–185. C: p. 146
926. M. Ohkusu, On the motion of multihull ships in waves (I). *Reports of Research Institute for Applied Mechanics*, Kyushu University **18**, No. 60 (1970) 33–60. C: p. 146
927. M. Ohkusu, Hydrodynamic forces on multiple cylinders in waves. In: [102], pp. 107–112. C: pp. 4, 148
928. S. Ohmatsu, On the irregular frequencies in the theory of oscillating bodies in a free surface. *Papers of Ship Research Institute*, Tokyo, No. 48, 1975. C: p. 221
929. M.A. Oien & Y.-H. Pao, Scattering of compressional waves by a rigid spheroidal inclusion. *JAM* **40** (1973) 1073–1077. C: p. 150
930. N. Okamoto, Electromagnetic scattering by many gyrotropic cylinders of different kinds. *IEAP* **22** (1974) 701–707. C: p. 280
931. N. Okamoto, Scattering of obliquely incident plane waves from a finite periodic structure of ferrite cylinders. *IEAP* **27** (1979) 317–323. C: p. 280
932. G.O. Olaofe, Scattering by two cylinders. *RS* **5** (1970) 1351–1360. C: p. 129
933. G.O. Olaofe, Scattering by an arbitrary configuration of parallel circular cylinders. *JOSA* **60** (1970) 1233–1236. C: p. 129
934. G.O. Olaofe, Scattering cross section for two cylinders. *IEAP* **18** (1970) 823–825. C: p. 129
935. M. O'Leary, Radiation and scattering of surface waves by a group of submerged, horizontal, circular cylinders. *AOR* **7** (1985) 51–57. C: p. 146
936. P. Olsson, The rigid movable inclusion in elastostatics and elastodynamics. *WM* **7** (1985) 421–445. C: pp. 21, 251
937. P. Olsson, S.K. Datta & A. Bostrom, Elastodynamic scattering from inclusions surrounded by thin interface layers. *JAM* **57** (1990) 672–676. C: p. 251
938. D.F. Paget, The numerical evaluation of Hadamard finite-part integrals. *NM* **36** (1981) 447–453. C: p. 338
939. L. Pan, F. Rizzo & P.A. Martin, Some efficient boundary integral strategies for time-harmonic wave problems in an elastic halfspace. *CMAME* **164** (1998) 207–221. C: p. 202
940. W.K.H. Panofsky & M. Phillips, *Classical Electricity and Magnetism*. Reading: Addison-Wesley, 1955. C: p. 140
941. Y.-H. Pao, Betti's identity and transition matrix for elastic waves. *JASA* **64** (1978) 302–310. C: p. 248
942. Y.-H. Pao, The transition matrix for the scattering of acoustic waves and for elastic waves. In: *Modern Problems in Elastic Wave Propagation* (ed. J. Miklowitz & J.D. Achenbach) pp. 123–144. New York: Wiley, 1978. C: p. 248
943. Y.-H. Pao & C.C. Mow, Scattering of plane compressional waves by a spherical obstacle. *JAP* **34** (1963) 493–499. C: p. 142
944. Y.-H. Pao & C.-C. Mow, *Diffraction of Elastic Waves and Dynamic Stress Concentrations*. New York: Crane, Russak & Co., 1973. C: pp. 124, 141, 142, 149
945. Y.-H. Pao & V. Varatharajulu, Huygens' principle, radiation conditions, and integral formulas for the scattering of elastic waves. *JASA* **59** (1976) 1361–1371. C: pp. 199, 201
946. C.H. Papas, Diffraction by a cylindrical obstacle. *JAP* **21** (1950) 318–325. C: p. 177
947. N.F. Parsons & P.A. Martin, Trapping of water waves by submerged plates using hypersingular integral equations. *JFM* **284** (1995) 359–375. C: p. 208
948. V.Z. Parton & P.I. Perlin, *Integral Equations in Elasticity*. Moscow: Mir, 1982. C: pp. 196, 201

949. D.N. Pattanayak & E. Wolf, Scattering states and bound states as solutions of the Schrödinger equation with nonlocal boundary conditions. *PR* D **13** (1976) 913–923. C: p. 186
950. C. Patterson & M.A. Sheikh, A regular boundary element method for fluid flow. *IJNMF* **2** (1982) 239–251. C: p. 244
951. J. Pearce & D.M. Mittleman, Scale model experimentation: using terahertz pulses to study light scattering. *Physics in Medicine & Biology* **47** (2002) 3823–3830. C: p. 320
952. J. Pearce & D.M. Mittleman, Using terahertz pulses to study light scattering. *Physica* B **338** (2003) 92–96. C: p. 320
953. D.B. Pearson, *Quantum Scattering and Spectral Theory*. London: Academic Press, 1988. C: p. 186
954. R. Peierls, *Surprises in Theoretical Physics*. Princeton: Princeton University Press, 1979. C: pp. 89, 105
955. A.Y. Perelman, Scattering by particles with radially variable refractive indices. *AO* **35** (1996) 5452–5460. C: p. 140
956. E. Perrey-Debain, J. Trevelyan & P. Bettess, Plane wave interpolation in direct collocation boundary element method for radiation and wave scattering: numerical aspects and applications. *JSV* **261** (2003) 839–858. C: p. 178
957. E. Perrey-Debain, J. Trevelyan & P. Bettess, Wave boundary elements: a theoretical overview presenting applications in scattering of short waves. *EABE* **28** (2004) 131–141. C: p. 178
958. A.F. Peterson, The "interior resonance" problem associated with surface integral equations of electromagnetics: numerical consequences and a survey of remedies. *Electromagnetics* **10** (1990) 293–312. C: p. 209
959. A.F. Peterson, S.L. Ray & R. Mittra, *Computational Methods for Electromagnetics*. New York: IEEE Press, 1998. C: pp. 195, 209
960. B. Peterson, Numerical computations of acoustic scattering from two spheres. Report T.M.F. 73-2, Institute of Theoretical Physics, Chalmers University of Technology, Göteborg, Sweden, 1973. C: p. 280
961. B. Peterson, Multiple scattering of waves by an arbitrary lattice. *PR* A **16** (1977) 1363–1370. C: p. 280
962. B. Peterson & S. Ström, *T* matrix for electromagnetic scattering from an arbitrary number of scatterers and representations of E(3). *PR* D **8** (1973) 3661–3678. C: pp. 242, 281
963. B. Peterson & S. Ström, Matrix formulation of acoustic scattering from an arbitrary number of scatterers. *JASA* **56** (1974) 771–780. C: pp. 242, 280, 280, 282 & §7.10.1 (p. 276)
964. B. Peterson & S. Ström, *T*-matrix formulation of electromagnetic scattering from multilayered scatterers. *PR* D **10** (1974) 2670–2684. C: pp. 242, 273
965. B. Peterson & S. Ström, Matrix formulation of acoustic scattering from multilayered scatterers. *JASA* **57** (1975) 2–13. C: pp. 242, 273
966. B. Peterson, V.K. Varadan & V.V. Varadan, Scattering of elastic waves by a fluid inclusion. *JASA* **73** (1983) 1487–1493. C: p. 274
967. B. Peterson, V.V. Varadan & V.K. Varadan, On the multiple scattering of waves from obstacles with solid–fluid interfaces. In: [1248], pp. 225–252. C: p. 281
968. B. Peterson, V.V. Varadan & V.K. Varadan, Scattering of acoustic waves by layered elastic and viscoelastic obstacles in water. *JASA* **68** (1980) 673–685. C: p. 274
969. B.A. Peterson, V.V. Varadan & V.K. Varadan, *T*-matrix approach to study the vibration frequencies of elastic bodies in fluids. *JASA* **74** (1983) 1051–1056. C: p. 274

970. B.Å. Peterson, V.V. Varadan & V.K. Varadan, Scattering of acoustic waves by elastic and viscoelastic obstacles immersed in a fluid. *WM* **2** (1980) 23–38. C: p. 274
971. D.D. Phanord, Elastic scattering and energy theorems for a doubly periodic planar array of elastic obstacles. *PR* E **48** (1993) 555–562. C: p. 235
972. D.D. Phanord & N.E. Berger, Multiple scattering of elastic waves by a distribution of identical spheres. *JMP* **31** (1990) 2619–2625. C: pp. 235, 320
973. D.D. Phanord & N.E. Berger, Reciprocity relations and energy theorems for multiple scattering of elastic waves. *JMP* **31** (1990) 2626–2632. C: p. 235
974. A.D. Pierce, *Acoustics*. New York: Acoustical Society of America, 1989. C: pp. 12, 13, 181
975. A.D. Pierce, Wave equation for sound in fluids with unsteady inhomogeneous flow. *JASA* **87** (1990) 2292–2299. C: p. 14
976. A.D. Pierce, Variational formulations in acoustic radiation and scattering. In: *Physical Acoustics*, vol. 22 (eds. A.D. Pierce & R.N. Thurston) pp. 195–371. San Diego: Academic Press, 1993, pp. 195–371. C: p. 351
977. R. Pike & P. Sabatier, *Scattering*, vol. 1. San Diego: Academic Press, 2002. C: p. 10
978. A.C. Pipkin, *A Course on Integral Equations*. New York: Springer, 1991. C: p. 168
979. A.J. Poggio & E.K. Miller, Integral equation solutions of three-dimensional scattering problems. In: *Computer Techniques for Electromagnetics* (ed. R. Mittra) pp. 159–264. Oxford: Pergamon Press, 1973. C: pp. 189, 193, 195, 219
980. T. Pointer, E. Liu & J.A. Hudson, Numerical modelling of seismic waves scattered by hydrofractures: application of the indirect boundary element method. *GJI* **135** (1998) 289–303. C: p. 201
981. D. Porter & D.S.G. Stirling, *Integral Equations*. Cambridge: Cambridge University Press, 1990. C: p. 168
982. R. Porter & D.V. Evans, Complementary approximations to wave scattering by vertical barriers. *JFM* **294** (1995) 155–180. C: p. 148
983. R. Porter & D.V. Evans, The trapping of surface waves by multiple submerged horizontal cylinders. *JEM* **34** (1998) 417–433. C: p. 146
984. C.G. Poulton, A.B. Movchan, R.C. McPhedran, N.A. Nicorovici & Y.A. Antipov, Eigenvalue problems for doubly periodic elastic structures and phononic band gaps. *PRSA* **456** (2000) 2543–2559. C: pp. 9, 142
985. M.J.W. Povey, *Ultrasonic Techniques for Fluids Characterization*. San Diego: Academic Press, 1997. C: pp. 143, 312, 319
986. W. Prager, Die Druckverteilung an Körpern in ebener Potentialströmung. *Physikalische Zeitschrift* **29** (1928) 865–869. C: p. 177
987. Ch. Provatidis & N. Zafiropoulos, On the 'interior Helmholtz integral equation formulation' in sound radiation problems. *EABE* **26** (2002) 29–40. C: p. 244
988. I.E. Psarobas, N. Stefanou & A. Modinos, Scattering of elastic waves by periodic arrays of spherical bodies. *PR* B **62** (2000) 278–291. C: p. 143
989. J. Qian & D.E. Beskos, Harmonic wave response of two 3-D rigid surface foundations. *SDEE* **15** (1996) 95–110. C: p. 202
990. M. Quinten & U. Kreibig, Absorption and elastic scattering of light by particle aggregates. *AO* **32** (1993) 6173–6182. C: p. 141
991. A. Quirantes, F. Arroyo & J. Quirantes-Ros, Multiple light scattering by spherical particle systems and its dependence on concentration: a T-matrix study. *J. Colloid & Interface Sci.* **240** (2001) 78–82. C: p. 281
992. A. Quirantes & Á. Delgado, Cross section calculations of randomly oriented bispheres in the small particle regime. *JQSRT* **78** (2003) 179–186. C: p. 281

993. G. Racah, Theory of complex spectra. II. *PR* **62** (1942) 438–462. C: p. 326

994. R.P. Radlinski, Comments on "Multi-obstacle acoustic scattering in the simple source formulation with accuracy checks" [J. Acoust. Soc. Am. 66, 1536–1541 (1979)] (reference [1131]). *JASA* **68** (1980) 989–992. C: p. 172

995. R.P. Radlinski & T.J. Meyers, Radiation patterns and radiation impedances of a pulsating cylinder surrounded by a circular cage of parallel cylindrical rods. *JASA* **56** (1974) 842–848. C: pp. 129, 130

996. H.A. Ragheb & M. Hamid, Scattering by N parallel conducting circular cylinders. *Int. J. Electronics* **59** (1985) 407–421. C: pp. 129, 310

997. H.A. Ragheb & M. Hamid, Simulation of a cylindrical reflector by conducting circular cylinders. *IEAP* **35** (1987) 349–353. C: p. 129

998. H.A. Ragheb & M. Hamid, Scattering by two conducting strips with parallel edges. *CJP* **66** (1988) 376–383. C: p. 149

999. H.A. Ragheb, L. Shafai & M. Hamid, Plane wave scattering by a conducting elliptic cylinder coated by a nonconfocal dielectric. *IEAP* **39** (1991) 218–223. C: p. 149

1000. J. Rahola, Diagonal forms of the translation operators in the fast multipole algorithm for scattering problems. *BIT* **36** (1996) 333–358. C: pp. 80, 240

1001. A.G. Ramm, Convergence of the T-matrix approach to scattering theory. *JMP* **23** (1982) 1123–1125. C: p. 264

1002. A.G. Ramm, *Iterative Methods for Calculating Static Fields and Wave Scattering by Small Bodies*. New York: Springer, 1982. C: pp. 288, 290

1003. A.G. Ramm, Scattering by a penetrable body. *JMP* **25** (1984) 469–471. C: p. 187

1004. A.G. Ramm, Wave scattering by small bodies. *Reports on Mathematical Physics* **21** (1985) 69–77. C: §8.3.2 (p. 301)

1005. A.G. Ramm, *Scattering by Obstacles*. Dordrecht: Reidel, 1986. C: pp. 264, 266, 266

1006. A.G. Ramm, *Multidimensional Inverse Scattering Problems*. Harlow: Longman, 1992. C: pp. 291, 296

1007. A.G. Ramm, Equations for the self-consistent field in random medium. *Physics Letters* A **312** (2003) 256–261. C: §8.3.2 (p. 301)

1008. S.M. Rao & P.K. Raju, Application of the method of moments to acoustic scattering from multiple bodies of arbitrary shape. *JASA* **86** (1989) 1143–1148. C: p. 172

1009. Lord Rayleigh, On the influence of obstacles arranged in rectangular order upon the properties of a medium. *Phil. Mag.*, Series 5, **34** (1892) 481–502. C: pp. 7, 10 & §1.1.4 (p. 5)

1010. Lord Rayleigh (J.W. Strutt), *The Theory of Sound*, vol. II, second edition. London: Macmillan & Co., 1896. Reprinted by Dover, New York, 1945. C: pp. 69, 82, 124, 134, 288

1011. Lord Rayleigh, On the passage of waves through apertures in plane screens, and allied problems. *Phil. Mag.*, Series 5, **43** (1897) 259–272. C: p. 152

1012. Lord Rayleigh, On the passage of waves through fine slits in thin opaque screens. *PRSA* **89** (1914) 194–219. C: p. 152

1013. M. Razavy, Scattering by two impenetrable spheres. *CJP* **51** (1973) 1850–1860. C: p. 296

1014. J.M. Reese & W. Thompson, Jr, Acoustic coupling between two finite-sized spheres; $n = 2$ mode. *JASA* **69** (1981) 1587–1590. C: p. 137

1015. J.J. do Rêgo Silva, *Acoustic and Elastic Wave Scattering using Boundary Elements*. Southampton: Computational Mechanics Publications 1994. C: pp. 168, 203

1016. J.J. Rehr & R.C. Albers, Scattering-matrix formulation of curved-wave multiple-scattering theory: application to x-ray-absorption fine structure. *PR* B **41** (1990) 8139–8149. C: pp. 104, 105
1017. E. Reinoso, *Scattering of Seismic Waves: Applications to the Mexico City Valley*. Southampton: WIT Press, 2002. C: p. 202
1018. E. Reinoso, L.C. Wrobel & H. Power, Two-dimensional scattering of P, SV and Rayleigh waves: preliminary results for the Valley of Mexico. *EESD* **26** (1997) 595–616. C: p. 202
1019. M. Rezayat, D.J. Shippy & F.J. Rizzo, On time-harmonic elastic-wave analysis by the boundary element method for moderate to high frequencies. *CMAME* **55** (1986) 349–367. C: pp. 202, 221
1020. F. Riesz & B. Szőkefalvi-Nagy, *Functional Analysis*. New York: Ungar Publishing Co., 1955. C: p. 160
1021. F.J. Rizzo, An integral equation approach to boundary value problems of classical elastostatics. *QAM* **25** (1967) 83–95. C: p. 201
1022. F.J. Rizzo, The boundary element method. Some early history – a personal view. In: *Boundary Element Methods in Structural Analysis* (ed. D.E. Beskos) pp. 1–16. New York: ASCE, 1989. C: pp. 166, 166, 201
1023. F.J. Rizzo, D.J. Shippy & M. Rezayat, A boundary integral equation method for radiation and scattering of elastic waves in three dimensions. *IJNME* **21** (1985) 115–129. C: p. 202
1024. S. Robert, J.-M. Conoir, H. Franklin & F. Luppé, Resonant elastic scattering by a finite number of cylindrical cavities in an elastic matrix. *WM* **40** (2004) 225–239. C: p. 142
1025. V. Rokhlin, Solution of acoustic scattering problems by means of second kind integral equations. *WM* **5** (1983) 257–271. C: p. 180
1026. V. Rokhlin, Rapid solution of integral equations of scattering theory in two dimensions. *JCP* **86** (1990) 414–439. C: pp. 235, 240
1027. V. Rokhlin, Diagonal forms of translation operators for the Helmholtz equation in three dimensions. *Applied & Computational Harmonic Analysis* **1** (1993) 82–93. C: p. 240
1028. M.E. Rose, *Multipole Fields*. New York: Wiley, 1955. C: pp. 84, 86, 328 & Table B.1 (p. 327)
1029. M.E. Rose, *Elementary Theory of Angular Momentum*. New York: Wiley, 1957. Reprinted by Dover, New York, 1995. C: pp. 84, 330, 330 & Table B.1 (p. 327)
1030. M.E. Rose, The electrostatic interaction of two arbitrary charge distributions. *J. Maths & Phys.* **37** (1958) 215–222. C: p. 89
1031. M. Roseau, *Vibrations in Mechanical Systems*. Berlin: Springer, 1987. C: p. 198
1032. R. Rösel, Integer finite-part integral transforms. *Proc. South African Math. Soc.* **4** (1974) 209–217. C: p. 339
1033. F. Rouleau, Electromagnetic scattering by compact clusters of spheres. *Astronomy & Astrophysics* **310** (1996) 686–698. C: p. 141
1034. J.A. Roumeliotis, A.-G.P. Ziotopoulos & G.C. Kokkorakis, Acoustic scattering by a circular cylinder parallel with another of small radius. *JASA* **109** (2001) 870–877. C: p. 129
1035. R.V. Row, Theoretical and experimental study of electromagnetic scattering by two identical conducting cylinders. *JAP* **26** (1955) 666–675. C: pp. 129, 177, 206 & §4.5.3 (p. 131)
1036. A.J. Rudgers, Application of a Neumann-series method to two problems in acoustic radiation theory that are formulated in terms of Green's functions. *JASA* **79** (1986) 1211–1222. C: p. 228
1037. M. Rusek & A. Orłowski, Analytical approach to localization of electromagnetic waves in two-dimensional random media. *PR* E **51** (1995) R2763–R2766. C: p. 300

1038. M. Rusek, A. Orłowski & J. Mostowski, Localization of light in three-dimensional random dielectric media. *PR* E **53** (1996) 4122–4130. C: p. 300

1039. M. Rusek, A. Orłowski & J. Mostowski, Band of localized electromagnetic waves in random arrays of dielectric cylinders. *PR* E **56** (1997) 4892–4895. C: p. 300

1040. B.P. Rynne, Instabilities in time marching methods for scattering problems. *Electromagnetics* **6** (1986) 129–144. C: p. 11

1041. F.J. Sabina, V.P. Smyshlyaev & J.R. Willis, Self-consistent analysis of waves in a matrix-inclusion composite – I. Aligned spheroidal inclusions. *JMPS* **41** (1993) 1573–1588. C: p. 322

1042. F.J. Sabina & J.R. Willis, A simple self-consistent analysis of wave propagation in particulate composites. *WM* **10** (1988) 127–142. C: p. 322

1043. R.A. Sack, Three-dimensional addition theorem for arbitrary functions involving expansions in spherical harmonics. *JMP* **5** (1964) 252–259. C: pp. 87, 90

1044. R.A. Sack, Two-center expansion for the powers of the distance between two points. *JMP* **5** (1964) 260–268. C: p. 96

1045. R.A. Sack, Generating functions for spherical harmonics. Part I: Three-dimensional harmonics. *SIMA* **5** (1974) 774–796. C: pp. 89, 91, 116

1046. K. Særmark, A note on addition theorems for Mathieu functions. *J. Appl. Math. & Phys.* (ZAMP) **10** (1959) 426–428. C: pp. 53, 149

1047. A. Sáez & J. Domínguez, BEM analysis of wave scattering in transversely isotropic solids. *IJNME* **44** (1999) 1283–1300. C: p. 202

1048. A. Şahin & E.L. Miller, Recursive T-matrix methods for scattering from multiple dielectric and metallic objects. *IEAP* **46** (1998) 672–678. C: p. 283

1049. M. Saillard & A. Sentenac, Rigorous solutions for electromagnetic scattering from rough surfaces. *Waves in Random Media* **11** (2001) R103–R137. C: p. 11

1050. J. Sanchez Hubert & E. Sanchez Palencia, *Vibration and Coupling of Continuous Systems*. Berlin: Springer, 1989. C: p. 198

1051. F.J. Sánchez-Sesma, Diffraction of elastic waves by three-dimensional surface irregularities. *BSSA* **73** (1983) 1621–1636. C: p. 246

1052. F.J. Sánchez-Sesma & M. Campillo, Diffraction of P, SV, and Rayleigh waves by topographic features: a boundary integral formulation. *BSSA* **81** (1991) 2234–2253. C: p. 202

1053. A.S. Sangani, A pairwise interaction theory for determining the linear acoustic properties of dilute bubbly liquids. *JFM* **232** (1991) 221–284. C: pp. 137, 316

1054. A.S. Sangani & R. Sureshkumar, Linear acoustic properties of bubbly liquids near the natural frequency of the bubbles using numerical simulations. *JFM* **252** (1993) 239–264. C: pp. 137, 316

1055. J. Saranen & G. Vainikko, *Periodic Integral and Pseudodifferential Equations with Numerical Approximation*. Berlin: Springer, 2002. C: pp. 161, 168, 168, 291

1056. J. Sarkar, S.C. Mandal & M.L. Ghosh, Interaction of elastic waves with two coplanar Griffith cracks in an orthotropic medium. *EFM* **49** (1994) 411–423. C: p. 208

1057. A. Sarkissian, A.G. Dallas & R.E. Kleinman, Numerical search for a breakdown of the Waterman algorithm for approximating the T matrix of a rigid obstacle. *JASA* **80** (1986) S74. C: p. 266

1058. T. Sarpkaya & M. Isaacson, *Mechanics of Wave Forces on Offshore Structures*. New York: Van Nostrand Reinhold, 1981. C: pp. 144, 172, 205

1059. H. Sato, M. Kitahara & T. Shoji, Boundary element analysis of multiple scattering waves in high performance concretes. *JAM* **72** (2005) 165–171. C: p. 321

1060. H. Sato & Y. Shindo, Multiple scattering of plane elastic waves in a fiber-reinforced composite medium with graded interfacial layers. *IJSS* **38** (2001) 2549–2571. C: p. 142

1061. H. Sato & Y. Shindo, Multiple scattering of plane elastic waves in a particle-reinforced-composite medium with graded interfacial layers. *MM* **35** (2003) 83–106. C: p. 143

1062. P. Sayer, The long-wave behaviour of the virtual mass in water of finite depth. *PRSA* **372** (1980) 65–91. C: p. 145

1063. P. Sayer & H.K. Liang, Wave interactions between barges. *JWPCOE* **112** (1986) 402–417. C: p. 205

1064. P. Sayer & R. Spencer, The wave-induced motions of adjacent floating vessels. *Proc. Hydrodynamics in Ocean Engineering*, pp. 781–798. Trondheim: Norwegian Institute of Technology, 1981. C: p. 146

1065. C.M. Sayers, On the propagation of ultrasound in highly concentrated mixtures and suspensions. *JPDAP* **13** (1980) 179–184. C: p. 319

1066. M.N. Sayhi, Y. Ousset & G. Verchery, Solution of radiation problems by collocation of integral formulations in terms of single and double layer potentials. *JSV* **74** (1981) 187–204. C: p. 172

1067. P.J. Schafbuch, F.J. Rizzo & R.B. Thompson, Boundary element method solutions for elastic wave scattering in 3D. *IJNME* **36** (1993) 437–455. C: pp. 202, 202

1068. P.J. Schafbuch, R.B. Thompson & F.J. Rizzo, Application of the boundary element method to elastic wave scattering by irregular defects. *J. Nondestructive Evaluation* **9** (1990) 113–127. C: p. 202

1069. P.J. Schafbuch, R.B. Thompson & F.J. Rizzo, Elastic scatterer interaction via generalized Born series and far-field approximations. *JASA* **93** (1993) 295–307. C: pp. 202, 226, 227

1070. H.A. Schenck, Improved integral formulation for acoustic radiation problems. *JASA* **44** (1968) 41–58. C: §6.11 (p. 219)

1071. H.A. Schenck, Helmholtz integral formulation of the sonar equations. *JASA* **79** (1986) 1423–1433. C: pp. 228, 229

1072. L.H. Schick, Multiple-scattering analysis of the three-body scattering problem. *RMP* **33** (1961) 608–622. C: p. 137

1073. M. Schleiff, Singuläre Integraloperatoren in Hilbert-Räumen mit Gewichtsfunktion. *Math. Nachr.* **42** (1969) 145–155. C: p. 339

1074. B. Schlicht, K.F. Wall, R.K. Chang & P.W. Barber, Light scattering by two parallel glass fibers. *JOSA* A **4** (1987) 800–809. C: p. 129

1075. J. Schneider, J. Brew & I.C. Peden, Electromagnetic detection of buried dielectric targets. *IEGRS* **29** (1991) 555–562. C: p. 251

1076. J.B. Schneider & I.C. Peden, Differential cross section of a dielectric ellipsoid by the T-matrix extended boundary condition method. *IEAP* **36** (1988) 1317–1321. C: p. 251

1077. J.T. Schnute, The scattering of surface waves by two submerged cylinders. *PCPS* **69** (1971) 201–215. C: p. 206

1078. K.R. Schroll & S.L. Cheng, Dynamic stresses around elliptical discontinuities. *JAM* **39** (1972) 133–138. C: p. 149

1079. K. Schulten & R.G. Gordon, Exact recursive evaluation of $3j$- and $6j$-coefficients for quantum-mechanical coupling of angular momenta. *JMP* **16** (1975) 1961–1970. C: p. 93

1080. F.M. Schulz, K. Stamnes & J.J. Stamnes, Scattering of electromagnetic waves by spheroidal particles: a novel approach exploiting the $\mathcal{T}$ matrix computed in spheroidal coordinates. *AO* **37** (1998) 7875–7896. C: p. 274

1081. G.T. Schuster, A hybrid BIE + Born series modeling scheme: generalized Born series. *JASA* **77** (1985) 865–879. C: p. 226

1082. G.T. Schuster & L.C. Smith, A comparison among four direct boundary integral methods. *JASA* **77** (1985) 850–864. C: p. 180
1083. C. Schwab & W.L. Wendland, Kernel properties and representations of boundary integral operators. *Math. Nachr.* **156** (1992) 187–218. C: p. 349
1084. L. Schwartz, *Théorie des Distributions*. Paris: Hermann, 1966. C: pp. 336, 349
1085. L. Schwartz & T.J. Plona, Ultrasonic propagation in close-packed disordered suspensions. *JAP* **55** (1984) 3971–3977. C: p. 322
1086. L.M. Schwartz & D.L. Johnson, Long-wavelength acoustic propagation in ordered and disordered suspensions. *PR* B **30** (1984) 4302–4313. C: p. 322
1087. A. Sebak, Transverse magnetic scattering by parallel conducting elliptic cylinders. *CJP* **69** (1991) 1233–1241. C: p. 149
1088. A. Sebak & Y. Antar, Multiple scattering by parallel conducting elliptic cylinders: TE case. *IEE-H* **142** (1995) 178–182. C: p. 149
1089. A.R. Sebak, Electromagnetic scattering by two parallel dielectric elliptic cylinders. *IEAP* **42** (1994) 1521–1527. C: p. 149
1090. D. Sébilleau, New recurrence relations for matrix elements of the propagator. *J. Phys.: Condensed Matter* **7** (1995) 6211–6220. C: p. 105
1091. A.F. Seybert & D.K. Casey, An integral equation method for coupled fluid/fluid scattering in three dimensions. *JASA* **84** (1988) 379–384. C: p. 180
1092. A.F. Seybert & T.K. Rengarajan, The use of CHIEF to obtain unique solutions for acoustic radiation using boundary integral equations. *JASA* **81** (1987) 1299–1306. C: p. 221
1093. A.F. Seybert, B. Soenarko, F.J. Rizzo & D.J. Shippy, An advanced computational method for radiation and scattering of acoustic waves in three dimensions. *JASA* **77** (1985) 362–368. C: p. 221
1094. A.F. Seybert, B. Soenarko, F.J. Rizzo & D.J. Shippy, A special integral equation formulation for acoustic radiation and scattering for axisymmetric bodies and boundary conditions. *JASA* **80** (1986) 1241–1247. C: p. 178
1095. A.F. Seybert & T.W. Wu, Modified Helmholtz integral equation for bodies sitting on an infinite plane. *JASA* **85** (1989) 19–23. C: p. 178
1096. K. Sezawa, Scattering of elastic waves and some allied problems. *Bulletin of the Earthquake Research Institute*, Tokyo University **3** (1927) 19–41. C: pp. 141, 149
1097. A.H. Shah, K.C. Wong & S.K. Datta, Single and multiple scattering of elastic waves in two dimensions. *JASA* **74** (1983) 1033–1043. C: p. 142
1098. R. Shail, Two centred expansion of the retarded Green's function. *Mathematika* **6** (1959) 108–117. C: p. 96
1099. D.L. Sharma, Scattering of steady elastic waves by surfaces of arbitrary shape. *BSSA* **57** (1967) 795–812. C: p. 201
1100. P. Sheng, *Introduction to Wave Scattering, Localization, and Mesoscopic Phenomena*. San Diego: Academic Press, 1995. C: p. 11
1101. S.E. Sherer, Scattering of sound from axisymmetric sources by multiple circular cylinders. *JASA* **115** (2004) 488–496. C: p. 129
1102. Y. Shindo & R. Togawa, Multiple scattering of antiplane shear waves in a piezoelectric fibrous composite medium with slip at interfaces. *WM* **30** (1999) 225–238. C: p. 130
1103. M.V. Simkin & V.P. Roychowdhury, Read before you cite! *Complex Systems* **14** (2003) 269–274. C: p. 28
1104. M.J. Simon, Multiple scattering in arrays of axisymmetric wave-energy devices. Part 1. A matrix method using a plane-wave approximation. *JFM* **120** (1982) 1–25. C: p. 312
1105. V.A. Singh, Analytic (unitarity-preserving) approximation for the electronic structure of amorphous systems. *PR* B **24** (1981) 4852–4853. C: p. 322

1106. B.P. Sinha & R.H. MacPhie, Translational addition theorems for spheroidal scalar and vector wave functions. *QAM* **38** (1980) 143–158. C: pp. 150, 150
1107. B.P. Sinha & R.H. MacPhie, Electromagnetic plane wave scattering by a system of two parallel conducting prolate spheroids. *IEAP* **31** (1983) 294–304. C: p. 150
1108. P.R. Siqueira & K. Sarabandi, Method of moments evaluation of the two-dimensional quasi-crystalline approximation. *IEAP* **44** (1996) 1067–1077. C: p. 320
1109. P.R. Siqueira & K. Sarabandi, T-matrix determination of effective permittivity for three-dimensional dense random media. *IEAP* **48** (2000) 317–327. C: pp. 283, 320
1110. D.V. Sivukhin, Diffraction of plane sound waves by a spherical cavity. *SPA* **1** (1955) 82–93. C: p. 141
1111. N.C. Skaropoulos, H.D. Yagridou & D.P. Chrissoulidis, Interactive resonant scattering by a cluster of air bubbles in water. *JASA* **113** (2003) 3001–3011. C: p. 139
1112. D.C. Skigin, V.V. Veremey & R. Mittra, Superdirective radiation from finite gratings of rectangular grooves. *IEAP* **47** (1999) 376–383. C: p. 148
1113. V. Sladek & J. Sladek (eds.), *Singular Integrals in Boundary Element Methods*. Southampton: Computational Mechanics Publications, 1998. C: p. 338
1114. O.D. Sledge, Scattering of a plane electromagnetic wave by a linear array of cylindrical elements. *IEAP* **17** (1969) 169–174. C: p. 189
1115. B.D. Sleeman, The scalar scattering of a plane wave by an ellipsoid. *J. Inst. Maths. Applics.* **3** (1967) 4–15. C: p. 150
1116. B.D. Sleeman, Acoustic scattering by inhomogeneous media. *QJMAM* **33** (1980) 373–383. C: p. 138
1117. I.H. Sloan & R.S. Womersley, Constructive polynomial approximation on the sphere. *J. Approximation Theory* **103** (2000) 91–118. C: p. 263
1118. V.I. Smirnov, *A Course of Higher Mathematics*, vol. IV, *Integral Equations and Partial Differential Equations*. Oxford: Pergamon Press, 1964. C: pp. 160, 171
1119. F. Smithies, *Cauchy and the Creation of Complex Function Theory*. Cambridge: Cambridge University Press, 1997. C: pp. 334, 335
1120. V.P. Smyshlyaev, J.R. Willis & F.J. Sabina, Self-consistent analysis of waves in a matrix-inclusion composite – II. Randomly oriented spheroidal inclusions. *JMPS* **41** (1993) 1589–1598. C: p. 322
1121. V.P. Smyshlyaev, J.R. Willis & F.J. Sabina, Self-consistent analysis of waves in a matrix-inclusion composite – III. A matrix containing cracks. *JMPS* **41** (1993) 1809–1824. C: p. 322
1122. R. Snieder, General theory of elastic wave scattering. In: [977], pp. 528–542. C: p. 201
1123. R. Snieder, Scattering of surface waves. In: [977], pp. 562–577. C: p. 22
1124. H. So & J.Y. Huang, Determination of dynamic stress intensity factors of two finite cracks at arbitrary positions by dislocation method. *IJES* **26** (1988) 111–119. C: p. 208
1125. S. Soboleff, Méthode nouvelle à résoudre le problème de Cauchy pour les équations linéaires hyperboliques normales. *Matematičeskiĭ Sbornik. Recueil Mathématique*, New Series, **1** (1936) 39–72. C: p. 177
1126. S.L. Sobolev, *Applications of Functional Analysis in Mathematical Physics*. Providence: American Mathematical Society, 1963. C: p. 177
1127. S.L. Sobolev, *Partial Differential Equations of Mathematical Physics*. Oxford: Pergamon Press, 1964. Reprinted by Dover, New York, 1989. C: p. 177
1128. A. Sommerfeld, *Partial Differential Equations in Physics*. New York: Academic Press, 1949. C: p. 230
1129. A. Sommerfeld, *Optics* (*Lectures in Theoretical Physics*, vol. IV). New York: Academic Press, 1954. C: pp. 157, 165, 195

1130. J.M. Song & W.C. Chew, Multilevel fast multipole algorithm for solving combined field integral equation of electromagnetic scattering. *MOTL* **10** (1995) 14–19. C: p. 240
1131. L. Sorensen, Multi-obstacle acoustic scattering in the simple source formulation with accuracy checks. *JASA* **66** (1979) 1536–1541. C: p. 172
1132. C.M. Soukoulis (ed.), *Photonic Band Gaps and Localization*. New York: Plenum Press, 1993. C: p. 11
1133. P.D.M. Spelt, M.A. Norato, A.S. Sangani, M.S. Greenwood & L.L. Tavlarides, Attenuation of sound in concentrated suspensions: theory and experiments. *JFM* **430** (2001) 51–86. C: p. 312
1134. P.D.M. Spelt, M.A. Norato, A.S. Sangani & L.L. Tavlarides, Determination of particle size distributions from acoustic wave propagation measurements. *Physics of Fluids* **11** (1999) 1065–1080. C: p. 312
1135. R.D. Spence & S. Granger, The scattering of sound from a prolate spheroid. *JASA* **23** (1951) 701–706. C: p. 150
1136. B.H. Spring & P.L. Monkmeyer, Interaction of plane waves with vertical cylinders. *Proc. 14th Coastal Engineering Conf.*, pp. 1828–1847. New York: ASCE, 1974. C: p. 129
1137. M.A. Srokosz, The submerged sphere as an absorber of wave power. *JFM* **95** (1979) 717–741. C: p. 145
1138. M.A. Srokosz & D.V. Evans, A theory for wave-power absorption by two independently oscillating bodies. *JFM* **90** (1979) 337–362. C: p. 308
1139. J.J. Stamnes, Exact two-dimensional scattering by perfectly reflecting elliptical cylinders, strips and slits. *Pure & Appl. Optics* **4** (1995) 841–855. C: p. 149
1140. J.J. Stamnes & B. Spjelkavik, New method for computing eigenfunctions (Mathieu functions) for scattering by elliptical cylinders. *Pure & Appl. Optics* **4** (1995) 251–262. C: p. 49
1141. R.G. Standing, Applications of wave diffraction theory. *IJNME* **13** (1978) 49–72. C: p. 205
1142. R.G. Standing, Use of potential flow theory in evaluating wave forces on offshore structures. In: [241], pp. 175–212. C: p. 205
1143. S. Stein, Addition theorems for spherical wave functions. *QAM* **19** (1961) 15–24. C: pp. 87, 90, 107
1144. J.J. Stoker, On radiation conditions. *CPAM* **9** (1956) 577– 595. C: p. 13
1145. J.J. Stoker, *Water Waves*. New York: Interscience, 1957. C: p. 23
1146. J.J. Stoker, *Differential Geometry*. New York: Wiley- Interscience, 1969. C: pp. 160, 347
1147. J.E. Storer & J. Sevick, General theory of plane-wave scattering from finite, conducting obstacles with application to the two-antenna problems. *JAP* **25** (1954) 369–376. C: p. 195
1148. B. Stout, J.-C. Auger & J. Lafait, A transfer matrix approach to local field calculations in multiple-scattering problems. *J. Modern Optics* **49** (2002) 2129–2152. C: p. 283
1149. J.A. Stratton, *Electromagnetic Theory*. New York: McGraw-Hill, 1941. C: pp. 13, 63, 65, 107, 140, 177, 189
1150. J.A. Stratton & L.J. Chu, Diffraction theory of electromagnetic waves. *PR* **56** (1939) 99–107. C: pp. 189, 195
1151. S. Ström, *T* matrix for electromagnetic scattering from an arbitrary number of scatterers with continuously varying electromagnetic properties. *PR* D **10** (1974) 2685–2690. C: p. 281
1152. S. Ström, On the integral equations for electromagnetic scattering. *Amer. J. Phys.* **43** (1975) 1060–1069. C: pp. 189, 192, 195, 251

1153. S. Ström, Quantum-mechanical scattering from an assembly of nonover-lapping potentials. *PR* D **13** (1976) 3485–3490. C: p. 280
1154. S. Ström & W. Zheng, Basic features of the null field method for dielectric scatterers. *RS* **22** (1987) 1273–1281. C: pp. 264, 273
1155. S. Ström & W. Zheng, Null-field computations of radar cross sections of composite objects. *Proc. IEEE* **77** (1989) 761–769. C: p. 273
1156. D. Stroud & F.P. Pan, Self-consistent approach to electromagnetic wave propagation in composite media: application to model granular metals. *PR* B **17** (1978) 1602–1610. C: p. 321
1157. R.J. Strutt, *Life of John William Strutt, Third Baron Rayleigh*, augmented edition. Madison: University of Wisconsin Press, 1968. C: p. 1
1158. B. Stupfel, A. Lavie & J.N. Decarpigny, Combined integral equation formulation and null-field method for the exterior acoustic problem. *JASA* **83** (1988) 927–941. C: p. 221
1159. J.-H. Su, V.V. Varadan, V.K. Varadan & L. Flax, Acoustic wave scattering by a finite elastic cylinder in water. *JASA* **68** (1980) 686–691. C: p. 274
1160. Yu.V. Svishchev, Computing the coefficients in addition theorems for spherical wave functions and Debye potentials. *Differential Equations* **28** (1992) 1294–1300. C: p. 93
1161. D.R. Swatek & I.R. Ciric, A recursive single-source surface integral equation analysis for wave scattering by heterogeneous dielectric bodies. *IEAP* **48** (2000) 1175–1185. C: p. 180
1162. G.T. Symm, Integral equation methods in potential theory. II. *PRSA* **275** (1963) 33–46. C: p. 177
1163. J.L. Synge, The general problem of antenna radiation and the fundamental integral equation, with application to an antenna of revolution—part II. *QAM* **6** (1948) 133–156. C: p. 244
1164. A. Tadeu, L. Godinho & P. Santos, Wave motion between two fluid-filled boreholes in an elastic medium. *EABE* **26** (2002) 101–117. C: p. 202
1165. K. Takakuda, On integral equation methods for elastic boundary value problems (3rd report, direct methods for dynamic problems). *Bull. Japan Soc. Mechanical Engineers* **27** (1984) 625–632. C: p. 219
1166. T.H. Tan, Scattering of elastic waves by elastically transparent obstacles (integral-equation method). *Applied Scientific Res.* **31** (1975) 29–51. C: p. 201
1167. T.H. Tan, Diffraction of time-harmonic elastic waves by a cylindrical obstacle. *Applied Scientific Res.* **32** (1976) 97–144. C: pp. 22, 201
1168. M. Tanaka, V. Sladek & J. Sladek, Regularization techniques applied to boundary element methods. *AMR* **47** (1994) 457–499. C: p. 338
1169. R. Tazzioli, Green's function in some contributions of 19th century mathematicians. *Historia Mathematica* **28** (2001) 232–252. C: p. 223
1170. T. Terai, On calculation of sound fields around three dimensional objects by integral equation methods. *JSV* **69** (1980) 71–100. Addendum: **70** (1980) 140–141. C: pp. 219, 351
1171. J.M. Terrón, A. Sánchez-Lavega & A. Salazar, Multiple scattering effects of thermal waves by two subsurface cylinders. *JAP* **87** (2000) 2600–2607. C: p. 129
1172. L.G. Tham, J. Qian & Y.K. Cheung, Dynamic response of a group of flexible foundations to incident seismic waves. *SDEE* **17** (1998) 127–137. C: p. 202
1173. G.P. Thomas & D.V. Evans, Arrays of three-dimensional wave- energy absorbers. *JFM* **108** (1981) 67–88. C: p. 145
1174. W. Thompson, Jr, Acoustic coupling between two finite-sized spherical sources. *JASA* **62** (1977) 8–11. C: p. 137

1175. W. Thomson & P.G. Tait, *Treatise on Natural Philosophy*, Part 1. Cambridge: Cambridge University Press, 1912. Reprinted as *Principles of Mechanics and Dynamics*. New York: Dover, 1962. C: pp. 30, 62, 65

1176. R.C. Thorne, Multipole expansions in the theory of surface waves. *PCPS PCPS* **49** (1953) 707–716. C: pp. 55, 110, 115, 145

1177. W. Tobocman, Comparison of the T-matrix and Helmholtz integral equation methods for wave scattering calculations. *JASA* **77** (1985) 369–374. Addendum and erratum: **78** (1985) 805–808. C: pp. 263, 273

1178. W. Tobocman, Extension of the Helmholtz integral equation method to shorter wavelengths. *JASA* **80** (1986) 1828–1837. C: p. 219

1179. W. Tobocman, Extension of the Helmholtz integral equation method to shorter wavelengths II. *JASA* **82** (1987) 704–706. C: p. 219

1180. K.C. Toh & S. Mukherjee, Hypersingular and finite part integrals in the boundary element method. *IJSS* **31** (1994) 2299–2312. C: p. 208

1181. I. Tolstoy, Comment on "Multiple scattering by finite regular arrays of resonators" [J. Acoust. Soc. Am. 87, 25–41 (1990)] (reference [1216]). *JASA* **88** (1990) 1178–1179. C: p. 299

1182. A. Tourin, M. Fink & A. Derode, Multiple scattering of sound. *Waves in Random Media* **10** (2000) R31–R60. C: pp. 10, 182

1183. F.G. Tricomi, *Integral Equations*. New York: Interscience, 1957. Reprinted by Dover, New York, 1985. C: pp. 168, 257

1184. W. Trinks, Zur Vielfachstreuung an kleinen Kugeln. *Annalen der Physik*, 5 Folge, **22** (1935) 561–590. C: p. 140

1185. R. Truell, C. Elbaum & B.B. Chick, *Ultrasonic Methods in Solid State Physics*. New York: Academic Press, 1969. C: pp. 141, 319

1186. L. Tsang & J.A. Kong, Multiple scattering of electromagnetic waves by random distributions of discrete scatterers with coherent potential and quantum mechanical formulism. *JAP* **51** (1980) 3465–3485. C: p. 321

1187. L. Tsang & J.A. Kong, Multiple scattering of acoustic waves by random distributions of discrete scatterers with the use of quasicrystalline-coherent potential approximation. *JAP* **52** (1981) 5448–5458. C: p. 322

1188. L. Tsang & J.A. Kong, Effective propagation constants for coherent electromagnetic wave propagation in media embedded with dielectric scatters. *JAP* **53** (1982) 7162–7173. C: p. 322

1189. L. Tsang & J.A. Kong, *Scattering of Electromagnetic Waves: Advanced Topics*. New York: Wiley, 2001. C: pp. 320, 321

1190. L. Tsang, J.A. Kong & K.-H. Ding, *Scattering of Electromagnetic Waves: Theories and Applications*. New York: Wiley, 2000. C: pp. 10, 140, 195, 251

1191. L. Tsang, J.A. Kong, K.-H. Ding & C.O. Ao, *Scattering of Electromagnetic Waves: Numerical Simulations*. New York: Wiley, 2001. C: pp. 5, 194, 195, 195

1192. L. Tsang, J.A. Kong & T. Habashy, Multiple scattering of acoustic waves by random distribution of discrete spherical scatterers with the quasicrystalline and Percus–Yevick approximation. *JASA* **71** (1982) 552–558. C: pp. 320, 322

1193. L. Tsang, J.A. Kong & R.T. Shin, *Theory of Microwave Remote Sensing*. New York: Wiley, 1985. C: p. 5

1194. L. Tsang, C.E. Mandt & K.H. Ding, Monte Carlo simulations of the extinction rate of dense media with randomly distributed dielectric spheres based on solution of Maxwell's equations. *OL* **17** (1992) 314–316. C: p. 322

1195. T.-G. Tsuei & P.W. Barber, Multiple scattering by two parallel dielectric cylinders. *AO* **27** (1988) 3375–3381. C: p. 129
1196. V. Twersky, Multiple scattering of radiation by an arbitrary configuration of parallel cylinders. *JASA* **24** (1952) 42–46. C: pp. 3, 130
1197. V. Twersky, Multiple scattering of radiation by an arbitrary planar configuration of parallel cylinders and by two parallel cylinders. *JAP* **23** (1952) 407–414. C: p. 130
1198. V. Twersky, On multiple scattering of waves. *J. Res. National Bureau of Standards* **64D** (1960) 715–730. C: pp. 4, 10, 172
1199. V. Twersky, Multiple scattering by arbitrary configurations in three dimensions. *JMP* **3** (1962) 83–91. C: p. 235
1200. V. Twersky, On scattering of waves by random distributions. I. Free-space scatterer formalism. *JMP* **3** (1962) 700–715. C: pp. 10, 316, 319, 319, 320, 321
1201. V. Twersky, On scattering of waves by random distributions. II. Two-space scatterer formalism. *JMP* **3** (1962) 724–734. C: pp. 10, 321
1202. V. Twersky, On scattering of waves by the infinite grating of circular cylinders. *IRE Trans. on Antennas & Propag.* **10** (1962) 737–765. C: p. 235
1203. V. Twersky, Scattering of waves by two objects. In: *Electromagnetic Waves* (ed. R.E. Langer) pp. 361–389. Madison: University of Wisconsin Press, 1962. C: §6.13 (p. 229)
1204. V. Twersky, Acoustic bulk parameters of random volume distributions of small scatterers. *JASA* **36** (1964) 1314–1329. C: pp. 320, 321
1205. V. Twersky, Rayleigh scattering. *AO* **3** (1964) 1150–1162. C: pp. 7, 134, 296
1206. V. Twersky, Multiple scattering of electromagnetic waves by arbitrary configurations. *JMP* **8** (1967) 589–610. C: p. 235
1207. V. Twersky, Multiple scattering of sound by a periodic line of obstacles. *JASA* **53** (1973) 96–112. C: p. 235
1208. V. Twersky, Multiple scattering of waves by the doubly periodic planar array of obstacles. *JMP* **16** (1975) 633–643. C: p. 235
1209. V. Twersky, Lattice sums and scattering coefficients for the rectangular planar array. *JMP* **16** (1975) 644–657. C: p. 235
1210. V. Twersky, Low frequency coupling in the planar rectangular lattice. *JMP* **16** (1975) 658–666. C: p. 235
1211. V. Twersky, Coherent scalar field in pair-correlated random distributions of aligned scatterers. *JMP* **18** (1977) 2468–2486. C: p. 320
1212. V. Twersky, Acoustic bulk parameters in distributions of pair-correlated scatterers. *JASA* **64** (1978) 1710–1719. C: p. 320
1213. V. Twersky, Coherent electromagnetic waves in pair-correlated random distributions of aligned scatterers. *JMP* **19** (1978) 215–230. C: p. 320
1214. V. Twersky, Multiple scattering of sound by correlated monolayers. *JASA* **73** (1983) 68–84. C: p. 320
1215. V. Twersky, Comments on resonant systems of scatterers. *JASA* **88** (1990) 1179–1180. C: p. 299
1216. V. Twersky, Multiple scattering by finite regular arrays of resonators. *JASA* **87** (1990) 25–41. C: p. 299
1217. V.V. Tyutekin, Scattering of plane waves by a cylindrical cavity in an isotropic elastic medium. *SPA* **5** (1959) 105–109. C: p. 141
1218. Y.C. Tzeng & A.K. Fung, T-matrix approach to multiple scattering of EM waves from N-spheres. *JEWA* **8** (1994) 61–84. C: p. 282

1219. H. Überall (ed.), *Acoustic Resonance Scattering*. Philadelphia: Gordon & Breach, 1992. C: p. 141
1220. R.J. Urick, *Principles of Underwater Sound*, third edition. New York: McGraw-Hill, 1983. C: p. 228
1221. R.J. Urick & W.S. Ament, The propagation of sound in composite media. *JASA* **21** (1949) 115–119. C: p. 319
1222. F. Ursell, On the heaving motion of a circular cylinder on the surface of a fluid. *QJMAM* **2** (1949) 218–231. C: pp. 55, 144
1223. F. Ursell, On the rolling motion of cylinders in the surface of a fluid. *QJMAM* **2** (1949) 335–353. C: pp. 55, 55, 144
1224. F. Ursell, Surface waves on deep water in the presence of a submerged circular cylinder. I. *PCPS* **46** (1950) 141–152. C: p. 145
1225. F. Ursell, Surface waves on deep water in the presence of a submerged circular cylinder. II. *PCPS* **46** (1950) 153–158. C: pp. 27, 145
1226. F. Ursell, On the exterior problems of acoustics. *PCPS* **74** (1973) 117–125. C: pp. 210, 211
1227. F. Ursell, On the exterior problems of acoustics. II. *MPCPS* **84** (1978) 545–548. C: pp. 210, 212
1228. F. Ursell, Irregular frequencies and the motion of floating bodies. *JFM* **105** (1981) 143–156. C: pp. 56, 217
1229. F. Ursell, Mathematical observations on the method of multipoles. *Schiffstechnik* **33** (1986) 113–128. C: pp. 55, 144
1230. J. Van Bladel, Low-frequency scattering by cylindrical bodies. *Applied Scientific Res.*, Series B, **10** (1963) 195– 202. C: pp. 293, 296
1231. J. Van Bladel, *Singular Electromagnetic Fields and Sources*. Oxford: Oxford University Press, 1991. C: pp. 195, 352
1232. A.L. Van Buren & B.J. King, Acoustic radiation from two spheroids. *JASA* **52** (1972) 364–372. C: p. 150
1233. H.C. van de Hulst, *Light Scattering by Small Particles*. New York: Wiley, 1957. Reprinted by Dover, New York, 1981. C: pp. 1, 2, 140
1234. A.E. van de Vooren, Unsteady airfoil theory. *Advances in Applied Mechanics* **5** (1958) 35–89. C: p. 207
1235. P.M. van den Berg, Transition matrix in acoustic scattering by a strip. *JASA* **70** (1981) 615–619. C: p. 274
1236. P.M. van den Berg, Scattering of two-dimensional elastodynamic waves by a rigid plane strip or a plane crack of finite width: the transition matrix. *JASA* **72** (1982) 1038–1045. C: p. 274
1237. P.M. van den Berg, Iterative schemes based on the minimization of the error in field problems. *Electromagnetics* **5** (1985) 237–262. C: p. 263
1238. B. van der Pol, A generalization of Maxwell's definition of solid harmonics to waves in n dimensions. *Physica* **3** (1936) 393–397. C: p. 71
1239. G. van Oortmerssen, The motions of a ship in shallow water. *OE* **3** (1976) 221–255. C: p. 205
1240. G. van Oortmerssen, Hydrodynamic interaction between two structures, floating in waves. *Proc. 2nd Int. Conf. on Behaviour of Off-Shore Structures* (BOSS'79), vol. 1, pp. 339–356. Cranfield: BHRA Fluid Engineering, 1979 C: p. 205
1241. B.A. van Tiggelen, A. Lagendijk & A. Tip, Multiple-scattering effects for the propagation of light in 3D slabs. *J. Phys.: Condensed Matter* **2** (1990) 7653–7677. C: p. 305

1242. L. van Wijngaarden, One-dimensional flow of liquids containing small gas bubbles. *ARFM* **4** (1972) 369–396. C: p. 316
1243. V.K. Varadan, Scattering of elastic waves by randomly distributed and oriented scatterers. *JASA* **65** (1979) 655–657. C: p. 320
1244. V.K. Varadan, V.N. Bringi & V.V. Varadan, Coherent electromagnetic wave propagation through randomly distributed dielectric scatterers. *PR* D **19** (1979) 2480–2489. C: pp. 282, 320
1245. V.K. Varadan, V.N. Bringi, V.V. Varadan & Y. Ma, Coherent attenuation of acoustic waves by pair-correlated random distribution of scatterers with uniform and Gaussian size distributions. *JASA* **73** (1983) 1941–1947. C: p. 320
1246. V.K. Varadan, Y. Ma & V.V. Varadan, A multiple scattering theory for elastic wave propagation in discrete random media. *JASA* **77** (1985) 375–385. C: p. 320
1247. V.K. Varadan, Y. Ma & V.V. Varadan, Multiple scattering of compressional and shear waves by fiber-reinforced composite materials. *JASA* **80** (1986) 333–339. C: pp. 142, 320
1248. V.K. Varadan & V.V. Varadan (eds.), *Acoustic, Electromagnetic and Elastic Wave Scattering – Focus on the T-matrix Approach*. New York: Pergamon, 1980. C: p. 242
1249. V.K. Varadan, V.V. Varadan, L.R. Dragonette & L. Flax, Computation of rigid body scattering by prolate spheroids using the T-matrix approach. *JASA* **71** (1982) 22–25. C: p. 273
1250. V.K. Varadan, V.V. Varadan & Y. Ma, Multiple scattering of elastic waves by cylinders of arbitrary cross section. II. Pair-correlated cylinders. *JASA* **78** (1985) 1874–1878. C: p. 320
1251. V.K. Varadan, V.V. Varadan & Y.-H. Pao, Multiple scattering of elastic waves by cylinders of arbitrary cross section. I. SH waves. *JASA* **63** (1978) 1310–1319. C: p. 320
1252. V.K. Varadan, V.V. Varadan, S.J. Tsao & W.G. Neubauer, Scattering of acoustic waves by rigid cylindrical obstacles with sharp corners. *JASA* **72** (1982) 1957–1964. C: p. 250
1253. V.V. Varadan, Scattering matrix for elastic waves. II. Application to elliptic cylinders. *JASA* **63** (1978) 1014–1024. C: p. 274
1254. V.V. Varadan, A. Lakhtakia & V.K. Varadan, Comments on recent criticism of the T-matrix method. *JASA* **84** (1988) 2280–2284. C: pp. 242, 265, 269, 320
1255. V.V. Varadan & V.K. Varadan, Low-frequency expansions for acoustic wave scattering using Waterman's T-matrix method. *JASA* **66** (1979) 586–589. C: pp. 280, 296
1256. V.V. Varadan & V.K. Varadan, Scattering matrix for elastic waves. III. Application to spheroids. *JASA* **65** (1979) 896–905. C: p. 274
1257. V.V. Varadan & V.K. Varadan, Multiple scattering of electromagnetic waves by randomly distributed and oriented dielectric scatterers. *PR* D **21** (1980) 388–394. C: pp. 282, 320
1258. V.V. Varadan & V.K. Varadan, Configurations with finite numbers of scatterers – A self-consistent T-matrix approach. *JASA* **70** (1981) 213–217. C: pp. 281, 282
1259. V.V. Varadan, D.J.N. Wall, S.J. Tsao & V.K. Varadan, Elastic wave scattering from surface breaking cylindrical cracks: SH waves. *JASA* **71** (1982) 1384–1390. C: p. 250
1260. V. Varatharajulu & Y.-H. Pao, Scattering matrix for elastic waves, I. Theory. *JASA* **60** (1976) 556–566. C: pp. 251, 274
1261. N.D. Veksler, *Resonance Acoustic Spectroscopy*. Berlin: Springer, 1993. C: p. 141
1262. S. Venakides, M.A. Haider & V. Papanicolaou, Boundary integral calculations of two-dimensional electromagnetic scattering by photonic crystal Fabry–Perot structures. *SIAP* **60** (2000) 1686–1706. C: p. 180

1263. G. Videen, Light scattering from a particle on or near a perfectly conducting surface. *OC* **115** (1995) 1–7. C: p. 273
1264. S. Vijayakumar & D.E. Cormack, An invariant imbedding method for singular integral evaluation on finite domains. *SIAP* **48** (1988) 1335–1349. C: p. 349
1265. I.A. Viktorov, *Rayleigh and Lamb Waves*. New York: Plenum Press, 1967. C: pp. 23, 143
1266. W.M. Visscher, A new way to calculate scattering of acoustic and elastic waves I. Theory illustrated for scalar waves. *JAP* **51** (1980) 825–834. C: p. 263
1267. O. Von Estorff (ed.), *Boundary Elements in Acoustics: Advances and Applications*. Southampton: Computational Mechanics Publications, 2000. C: p. 168
1268. A.G. Voronovich, *Wave Scattering from Rough Surfaces*. Berlin: Springer, 1994. C: p. 11
1269. N.V. Voshchinnikov & V.G. Farafonov, Optical properties of spheroidal particles. *Astrophysics & Space Science* **204** (1993) 19–86. C: p. 150
1270. G.T. Walker, Some formulae for transforming the origin of reference of Bessel's functions. *Messenger of Mathematics* **25** (1896) 76–80. C: p. 38
1271. D.J.N. Wall, Methods of overcoming numerical instabilities associated with the T-matrix method. In: [1248], pp. 269–286. C: pp. 258, 265
1272. D.J.N. Wall, A uniqueness theorem for the reduced wave equation governing the acoustic wave in a heterogeneous medium. *QJMAM* **41** (1988) 141–153. C: p. 181
1273. D.J.N. Wall, Uniqueness theorems for the inverse problem of elastodynamic boundary scattering. *IMAP* **44** (1990) 221–241. C: p. 199
1274. D.J.N. Wall, V.V. Varadan & V.K. Varadan, Dynamic stress concentrations of cylindrical cavities with sharp and smooth boundaries: I. SH waves. *WM* **3** (1981) 203–213. C: p. 250
1275. C.-Y. Wang & J.D. Achenbach, Elastodynamic fundamental solutions for anisotropic solids. *GJI* **118** (1994) 384–392. C: p. 202
1276. C.-Y. Wang & J.D. Achenbach, Three-dimensional time-harmonic elastodynamic Green's functions for anisotropic solids. *PRSA* **449** (1995) 441–458. C: p. 202
1277. J.J.H. Wang, *Generalized Moment Methods in Electromagnetics*. New York: Wiley, 1991. C: p. 195
1278. K.-H. Wang & Q. Shen, Wave motion over a group of submerged horizontal plates. *IJES* **37** (1999) 703–715. C: p. 148
1279. R.T. Wang & H.C. van de Hulst, Rainbows: Mie computations and the Airy approximation. *AO* **30** (1991) 106–117. C: p. 140
1280. S. Wang, Wave radiation due to oscillation of two parallel spaced cylinders. *OE* **8** (1981) 599–621. C: p. 146
1281. S. Wang, Motions of a spherical submarine in waves. *OE* **13** (1986) 249–271. C: p. 145
1282. S. Wang & R. Wahab, Heaving oscillations of twin cylinders in a free surface. *JSR* **15** (1971) 33–48. C: p. 146
1283. Y.M. Wang & W.C. Chew, A recursive T-matrix approach for the solution of electromagnetic scattering by many spheres. *IEAP* **41** (1993) 1633–1639. C: p. 282
1284. Z.L. Wang, L. Hu & W.G. Lin, A modified T-matrix formulation for multiple scattering of electromagnetic waves. *JPDAP* **27** (1994) 447–451. C: p. 321
1285. G.N. Ward, *Linearized Theory of Steady High-Speed Flow*. Cambridge: Cambridge University Press, 1955. C: p. 336
1286. K.F. Warnick & W.C. Chew, Numerical simulation methods for rough surface scattering. *Waves in Random Media* **11** (2001) R1–R30. C: p. 11
1287. P.C. Waterman, Matrix formulation of electromagnetic scattering. *Proc. IEEE* **53** (1965) 805–812. C: pp. 242, 247, 247, 251, 264

1288. P.C. Waterman, New formulation of acoustic scattering. *JASA* **45** (1969) 1417–1429. C: pp. 250, 264, 273 & §7.8.1 (p. 258)
1289. P.C. Waterman, Scattering by dielectric obstacles. *Alta Frequenza* **38** (1969) 348–352. C: p. 251
1290. P.C. Waterman, Symmetry, unitarity, and geometry in electromagnetic scattering. *PR* D **3** (1971) 825–839. C: pp. 269, 273
1291. P.C. Waterman, Matrix theory of elastic wave scattering. *JASA* **60** (1976) 567–580. C: pp. 142, 251, 274
1292. P.C. Waterman, Matrix methods in potential theory and electromagnetic scattering. *JAP* **50** (1979) 4550–4566. C: pp. 242, 248
1293. P.C. Waterman, Survey of T-matrix methods and surface field representations. In: [1248], pp. 61–78. C: p. 258
1294. P.C. Waterman, Surface fields in potential theory and acoustic scattering. *JASA* **76** (1984) 1215–1226. C: p. 258
1295. P.C. Waterman, Surface fields and the T matrix. *JOSA* A **16** (1999) 2968–2977. C: pp. 258, 264
1296. P.C. Waterman & N.E. Pedersen, Electromagnetic scattering by periodic arrays of particles. *JAP* **59** (1986) 2609–2618. C: p. 281
1297. P.C. Waterman & R. Truell, Multiple scattering of waves. *JMP* **2** (1961) 512–537. C: pp. 315, 316, 318, 319
1298. G.N. Watson, *A Treatise on the Theory of Bessel Functions*, second edition. Cambridge: Cambridge University Press, 1944. C: pp. 29, 34, 36, 37, 38, 39, 77, 80, 82, 110, 127, 135, 147, 230
1299. K.M. Watson, Multiple scattering of electromagnetic waves in an underdense plasma. *JMP* **10** (1969) 688–702. C: p. 299
1300. R.L. Weaver & Y.-H. Pao, Application of the transition matrix to a ribbon-shaped scatterer. *JASA* **66** (1979) 1199–1206. C: p. 274
1301. A.G. Webster, *Partial Differential Equations of Mathematical Physics*. New York: Dover, 1955. C: p. 223
1302. J.V. Wehausen, The motion of floating bodies. *ARFM* **3** (1971) 237–268. C: p. 205
1303. J.V. Wehausen & E.V. Laitone, Surface waves. In: *Encyclopedia of Physics*, vol. IX (ed. S. Flügge & C. Truesdell) pp. 446–778. Berlin: Springer, 1960. Available at http://www.coe.berkeley.edu/SurfaceWaves/ with index and errata. C: pp. 23, 24, 56, 114, 144, 145, 204
1304. P.S.P. Wei, A.W. Reed & C.N. Ericksen, Two-sphere interference in RCS measurements and imaging. *IEAPM* **46**, 2 (2004) 166– 172. C: p. 141
1305. W. Wendland, Die Behandlung von Randwertaufgaben im R_3 mit Hilfe von Einfach- und Doppelschichtpotentialen. *NM* **11** (1968) 380–404. C: p. 169
1306. W.L. Wendland, Strongly elliptic boundary integral equations. In: *The State of the Art in Numerical Analysis* (eds. A. Iserles & M.J.D. Powell) pp. 511–562. Oxford: Oxford University Press, 1987. C: pp. 168, 177
1307. W.L. Wendland, Boundary element methods for elliptic problems. In: *Mathematical Theory of Finite and Boundary Element Methods* (A.H. Schatz, V. Thomée & W.L. Wendland) pp. 219–276. Basel: Birkhäuser, 1990. C: p. 168
1308. E.J. Weniger & E.O. Steinborn, A simple derivation of the addition theorems of the irregular solid harmonics, the Helmholtz harmonics, and the modified Helmholtz harmonics. *JMP* **26** (1985) 664–670. C: p. 90
1309. M.F. Werby & R.B. Evans, Scattering from objects submerged in unbounded and bounded oceans. *IEEE J. Oceanic Engineering* **12** (1987) 380–394. C: p. 280

1310. M.F. Werby & L.H. Green, An extended unitary approach for acoustical scattering from elastic shells immersed in a fluid. *JASA* **74** (1983) 625–630. C: p. 269
1311. P. Werner, Zur mathematischen Theorie akustischer Wellenfelder. *ARMA* **6** (1960) 231–260. C: p. 181
1312. P. Werner, Beugungsprobleme der mathematischen Akustik. *ARMA* **12** (1963) 155–184. C: pp. 181, 185
1313. V.H. Weston, Multifrequency inverse problem for the reduced wave equation with sparse data. *JMP* **25** (1984) 1382–1390. C: p. 186
1314. V.H. Weston, Formulation of the inverse problem for the reduced wave equation in momentum space. *JMP* **27** (1986) 1772–1781. C: p. 186
1315. R.M. White, Elastic wave scattering at a cylindrical discontinuity in a solid. *JASA* **30** (1958) 771–785. C: p. 141
1316. E.T. Whittaker, *A Treatise on the Analytical Dynamics of Particles and Rigid Bodies*, fourth edition. Cambridge: Cambridge University Press, 1937. C: p. 330
1317. E.T. Whittaker & G.N. Watson, *A Course of Modern Analysis*, fourth edition. Cambridge: Cambridge University Press, 1927. C: pp. 63, 65
1318. E.P. Wigner, *Group Theory and its Application to the Quantum Mechanics of Atomic Spectra*. New York: Academic Press, 1959. C: p. 326
1319. E.P. Wigner, On the matrices which reduce the Kronecker products of representations of S.R. groups. In: *Quantum Theory of Angular Momentum* (eds. L.C. Biedenharn & H. van Dam) pp. 87–133. New York: Academic Press, 1965. C: p. 326
1320. C.H. Wilcox, A variational method for computing the echo area of a lamina. In: *Boundary Problems in Differential Equations* (ed. R.E. Langer) pp. 301–314. Madison: University of Wisconsin Press, 1960. C: p. 351
1321. A.N. Williams, Wave diffraction by elliptical breakwaters in shallow water. *OE* **12** (1985) 25–43. C: p. 149
1322. A.N. Williams & W. Li, Water wave interaction with an array of bottom-mounted surface-piercing porous cylinders. *OE* **27** (2000) 841–866. C: p. 129
1323. J.R. Willis, A polarization approach to the scattering of elastic waves – I. Scattering by a single inclusion. *JMPS* **28** (1980) 287–305. C: p. 202
1324. J.R. Willis, A polarization approach to the scattering of elastic waves – II. Multiple scattering from inclusions. *JMPS* **28** (1980) 307–327. C: p. 320
1325. D.R. Wilton, Computational methods. In: [977], pp. 316– 365. C: p. 195
1326. D.E. Winch & R.W. James, Computations with spherical harmonics and Fourier series in geomagnetism. In: *Methods in Computational Physics*, vol. 13 (ed. B.A. Bolt) pp. 93–132. New York: Academic Press, 1973. C: pp. 84, 93, 333
1327. A. Wirgin, Green function theory of the scattering of electromagnetic waves from a cylindrical boundary of arbitrary shape. *OC* **7** (1973) 65–69. C: p. 228
1328. A. Wirzba, Quantum mechanics and semiclassics of hyperbolic n-disk scattering systems. *Physics Reports* **309** (1999) 1–116. C: p. 129
1329. A. Wirzba & P.E. Rosenqvist, Three disks in a row: a two- dimensional scattering analog of the double-well problem. *PR* A **54** (1996) 2745–2754. Erratum: **55** (1997) 1555. C: p. 129
1330. W.J. Wiscombe & A. Mugnai, Single scattering from nonspherical Chebyshev particles: a compendium of calculations. NASA Reference Publication 1157, 1986. C: pp. 16, 251
1331. W.J. Wiscombe & A. Mugnai, Scattering from nonspherical Chebyshev particles. 2: Means of angular scattering patterns. *AO* **27** (1988) 2405–2421. C: p. 251
1332. R.C. Wittmann, Spherical wave operators and the translation formulas. *IEAP* **36** (1988) 1078–1087. C: p. 86

1333. A. Wolf, Motion of a rigid sphere in an acoustic wave field. *Geophysics* **10** (1945) 91–109. C: p. 142
1334. H.L. Wong, Effect of surface topography on the diffraction of *P*, *SV*, and Rayleigh waves. *BSSA* **72** (1982) 1167–1183. C: p. 246
1335. H.L. Wong & M.D. Trifunac, Two-dimensional, antiplane, building-soil-building interaction for two or more buildings and for incident plane *SH* waves. *BSSA* **65** (1975) 1863–1885. C: p. 129
1336. T. Wriedt & A. Doicu, Comparison between various formulations of the extended boundary condition method. *OC* **142** (1997) 91–98. C: p. 265
1337. G.X. Wu, The interaction of water waves with a group of submerged spheres. *AOR* **17** (1995) 165–184. C: p. 146
1338. G.X. Wu & R. Eatock Taylor, The exciting force on a submerged spheroid in regular waves. *JFM* **182** (1987) 411–426. C: p. 206
1339. G.X. Wu & R. Eatock Taylor, On the radiation and diffraction of surface waves by submerged spheroids. *JSR* **33** (1989) 84–92. C: p. 206
1340. T.W. Wu (ed.), *Boundary Element Acoustics: Fundamentals and Computer Codes.* Southampton: Computational Mechanics Publications, 2000. C: p. 168
1341. T.W. Wu & A.F. Seybert, A weighted residual formulation for the CHIEF method in acoustics. *JASA* **90** (1991) 1608–1614. C: p. 221
1342. X.-J. Wu & W.G. Price, A multiple Green's function expression for the hydrodynamic analysis of multi-hull structures. *AOR* **9** (1987) 58–66. C: p. 217
1343. P.J. Wyatt, Scattering of electromagnetic plane waves from inhomogeneous spherically symmetric objects. *PR* **127** (1962) 1837–1843. Errata: **134** (1964) AB1. C: p. 140
1344. X.M. Xu & Q.H. Liu, Fast spectral-domain method for acoustic scattering problems. *IEEE Trans. on Ultrasonics, Ferroelectrics, & Frequency Control* **48** (2001) 522–529. C: p. 186
1345. Y.-L. Xu, Electromagnetic scattering by an aggregate of spheres. *AO* **34** (1995) 4573–4588. Errata: **37** (1998) 6494 and **40** (2001) 5508. C: p. 141
1346. Y.-L. Xu, Fast evaluation of Gaunt coefficients. *Mathematics of Computation* **65** (1996) 1601–1612. C: p. 93
1347. Y.-L. Xu, Calculation of the addition coefficients in electromagnetic multisphere-scattering theory. *JCP* **127** (1996) 285–298. Erratum: **134** (1997) 200. C: pp. 94, 107
1348. Y.-L. Xu, Efficient evaluation of vector translation coefficients in multiparticle light-scattering theories. *JCP* **139** (1998) 137–165. C: pp. 93, 107, 141
1349. A.D. Yaghjian, Augmented electric- and magnetic-field integral equations. *RS* **16** (1981) 987–1001. C: p. 219
1350. R.-B. Yang, A dynamic generalized self-consistent model for wave propagation in particulate composites. *JAM* **70** (2003) 575–582. C: p. 321
1351. R.-B. Yang & A.K. Mal, Multiple scattering of elastic waves in a fiber-reinforced composite. *JMPS* **42** (1994) 1945–1968. C: p. 321
1352. R.-B. Yang & A.K. Mal, Elastic waves in a composite containing inhomogeneous fibers. *IJES* **34** (1996) 67–79. C: p. 321
1353. S.A. Yang, An integral equation approach to three-dimensional acoustic radiation and scattering problems. *JASA* **116** (2004) 1372–1380. C: pp. 178, 221
1354. Z. Ye & A. Alvarez, An analytic approach to sound scattering from a pair of scatterers with application to rigid spheres. *J. Computational Acoustics* **9** (2001) 311–325. C: p. 304
1355. Z. Ye & L. Ding, Acoustic dispersion and attenuation relations in bubbly mixture. *JASA* **98** (1995) 1629–1636. C: pp. 316, 319

1356. M.S. Yeung, Single integral equation for electromagnetic scattering by three-dimensional homogeneous dielectric objects. *IEAP* **47** (1999) 1615–1622. C: p. 195
1357. R.W. Yeung, Added mass and damping of a vertical cylinder in finite-depth waters. *AOR* **3** (1981) 119–133. C: p. 148
1358. R.W. Yeung, Numerical methods in free-surface flows. *ARFM* **14** (1982) 395–442. C: p. 206
1359. O. Yılmaz, Hydrodynamic interactions of waves with group of truncated vertical cylinders. *JWPCOE* **124** (1998) 272–279. C: p. 283
1360. O. Yılmaz, An iterative procedure for the diffraction of water waves by multiple cylinders. *OE* **31** (2004) 1437–1446. C: p. 132
1361. O. Yılmaz & A. Incecik, Analytical solutions of the diffraction problem of a group of truncated vertical cylinders. *OE* **25** (1998) 385–394. C: p. 283
1362. W.-Y. Yin, L.-W. Li, T.-S. Yeo & M.-S. Leong, Multiple scattering by an array of finite cylindrically curved thin screens. *JEWA* **14** (2000) 1449–1464. C: p. 130
1363. C.F. Ying & R. Truell, Scattering of a plane longitudinal wave by a spherical obstacle in an isotropically elastic solid. *JAP* **27** (1956) 1086–1097. C: p. 141
1364. E. Yogeswaren & J.G. Harris, A model of a confocal ultrasonic inspection system for interfaces. *JASA* **96** (1994) 3581– 3592. C: p. 208
1365. J.W. Young & J.C. Bertrand, Multiple scattering by two cylinders. *JASA* **58** (1975) 1190–1195. C: p. 129
1366. R.M. Young, *An Introduction to Nonharmonic Fourier Series*. New York: Academic Press, 1980. C: p. 254
1367. H.A. Yousif & S. Köhler, Scattering by two penetrable cylinders at oblique incidence. I. The analytical solution. *JOSA* A **5** (1988) 1085–1096. C: p. 129
1368. J.-W. Yu, W.J. Byun & N.-H. Myung, Multiple scattering from two dielectric-filled semi-circular channels in a conducting plane: TM case. *IEAP* **50** (2002) 1250–1253. C: p. 130
1369. Y.S. Yu & F. Ursell, Surface waves generated by an oscillating circular cylinder on water of finite depth: theory and experiment. *JFM* **11** (1961) 529–551. C: pp. 55, 145
1370. E.V. Zakharov & I.V. Sobyanina, On one-dimensional integro-differential equations for problems of diffraction by screens. *USSR Comput. Maths. Math. Phys.* **26**, 2 (1986) 194–198. C: p. 161
1371. R.N. Zare, *Angular Momentum*. New York: Wiley, 1988. C: pp. 63, 326, 330, 330, 332 & Table B.1 (p. 327)
1372. F. Záviška, Über die Beugung elektromagnetischer Wellen an parallelen, unendlich langen Kreiszylindern. *Annalen der Physik*, 4 Folge, **40** (1913) 1023–1056. C: pp. 122, 129
1373. Ch. Zhang & D. Gross, The solution of plane problems of wave loaded cracks by an integral equation method. *ZAMM* **68** (1988) 299–305. C: p. 208
1374. Ch. Zhang & D. Gross, *On Wave Propagation in Elastic Solids with Cracks*. Southampton: Computational Mechanics Publications, 1998. C: p. 207
1375. S. Zhang & A.N. Williams, Wave scattering by submerged elliptical disk. *JWPCOE* **122** (1996) 38–45. C: p. 149
1376. W. Zheng & S. Ström, The null-field approach to electromagnetic resonance of composite objects. *Computer Physics Comm.* **68** (1991) 157–174. C: p. 282
1377. X. Zheng & R.E. Apfel, Acoustic interaction forces between two fluid spheres in an acoustic field. *JASA* **97** (1995) 2218–2226. C: p. 137
1378. S. Zhu & G. Moule, An efficient numerical calculation of wave loads on an array of vertical cylinders. *Appl. Math. Modelling* **20** (1996) 26–33. C: p. 178

1379. J.M. Ziman, *Models of Disorder*. Cambridge: Cambridge University Press, 1979. C: pp. 1, 9, 321
1380. C. Zimmerman & M. Stern, Scattering of plane compressional waves by spherical inclusions in a poroelastic medium. *JASA* **94** (1993) 527–536. C: p. 143
1381. N.R. Zitron, Multiple scattering of elastic waves by two arbitrary cylinders. *JASA* **42** (1967) 394–402. C: p. 2
1382. N. Zitron & S.N. Karp, Higher-order approximations in multiple scattering. I. Two-dimensional scalar case. *JMP* **2** (1961) 394–402. C: pp. 2, 310, 311
1383. N. Zitron & S.N. Karp, Higher-order approximations in multiple scattering. II. Three-dimensional scalar case. *JMP* **2** (1961) 402–406. C: pp. 2, 310
1384. F. Zolla & R. Petit, Method of fictitious sources as applied to the electromagnetic diffraction of a plane wave by a grating in conical diffraction mounts. *JOSA* A **13** (1996) 796–802. C: p. 245
1385. F. Zolla, R. Petit & M. Cadilhac, Electromagnetic theory of diffraction by a system of parallel rods: the method of fictitious sources. *JOSA* A **11** (1994) 1087–1096. C: p. 245
1386. L.M. Zurk, L. Tsang, K.H. Ding & D.P. Winebrenner, Monte Carlo simulations of the extinction rate of densely packed spheres with clustered and nonclustered geometries. *JOSA* A **12** (1995) 1772–1781. C: pp. 28, 141, 322

Citation index

Subject index